**Corrosion Resistance of Steels
Against Lyes and Organic Acids**

Edited by
Michael Schütze, Thomas Ladwein
and Roman Bender

Corrosion Resistance of Steels Against Lyes and Organic Acids

WILEY-VCH Verlag GmbH & Co. KGaA

Editors

Prof. Dr.-Ing. Michael Schütze
DECHEMA-Forschungsinstitut
Chairman of the Executive Board
Theodor-Heuss-Allee 25
60486 Frankfurt am Main
Germany

Prof. Dr. Thomas Ladwein
Aalen University of Applied Sciences
Surface and Materials Science
73428 Aalen
Germany

Dr. rer. nat. Roman Bender
Chief Executive of GfKORR e. V.
Society for Corrosion Protection
Theodor-Heuss-Allee 25
60486 Frankfurt am Main
Germany

Cover Illustration
Source: DECHEMA-Forschungsinstitut,
Frankfurt (Main), Germany

Warranty Disclaimer

This book has been compiled from literature data with the greatest possible care and attention. The statements made only provide general descriptions and information.

Even for the correct selection of materials and correct processing, corrosive attack cannot be excluded in a corrosion system as it may be caused by previously unknown critical conditions and influencing factors or subsequently modified operating conditions.

No guarantee can be given for the chemical stability of the plant or equipment. Therefore, the given information and recommendations do not include any statements, from which warranty claims can be derived with respect to DECHEMA e. V. or its employees or the authors.

The DECHEMA e. V. is liable to the customer, irrespective of the legal grounds, for intentional or grossly negligent damage caused by their legal representatives or vicarious agents.

For a case of slight negligence, liability is limited to the infringement of essential contractual obligations (cardinal obligations). DECHEMA e. V. is not liable in the case of slight negligence for collateral damage or consequential damage as well as for damage that results from interruptions in the operations or delays which may arise from the deployment of this book.

■ This book was carefully produced. Nevertheless, editors, authors and publisher do not warrant the information contained therein to be free of errors. Readers are advised to keep in mind that statements, data, illustrations, procedural details or other items may inadvertently be inaccurate.

Library of Congress Card No.: Applied for.

British Library Cataloguing-in-Publication Data:
A catalogue record for this book is available from the British Library.

Bibliographic information published by Die Deutsche Bibliothek
Die Deutsche Bibliothek lists this publication in the Deutsche Nationalbibliografie; detailed bibliographic data is available in the Internet at
<http://dnb.ddb.de>.

© 2013 DECHEMA e. V., Society for Chemical Engineering and Biotechnology, 60486 Frankfurt (Main), Germany

All rights reserved (including those of translation into other languages). No part of this book may be reproduced in any form – nor transmitted or translated into machine language without written permission from the publishers. Registered names, trademarks, etc. used in this book, even when not specifically marked as such, are not to be considered unprotected by law.

Printed in the Federal Republic of Germany
Printed on acid-free paper

Typesetting Kühn & Weyh, Satz und Medien, Freiburg
Printing Strauss GmbH, Mörlenbach
Binding Strauss GmbH, Mörlenbach
Cover Design Graphik-Design Schulz, Fußgönheim

ISBN: 978-3-527-33679-1

Contents

Preface IX

How to use the Handbook XI

Warranty disclaimer 1

Acetic Acid 3
Unalloyed steels and cast steel 3
Unalloyed cast iron 12
High-alloy cast iron, high-silicon cast iron 13
Structural steels with up to 12 % chromium 19
Ferritic chromium steels with more than 12 % chromium 19
Ferritic-austenitic steels with more than 12 % chromium 19
Austenitic chromium-nickel steels 32
Austenitic chromium-nickel-molybdenum steels 51
Austenitic chromium-nickel steels with special alloying additions 68
Special iron-based alloys 73
Bibliography 77

Alkanecarboxylic Acids 95
Unalloyed steels and cast steel 95
Unalloyed cast iron 95
Austenitic chromium-nickel-molybdenum steels 100
Austenitic chromium-nickel steels with special alloying additions 100
Special iron-based alloys 100
High-alloy cast iron, high-silicon cast iron 101
Structural steels with up to 12 % chromium 102
Ferritic chromium steels with more than 12 % chromium 103
Ferritic-austenitic steels with more than 12 % chromium 103
Austenitic chromium-nickel steels 104

Austenitic chromium-nickel-molybdenum steels *107*
Austenitic chromium-nickel steels with special alloying additions *107*
Special iron-based alloys *107*
Bibliography *116*

Carbonic Acid *119*
Unalloyed and low-alloy steels/cast steel *119*
Unalloyed cast iron and low-alloy cast iron *119*
Ferritic chrome steels with < 13 % Cr *146*
Ferritic chromium steels with ≥ 13 % Cr *147*
High-alloy multiphase steels *147*
Ferritic/perlitic-martensitic steels *147*
Austenitic chromium-nickel steels *151*
Austenitic CrNiMo(N) steels *169*
Austenitic CrNiMoCu(N) steels *169*
Bibliography *172*

Formic Acid *175*
Unalloyed steels and cast steel *175*
Unalloyed cast iron *175*
High-alloy cast iron, high-silicon cast iron *175*
Structural steels with up to 12 % chromium *176*
Ferritic chromium steels with more than 12 % chromium *176*
Ferritic-austenitic steels with more than 12 % chromium *179*
Austenitic chromium-nickel steels *180*
Austenitic chromium-nickel-molybdenum steels *180*
Austenitic chromium-nickel steels with special alloying additions *180*
Special iron-based alloys *183*
Bibliography *185*

Sulfonic Acids *187*
Unalloyed steels and cast steel *187*
High-alloy cast iron, high-silicon cast iron *191*
Ferritic chromium steels with more than 12 % chromium *191*
Austenitic chromium-nickel steels *191*
Austenitic chromium-nickel-molybdenum steels *194*
Austenitic chromium-nickel steels with special alloying additions *196*
Bibliography *197*

Alkaline Earth Hydroxides *199*
Unalloyed steels and cast steel   *199*
Unalloyed cast iron   *218*
High-alloy cast iron, high-silicon cast iron   *218*
Structural steels with up to 12 % chromium   *219*
Ferritic chromium steels with more than 12 % chromium   *219*
Ferritic-austenitic steels with more than 12 % chromium   *219*
Austenitic chromium-nickel steels   *219*
Austenitic chromium-nickel-molybdenum steels   *219*
Austenitic chromium-nickel steels with special alloying additions   *219*
Bibliography   *220*

Ammonia and Ammonium Hydroxide   *223*
Unalloyed and low-alloy steels/cast steel   *223*
Unalloyed and low-alloyed cast iron   *223*
Highly alloyed cast iron   *244*
Cast ferrosilicon   *244*
Austenitic cast iron (amongst others)   *244*
Ferritic chrome steels with < 13 % Cr   *244*
Ferritic chrome steels with ≥ 13 % Cr   *244*
Ferritic/perlitic-martensitic steels   *244*
Ferritic-austenitic steels/duplex steels   *244*
Austenitic chromium-nickel steels   *244*
Austenitic chromium-nickel-molybdenum(N) steels and CrNiMoCu(N) steels   *244*
Bibliography   *247*

Lithium Hydroxide   *251*
Unalloyed steels and cast steel   *251*
Unalloyed cast iron   *257*
High-alloy cast iron, high-silicon cast iron   *257*
Ferritic chromium steel with more than 12 % chromium   *257*
Austenitic chromium-nickel steels   *258*
Austenitic chromium-nickel-molybdenum steels   *258*
Bibliography   *261*

Potassium Hydroxide   *263*
Unalloyed steels and low-alloy steels/cast steel   *263*
Unalloyed cast iron and low-alloy cast iron   *269*
High-alloy cast iron   *269*

Cast ferrosilicon *269*
Austenitic cast iron (among other things) *269*
Ferritic chrome steels with < 13 % Cr *270*
Ferritic chrome steels with ≥ 13 % Cr *271*
Ferritic/Perlitic-martensitic steels *271*
Ferritic-austenitic steels/duplex steels *271*
Austenitic chromium-nickel steels *274*
Austenitic CrNiMo(N) steels and CrNiMoCu(N) steels *277*
Bibliography *286*

Sodium Hydroxide *293*
Unalloyed and low-alloy steels/cast steel *293*
Unalloyed cast iron and low-alloy cast iron *316*
High-alloy cast iron *319*
Silicon cast iron *319*
Austenitic cast iron (etc.) *319*
Ferritic chromium steels with < 13 % Cr *319*
Ferritic chromium steels with ≥ 13 % Cr *329*
High-alloy multiphase steels *334*
Ferritic/perlitic-martensitic steels *334*
Ferritic-austenitic steels/duplex steels *334*
Austenitic chromium-nickel steels *342*
Austenitic chromium-nickel-molybdenum(N) steels *349*
Austenitic CrNiMoCu(N) steels *361*
Special iron-based alloys *362*
Bibliography *363*

Key to materials compositions *371*

Index of materials *411*

Subject index *421*

Preface

Practically all industries face the problem of corrosion – from the micro-scale of components for the electronics industries to the macro-scale of those for the chemical and construction industries. This explains why the overall costs of corrosion still amount to about 2 to 4% of the gross national product of industrialized countries despite the fact that zillions of dollars have been spent on corrosion research during the last few decades.

Much of this research was necessary due to the development of new technologies, materials and products, but it is no secret that a considerable number of failures in technology nowadays could, to a significant extent, be avoided if existing knowledge were used properly. This fact is particularly true in the field of corrosion and corrosion protection. Here, a wealth of information exists, but unfortunately in most cases it is scattered over many different information sources. However, as far back as 1953, an initiative was launched in Germany to compile an information system from the existing knowledge of corrosion and to complement this information with commentaries and interpretations by corrosion experts. The information system, entitled "DECHEMA-WERKSTOFF-TABELLE" (DECHEMA Corrosion Data Sheets), grew rapidly in size and content during the following years and soon became an indispensable tool for all engineers and scientists dealing with corrosion problems. This tool is still a living system today: it is continuously revised and updated by corrosion experts and thus represents a unique source of information. Currently, it comprises more than 12,000 pages with approximately 110,000 corrosion systems (i.e., all relevant commercial materials and media), based on the evaluation of over 100,000 scientific and technical articles which are referenced in the database.

Last century, an increasing demand for an English version of the DECHEMA-WERKSTOFF-TABELLE arose in the 80s; accordingly the first volume of the DECHEMA Corrosion Handbook was published in 1987. This was a slightly condensed version of the German edition and comprised 12 volumes. Before long, this handbook had spread all over the world and become a standard tool in countless laboratories outside Germany. The second edition of the DECHEMA Corrosion Handbook was published in 2004. Together the two editions covered 24 volumes.

The present book compiles all information on the corrosion behaviour of steels against lyes and organic acids that was compiled in the volumes of the corrosion handbook. This compilation is an indispensable tool for all engineers and scientists dealing with corrosion problems of Steels in contact with lyes and organic acids.

Steel is one of the most widely used construction material with more than 1.3 billion tons produced each year. Buildings, industrial plants, machines, tools, pipelines, vessels and tanks are only a few of its applications in our daily life. Steel is an alloy made of iron and additional elements like carbon, chromium, manganese, vanadium and tungsten, and its quality, ductility, hardness and strength vary with the amount of the alloying element.

As steels corrode in various atmospheres, in water and in soil its corrosion resistance against lyes and organic acids is essential and a crucial financial factor for many industries. These chemicals are present in nearly every industrial production process such as metal manufacturing but also explosives, food, dyes, leather, paper and fertilizers, to name only a few.

Understanding how to strengthen the corrosion resistance of steels as reaction, transport and storage devices against these omnipresent chemicals is key for all industries involved. This book is therefore a must-have for all mechanical, civil and chemical engineers, material scientists and chemists working with steel in alkaline or acidic media.

This handbook highlights the limitations of steels in lyes and organic acids and provides vital information on corrosion protection measures.

The chapters are arranged by the agents leading to individual corrosion reactions, and a vast number of steels are presented in terms of their behaviour in these agents. The key information consists of quantitative data on corrosion rates coupled with commentaries on the background and mechanisms of corrosion behind these data, together with the dependencies on secondary parameters, such as flow-rate, pH, temperature, etc. This information is complemented by more detailed annotations where necessary, and by an immense number of references listed at the end of each chapter.

An important feature of this handbook is that the data was compiled for industrial use. Therefore, particularly for those working in industrial laboratories or for industrial clients, the book will be an invaluable source of rapid information for day-to-day problem solving. The handbook will have fulfilled its task if it helps to avoid the failures and problems caused by corrosion simply by providing a comprehensive source of information summarizing the present state-of-the-art. Last but not least, in cases where this knowledge is applied, there is a good chance of decreasing the costs of corrosion significantly.

Finally the editors would like to express their appreciation to Gudrun Walter of Wiley-VCH for their valuable assistance during all stages of the preparation of this book.

Michael Schütze, Thomas Ladwein and Roman Bender

How to use the Handbook

The Handbook provides information on the chemical resistance and the corrosion behavior of steels in lyes and organic acids.

The user is given information on the range of applications and corrosion protection measures.

Research results and operating experience reported by experts allow recommendations to be made for the selection of steels and to provide assistance in the assessment of damage.

The objective is to offer a comprehensive and concise description of the behavior of steels in contact with a particular acid or lye.

The information on resistance is given as text, tables, and figures. The literature used by the authors is cited at the corresponding point. There is an index of materials as well as a subject index at the end of the book so that the user can quickly find the information given for a particular keyword.

The Handbook is thus a guide that leads the reader to steels that have already been used in certain cases, that can be used or that are not suitable owing to their lack of resistance.

The resistance is coded with three evaluation symbols in order to compress the information. Uniform corrosion is evaluated according to the following criteria:

Symbol	Meaning	Area-related mass loss rate[1]		Corrosion rate
		$g/(m^2 \, h)$	$g/(m^2 \, d)$	mm/a
+	resistant	≤ 0.1	≤ 2.4	≤ 0.1[2]
⊕	fairly resistant	> 0.1 to ≤ 1.0	> 2.4 to ≤ 24.0	> 0.1 to ≤ 1.0
−	not resistant	> 1.0	> 24.0	> 1.0

[1] valid for steels; for Al, Mg, and its alloys, 1/3 of the value must be used
[2] the values for Ta, Ti, and Zr are too high (possible embrittlement due to hydrogen absorption in the event of corrosion! Therefore, corrosion rate = 0.01 mm/a, see the individual cases)

The evaluation of the corrosion resistance of metallic materials is given

- for uniform corrosion or local penetration rate, in: mm/a and mpy
- or if the density of the material is not known, in: $g/(m^2 \, h)$ or $g/(m^2 \, d)$.

Pitting corrosion, crevice corrosion, and stress corrosion cracking or non-uniform attack are particularly highlighted.

The following equations are used to convert mass loss rates, x, into the corrosion rate, y:

from x_1 into $g/(m^2\,h)$ from x_2 into $g/(m^2\,d)$ where

$$\frac{x_1 \cdot 365 \cdot 24}{\rho \cdot 1{,}000} = y\ (mm/a) \qquad \frac{x_2 \cdot 365}{\rho \cdot 1{,}000} = y\ (mm/a)$$

x_1: value in $g/(m^2\,h)$
x_2: value in $g/(m^2\,d)$
ρ: density of material in g/cm^3
y: value in mm/a
d: days
h: hours

In those media in which uniform corrosion can be expected, if possible, isocorrosion curves (corrosion rate = 0.1 mm/a (3.94 mpy)) are given.

Unless stated otherwise, the data was measured at atmospheric pressure and room temperature.

The resistance data should not be accepted by the user without question, and the materials for a particular purpose should not be regarded as the only ones that are suitable. To avoid wrong conclusions being drawn, it must be always taken into account that the expected material behavior depends on a variety of factors that are often difficult to recognize individually and which may not have been taken deliberately into account in the investigations upon which the data is based. Under certain circumstances, even slight deviations in the chemical composition of the medium, in the pressure, in the temperature or, for example, in the flow rate are sufficient to have a significant effect on the behavior of the materials. Furthermore, impurities in the medium or mixed media can result in a considerable increase in corrosion.

The composition or the pretreatment of the material itself can also be of decisive importance for its behavior. In this respect, welding should be mentioned. The suitability of the component's design with respect to corrosion is a further point which must be taken into account. In case of doubt, the corrosion resistance should be investigated under operating conditions to decide on the suitability of the selected materials.

Warranty disclaimer

This book has been compiled from literature data with the greatest possible care and attention. The statements made in this book only provide general descriptions and information.

Even for the correct selection of materials and correct processing, corrosive attack cannot be excluded in a corrosion system as it may be caused by previously unknown critical conditions and influencing factors or subsequently modified operating conditions.

No guarantee can be given for the chemical stability of the plant or equipment. Therefore, the given information and recommendations do not include any statements, from which warranty claims can be derived with respect to DECHEMA e.V. or its employees or the authors.

The DECHEMA e.V. is liable to the customer, irrespective of the legal grounds, for intentional or grossly negligent damage caused by their legal representatives or vicarious agents.

For a case of slight negligence, liability is limited to the infringement of essential contractual obligations (cardinal obligations). DECHEMA e.V. is not liable in the case of slight negligence for collateral damage or consequential damage as well as for damage that results from interruptions in the operations or delays which may arise from the deployment of this book.

Acetic Acid

Unalloyed steels and cast steel

– *Acetic acid* –

Iron and steel [1, 2] are attacked by acetic acid more or less severely depending on concentration and temperature (Table 27) [3–7].

It is assumed that dissolution in the cathodic region proceeds predominantly chemically, while in the anodic region a chemical and electrochemical dissolution mechanism coexist. In 1 to 90 % aqueous acetic acid solution, for example, iron corrodes by a mixed chemical-electrochemical mechanism, in which the proportion of the chemical reaction is 15 to 20 % at low and medium concentrations [8–10].

The dissolution mechanism of Armco® iron in aqueous acetic acid solution is dependent on the concentration, thus on the degree of dissociation and the water content of the acid, which is the reason that the corrosion current goes through a maximum at about 30 % [11].

Some of the hydrogen formed at the acid concentration of up to 95 % is absorbed by the metal; at concentrations above 95 %, no more gas evolution takes place, which indicates a change in the dissolution mechanism, which, however, has not yet been completely elucidated; in any case, the lack of free water is probably the decisive factor [12].

The dissolution behavior of iron in deaerated 95 % acetic acid solution and its passivation by anodic polarization was studied in [13]. Comparable investigations, but with an addition of 0.5 mol/l sodium acetate to the acetic acid, in a N_2 atmosphere showed that anodic polarization leads to oxygen passivation of Fe [14, 15]. In the presence of O_2 the formation of a passivating iron acetate film took place and a gradual decrease in the corrosion rate in the first few hours has been observed. An explanation for this mechanism is proposed in [16].

Based on tests using Armco® iron in oxygen-free sodium acetate and sodium chloride solutions acidified with acetic acid, acetic acid acts as proton donor [17]. Acetic acid also considerably increases the cavitation loss of Armco® iron in sodium chloride solution [18].

As for older theoretical works on the dissolution of iron in acetic acid, see [19–21].

Acetic acid dissolved in organic solvents attacks iron [22, 23].

On increase of the temperature to 373 K (100°C), the corrosion loss of iron increases abruptly in aqueous 1 to 15 % acetic acid but is independent of the acetic acid concentration [24].

The interaction of hydrogen sulfide (0.07 g/l) with iron and C-steel is supported by small amounts (5 to 150 mg/l) of organic acids such as acetic, formic, butyric acid; this effect does not occur at higher hydrogen sulfide concentrations [25].

C-steel is not resistant to aqueous 10 % acetic acid solution [26]. The low-alloyed iron grades have a similar behavior to unalloyed iron in acetic acid [27].

The unalloyed steel SS 41 (Fe-0.04C-0.30Mn-0.020P-0.017P-Si-traces) was investigated in acetic acid of concentrations up to 99.7 % by volume at room temperature (test duration 60 min) and at the boiling point (7 d). The corrosion rate at room temperature increases with the concentration of acetic acid to a maximum of 68.5 mm/a (2,697 mpy) at 30 %. The corrosion rate for nearly water-free acetic acid was about 25.7 mm/a (1,012 mpy). At the boiling point the corrosion rates are hundred times higher than at room temperature [28].

According to a Soviet publication, unalloyed steel can be protected against corrosion in 20 to 90 % acetic acid by means of a protective current density of 40 to 70 mA/dm^2 at 800 to 900 mV$_{SCE}$ [29].

These findings were confirmed by other results, according to which the optimum protective current density for St 3 (1.0333) is 40 to 70 mA/dm^2 at 293 to 323 K (20 to 50°C) in concentrations between 20 and 93 % (degree of protection 93 to 99 %); at 323 to 368 K (50 to 95°C), the degrees of protection obtained in acid of the concentration range mentioned are 95 to 98 %. Therefore, when cellulose acetate is prepared, cathodically protected steel containers can be used instead of containers made of expensive stainless steels.

The most suitable material for protective electrodes is an FeSi-alloy containing 18 to 20 % silicon [30].

In a plant for the production of acetic acid/acetic anhydride mixtures by oxidation of acetaldehyde, the corrosion of CrNiMo-steel 18 8 3 was caused by the water content and by entrained catalysts. Remedial action was taken by preventing catalyst entrainment and accelerating the evaporation [31].

In a steel bubble-cap column for the distillation of a water-saturated mixture of acetic acid + ethyl acetate, severe corrosion at the bubble caps took place, because the mixture also contained traces of formic acid; the steel column was therefore replaced by a column made of CrNiMo-steel 18 8 3 [32].

In a C-steel container for pyroligneous acid solutions consisting of the mixture obtained in the dry distillation of wood and of about (%) 8 acetic acid, 4 methanol, 7 other organic components and 81 water, C-steel has a material consumption rate of 48 mg/dm^2 d (0.218 mm/a (8.58 mpy)). By using graphite anodes, the corrosion rate at 55.5 mA/dm^2 is reduced to 5 mg/dm^2 d (0.022 mm/a (0.87 mpy)) [33].

According to investigations in Soviet natural gas deposits, the severe corrosion of steel (St 45; cf. E255, SAE 1020) in the gas condensate is caused by the presence of organic acids therein. The material consumption rates reach, for example, at 100 mg/l acetic acid 0.026 g/m^2 h and at 1,000 mg/l acetic acid 0.167 g/m^2 h (0.028 and 0.18 mm/a (1.10 and 7.09 mpy)) [34].

According to data which should be viewed with reservations (maybe a translation error, since it is a Korean article), the susceptibility to stress corrosion cracking in 0.5 % acetic acid saturated with hydrogen sulfide is said to increase with tensile load, but surprisingly in the order soft steel – stainless steel – hard steel – high-strength steel [35].

Under these conditions, the hydrogen overvoltage is increased and hydrogen is absorbed by the metal (welded high-strength steel) so that crack formation can be attributed to embrittlement [36].

At 308 K (35°C), high-strength steels are, as expected, highly susceptible to stress corrosion cracking; the threshold stress leading to crack formation is independent of whether the steels are welded or not; neither does the welding process affect the behavior [37].

The resistance of the low-alloyed API steels N-80 (1.0564), P-105 (1.0670) and P-110 (1.0671) in solutions with 10 % acetic acid and 35 g/l sodium chloride saturated with hydrogen sulfide depends on the temperature of the heat treatment: quenched steels are more resistant than steels hardened by normal heat treatment. The resistance to stress corrosion cracking decreases, as expected, with increasing yield point of the steels. The crack formation occured transgranularly in all steels investigated [38].

The corrosion of a high-strength steel in 0.5 % acetic acid saturated with hydrogen sulfide could not be prevented by means of sodium dichromate and sodium molybdate; however, an oleylamine, which inhibits the hydrogen absorption, proved to be effective. The same effect is also exhibited by a zinc-pigmented inorganic varnish, while a polyurethane coating did not offer effective protection because of swelling [39].

Data on the behavior of iron and steel in acetic acid are summarized in Table 1; see also [40–43].

Possible alternative fuels like methanol and ethanol can cause corrosion problems because of their water contents and the formation of acids by oxidation on long exposure to atmospheric oxygen. Therefore, tests with ethanol and 500 ppm acetic acid were performed, which showed that a noticeable increase in the corrosion rates of unalloyed steels compared with their behavior in pure alcohol occurs. Inhibitors like morpholine, piperazine, and hexamethylenediamine can lower the corrosion rates to acceptable values [44]. Similar corrosion tests of unalloyed steels in acetic acid containing alcohols confirm the results mentioned above [45, 46].

The influence of carbon and chromium content, the manufacturing conditions, and the heat treatment on the SCC susceptibility of unalloyed steels in acetic acid containing ethanol was studied in constant strain rate tests. It was observed that Cr-contents (0.04 %) improved the resistance to SCC while temper-rolling and lower coiling temperature led to a decrease. The C-content showed no influence [47].

The presence of hydrocarbons in 0.0066 mol/l acetic acid leads to a doubling of the corrosion rates of St 3 (cf. 1.0333), while the H-uptake by the steel is decreased [48].

Furthermore do amounts of Cl^-- and NO_3^--ions ($> 10^{-4}$ mol/l) in deaerated 0.05 mol/l acetic acid accelerate the anodic dissolution of unalloyed steels. NO_2^--ions tended to promote passivation [49].

The corrosion of iron in aqueous acetic acid solutions of 358 K (85°C) is inhibited by iodide ions on addition of potassium iodide [50].

To inhibit the corrosion of iron, steel, Fe-alloys and nickel in non-oxidizing acids (inter alia acetic acid), a mixture of 95 % by volume of propargyl alcohol and 5 % by volume of dipropargyl formal (dosage 0.001 to 1.0 % by volume, relative to the volume of the acid at an acid concentration of 1 to 15 %) has been proposed in a patent.

The two components of the inhibitor have a synergistic effect with respect to one another [51].

Material	Acetic Acid %	Temperature K (°C)	Corrosion rate mm/a (mpy)	Remarks	Literature
Pure iron	33	boiling point	0.145 (5.71)		[52]
Armco® iron	33	boiling point	0.16 (6.30)		[53]
Steel	33	boiling point	2.33 (91.7)		[53]
Steel	5 to 20	room temp.	1.14 to 1.37 (44.9 to 53.9)		[52]
	glacial acetic acid	room temp.	0.36 (14.2)	hydrogen bubbled through	[52]
	glacial acetic acid	room temp.	17.3 (681)	oxygen bubbled through	[52]
	33	boiling point	3.28 (129)		[52]
Unalloyed steel	5	293 (20)	0.79 (31.1)		[53]
	15	293 (20)	1.23 (48.4)		[53]
	33	293 (20)	1.34 (52.8)		[53]
Steel	5 to 100	297 (24)	> 1.25 (> 50)	aerated	[54, 55]
	5 to 100	297 (24)	> 1.25 (> 50)	air-free	[54, 55]
	50	373 (100)	> 1.25 (> 50)	acetic acid vapor	[54, 55]
	100	373 (100)	> 1.25 (> 50)	acetic acid vapor	[54, 55]
	100	297 (24)	> 1.25 (> 50)	acetic acid + acetic anhydride + peracetic acid	[54, 55]
	100	297 (24)	> 1.25 (> 50)	acetic acid + acetone + carbon tetrachloride	[54, 55]
	100	297 (24)	> 1.25 (> 50)	acetic acid + ether	[54, 55]
	100	297 (24)	> 1.25 (> 50)	acetic acid + ethyl alcohol	[54, 55]
Steel	100	297 (24)	> 1.25 (> 50)	acetic acid + formic acid	[54, 55]

Table 1: Corrosion behavior of iron and steel in acetic acid of various concentrations and temperatures

Table 1: Continued

Material	Acetic Acid %	Temperature K (°C)	Corrosion rate mm/a (mpy)	Remarks	Literature
	100	297 (24)	> 1.25 (> 50)	acetic acid + hydrobromic acid	[54, 55]
	100	297 (24)	> 1.25 (> 50)	acetic acid + hydrochloric acid	[54, 55]
	100	297 (24)	> 1.25 (> 50)	acetic acid + mercury salts	[54, 55]
	100	297 (24)	> 1.25 (> 50)	acetic acid + sulfuric acid	[54, 55]
Cu-Steel	3	311 (38)	1.21 (47.6)		[53]
Mn-Steel	33	boiling point	1.18 (46.6)		[53]

Table 1: Corrosion behavior of iron and steel in acetic acid of various concentrations and temperatures

Since sorbic acid is used as preservative in the packing industry, its effect on the corrosion of unalloyed steel was investigated in 0.1 mol/l acetic acid solutions. It was shown that 1 g/l sorbic acid inhibits the corrosion to an extent of 18 % in the presence of oxygen and of 41 % in the absence of oxygen in the solution. The test duration was 20 days. Smaller amounts of sorbic acid (for example 0.1 or 0.01 g/l) cause stimulation of corrosion [56].

Red rhodamine 6Zh KDM showed good inhibiting efficiency for the steels St 2 (1.0330, cf. SAE 1008) and St 3 (1.0333) in 1 to 5 mol/l acetic acid solutions at 298 K (25°C) but was less effective at higher temperatures [57]. The inhibiting efficiency for St 2 at an inhibitor concentration of 0.6 % in 1 mol/l acetic acid was 97.2 % at 298 K (25°C).

Dithiocarbamates increase the corrosion rates of low carbon steels (St 3, St 10) in acetic acid solutions by 1.3 to 1.7 times but decrease them in the presence of N-containing corrosion inhibitors such as Katapin BPV, ChM-r, PKU-M or I-1-A [58].

In 10 % acetic acid solution, the effect of the addition of piperidine, Katapin, and KPI-4 was investigated with additional cathodic protection. The corrosion inhibitors suppressed the electrochemical corrosion of Fe, but had no effect on its chemical dissolution [59].

Testing of some urotropin derivates showed that benzyl-hexamethylene-tetraamineiodide at 0.003 mol/l had the best inhibition efficiency for St 3 in acetic acid solutions [60]. For technical use the inhibition effect of the substance proved to be to small.

It is theoretically shown that a relationship between the inhibition of weight loss and hydrogen induced corrosion cannot be generally expected. Experiments were performed with ten commercially available inhibitors for pickling and sour gas service in various solutions of 6 ml/l acetic acid H_2S-saturated at pH 3 either with or without the addition of 50 g/l NaCl or in the absence of H_2S with NaCl at pH 3 and bubbled through by N_2 [61].

The corrosion rate of steel ((%) 0.85 C-0.25 Si-0.002 S-0.01 P-0.08 Cr-0.14 Ni) in acetic acid is increased by 20 mg/l selenium dioxide. It is diminished by the addition of small amounts of aniline or pyridine [62]. Likewise, the corrosion of unalloyed steel in aqueous acetic acid solution decreases in the presence of water-soluble phenols (from slate) [63].

In simulating mixtures of the triacetate fiber production the influence of the water content on the corrosion of St 3 was investigated. The mixture consists of methylene chloride/ethanol (9.1) containing HCl, acetic acid and water. The corrosion rate increased with rising water concentration [64].

St 3 with anodic protection is recommended for plant components for the acetyl cellulose production. This was investigated in aqueous acetic acid solutions of H_2SO_4. An equation was derived for predicting the corrosion rate at any ratio of the concentrations. At a potential of higher than 1.0 V the material consumption rate was 0.5 g/m^2 h [65].

– *Acetic acid vapors* –

Hot acetic acid vapor is catalytically cleaved into carbon, water, acetone and many more on iron surfaces from about 673 K (400°C) upwards [52]. During the distillation of mixtures of acetic acid, benzene and water, pyrophoric iron was formed by corrosion in a boiler, producing a darting flame when the boiler was opened after the distillation [52].

Vapors which are given off from elastomers or plastics such as phenol-formaldehyde and urea-formaldehyde resins even at 308 K (35°C) and 100 % relative humidity have a very corrosive effect on iron and on metals such as aluminium, copper, cadmium, due to their contents of acetic acid, formaldehyde and formic acid; even at air humidities from 70 % upwards, corrosion has to be expected [66].

Wood contains acetic acid esters, which hydrolyse in the presence of water to free acetic acid. This is the reason for the accelerated corrosion of unalloyed steels in contact with wood and in the vapor space near wood surfaces. Moisture contents of wood and air, and temperature influence the corrosion rate [67].

As for the corrosion of iron in wooden casings caused by volatile acetic acid from the wood [68].

The used solvent which is formed when lubricating oil is purified by means of furfural to remove olefinic and naphthenic components is partially oxidized to acetic acid during the subsequent processing at 478 to 533 K (205 to 260°C) so that unalloyed steel, in particular in the vapor phase, is severely attacked; the recovery columns are therefore lined with Monel®, which ensures a long service life [69].

The damage on iron materials, which had occurred as a result of the fully demineralized boiler feedwater in power plants, consisted in destruction of the material in the region of high heat transfers, where organic and inorganic acids such as acetic acid, formic acid, hydrochloric acid, sulfuric acid and the like were formed by hydrothermal decomposition of organic substances [70].

When steel is exposed to acid vapors, for example of acetic acid, hydrochloric acid, sulfurous acid, followed by condensation of moisture, black horn-like magnetite efflorescences are formed on the steel. These types of corrosion were observed in partially filled emulsion paint containers and on steel which was exposed to the thermal decomposition products of chlorine-containing polymers [71].

Acetic acid vapors, for example from wood, plastics or elastomers, severely attack most metals (Table 2). However, steel and CuZn-alloys can be protected against

Material	Degree of purity %	Material consumption, g/dm^2 Concentration of acetic acid			
		in the vapor			blank
		0.5 ppm	5 ppm	50 ppm	
Mg	99.9	0.38	2.25	5.70	0.10
Al	99.9	0.01	0.02	0.12	0.02
Ti	99.9	< 0.001	< 0.001	< 0.001	< 0.001
Mn			0.92	8.60	0.05
V	99.97	0.035	0.040	0.020	0.045
Zn	99.95	1.16	3.73	0.72	0.01
Fe-sheet	99.99	0.20	2.88	3.50	0.001
Cd	99.50	0.089	2.48	3.05	0.012
Co	99.99	0.014	0.21	0.55	0.002
In	99.995	0.065	0.02	0.87	0.004
Ni	99.9	< 0.001	< 0.001	< 0.06	< 0.001
Mo		0.003	0.003	0.003	0.003
Sn	99.90	< 0.001	< 0.001	< 0.02	< 0.001
Pb-sheet	99.90	1.06	1.57	3.10	0.06
W, sintered		0.029	0.030	0.040	0.020
Cu-sheet	99.99	0.02	0.18	0.60	0.003
Ag-sheet	99.99	< 0.001	< 0.001	< 0.001	< 0.001

Table 2: Corrosion of metals by vapors above aqueous acetic acid solutions (303 K (30°C), 100% relative humidity, test duration 3 weeks) [72]

acetic acid vapor at least for some time by means of metal and alloy coatings (Table 3) if the coatings are anodic with respect to the basic material; ZnNi-, MnSe- and

CdSn-alloys provide the best protection. In the case of duplex coatings (Ni + Cr), which are usually resistant, care must be taken that they are free of pores, because attack of the basic material has been observed at these sites (Table 4) [72].

Acetic acid/lactic acid mixtures in the concentration of 0.1 mol/dm^3 achieved an inhibition efficiency of 70 % of St 3S (cf. 1.0333) in 0.5 mol/l sulfuric acid solution [73].

– Acetic anhydride –

It is possible to distill off acetic anhydride from mixtures of sodium acetate and sulfuryl chloride in iron vacuum thin-film evaporators (drum dryers) by taking advantage of Leidenfrost's phenomenon [74, 75].

In the thermal cracking of acetic acid to acetic anhydride, the crude anhydride produced is fractionated continuously in copper columns and the pure anhydride is condensed in silver condensers. The residue in the boiler of the crude anhydride column, which contains acetic anhydride, tarlike products and higher fatty acids from the crude acetic acid used for the cracking, is further processed in vacuum in boilers equipped with a stirrer made of unalloyed steel or cast iron [74].

In the preparation of di-tert.-butyl sulfide by heating tert.-butyl mercaptan with isobutane and acetic anhydride, iron autoclaves are used. An iron sulfide film is produced beforehand on the inside iron surface of the autoclave by treatment with soluble sodium sulfides in order to prevent unfavorable catalytic effects [74].

Coating	Material consumption of coated steel, g/dm^2 acetic acid concentration in air, ppm								
	0.005	0.05	0.5	2.0	3.5	5.0	20	35	50
Mn	0.03	0.04	0.5	*	*	*	*	*	*
Mn/Se (99:1)	0.06	0.05	0.04	0.12	0.17	0.33	*	*	*
Mn/Zn (35:65)	0.16	0.17	0.50	0.66	*	*	*	*	*
Zn	0.02	0.03	0.25	*	*	*	*	*	*
Zn/Fe (60:40)	0.03	0.04	0.51	*	*	*	0.53	*	*
Zn/Ni (84:16)	0.01	0.008	0.06	0.10	0.20	0.21	0.35	0.51	0.68
Zn/In (76:24)	0.04	0.06	0.05	*	*	*	*	*	*
Cd	0.03	0.04	0.09	*	*	*	*	*	*
Cd/In (48:52)	0.02	0.02	0.09	0.30	*	*	*	*	*
Cd/Sn (70:30)	0.01	0.02	0.04	0.18	0.35	0.56	0.64	0.60	0.69

* specimens completely corroded

Table 3: Corrosion of electrolytically prepared alloy coatings (10 micron) caused by the vapor above aqueous acetic acid solutions (303 K (30°C), 100 % relative humidity, test duration 3 weeks) [72]

Acetic acid in the vapor	Material consumption in the vapor space g/dm²			
	Ni + Cr on steel		Ni + Cr on CuZn-alloy	
	Layer thickness, μm			
ppm	10	20	7.5	15
0.5	0.01	0.001	0.0007	–
5.0	0.11	0.02	0.002	0.0005

Table 4: Corrosion of duplex coatings (Ni + Cr) on steel and CuZn-alloys caused by the vapor solutions (303 K (30°C), 100 % relative humidity, test duration 3 weeks) [72]

Iron and steel are resistant to acetic anhydride up to a maximum temperature of 298 K (25°C); but at 333 K (60°C), the material consumption is already quite extensive [76]. The values given in various volumes of tables differ considerably: thus, corrosion rates of more than 1.25 mm/a (49.2 mpy) [77] and less than 0.05 mm/a (1.97 mpy) [78] are given for steel in acetic anhydride, however, without any more detailed information about the conditions. In the presence of oxygen or oxidizing agents, iron is severely attacked, as expected, by acetic anhydride, oxidizing agents (for example sodium oxide) can lead to very violent reactions in the reaction [79, 80].

In the absence of moisture, iron is quite resistant to acetic anhydride and can be used in the preparation and processing, for example for pumps and tanks [81].

– Acetic anhydride vapor –

At high temperatures above 723 K (450°C), unalloyed iron causes catalytic decomposition of acetic anhydride and acetic acid [74].

– Vinegar –

As for the behavior of iron and steel in wine vinegar, see Table 5 [52].

Material	Conc., %	Corrosion rate mm/a (mpy)
Electrolytic iron	0.025	0.18 (7.09)
Steel	0.03	0.41 (16.1)
Steel	0.3	0.137 (5.39)
Steel	0.9	0.137 (5.39)
Cast iron (2.3 % Si)	2.5	7.29 (287)

Table 5: Behavior of iron and steel in unagitated wine vinegar at room temperature [52]

Despite relatively small losses, unalloyed iron cannot be used as construction material for vinegar fermentation plants, because even such small amounts of iron compounds impair the taste of vinegar [52, 82].

Starting from defects in the coating, steel can be attacked by the organic acids (acetic acid, citric acid, tartaric acid etc.), which are present in fruit juices and foods. Organic acids, such as acetic acid, rapidly dissolve the iron oxides formed, as a result of which the attack is further enhanced by complex formation. Since chromium oxide dissolves much more slowly, this subsurface rusting can be prevented by electrolytically deposited Cr-coatings having a surface layer of chromium oxide with an overall thickness about 10 nm [83]. In the canning industry, chromate-coated sheet metal is used as protection against 3 % acetic acid and other organic acids if it is desirable to economize on tin [84]. As for the effect of nitrite additions on the corrosion behavior of tin-plated iron (tinplate) in fruit juices and fruit acids such as acetic acid, citric acid, tartaric acid etc., where nitrite leads to extensive dissolution of iron in combination with pitting corrosion even in the case of coated containers; see [85]. According to this reference, an electrochemical accelerated test in nitrite-containing acetic acid is very suitable for testing the expected corrosion resistance of such containers.

The following observation is important for the shipping of steel parts: In wooden boxes, acetic acid vapor can be formed, which after some time of storage (2 months are sufficient) leads to severe corrosion of unprotected steel [52, 86].

Unalloyed cast iron

– Acetic acid –

The dry distillation of wood, during which acetic acid is formed, is carried out in cast iron boilers which contain a transition piece made of copper [52].

Unalloyed cast iron [87, 88] is attacked by acetic acid (see Table 6) [89].

According to corrosion tests in 5 % acetic acid, malleable cast iron has no better corrosion resistance than steel or cast iron [90].

– Acetic anhydride –

The attack on cast iron in the heat is considerably affected by the ratio of acetic anhydride: acetic acid and even more by the presence of air. Mixtures which are high in acetic anhydride content attack less than those containing less acetic anhydride. Thus, a mixture consisting of about 92 % acetic anhydride + 6 to 7 % acetic acid + 1 to 2 % ethylidene acetate, such as is formed as crude anhydride in the oxidation of acetaldehyde, can be economically fractionated in cast iron columns under reduced pressure. The forerun is condensed in a condenser made of copper and the main fraction in a silver condenser [91].

Unalloyed cast iron is resistant to acetic anhydride at 298 K (25 °C) [92]. This behavior cannot be improved by small alloying additions [93].

As for the corrosion behavior of unalloyed cast iron in acetic anhydride, see Table 7 [54, 55, 94, 95].

– *Acetic anhydride vapor* –

Cast iron is not resistant to acetic anhydride above 673 K (400°C) [96].

– *Vinegar* –

Cast iron has a corrosion rate of 0.45 to 0.50 mm/a (17.7 and 19.7 mpy) in about 6 % vinegar at room temperature [52].

High-alloy cast iron, high-silicon cast iron

– *Acetic acid* –

High-alloy cast iron (high-silicon cast iron) [97, 98] is not attacked by acetic acid, with the exception of a few cases (see Table 8) [52–55, 99].

The acid-resistant FeSi-alloys with 8 to 18 % (mostly 14 to 15 %) silicon are resistant to acetic acid [100]. High-silicon cast iron (> 14.5 % silicon) is resistant to acetic acid of all concentrations up to their boiling points [101]; pumps for 50 to 75 % acetic acid were in use for 2 years without failure, for example, at 411 K (138°C) [52, 102].

Material	Acetic acid %	Temperature K (°C)	Corrosion rate mm/a (mpy)	Remarks	Literature
Cast iron	20	293 (20)	18.5 (728)		[55]
	20	boiling	68.0 (2,677)		[53]
	60	293 (20)	24.5 (965)		[53]
	60	boiling	30.0 (1,181)		[53]
	100	293 (20)	1.1 (43.3)		[53]
	100	boiling	6.0 (236)		[53]
Cast iron	20 to 60	room temp.	9.1 to 22.8 (359 to 898)		[52]
	glacial acetic acid	room temp.	0.9 to 1.14 (35.4 to 44.9)		[52]
	glacial acetic acid	boiling point	5.7 (224)		[52]

Table 6: Corrosion behavior of cast iron in acetic acid of various concentrations and temperatures

Table 6: Continued

Material	Acetic acid %	Temperature K (°C)	Corrosion rate mm/a (mpy)	Remarks	Literature
Cast iron (gray cast iron)	5 to 100	297 (24)	> 1.25 (> 50)	aerated	[54, 55]
	5 to 100	297 (24)	> 1.25 (> 50)	air-free	[54, 55]
	50	373 (100)	> 1.25 (> 50)	acetic acid vapor	[54, 55]
	100	373 (100)	> 1.25 (> 50)	acetic acid vapor	[54, 55]
	100	297 (24)	> 1.25 (> 50)	acetic acid + acetic anhydride + peracetic acid	[54, 55]
	100	297 (24)	> 1.25 (> 50)	acetic acid + acetone + carbon tetrachloride	[54, 55]
	100	297 (24)	> 1.25 (> 50)	acetic acid + ether	[54, 55]
	100	297 (24)	> 1.25 (> 50)	acetic acid + ethyl alcohol	[54, 55]
	100	297 (24)	> 1.25 (> 50)	acetic acid + formic acid	[54, 55]
	100	297 (24)	> 1.25 (> 50)	acetic acid + hydrobromic acid	[54, 55]
	100	297 (24)	> 1.25 (> 50)	acetic acid + hydrochloric acid	[54, 55]
	100	297 (24)	> 1.25 (> 50)	acetic acid + mercury salts	[54, 55]
	100	297 (24)	> 1.25 (> 50)	acetic acid + sulfuric acid	[54, 55]

Table 6: Corrosion behavior of cast iron in acetic acid of various concentrations and temperatures

Material	Acetic acid %	Temperature K (°C)	Corrosion rate mm/a (mpy)	Remarks	Literature
Cast iron	90	297 (24)	> 1.25 (> 50)		[54]
	100	297 (24)	> 1.25 (> 50)		[54]
Cast iron (gray cast iron)	90	373 (100)	< 0.5 (< 20)	acetic anhydride in acetic acid	[55]

Table 7: Corrosion behavior of unalloyed cast iron in acetic anhydride of various concentrations and temperatures

Table 7: Continued

Material	Acetic acid %	Temperature K (°C)	Corrosion rate mm/a (mpy)	Remarks	Literature
Cast iron		298 (25)	0.72 (28.3)	acetic anhydride	[53, 103]
		boiling	3.95 (156)	: glacial acetic acid = 80:20	
		298 (25)	0.60 (23.6)	= 60:40	
		boiling	16.5 (650)		
		298 (25)	0.17 (6.7)	= 10:90	

Table 7: Corrosion behavior of unalloyed cast iron in acetic anhydride of various concentrations and temperatures

Material	Acetic acid %	Temperature K (°C)	Corrosion rate mm/a (mpy)	Remarks	Literature
Cast iron (Fe-2.5C2.3Si)	33	boiling	3.6 to 7.75 (142 to 305)		[52]
Cast iron (Fe-3C1Si)	1.2	293 (20)	0.1 (3.94)		[53]
(Fe-3C3Si)	1.2	293 (20)	0.17 (6.69)		[53]
(Fe-3C6Si)	1.2	293 (20)	0.09 (3.54)		[53]
Si-cast iron (> 14.5 % Si)	10	293 (20)	< 0.1 (< 4)		[53]
	10	boiling	< 0.1 (< 4)		[53]
	50	293 (20)	< 0.1 (< 4)		[53]
	50	boiling	< 0.1 (< 4)		[53]
	80	293 (20)	< 0.1 (< 4)		[53]
	80	boiling	< 0.1 (< 4)		[53]
	100	293 (20)	< 0.1 (< 4)		[53]
	100	boiling	< 0.1 (< 4)		[53]
	100	293 (20)	< 0.1 (< 4)		[53]
	100	boiling	< 1 (< 40)		[53]
		boiling	1.09 (42.9)	pure acetic acid + pure butyric acid (1:1)	[53]

* concentrated

Table 8: Corrosion behavior of high-alloy cast iron (high-silicon cast iron) in acetic acid of various concentrations and temperatures

Table 8: Continued

Material	Acetic acid %	Temperature K (°C)	Corrosion rate mm/a (mpy)	Remarks	Literature
Si-cast iron (> 14.5 % Si)		boiling	0.28 (11.0)	pure acetic acid + pure butyric acid (1:1) + 50 % water	[53]
	5 to 100	373 (100)	< 0.05 (< 2)	aerated	[54, 55]
	30	398 (125)	0.5 to 1.25 (19.7 to 49.2)	aerated	[55]
	30	398 and 423 (125 and 150)	0.5 to 1.25 (19.7 to 49.2)	air-free	[55]
	100	373 (100)	< 0.05 (< 2)	acetic acid vapor	[54, 55]
	90	373 (100)	< 0.05 (< 2)	90 % acetic acid with acetaldehyde + 7 % water	[55]
	30	323 (50)	< 0.05 (< 2)	90 % acetic acid with acetaldehyde + 7 % water	[55]
	100	373 (100)	< 0.05 (< 2)	acetic acid + acetic anhydride + peracetic acid	[54, 55]
	100	373 (100)	< 0.05 (< 2)	acetic acid + acetone + carbon tetrachloride	[54, 55]
	100	373 (100)	< 0.05 (< 2)	acetic acid + ether	[54, 55]
	100	373 (100)	< 0.05 (< 2)	acetic acid + ethyl alcohol	[54, 55]
	100	373 (100)	< 0.5 (< 20)	acetic acid + formic acid	[54, 55]
	100	297 (24)	> 1.25 (> 50)	acetic acid + hydrobromic acid	[54, 55]
	100	297 (24)	< 0.5 (< 20)	acetic acid + mercury salts	[54, 55]
	100	373 (100)	< 0.5 (< 20)	acetic acid + phenylacetic acid	[54]

* concentrated

Table 8: Corrosion behavior of high-alloy cast iron (high-silicon cast iron) in acetic acid of various concentrations and temperatures

Table 8: Continued

Material	Acetic acid %	Temperature K (°C)	Corrosion rate mm/a (mpy)	Remarks	Literature
Si-cast iron (> 14.5 % Si)	100	373 (100)	< 0.05 (< 2)	acetic acid + salicylic acid vapors	[54, 55]
	100	373 (100)	< 0.05 (< 2)	acetic acid + sulfuric acid	[54, 55]
	100	298 (25)	> 1.25 (> 50)	acetic acid + hydrochloric acid	[55]
	up to 50	398 (125)	< 0.05 (< 2)	acetic acid + formic acid + sulfuric acid	[55]
	60 to 100	398 (125)	< 0.5 (< 20)	acetic acid + formic acid + sulfuric acid	[55]
Si-cast iron (13.87 % Si)	conc.*	378 (105)	0.061 (2.40)		[52, 104]
Si-cast iron (18.22 % Si)	conc.*	378 (105)	0.086 (3.39)		[52, 104]
Cr-cast iron	5	293 (20)	< 0.1 (< 4)		[53]
	5	373 (100)	< 0.1 (< 4)		[53]
	10	293 (20)	< 0.1 (< 4)		[53]
	10	373 (100)	< 1.0 (< 4)		[53]
	60	293 (20)	< 0.1 (< 4)		[53]
	60	373 (100)	< 1.0 (< 4)		[53]
	100	293 (20)	< 0.1 (< 4)		[53]
	100	373 (100)	< 1.0 (< 4)		[53]
CrMo-cast iron	5	293 (20)	< 0.1 (< 4)	The addition of formic acid considerably increases the attack	[53]
	5	373 (100)	< 0.1 (< 4)		
	10	293 (20)	< 0.1 (< 4)		
	10	373 (100)	< 0.1 (< 4)		
	60	293 (20)	< 0.1 (< 4)		
	60	373 (100)	< 0.1 (< 4)		
	100	293 (20)	< 0.1 (< 4)		
	100	373 (100)	< 0.1 (< 4)		

* concentrated

Table 8: Corrosion behavior of high-alloy cast iron (high-silicon cast iron) in acetic acid of various concentrations and temperatures

Boilers made of high-silicon cast iron (> 14.5 % silicon) are used in the production of acetic acid by distillation of wood [52, 105] and by extraction with solvents [106].

High-silicon cast iron is also used in the production of acetic acid from calcium carbide for the reactor and the distillation columns for crude acetic acid [52].

FeSi-alloys, which are high in silicon content, are therefore used for valves, pipelines, pumps, columns, condensers, boilers, stopcocks etc., independent of acetic acid concentration, temperature and degree of aeration. Embrittlement, susceptibility to failure by heat shock and expensive fabrication methods are the most important factors which preclude widespread use of these materials [107, 108].

Valves and tubes made of high-silicon cast iron resisted for 2 years without any failure whatsoever the exposure to acid/ether mixtures at 377 K (104°C), mixtures of 30 % acetic acid + 70 % propionic acid and mixtures of acetic acid + formic acid [52, 109].

From the series of Tantiron® alloys which are high in silicon content, especially the standard type Tantiron® N, which contains 14.25 to 15.25 % silicon is used for handling acetic acid or sulfuric acid [110–112].

The resistance of C-steels to acetic acid is increased by silicon diffusion layers, improved by longer siliconization time and higher carbon content [113].

Cast iron alloyed with 25 to 30 % chromium is resistant to acetic acid of all concentrations up to the boiling point [106]. It is not suitable for a mixture of 5 % acetic acid + 10 % sulfuric acid at 393 K (120°C) [52, 106].

High-chromium cast iron (Guronit® GS2, DIN-Mat.-No. 1.4136, GX70CrMo29-2) is resistant to pure acetic acid of all concentrations and to mixtures of equal parts of acetic acid and formic acid at 373 K (100°C) [52, 106], therefore suitable for pumps and gate valves [52].

Mold-cast iron with 21 % chromium and 4 % aluminium is said to resist pure acetic acid even at 473 K (200°C). In a mixture of 87.23 % acetic acid + 3 % extraction solvent + 9.6 % water + 0.17 % salt (sodium acetate + sodium salicylate), a material consumption rate of 10 to 12 g/m^2 d (0.048 to 0.057 mm/a (1.89 to 2.24 mpy)) was observed at the boiling point in the liquid space; in the vapor space, the attack was considerably stronger [52].

– Acetic anhydride –

High-alloy cast iron (high-silicon cast iron) has varying resistance to acetic anhydride (see Table 9) [53–55, 114].

Columns and centrifugal pumps made of high-silicon cast iron have been used in the acetaldehyde oxidation process for the preparation of acetic anhydride [115].

High-silicon cast iron with 16 % silicon is recommended for pumps for pure acetic anhydride [53].

Structural steels with up to 12% chromium
Ferritic chromium steels with more than 12% chromium
Ferritic-austenitic steels with more than 12% chromium

– Acetic acid –

Stainless chromium steels (see also [116, 117]) are more or less attacked by acetic acid, depending on the concentration and temperature (see Table 10).

A heat-treated molybdenum-containing chromium steel ((%) Fe-0.66C-0.48Mn-1.15Si-10.9Cr-1.12Mo) has a corrosion rate of 0.5 to 1.1 mm/a (19.7 to 43.3 mpy) in 0.5% acetic acid at 303 K (30°C), while a molybdenum-free chromium steel ((%) Fe-0.96C-13.3Cr) has a corrosion rate of 41 to 98 mm/a (1,614 to 3,858 mpy) and a chromium steel ((%) Fe-0.6C- 14.1Cr) 1.3 to 5.5 mm/a (51.2 to 216 mpy) [118]. Accordingly, it is recommended to use the Mo-containing types which have a higher chromium content.

Stainless 13% chromium steel is resistant to acetic acid at moderate temperature, but is attacked by boiling acetic acid of any concentration [119].

The annealing temperature affects the corrosion behavior. The start of the steep increase of the corrosion rates and the position of the corrosion maximum of the steels which contain 13% chromium are shifted to longer tempering times with decreasing tempering temperature from 873 to 823 to 773 K (600 to 550 to 500°C), for example X20Cr13 (SAE 420) in 5%, X17CrNi16-2 (cf. SAE 431), X39CrMo17-1 (1.4122) in 20% boiling acetic acid [120, 121]. The addition of molybdenum has a favorable effect on the behavior in all cases: this is also confirmed by [122–124]. [122] mentions a minimum chromium content of 25% for molybdenum-free chromium steels in boiling acetic acid; it also describes that in the case of 12% chromium as little as 3% molybdenum are sufficient for stabilization. Titanium and nickel do not have any specific effect, and manganese and niobium are even impairing.

A chromium-molybdenum steel ((%) 2.25 Cr, 1 Mo) became embrittled by hydrogen in 0.5% acetic acid saturated with hydrogen sulfide [125]. The effect of temper embrittlement on hydrogen embrittlement was investigated on the same steel in the same medium in [126]. It was found that temper embrittlement enhances the susceptibility to H-embrittlement on grain boundaries; the threshold stress intensity factor decreases with fracture surface transition temperature, grain size, hydrogen content, and has a good correlation with the hydrogen embrittlement susceptibility factor. The relation between these two factors was used to determine the maximum of residual H-content in a pressure wall. A method for preventing hydrogen induced stress corrosion cracking during hydrostatic testing of pressure vessels after service was proposed [126]. Hardened martensitic steels are also susceptible to hydrogen embrittlement and can be subject to stress corrosion cracking in acetic acid + hydrogen sulfide. Cathodic protection by means of magnesium, zinc, aluminium or unalloyed steel is possible; likewise, chromium layers are suitable [127].

Material	Acetic acid %	Temperature K (°C)	Corrosion rate mm/a (mpy)	Remarks	Literature
Si-cast iron (without further details)	100	373 (100)	< 0.05 (< 2)		[54]
Si-cast iron (containing more than 14.5 % Si)	without details	333 (60)	0.33 (13.0)		[114]
Si-cast iron (without further details)	10	398 (125)	< 0.05 (< 2)	acetic anhydride in acetic acid	[54]
	40	398 (125)	< 0.05 (< 2)		
	60	398 (125)	< 0.5 (< 20)		
	80	398 (125)	> 1.25 (> 50)		
Si-cast iron (without further details)		298 (25)	0.16 (6.30)	acetic anhydride :glacial acetic acid = 60:40 10:90	[53, 114]
		boiling	3.75 (148)		
		298 (25)	0.22 (8.66)		
		boiling	0.42 (16.5)		
	pure	333 (60)	0.37 (14.6)		

Table 9: Corrosion behavior of high-silicon cast iron in acetic anhydride of various concentrations and temperatures

Material	Acetic acid %	Temperature K (°C)	Corrosion rate mm/a (mpy)	Remarks	Literature
Chromium steel (17 % Cr)	10	293 (20)	< 0.11 (< 4.33)		[82]
	10	boiling	< 1.1 (< 43.3)		
	50	293 (20)	< 0.11 (< 4.33)		
	50	boiling	< 11.0 (< 433)		
	glacial acetic acid	293 (20)	< 0.11 (< 4.33)		
	glacial acetic acid	boiling	< 1.1 (< 43.3)		

Table 10: Corrosion behavior of stainless chromium steels in acetic acid of various concentrations and temperatures

Table 10: Continued

Material	Acetic acid %	Temperature K (°C)	Corrosion rate mm/a (mpy)	Remarks	Literature
Chromium steel (without further details)	10	293 (20)	< 0.1 (< 4)		[53]
	10	boiling	< 1.0 (< 39.4)		
	50 % conc.	293 (20)	< 0.1 (< 4)		
	50 % conc.	boiling	> 10.0 (> 394)		
Chromium steel (12 % Cr)	10	297 (24)	< 0.5 (< 20)	aerated, avoid traces of HCl, H$_2$SO$_4$ and NaCl	[54]
	10	325 (52)	> 1.25 (> 50)		
	20 to 100	297 (24)	> 1.25 (> 50)		
	10	297 (24)	< 0.5 (< 20)	air-free, avoid traces of HCl, H$_2$SO$_4$ and NaCl	[54]
	10	325 (52)	> 1.25 (> 50)		
	20 to 90	297 (24)	> 1.25 (> 50)		
	100	297 (24)	< 0.5 (< 20)		
	100	373 (100)	> 1.25 (> 50)		
	50	373 (100)	> 1.25 (> 50)	acetic acid vapor	[54]
	100	373 (100)	> 1.25 (> 50)		
	100	373 (100)	> 1.25 (> 50)	acetic acid + acetic anhydride + peracetic acid	[54]
	100	373 (100)	> 1.25 (> 50)	acetic acid + ethyl alcohol	[54]
	100	373 (100)	> 1.25 (> 50)	acetic acid + formic acid	[54]
	100	297 (24)	> 1.25 (> 50)	acetic acid + hydrobromic acid	[54]
	100	297 (24)	> 1.25 (> 50)	acetic acid + hydrochloric acid	[54]
	100	297 (24)	> 1.25 (> 50)	acetic acid + sulfuric acid	[54]

Table 10: Corrosion behavior of stainless chromium steels in acetic acid of various concentrations and temperatures

Table 10: Continued

Material	Acetic acid %	Temperature K (°C)	Corrosion rate mm/a (mpy)	Remarks	Literature
Remanit® steels corresponding to X6CrAl13 (cf. SAE 405) X12CrS13 (cf. SAE 416) X12Cr13 (cf. SAE 410) X20Cr13 (cf. SAE 420)	10	293 (20)			[128]
	10	boiling	1.1–11.0 (43.3–433)		
	50	293 (20)	1.1–11.0 (43.3–433)		
	50	boiling	> 11.0 (> 433)		
	glacial acetic acid 100 %	293 (20)	0.11–1.1 (4.33–43.3)		
	glacial acetic acid 100 %	boiling	> 11.0 (> 433)		
	10	293 (20)	0.11–1.1 (4.33–43.3)	acetic acid + hydrogen peroxide	
	10	323 (50)	1.1–11.0 (43.3–433)		
	10	363 (90)	> 11.0 (> 433)		
	50	293 (20)	0.11–1.1 (4.33–43.3)		
	50	323 (50)	1.1–11.0 (43.3–433)		
	50	363 (90)	> 11.0 (> 433)		
Remanit® steel corresponding to X20CrMo13	10	293 (20)	< 0.11 (< 4.33)		[128]
	10	boiling	1.1–11.0 (43.3–433)		
	50	293 (20)	0.11–1.1 (4.33–43.3)		
	50	boiling	1.1–11.0 (43.3–433)		
	glacial acetic acid 100 %	293 (20)	< 0.11 (< 4.33)		
	glacial acetic acid 100 %	boiling	1.1–11.0 (43.3–433)	acetic acid + hydrogen peroxide	

Table 10: Corrosion behavior of stainless chromium steels in acetic acid of various concentrations and temperatures

Table 10: Continued

Material	Acetic acid %	Temperature K (°C)	Corrosion rate mm/a (mpy)	Remarks	Literature
Remanit® steel corresponding to X20CrMo13	10	293 (20)	< 0.11 (< 4.33)		
	10	323 (50)	< 0.11 (< 4.33)		
	10	363 (90)	1.1–11.0 (43.3–433)		
	50	293 (20)	< 0.11 (< 4.33)		
	50	323 (50)	< 0.11 (< 4.33)		
	50	363 (90)	1.1–11.0 (43.3–433)		
Cr-steel ((%) 13 Cr, 0.20 C, 0.60 Si, 0.60 Mn, 0.40 Ni)	6	room temp.	0.006 (0.24)		[129]
	6	boiling	2.48 (97.6)		
Cr-steel ((%) 13 Cr, 0.20 C, 0.60 Si, 0.60 Mn, 0.40 Ni, 1.15 Mo)	6	room temp.	0.028 (1.10)		[129]
	6	boiling	0.193 (7.6)		
Cr-steel ((%) 12.75 Cr, 0.42 C, 0.45 Si, 0.60 Mn, 0.50 Ni)	6	room temp.	0.005 (0.197)		[129]
	6	boiling	0.052 (2.05)		
Cr-steel ((%) 13.75 Cr, 0.55 C, 0.60 Si, 0.60 Mn, 0.55 Mo)	6	room temp.	0.019 (0.75)		[129]
	6	boiling	0.221 (8.70)		

Table 10: Corrosion behavior of stainless chromium steels in acetic acid of various concentrations and temperatures

Table 10: Continued

Material	Acetic acid %	Temperature K (°C)	Corrosion rate mm/a (mpy)	Remarks	Literature
Cr-steel ((%) 17 Cr, 0.89 C, 0.45 Si, 0.65 Mn, 0.18 Ni, 1.00 Mo, 0.10 V)	6	room temp.	0.005 (0.197)		[129]
	6	boiling	0.023 (0.91)		
Cr-steel ((%) 17.50 Cr, 1.10 C, 0.45 Si, 0.65 Mn, 0.50 Ni, 0,90 Mo)	6	room temp.	0.030 (1.18)		[129]
	6	boiling	0.132 (5.20)		
Cr-steel ((%) 17.25 Cr, 0.65 C, 0.55 Si, 0.60 Mn, 0.50 Ni, 0.50 Mo)	6	room temp.	0.003 (0.12)		[129]
	6	boiling	0.070 (2.76)		
Cr-steel ((%) 16.50 Cr, 0.38 C, 0.55 Si, 0.60 Mn, 0.80 Ni, 1.15 Mo)	6	room temp.	0.002 (0.08)		[129]
	6	boiling	0.009 (0.35)		
Cr-steel ((%) 16.70 Cr, 0.94 C, 0.64 Si, 0.42 Mn, 1.13 Mo, 0.18 V, 2.60 W, 2.54 Co)	6	room temp.	0.003 (0.12)		[129]
	6	boiling	0.094 (3.70)		

Table 10: Corrosion behavior of stainless chromium steels in acetic acid of various concentrations and temperatures

Table 10: Continued

Material	Acetic acid %	Temperature K (°C)	Corrosion rate mm/a (mpy)	Remarks	Literature
Cr-steel ((%) 19.85 Cr, 0.79 C, 0.47 Si, 0.65 Mn, 0.40 Ni, 0.80 Mo)	6	room temp.	0.001 (0.039))		[129]
	6	boiling	0.40 (1.57)		
Chromium Steel (17 % Cr)	10	373 (100)	< 0.05 (< 2)	aerated, also avoid traces of HCl, H$_2$SO$_4$ and NaCl	[54]
	20–50	373 (100)	< 0.5 (< 20)		
	30	477 (204)	> 1.25 (> 50)		
	60–100	352 (79)	0.5–1.25 (19.7–49.2)		
	60	373 (100)	> 1.25 (> 50)		
	80	373 (100)	> 1.25 (> 50)		
	100	373 (100)	> 1.25 (> 50)		
	10	297 (24)	< 0.5 (< 20)		
	10	325–373 (52–100)	0.5–1.25 (19.7–49.2)		
	20–60	373 (100)	0.5–1.25 (19.7–49.2)		
	70–100	352 (79)	0.5–1.25 (19.7–49.2)		
	70–100	373 (100)	> 1.25 (> 50)		
	100	297 (24)	< 0.5 (< 20)		
	50	373 (100)	> 1.25 (> 50)	acetic acid vapor	
	100	373 (100)	> 1.25 (> 50)		
	100	373 (100)	> 1.25 (> 50)	acetic acid + acetic anhydride + peracetic acid	

Table 10: Corrosion behavior of stainless chromium steels in acetic acid of various concentrations and temperatures

Table 10: Continued

Material	Acetic acid %	Temperature K (°C)	Corrosion rate mm/a (mpy)	Remarks	Literature
Chromium Steel (17 % Cr)	100	373 (100)	> 1.25 (> 50)	acetic acid + ethyl alcohol	
	100	373 (100)	> 1.25 (> 50)	acetic acid + formic acid	
	100	297 (24)	> 1.25 (> 50)	acetic acid + hydrobromic acid	
	100	297 (24)	> 1.25 (> 50)	acetic acid + hydrochloric acid	
	100	297 (24)	> 1.25 (> 50)	acetic acid + sulfuric acid	
Stainless martensitic steel SAE 405 and SAE 410, CrMn-steel 18 9	60	373 (100)	> 1.25 (> 50)	aerated	[54]
	10	293 (20)	< 0.1 (< 4)		[53]
	10	boiling	< 0.1 (< 4)		
	50	293 (20)	< 0.1 (< 4)		
	50	boiling	< 1.0 (< 39.4)		
	conc.	293 (20)	< 0.1 (< 4)		
	conc.	boiling	< 1.0 (< 39.4)		
	conc.	0.1 MPa/473	> 10.0 (> 394)		
CrMo-steel	10 up to conc.	293 (20)	< 0.1 (< 4)		[53]
	10 up to conc.	boiling	< 0.1 (< 4)		
CrSi-steel	25	293 (20)	0.26 (10.2)		[53]
			0.95 (37.4)		
			0.014 (0.55)		
			0.48 (18.9)		
			0.17 (6.69)		

Table 10: Corrosion behavior of stainless chromium steels in acetic acid of various concentrations and temperatures

The Soviet chromium steels Ch17T (cf. 1.4510, SAE 439) and Ch12N2 are resistant to acetic acid up to concentrations of 55 % and temperatures up to 343 K (70°C), even when welded [130]. The stainless cast steel Ch17M2TL has good resistance in boiling 74 % acetic acid at 373 K (100°C) and is suitable as replacement for CrNi-steels 18 8 [131]. Chromium steel Ch17T is slightly attacked in boiling 1 to 15 % acetic acid [132]. In tar water which is formed during peat processing and the main corrosive agent of which is acetic acid, 17 % chromium steel is considered sufficiently resistant. Investigations carried out in this context showed that this steel is also only slightly attacked by solutions which contain 1 to 15 % acetic acid, 5 to 40 % ammonium sulfate and 0.6 % ethanol and the corrosion reaches a constant rate after 50 h [132]. In crude acetic acid, the corrosion rate of stainless Cr-steels remains small: for CrTi-steel 28: 0.002 mm/a (0.079 mpy), CrMoTi-steel 28 (1 to 2 Mo): 0.003 mm/a (0.12 mpy), CrMoTi-steel 25 2: 0.008 mm/a (0.31 mpy), CrNiMoTi-steel 21 6 2: 0.001 mm/a (0.039 mpy).

In acetic acid at the boiling point the corrosion rate of the steel SAE 430 (X6Cr17) considerably increased as the amount of water decreased below 5 %. The corrosion rate at a water content of about 65 ppm was determined to be 1,136 mm/a (44,724 mpy) [133]. A further investigation of the water influence on the corrosion behavior of SAE 430 was carried out in [134]. Weight loss measurements were made at ambient temperature and at the solution boiling temperatures. High corrosion rates were particularly found in pure acetic acid, which confirm to the value mentioned above. The high conductivity of the acetic acid at the boiling point and the small Cr_2O_3 content of the surface layer were given as reasons for the high corrosion rates.

Similar corrosion investigations in acetic acid were performed for SAE 444 (X2CrMoTi18-2) in [135].

Ion-nitrided surface layers of SAE 430 showed inferior corrosion resistance to the original steel in 0.1 mol/l acetic acid [136].

However, in technical grade acid, the corrosion rates remain within acceptable limits only in the case of high chromium content (CrTi-steel 28). In a mixture of (%) 75 acetic acid, 5 formic acid, 2 propionic acid, 0.2 sulfuric acid, this steel loses 0.1 mm/a (3.94 mpy), and, in a mixture of (%) 9 acetic acid, 5 formic acid, 1 propionic acid, Cr-steel 28 loses 0.024 mm/a (0.94 mpy), CrMoTi-steel 25 (1 to 2 % Mo) 0.014 mm/a (0.55 mpy) and CrNiMoTi-steel 21 6 2 0.018 mm/a (0.71 mpy).

The stainless ferritic-austenitic steels 0Ch21N5T and 0Ch21N6M2T have good resistance in 6 % acetic acid and other organic acids at room temperature and at 333 K (60°C). As expected, the addition of chloride ions increases the weight loss. In acid solutions which have a high chloride content, ferritic-austenitic steels, in particular 0Ch21N5T, corrode faster than the austenitic steels [137].

Technical media for the production of acetic acid by the synthesis from methanol and carbon monoxide often contain 0.01 to 1 % hydrogen iodide. This iodide content was found to cause pitting corrosion in 70 % acetic acid solutions at 363 K (90°C) on steel 08Ch22N6T, and was promoted in the presence of oxidants. Sodium hypophosphite proved to hinder pitting as inhibiting agent [138].

For the production of 2-methyl-5-methoxybenzotriazole by reduction of 4.4′-dimethoxy-2.2′-dinitrophenyl sulfide by zinc powder in acetic acid, the steel 08Ch21N6M2T (X 8 CrNiMoTi 21 6 2) exhibits sufficient corrosion resistance in this medium [139].

The stainless titanium-stabilized steel Uniloy® 326 ((%) 26 Cr, 6.5 Ni, max. 0.05C, 0.2 Ti), a two-phase alloy with 30% austenite in a ferrite matrix, is superior to the stainless steels SAE 304 and 316 in its resistance to acetic acid and acetic anhydride, even with respect to stress corrosion cracking and crevice corrosion and comparable to NiCr-steels 18 16 [140].

The corrosion resistance of nickel-free austenitic Tenelon® steels ((%) 17 Cr, 14.4 Mn, 0.1 C, 0.4 N, 1 Si, 0.045 S or P) in 60% acetic acid at 343 K (70°C), test duration 96 h, which have a corrosion rate of about 0.04 mm/a (1.57 mpy), correspond to that of the steel SAE 302 [141].

On a test duration of 100 h at 323 K (50°C) the Ni-free CrMn-steels 12Ch13G18D and especially 10Ch14AG15 showed also good resistance in 1, 5, 10, and 15% acid solutions [142].

The ultra-low C and N high-chromium ferritic steels S 30-2 and SR 26-4 with 2 to 4% Mo were used for rectification towers for acetic acid plants [143].

According to data, from the manufacturer the CrMo-steel 18 2 is clearly superior to the CrNi-steel X5CrNi18-10 (1.4301, SAE 304) in boiling acetic acid and to the CrNiMo-steel X5CrNiMo17-12-2 (1.4401, SAE 316) in 20% acetic acid [144]; (Table 11).

Steels	Acetic acid concentration	
	20%	50%
	Material consumption rate mg/dm² d (mm/a)	
X5CrNi18-10 (SAE 304)	164 (0.8)	1,060 (4.5)
X5CrNiMo17-12-2 (SAE 316)	2 (0.01)	
18Cr2Mo	1 (0.05)	2 (0.01)

Table 11: Corrosion behavior of a CrMo-steel in comparison to that of CrNi-steels in boiling acetic acid of various concentrations [144]

Ferritic steels of the type 18Cr 2Mo is said to be extremely resistant in boiling acetic acid (20 to 80%); based on the test duration of 24 h, the corrosion rates are said to be only 0.005 mm/a (0.20 mpy) [145].

For the two CrMo-steels Remanit® 4521 (X2CrMoTi18-2, SAE 444) and Remanit® 4522 (X2CrMoNb18-2, SAE 443), corrosion rates of less than 0.11 mm/a (4.33 mpy) in 10 and 50% acetic acid at 293 and 323 K (20 and 50°C), glacial acetic acid at 293 K (20°C) and in acetic acid (10 and 50%) with hydrogen peroxide are mentioned; only in boiling glacial acetic acid are the values between 0.11 and 1.1 mm/a (4.33 and 43.3 mpy) [146].

As for the corrosion resistance of other Remanit® chromium steels, see Table 10.

For cutting synthetic fibers, knives made of chromium-molybdenum steels, which are resistant to acetic acid, can be used [147]; (see Table 10).

Midvale® 2024, an alloy containing (%) 26 Cr, 4 Mo, is said to be particularly resistant to acetic acid [52].

It is true that conventional ferritic steels are resistant to stress corrosion cracking, but they are difficult to weld. Superferrites such as 18Cr2Mo, 26Cr1Mo, 29Cr4Mo can be welded by the TIG and MIG processes and are even superior to austenitic steels in their corrosion resistance, inter alia, to acetic acid, and formic acid [148].

The same corrosion behavior is also exhibited by ferritic steels which contain 35 % chromium or (%) 28 Cr, 2 Mo, 0.2 and 0.4 Ni and 0.5 Cu, max. 0.006 N, which are superior to austenitic CrNiMo-steels which contain up to 2.8 % Mo and also to the steels of the type X3CrNiMo17-13-3 (1.4436, SAE 316), which are extremely resistant to pitting corrosion. In terms of resistance to pitting corrosion, the steel containing 28 % chromium is superior to that containing 35 % chromium [149].

The ferritic chromium steel with (%) 26 Cr, 1.0 Mo, 0.001 C, 0.010 N, 0.25 Si, 0.01 Mn, 0.01 S, 0.01 P, 0.12 Ni + Cu, commercial name E-Brite® 26-1 (UNS S44627) from Airco Vacuum Metals and Orion® 26-1 from Creusot-Loire, DIN-Mat. No. 1.4131 ((%) 26 Cr, 0.0020 C, 0.400 Mn, 0.400 Si, 1.00 Mo, 0.015 N, 0.500 Ni + Cu), which are prepared by electron beam remelting, are ductile and weldable in contrast to the previous ferritic steels, due to their very high purities. They are distinguished by excellent resistance to stress corrosion cracking in neutral chloride solutions and at the same time high resistance to uniform corrosion, in which it is superior to the stainless austenitic steels of the classical type 18 10 Mo. The absence of nickel leads to good behavior towards media which convert nickel into complex compounds (organic acids, hydrogen sulfide). As for the corrosion rate of E-Brite® 26-1 and Orion® 26-1 in boiling acetic acid, see Table 12 [150, 151].

There are many possible applications for the type of low interstitial steels (ASTM XM-27, (%) 26 Cr, 1 Mo) of ferritic structure in chemical plant construction for tubes and tube-like equipment for organic acids such as acetic acid, formic acid, oxalic acid, sulfide-containing media and chloride-containing water, which lead to stress corrosion cracking [150–154]. Tubes made of the stainless ferritic CrMo-steel 26 1 have successfully been used for the handling of hot crude acetic acid for 2 years; tubes made of the steel SAE 304, which had been corroded by stress corrosion cracking, were replaced by 26 1 tubes in a condenser for aqueous acetic acid [152, 153].

A ferritic CrMo-steel ((%) 28 Cr, 2 Mo, max. 0.010 C + N) when welded and sensibilization-annealed is not susceptible to intercrystalline corrosion and is highly resistant to acetic, formic and oxalic acid [155].

High chromium ferritic low-carbon steels ((%) 23.8 to 25 Cr, 0.6 to 4.7 Mo, 0.003 to 0.005 C) melted in vacuo and stabilized with niobium have good resistance in hot acetic acid solutions at a ratio Nb/(C + N) ≥ 20 to intercrystalline corrosion [156]. The CrMo-steel (X6CrMo17-1, cf. 1.4113, SAE 434) is also very resistant in the CASS test [157].

As for the behavior of stainless chromium steels towards acetic acid, see also Table 10.

– *Acetic acid vapor* –

The resistance of 17% chromium steel to acetic acid of all concentrations at all temperatures up to the boiling point is improved by 2% molybdenum. According to laboratory tests, 17% Cr steel is resistant to boiling 10% acetic acid even when it does not contain molybdenum. Additions of 1% molybdenum and 1% copper improve the corrosion resistance to vapors of 99.8% acetic acid considerably. Under these conditions, usually pitting corrosion takes place, which, however, is very slight in the presence of molybdenum + copper. The material consumption of 17% chromium steel with 1% Mo and 1% Cu was 0.084 g/m^2 (0.00097 mm/a (0.038 mpy)) after 96 h in comparison to 6.36 g/m^2 (0.074 mm/a (2.91 mpy)) for the molybdenum-free steel [158].

Attacking medium	Temperature	Corrosion rate, mm/a (mpy)				Literature
		E-Brite® 26-1	Orion® 26-1	SAE 316	SAE 304 L	
Acetic acid/acetic anhydride (1:1)	boiling	0.203 (7.99)		1.067 (42.0)		[151]
Acetic acid (95%) + formic acid (5%)	boiling	0.0254 (1.00)		0.229 (9.02)		[151]
Glacial acetic acid	boiling	0.0254 (1.00)		0.229 (9.02)		[151]
Acetic acid (99.7%)	boiling		0.0 (0.00)	0.090 (3.54)	0.170 (6.69)	[150]
Acetic acid (71%) + formic acid (7%)	boiling		0.004 (0.16)	2.000 (78.7)	3.000 (118)	[150]

Table 12: General corrosion rate of E-Brite® 26-1 (XM 27) and Orion® 26-1 in boiling acetic acid of various concentrations compared with the stainless steel types SAE 316 and 304 L

In chloride-containing vapors, all stainless steels, in particular the molybdenum-free steels, are attacked by pitting corrosion [159].

– *Acetic anhydride* –

For acetylation installations for the production of acetyl cellulose, these steels are not suitable. Exceptions are CrMo-steels which contain (%) 16 to 18 chromium, 1.5 to 2.0 molybdenum. These are virtually resistant even at the boiling point of the acetic anhydride [160].

Stainless chromium steels are attacked by acetic anhydride more or less, depending on concentration and temperature; see Table 13.

The ferritic and martensitic steels which contain 12 to 18% chromium (DIN-Mat. No. 1.4401, 1.4016, 1.4021, 1.4057, 1.4501) have good resistance to acetic anhydride at room temperature but lose 1.1 to > 11 mm/a (43.3 to > 433 mpy) at its boiling point [160].

– Vinegar –

Tests using various stainless steels (Cr-steels 12 to 18 Cr, CrNi-steels 18 8 and CrNiMo-steels 18 8 2) in 5 % acetic acid at 313 K and 373 K (40°C and 100°C) suggest that, when selecting materials which come into contact with food, care should always be taken to subject them to a certain initial stress: the reason for this is that in about the first 10 d ppm amounts of the individual alloy elements pass into the solution, while after this time virtually no more loss can be detected [161].

Material	Acetic anhydride %	Temperature K (°C)	Corrosion rate mm/a (mpy)	Remarks	Literature
Chromium steel (without specified details)	no data given	293 (20)	< 0.1 (< 4)		[53]
		boiling	> 10.0 (> 394)		
Chromium steel (12 % Cr)	30	352 (79)	> 1.25 (> 50)		[54]
	50	297 (24)	0.5 to 1.25 (19.7 to 49.2)		
	90	373 (100)	> 1.25 (> 50)		
	100	373 (100)	0.5 to 1.25 (19.7 to 49.2)		
Chromium steel (17 % Cr)	no data given	293 (20)		resistant	[114]
		boiling point	10.94 (431)		
Chromium steel (17 % Cr)	30	352 (79)	> 1.25 (> 50)		[54]
	50	297 (24)	> 1.25 (> 50)		
	90	297 (24)	0.5 to 1.25 (19.7 to 49.2)		
	90	325 (52)	> 1.25 (> 50)		
	100	297 to 373 (24 to 100)	> 1.25 (> 50)		
Remanit® steels corresponding to X12Cr13 cf. SAE 410 X15Cr13	no data given	293 (20)	< 0.11 (< 4.33)		[162]
		boiling	1.1 to 11.0 (43.3 to 433)		
CrMn-steel 18 9	no data given	293 (20)	< 0.1 (< 4)		[53]
		boiling	< 10.0 (< 394)		

Table 13: Corrosion behavior of stainless chromium steels towards acetic anhydride of various concentrations and temperatures

Table 13: Continued

Material	Acetic anhydride %	Temperature K (°C)	Corrosion rate mm/a (mpy)	Remarks	Literature
CrNiMn-steel	crude	389 (116)	2.61 (103)		[163, 164]
SAE 201 (%) 16 to 18 Cr, 3.5 to 5.5 Ni, 6.5 Mn, < 0.15 C	vapor	383 (110)	1.70 (66.9)		
CrMo-steel	no data given	293 (20)	< 0.1 (< 4)		[53]
		boiling	< 0.1 (< 4)		
CrMo-steel 17 2	no data given	boiling point		resistant	[114]
Remanit® steel corresponding to CrMo-steel with 13 % Cr	no data given	293 (20)	< 0.11 (< 4.33)		[165]
		boiling	0.11 to 1.1 (4.33 to 43.3)		

Table 13: Corrosion behavior of stainless chromium steels towards acetic anhydride of various concentrations and temperatures

Nitrogen-containing stainless chromium-manganese steels with (%) 20 to 22 Cr, 12 to 14 Mn, 0.5 to 1.0 N, max. 0.10 C are very corrosion-resistant to vinegar and citric acid and thus comparable to 18 8 steels [166].

Austenitic chromium-nickel steels

– Acetic acid –

According to an older source, stainless steel is said to be suitable for acetic acid of any concentration up to 495 K (222°C), based on observations in industrial plants [167]. This is contradicted by the fact that above 423 K (150°C) even the austenitic steels were destroyed in concentrated acetic acid [52]. In the case of completely stabilized austenite structure, the chromium content has in general no effect on the resistance to acetic acid [52, 168].

CrNi-steels are used in the reaction of methanol with carbon monoxide to give acetic acid at 30.0 to 40.0 MPa at temperatures of up to 773 K (500°C) [52, 169]. Likewise, autoclaves made of austenitic steel are used for gas mixtures of aqueous 80 % methanol with small amounts of cobalt acetate, hydrogen iodide and carbon disulfide, into which carbon monoxide is injected under pressure at 473 K (200°C), which gives mixtures of about 80 % free acetic acid and 20 % methyl acetate [52, 170].

Higher-boiling acids and substances, such as are formed during oxidation, as well as the presence of salicylic acid (workup of acids of the acetyl salicylic acid preparation) have a considerable effect on the resistance [52]; (see Table 14).

In the cellulose and paper industry, solutions containing sulfur dioxide and liberated organic acids – mainly acetic acid and formic acid – cause severe attack in the sulfite process at higher pressures and temperatures, which is the reason why acid-resistant materials are used in the sulfite plants, for example alloys based on NiMo or NiMoCr or steels with the DIN-Mat. No. 1.4580 (SAE 316 Cb), 1.4438 (SAE 317 L) for digesters, 1.4301 (SAE 304), 1.4550 (SAE 347), 1.4541 (SAE 321), 1.4401 (SAE 316), 1.4580 (SAE 316 Cb), 1.4571 (SAE 316 Ti) in cellulose processing before or after the chlorination. Equipment which are under chemical and thermal stress, such as digesters, evaporators and pressure containers, and the respective pipelines and fittings must be manufactured from steels which are resistant to intergranular corrosion (solid sheet or clad steel) [171].

A stainless steel distillation boiler for separating a mixture of acetaldehyde with 1 % acetic acid corroded at the unalloyed steel flanges and had stress corrosion cracks at the unalloyed steel/stainless steel contact areas; it is assumed that the failure was caused by a defect in the stainless steel [172].

Material	Corrosion rate, mm/a (mpy)				
	a	b	c	d	e
Monel®	2.05 (80.7)	2.25 (88.6)	1.65 (65.0)	1.025 (40.3)	8.0 (314)
Nickel	1.3 (51.2)	2.425[2] (95.5)	2.425[2] (95.5)	2.425[2] (95.5)	2.425[2] (95.5)
Inconel®	0.95 (37.4)	1.875 (73.8)	2.375[2] (93.5)	2.375[2] (93.5)	2.4[2] (94.5)
SAE 304	0.225 (8.85)	0.525 (20.7)	2.625[2] (103)	0.625 (24.6)	1.05 (41.3)
SAE 316	0.03 (1.18)	0.35 (13.8)	0.8 (31.5)	0.04 (1.57)	0.0175 (0.69)
SAE 317	0.0275 (1.08)	0.125 (4.92)	0.25 (9.84)	0.0325 (1.28)	0.03 (1.18)
Carpenter® 20	0.025 (0.98)	0.125 (4.92)	0.25 (9.84)	0.0425 (1.67)	0.005 (0.197)
Hastelloy® C	0.005 (0.197)	0.0075 (0.30)	0.01 (0.39)	0.01 (0.39)	0.0125 (0.49)
Copper		2.175[1] (85.6)	0.95[1] (37.4)	0.6[1] (23.6)	2.5[1] (98.4)
Aluminium bronze (10 % Al and 3.5 % Fe)		2.5[1] (98.4)	1.55[1] (61.0)	0.6[1] (23.6)	3.5[1] (138)

1) calculation of rates based on a test duration of 900 h
2) minimum rate – after 1,380 h specimens had been destroyed
a) Test under service conditions, duration 891 h in vapors above plate 8 of the column at 393 K (120°C). Composition of the test solution: 91.68 acetic acid; 0.86 salicylic acid; 7.46 water
b) Test under service conditions, duration 1,530 h in vapors above plate 17 and 391.9 K (118.9°C). Composition: 91.68 acetic acid; 0.95 salicylic acid; balance water
c) Test under service conditions, duration 1,380 h in vapors above plate 24 at 380.7 K (107.7°C). Composition: 74.66 acetic acid; 0.27 salicylic acid; balance water
d) Test under service conditions, duration 1,380 h above plate 32 in vapors of 376.9 K (103.9°C). Composition: 55.5 acetic acid; 0.016 salicylic acid; balance water
e) Test under service conditions, duration 1,380 h above plate 40 in vapors of 373 K (100°C). Composition: 23.44 acetic acid; 0.011 salicylic acid; balance water

Table 14: Results of service relevant corrosion tests using various materials under the conditions of acetic acid distillation [173]

Stirrers made of CrNi-steel 18 8 are very suitable for the nitration of 2-methylnaphthalene, in which acetic acid alone or in a mixture with acetic anhydride is used as solvent [52, 174].

The results obtained in reaction media during the terephthalic acid preparation by oxidation of xylene lead to the observation that in 98 % chemically pure boiling acetic acid, not only CrNi-steel 12Ch18N10T (cf. SAE 321) but also CrNiMo-steels 10Ch17N13M2T (cf. SAE 316 Ti), 10Ch17N13M3T (cf. SAE 316 Ti) and 03Ch21N21M4B and the NiCrMoV-alloy ChN65MV are resistant.

In boiling 99.94 % synthetic acetic acid with 0.5 % formic acid, resistance can only be guaranteed for the Ni-alloy, which can therefore be used and is recommended for rectification columns. The CrNi-steel 18 10 is only suitable up to 323 K (50°C). For these steels 0.01 to 0.06 mol/l bromide ions lead to pitting corrosion and material consumption rates of more than 1 g/m^2 h [175].

In general austenitic CrNi-steels are resistant to acetic acid solutions of any concentration [176]. High corrosion attack, which was observed on these steels, was referred to originate from the chloride contamination of the acetic acid.

Therefore, corrosion tests were performed with welded samples and looped specimens of 1.4541 in nearly anhydrous acetic acid of 99.5 to 99.95 % with chloride contents of less than 1 up to 100 ppm. The test temperature was 391 K (118°C) at normal pressure or 423 K (150°C) under pressure for a test duration of up to 246 d. Pitting corrosion was found in all cases also at chloride contents less than 1 ppm. Under access of oxygen high corrosion rates can occur and pitting corrosion and general corrosion may overlap. Susceptibility to chloride induced transcrystalline stress corrosion cracking was not found under these conditions. For further details on test conditions, material composition and test results see [177].

The same is true for the reaction mixture containing 88 % acetic acid, 1.3 g/l cobalt ions and 1.42 g/l bromide ions at 373 K (100°C) and for 98 % acetic acid containing cobalt and bromide ions; the corrosion of the CrNi-steel 12Ch18N10T (cf. SAE 321 and the CrNiMo-steel 10Ch17N13M2T (cf. SAE 316 Ti) is somewhat diminished by cobalt and bromide ions mentioned, while that of the other three materials is not affected.

Stainless steel linings in the BASF high-pressure process for the production of acetic acid from methanol and carbon monoxide corroded in the autoclaves under the synthesis conditions within a very short time [178].

The austenitic chromium nickel steels which are low in carbon (SAE 304, 304 L, 310, 316, 316 L, 347) can be used at normal temperatures for acetic acid of all concentrations. The molybdenum-containing steels, for example Type SAE 316, are particularly suitable, even if the acid contains small amounts of chlorides, while the CrNi-steel SAE 304 was resistant at 5.6 ppm chlorides but not at 20 ppm chloride concentration [179–181].

In general, CrNi-steels without Mo are more susceptible to pitting corrosion in presence of chloride ions than CrNiMo-steels.

Under oxidizing conditions, the steels can be used without problems due to the formation of a passivating oxide film. Under reducing conditions and in the presence of chlorides, there is a risk of the passive film being destroyed and the metal underneath is rapidly attacked.

High tensile stress in combination with exposure to chloride ions causes stress corrosion cracking.

Severe intercrystalline corrosion was observed at the heat-affected zones in the region of weldings in stainless austenitic steels on exposure to acetic acid [182].

Stainless steels are used to manufacture columns, e.g. distillation columns, tanks, pipelines, pumps, and valves for acetic acid, preferably in the form of welded and flanged constructions. The steel SAE 304 is often used for storage tanks and tank cars [182].

The corrosion maximum of SAE 304 was found at 80% by volume acetic acid solutions [183].

Ion-nitriding of steel surfaces of SAE 304 L and SAE 310 led to reduced corrosion resistance compared to that of the base alloys in 0.1 mol/l acetic acid solutions [184].

When low-carbon austenitic stainless steels are welded, inclusions such as Fe- and Cr-carbides reduce the corrosion resistance to acetic acid [185].

As for the behavior of austenitic steels towards pure acetic acid, see also [52, 186–191].

The service life of austenitic stainless steel in glacial acetic acid is longer than in acetic acid/formic acid mixtures. For hot (above 393 K (120°C)) mixtures of acetic acid with more than 2% formic acid, stainless steel is not as suitable as copper alloys or Monel®; however, for dilute acetic acid/formic acid solutions, Mo-containing CrNi-steels are preferred to copper or nickel alloys [182].

The usefulness of austenitic steels for handling acetic acid depends largely on the admixtures contained in the acid. As for the effect of formic acid (mixtures of both acids are used in the textile industry), acetic anhydride, salicylic acid etc., see Tables 26 and 29 [192].

Stainless molybdenum-free steels are attacked by pitting corrosion in media which contain acetic acid and are also susceptible to galvanic corrosion in certain cases. In 98 to 99% acetic acid and in acetic acid with 0.5 to 5% acetic anhydride, stainless steels and titanium and zirconium are attacked not uniformly and in some cases even in the form of pitting corrosion [193]. Small amounts of sodium chloride in the acetic acid already increase the corrosion considerably and lead to pitting [52].

Stainless steel 18 8 is considered resistant to acetic acid of all concentrations up to the boiling point; however, slight attack must be expected in boiling 100% acetic acid [194, 195].

CrNi-steel 18 8 is attacked under galvanostatic test conditions in a methanolic solution of organic acids such as acetic acid, formic acid, at low temperature from 243 to 263 K (–30 to –10°C) in the form of pitting corrosion; the local corrosion attack is inhibited by water [196].

The CrNi-steel 1.4301 is resistant to methanol and ethanol containing small amounts of acetic acid and water at room temperature [197].

According to an older reference, the austenitic 18 8 steels are resistant to acetic acid of concentrations above 20% up to the boiling point, at higher concentration and in glacial acetic acid up to 353 K (80°C).

In boiling glacial acetic acid the corrosion rates given in Table 15 are found.

Steel	Corrosion rate, mm/a (mpy)
CrNi 18 10	0.5 (19.7)
CrNi 18 10 Ti	1.2 (47.2)
CrNi 18 10 Nb	1.1 (43.3)
CrNi 18 10 Mo	0.15 (5.91)

Table 15: Corrosion rates of various CrNi-steels in boiling glacial acetic acid

It is recommended to choose steels which contain 17.5 % chromium and 2.5 % molybdenum for this type of apparatus. Molybdenum in particular suppresses the pitting corrosion which occurs often with CrNi-steels [198, 199]. Small amounts of mineral acids considerably increase the corrosion attack; NiCrMoCu-steels 25 20 must be used in this case [199].

Glacial acetic acid and completely anhydrous acetic acid differ considerably in their aggressivity on CrNi- and CrNiMo-steels. Especially at temperatures above the boiling point under pressure, the action of glacial acetic acid is often very strong because it destroys the protective passivating film [52, 200, 201].

According to electrochemical measurements, 18 10 steel has very good resistance to acetic acid even in high concentrations in the presence of dissolved oxygen; the reduced resistance at low concentrations is attributed in particular to the higher dissociation rate of acetic acid [202]. As for the effect of oxygen on the corrosion of stainless steels, for example CrNi 18 10, CrNiMo 17 13 3, CrNiMo 21 21 4, and the nickel-based alloy EP-567 in acetic acid, see [203]; this report also provides limits for temperature and acid concentration. Thus, corrosion is increased by gassing using nitrogen, while upon gassing using air oxygen the corrosion rates are in most cases 60 to 70 % lower even at high concentrations such as 90 %, 98 % and glacial acetic acid as well as at temperatures of 391, 423, 474 and 523 K (118, 150, 201 and 250°C).

The Soviet steels Ch18N10T (cf. SAE 347), Ch17N13M2T (cf. SAE 316 Ti), Ch17N13M3T (cf. SAE 316 Ti), Ch17T (cf. SAE 439) and titanium are not attacked in 5 to 98 % acetic acid up to 373 K (100°C). If 0.3 or 5 % formic acid is added to acetic acid of all concentrations, the corrosion resistance of the various stainless steels and of titanium is supposedly not affected, although a slight increase in the corrosion loss can occur at higher temperatures [204, 205].

According to potentiostatic measurements, steels which are lower in nickel or nickel-free are passive in aerated acetic acid solutions in concentrations of (%) 15, 30, 50, 70, 98 at 303 K (30°C). The stainless steel Ch18N9T (1.4541, SAE 321) can therefore be replaced under these conditions by stainless steels which are lower in nickel or nickel-free [206]. CrNi-steel ((%) 0.1 C-18 Cr-9.62 Ni), CrNiMo-steel ((%) 18.62 Cr-12.59 Ni-2.5 Mo), and CrNiMn-steel ((%) 13.95 Cr-1.43 Ni-13.48 Mn) were sufficiently resistant in tar water which is formed during peat processing and the main corrosive agent of which is acetic acid [207].

Investigations on the corrosion behavior of the Polish steels 1H18N9T (1.4541, SAE 321) and H17N13M2T (1.4571, SAE 316 Ti) in solutions of 40 to 80 % acetic acid at 313 K (40°C) to boiling point were performed in [208]. Application ranges were established.

The use of cobalt-containing nickel for the preparation of stainless steels affects the corrosion resistance, for example, of steel SAE 302 ((%) 17 to 19 Cr, 8 to 10 Ni, 2 Mn, max. 1 Si, 0.15 C) in acetic acid negatively, while the resistance of the steel SAE 309 ((%) 22 to 24 Cr, 12 to 15 Ni, 2 Mn, max. 1 Si, 0.20 C) is, on the other hand, hardly affected [209].

The corrosion resistance of Carpenter® Custom 455 steel ((%) 0.03 C-0.5 Mn-0.5 Si-11.8 Cr-9 Ni-1.2 Ti-2.3Cu-0.30 Nb + Ta) in boiling 10% acetic acid is higher than that of the stainless Cr-steel SAE 410 ((%) 11.5 to 13.5 Cr, 1 Mn, max. 1 Si, max. 0.15 C) and equal to that of the Cr-steel SAE 430 ((%) 14 to 18 Cr, 1 Mn, max. 1 Si, max. 0.12 C) [210].

Soviet ferritic-austenitic steels, for example Ch21N5, Ch21N5T, Ch21N6M2, Ch21N6M2T, which were developed especially to save nickel, exhibited intercrystalline cracks after bending in acetic acid for 100 h [211]. The steel 1Ch21N5T loses 0.01 mm/a (0.39 mpy) in acetic acid of all concentrations up to 373 K (100°C). The steels 0Ch21N6M2T and 0Ch21N5T are highly resistant in the media present during acetic acid production [212]. However, in boiling technical grade acetic acid, a corrosion rate of 0.6 mm/a (23.6 mpy) was found for 0Ch21N5T [213].

The stainless CrNiSi-steel 18 8 2 with 0.06% carbon and small amounts of phosphorus and molybdenum is resistant to stress corrosion cracking in cold and hot chloride solutions containing acetic acid [214].

A firmly adhering oxide film for metallic glass materials on wire is obtained by exposing CrNi-steel 40 5 to an aqueous solution containing 30 to 50% nitric acid, 12 to 20% hydrochloric acid, 3 to 5% acetic acid at about 335 K (62°C), followed by annealing in a gas flame [215].

The low-nickel austenitic steel SAE 201 ((%) 16 to 18 Cr, 3.5 to 5.5 Ni, 5.5 to 7.5 Mn, max. 0.15 C, max. 0.25 N) loses 0.03 mm/a (1.18 mpy) in aerated 60% acetic acid at 343 K (70°C) over a period of 96 h [216, 217] and is thus significantly less resistant than the above-mentioned steel SAE 302 and the steel USS Tenelon® ((%) 17 Cr-14.5Mn-0.1C-0.04N), which lose around 0.003 mm/a (0.12 mpy) under these conditions [218]. Attempts to replace chromium by manganese have not been very successful so far; thus, adding manganese to a CrNi-steel ((%) 17 to 18 Cr, 4 to 5 Ni) reduces its resistance to acetic acid [219].

The Soviet stainless CrNiMn-steel ((%) 17 Cr, 4 Ni, 8 Mn) is resistant to acetic acid of all concentrations up to 333 K (60°C) and in 30% acid up to the boiling point; as expected, the addition of formic acid reduces the corrosion resistance [220, 221].

The Czech steel ((%) 0.14 C, 1.27 Mn, 0.56 Si, 0.036 P, 0.27 S, 19.6 Cr, 8.65 Ni) according to CSN 17241 is susceptible to pitting corrosion in a solution of 1.8% acetic acid +0.4% citric acid +0.8% sugar with and without 3% sodium chloride, since chloride ions act as activator [222].

In acetic acid test solutions with hydrogen sulfide (pH 2), specimens consisting of Remanit® 1860 M ((%) 0.06 C, 0.54 Si, 8.60 Mn, 0.031 P, 0.012 S, 17.53 Cr, 5.87 Ni, 0.143 N) and Remanit® 1860 MS ((%) 0.03 C, 0.42 Si, 8.95 Mn, 0.023 P, 0.010 S, 17.41 Cr, 5.98 Ni, 0.146 N), when subjected to a stress around the minimum yield point, broke after 33 to 110 h at room temperature as a result of transcrystalline crack formation. The cracks are caused by hydrogen, which evolves at the cathode and – facilitated by the sulfide ion – diffuses into the metal. The sensitivity of the CrNiMn-austenites is in contrast to the experience with CrNi-steels 18 8, which show no tendency for stress corrosion cracking in an environment which contains hydrogen sulfide, unless they are highly cold-formed [223].

As for stress corrosion cracking in acetic acid solutions which contain hydrogen sulfide, see also Table 16 [52, 224].

The tests were carried out on horseshoe-shaped specimens, which were subjected to a stress of 7,030 kg/cm^2. The specimens were hardened and tempered in a muffle furnace in air and either quenched to room temperature in an oil bath or hardened in a salt bath at 580 K (307°C) for 10 min and cooled in air.

Acetic acid can be stored and transported in sealed containers made of stainless steel [225].

As for the behavior of stainless CrNi-steels towards acetic acid, see Table 17 and the respective literature.

According to older data, CrNi- and CrNiMo-steels (1.4541 cf. SAE 321 and 1.4571 cf. SAE 316 Ti) are attacked in boiling glacial acetic acid at a material consumption rate of 12 and 14 g/m^2 d (0.6 mm/a (23.6 mpy)) the addition of 1 % mercuric acetate nearly stops the attack (0.0 g/m^2 d) [52].

Ferric salts (nitrate, sulfate, oxalate) above a certain critical concentration have such an inhibitory effect on CrNi-steels 18 8 in organic acids such as acetic acid, formic acid and the like that the characteristic gloss of these steels is maintained in boiling acids even on long test durations; their concentration is, for example, 0.11 g/l iron for steel SAE 304 in 50 % acetic acid [226]. If small amounts of sodium dichromate are added to boiling concentrated acetic acid, the material consumption rate of CrNi-steels 18 8 is reduced to 2.5 to 6 g/m^2 d (0.11 to 0.27 mm/a (4.33 to 10.6 mpy)) [52].

Wear resistance and corrosion resistance of CrNi-steels can be improved by coating with a special Ni-alloy PG-Sr 3 [227].

– *Acetic acid vapor* –

Under heat transfer conditions, the steel SAE 304 L ((%) 0.05 C, 1.45 Mn, 0.62 Si, 18.22 Cr, 10.26 Ni) loses more than 1.25 mm/a (49.2 mpy) in 56 % acetic acid which contains 20 ppm chloride plus 26 ppm sulfate (specimen temperature 373 K (100°C), solution temperature 351 K (78°C) after a test duration of 96 h), but in acid without additions of salt even less than 0.015 mm/a (0.59 mpy) [228–230].

For the steel SAE 316 ((%) 0.07 C, 1.70 Mn, 0.49 Si, 17.47 Cr, 13.22 Ni, 2.15 Mo), the corrosion rates are in both cases below 0.015 mm/a (0.59 mpy) [229, 230].

In 56 % acetic acid with 1 or 5 % of sulfuric or 4 % formic acid (specimen temperature 383 K (110°C), solution temperature 373 K (100°C), test duration 96 h), the corrosion rates listed in Table 18 were found [229–231].

In dry acetic acid vapors above the dew point, the CrNi-steels 18 10 have good resistance at 873 K (600°C); however, this is only true for pure acetic acid. The acetic acid solutions used in industry often contain organic materials as impurities, in particular formic acid, which severely attacks these steels [232].

Type of steel	C-content %	Standard analysis %	Heat treatment	Rockwell hardness cone (HRC)	Behavior of 3 specimens each in aqueous solutions of	
					0.5 % Acetic acid + 6 % NaCl + H$_2$S	0.5 % Acetic acid + H$_2$S
440 C	1.0	18 Cr	1,313 K (1,040°C)/oil + 588 K (315°C)/air	54 to 55	all specimens spontaneously broken after 1 to 2 h	all specimens spontaneously broken after 1 to 2 d
440 A	0.7	18 Cr	1,313 K (1,040°C)/oil + 588 K (315°C)/air	52 to 53	all specimens spontaneously broken after 1 to 2 h	all specimens spontaneously broken after 5 d
420	0.4	13 Cr	1,313 K (1,040°C)/oil + 588 K (315°C)/air	50 to 52	all specimens spontaneously broken after 2 to 5 h	all specimens spontaneously broken after 2 to 3 d
17-7PH	0.07	17Cr, 7Ni, 1Al	annealed + 1,033 K (760°C)/air + 783 K (510°C)/air	44 to 45		all specimens spontaneously broken after 1 to 2 d
17-7PH	0.07	17Cr, 7Ni, 1Al	annealed + 1,033 K (760°C)/air + 783 K (510°C)/air	44 to 45	all specimens spontaneously broken after 1 to 2 h	all specimens spontaneously broken after 4 to 24 d
17-4PH	0.05	17Cr, 4Ni, 4Cu	annealed + 753 K (480°C)/air	41 to 43		1 specimen spontaneously broken after 2 d, 2 unchanged after 50 d
17-4PH	0.05	17Cr, 4Ni, 4Cu	annealed + 753 K (480°C)/air	44 to 45	all specimens spontaneously broken after 6 to 46 h	2 specimens spontaneously broken after 12 to 38 d, 1 unchanged after 62 d
431	0.15	16Cr, 2Ni	1,253 K (970°C)/oil + 588 K (315°C)/air	42 to 43	all specimens spontaneously broken after 30 to 46 h	2 specimens spontaneously broken after 57 d, 1 unchanged after 62 d

* cold drawn, 56 %

Table 16: Stress corrosion cracking of hardened steels in acetic acid solutions which contain hydrogen sulfide [233]

Table 16: Continued

Type of steel	C-content %	Standard analysis %	Heat treatment	Rockwell hardness cone (HRC)	Behavior of 3 specimens each in aqueous solutions of	
					0.5% Acetic acid + 6% NaCl + H$_2$S	0.5% Acetic acid + H$_2$S
410	0.1	12.5Cr, 0.5Mo	1,253 K (970°C)/oil + 588 K (315°C)/air	41 to 43	all specimens spontaneously broken after 6 to 21 h	all specimens spontaneously broken after 3 to 8 d
416	0.1	12.5Cr	1,253 K (970°C)/oil + 588 K (315°C)/air	40 to 42	all specimens spontaneously broken after 30 to 46 h	all specimens spontaneously broken after 3 to 8 d
17-7PH	0.07	17Cr, 7Ni, 1Al	annealed + 1,033 K (760°C)/air + 838 K (565°C)/air	41 to 43	all specimens spontaneously broken after 1 to 5 h	all specimens spontaneously broken after 1 to 5 d
17-4PH	0.05	17Cr, 4Ni, 4Cu	annealed + 923 K (650°C)/air	29 to 31	2 specimens spontaneously broken after 282 h. 1 cracked after 504 h, but not broken into two	all specimens cracked after 62 d, but not broken into two
304	0.05	19 Cr, 9 Ni	*	42 to 43	all specimens cracked after 504 h, but not broken into two	all specimens unchanged after 62 d
305	0.07	19 Cr, 12 Ni	*	34 to 35	all specimens cracked after 504 h, but not broken into two	all specimens unchanged after 62 d
17-10P	0.12	17 Cr, 11 Ni	annealed + 973 K (700°C)/air	30 to 31	2 specimens spontaneously broken after 101 to 432 h, 1 cracked after 504 h, but not broken into two	all specimens unchanged after 62 d

* cold drawn, 56 %

Table 16: Stress corrosion cracking of hardened steels in acetic acid solutions which contain hydrogen sulfide [233]

Table 16: Continued

Type of steel	C-content %	Standard analysis %	Heat treatment	Rockwell hardness cone (HRC)	Behavior of 3 specimens each in aqueous solutions of	
					0.5 % Acetic acid + 6 % NaCl + H$_2$S	0.5 % Acetic acid + H$_2$S
SAE 4130	0.27	1Cr, 0.2Mo	588 K (315°C) + 588 K (315°C)/air + 753 K (480°C)/air	27 to 28		all specimens spontaneously broken after 2 to 7 d
SAE 4130	0.27	1Cr, 0.2Mo	1,143 K (870°C) + 588 K (315°C)/air + 863 K (590°C)/air	21 to 22		all specimens unchanged after 42 d
SAE 4340	0.39	1Cr, 2Ni, 0.3Mo	1,088 K (815°C) + 613 K (340°C)/air + 753 K (480°C)/air	36 to 37		all specimens spontaneously broken after 1 day

* cold drawn, 56 %

Table 16: Stress corrosion cracking of hardened steels in acetic acid solutions which contain hydrogen sulfide [233]

CrNi-steel	Acetic acid %	Temperature K (°C)	Corrosion rate mm/a (mpy)	Remarks	Literature
18 Cr, 8 Ni	10	293 (20)	< 0.1 (< 4)		[53]
	10	boiling point	< 0.1 (< 4)		
	50	293 (20)	< 0.1 (< 4)		
	50	boiling point	< 0.1 (< 4)		
	80	293 (20)	< 0.1 (< 4)		
	80	boiling point	< 3.0 (< 118)		
	conc.	293 (20)	< 0.1 (< 4)		
	conc.	boiling point	< 1.0 (< 39.4)		

Table 17: Corrosion behavior of stainless chromium-nickel steels towards acetic acid of various concentrations and temperatures [52–55, 82, 234, 238]

Table 17: Continued

CrNi-steel	Acetic acid %	Temperature K (°C)	Corrosion rate mm/a (mpy)	Remarks	Literature
18 to 20 Cr, 8 to 11 Ni, stabilized	all conc.	293 (20)		resistant	[82]
	50	boiling point	< 1.1 (< 43.3)		
	80	373 (100)	< 1.0 (< 39.4)		
	glacial acetic acid	373 (100)	< 1.1 (< 43.3)		
	glacial acetic acid	473 (200)	< 1.1 (< 43.3)		
SAE 304 (19 Cr, 8 Ni)	24	308 (35)	< 0.0025 (< 0.098)	storage tests/115 d	[52]
	98	308 (35)	< 0.0025 (< 0.098)	storage tests/157 d	[52]
	10	no temperature given	0.005 (0.197)	laboratory tests/ 168 d	[52]
	50	no temperature given	0.01 (0.39)		
	75	no temperature given	0.83 (32.7)		
	glacial acetic acid	no temperature given	0.68 (26.8)		
	99	boiling point	0.45 (17.7)	plant test	[52]
SAE 321 (18 Cr, 10 Ni, Ti-stabilized)	99	boiling point	1.18 (46.5)	plant test	[52]
SAE 347 (19 Cr, 11 Ni, Nb-stabilized)	99	boiling point	1.03 (40.55)	plant test	[52]
SAE 308 (21 Cr, 11 Ni)	99	boiling point	1.33 (52.3)	plant test	[52]
SAE 310 (24 Cr, 18 Ni)	99	boiling point	0.98 (38.6)	plant test	[52]
SAE 304 (19 Cr, 8 Ni)	99.7	393 (120)	0.13 (5.12)		[52]
SAE 347 (19 Cr, 11 Ni, Nb-stabilized)	99.7	393 (120)	0.18 (7.09)		[52]

Table 17: Corrosion behavior of stainless chromium-nickel steels towards acetic acid of various concentrations and temperatures [52–55, 82, 234, 238]

Table 17: Continued

CrNi-steel	Acetic acid %	Temperature K (°C)	Corrosion rate mm/a (mpy)	Remarks	Literature
Cast steel (19 Cr, 10 Ni)	glacial acetic acid	383 (110)	1.6 (63.0)		[82]
	glacial acetic acid	450 (177)	17.3 (681)		
	30	383 (110)	0.03 (1.18)		
	30	450 (177)	0.1–3.5 (3.94–138)		
SAE 304 (19 Cr, 8 Ni)	99	393 (120)	1.65 (65.0)	acetic acid vapor plant test/82 d	[52]
21 Cr, 5 Ni, Ti-stabilized	5	438 (165)	0 (0.0)		[234]
	25	438 (165)	0.02 (0.79)		
	50	413 (140)	2.34 (92.1)		
	98	438 (165)	3.83 (151)		
18 Cr, 10 Ni, Ti-stabilized	5	438 (165)	0 (0.0)		[234]
	25	438 (165)	0.13 (5.12)		
	50	438 (165)	3.46 (136)		
	98	438 (165)	2.3 (90.6)		
18 Cr, 10 Ni	50	375 (102)	3.30 (130)	without heat transfer	[234]
	50	383 (110)	5.33 (210)	with heat transfer	
	50	398 (125)	5.59 (220)	with heat transfer	
	50	413 (140)	6.35 (250)	with heat transfer	
	99.6	391 (118)	1.75 (68.9)	without heat transfer	
	99.6	383 (110)	6.60 (260)	with heat transfer	
	99.6	398 (125)	8.64 (340)	with heat transfer	
	99.6	413 (140)	0.51 (20.1)	with heat transfer	

Table 17: Corrosion behavior of stainless chromium-nickel steels towards acetic acid of various concentrations and temperatures [52–55, 82, 234, 238]

Table 17: Continued

CrNi-steel	Acetic acid %	Temperature K (°C)	Corrosion rate mm/a (mpy)	Remarks	Literature
Remanit® steels corresponding to X5CrNi18-10 (SAE 304) X2CrNi19-11 (SAE 304 L) X6CrNiTi18-10 (SAE 321)	10	293 (20)	< 0.11 (< 4.33)		[235]
	10	boiling point	< 0.11 (< 4.33)		
	50	293 (20)	< 0.11 (< 4.33)		
	50	boiling point	0.11–1.1 (4.33–43.3)		
	glacial acetic acid, 100	293 (20)	< 0.11 (< 4.33)		
	glacial acetic acid, 100	boiling point	0.11–1.1 (4.33–43.3)		
	10	293 (20)	< 0.11 (< 4.33)	acetic acid + hydrogen peroxide	[236]
	50	323 (50)	< 0.11 (< 4.33)		
	50	363 (90)	< 0.11 (< 4.33)		
SAE 302, 321, 347	10	373 (100)	< 0.05 (< 2)		[55]
	20 – 50	373 (100)	< 0.5 (< 20)		
	60 – 100	348 (75)	0.5–1.25		
	30	398 (125)	< 0.5 (< 20)		
	30	423 (150)	> 1.25 (> 50)	aerated, no HCl, H_2SO_4, NaCl	
	50	398 (125)	> 1.25 (> 50)		
	60	373 (100)	> 1.25 (> 50)		
	80	373 (100)	> 1.25 (> 50)		
	100	373 (100)	> 1.25 (> 50)		
	10 – 90	348 (75)	0.5–1.25 (19.7–49.2)	air free; intergranular corrosion, no HCl, H_2SO_4, NaCl	[55]
	10 – 50	373 (100)	0.5–1.25 (19.7–49.2)		
	60 – 100	373 (100)	> 1.25 (> 50)		
	30	323 (50)	< 0.05 (< 2)	90 % acetic acid with acetaldehyde + 7 % water	[55]

Table 17: Corrosion behavior of stainless chromium-nickel steels towards acetic acid of various concentrations and temperatures [52–55, 82, 234, 238]

Table 17: Continued

CrNi-steel	Acetic acid %	Temperature K (°C)	Corrosion rate mm/a (mpy)	Remarks	Literature
18 Cr, 8 Ni	10 – 100	297 (24)	< 0.05 (< 2)	aerated, avoid traces of HCl, H_2SO_4, NaCl	[54]
	10 – 50	373 (100)	< 0.05 (< 2)		
	30	422 (149)	< 0.5 (< 20)		
	30	477 (204)	> 1.25 (> 50)		
	50	422 (149)	> 1.25 (> 50)		
	60	373 (100)	< 0.5 (< 20)		
	70 – 100	352 (79)	< 0.5 (< 20)		
	70 – 100	373 (100)	0.5–1.25 (19.7–49.2)		
	100	422 (149)	> 1.25 (> 50)		
	10 – 90	352 (79)	< 0.5 (< 20)	air-free, avoid traces of HCl, H_2SO_4, NaCl; intergranular corrosion	
	50	325 (52)	0.5–1.25 (19.7–49.2)		
	100	325 (52)	0.5–1.25 (19.7–49.2)		
	10 – 20	373 (100)	< 0.5 (< 20)		
	30 – 60	373 (100)	0.5–1.25 (19.7–49.2)		
	70 – 100	373 (100)	> 1.25 (> 50)		
	20	373 (100)	0.5–1.25 (19.7–49.2)	acetic acid vapor	[54, 55]
	40	373 (100)	> 1.25 (> 50)		
	100	373 (100)	> 1.25 (> 50)		
	30	325 (52)	< 0.05 (1.97)	acetic acid + acetaldehyde	[54]
	100	373 (100)	> 1.25 (> 50)	acetic acid + acetic anhydride + peracetic acid	[54, 55]
	100	373 (100)	> 1.25 (> 50)	acetic acid + ethyl alcohol	[54, 55]

Table 17: Corrosion behavior of stainless chromium-nickel steels towards acetic acid of various concentrations and temperatures [52–55, 82, 234, 238]

Table 17: Continued

CrNi-steel	Acetic acid %	Temperature K (°C)	Corrosion rate mm/a (mpy)	Remarks	Literature
18 Cr, 8 Ni	10	373 (100)	0.5–1.25 (19.7–49.2)	acetic acid + formic acid	[54, 55]
	100	373 (100)	> 1.25 (> 50)	acetic acid + formic acid, pitting corrosion	
	100	297 (24)	> 1.25 (> 50)	acetic acid + hydrobromic acid	[54, 55]
	100	297 (24)	> 1.25 (> 50)	acetic acid + hydrochloric acid	[54, 55]
	100	373 (100)	0.5–1.25 (19.7–49.2)	95 % acetic acid + 5 % propionic acid, stress corrosion cracking	[55]
	100	422 (149)	> 1.25 (> 50)	90 % acetic acid + 10 % salicylic acid	[54, 55]
	100	348 (75)	< 0.5 (< 20)	acetic acid + salicylic acid	[54, 55]
	100	373 (100)	> 1.25 (> 50)	vapors, intergranular corrosion	
	100	297 (24)	1.25 (49.2)	acetic acid + sulfuric acid	[54]
19 Cr, 9 Ni, 0.06 C	100	boiling point	0.0825 (3.25)	90 % acetic acid + 10 % acetic anhydride	[52, 237]
				reflux, laboratory tests	
	99	boiling point	2.450 (96.5)	99 % acetic acid + 1 % NaCl	
18 Cr, 8 Ni	56	373 (100)	0.0175 (0.69)	operational test; condensate consisting of 6 parts of 56 % acetic acid and 1 part of 90 % formic acid; 150 h, severe pitting corrosion	[52, 237]
	56	378 (105)	0.15 (5.91)	as above; vapor; 150 h	

Table 17: Corrosion behavior of stainless chromium-nickel steels towards acetic acid of various concentrations and temperatures [52–55, 82, 234, 238]

Table 17: Continued

CrNi-steel	Acetic acid %	Temperature K (°C)	Corrosion rate mm/a (mpy)	Remarks	Literature
SAE 302 (18 Cr, 9 Ni, 0.2 C)	50	391 (118)	destroyed	50 % acetic acid with 2 % formic acid	[52]
	25	383 (110)	destroyed	25 % acetic acid with 4 % formic acid	[52]
	30	408 (135)	destroyed	30 % acetic acid with 8 % formic acid	[52]
	25	377 (104)	0.225 (8.86)	25 % acetic acid with 1.5 % formic acid, operational test	[52, 237]
	30 – 50	377 (104)	destroyed	30–50 % acetic acid with 2–10 % formic acid, operational test	
	86–91	399.7 (126.7)	3.425 (135)	86 – 91 % acetic acid with 8.4 % salicylic acid vapors; operational test, 891 h	[52, 237]
	99.5	385.2 (112.2)	0.375 (14.8)	99.5 % acetic acid + 0.05 % salicylic acid vapors; 1,022 h	
SAE 347 (18 Cr, 10 Ni, Nb-stabilized)	50	391 (118)	destroyed	50 % acetic acid with 2 % formic acid	[52]
	25	383 (110)	destroyed	25 % acetic acid with 4 % formic acid	[52]
	30	408 (135)	1.05 (41.3)	30 % acetic acid with 8 % formic acid	[52]
	25	377 (104)	0.25 (9.84)	25 % acetic acid with 1.5 % formic acid	[52, 237]
	30 – 50	377 (104)	destroyed	30–50 % acetic acid with 2–10 % formic acid, operational test	

Table 17: Corrosion behavior of stainless chromium-nickel steels towards acetic acid of various concentrations and temperatures [52-55, 82, 234, 238]

In acetic acid cracking furnaces, the alloy Incoloy® 800 H ((%) 32.0 Ni, 21.0 Cr, 45.0 Fe, 0.3 Al, 0.4 Ti, 0.08 C) is increasingly used for furnace coils, since the customary stainless steels of the series 300 precipitate prematurely in the furnaces as a result of the formation of the sigma phase [239, 240].

– Acetic anhydride –

Austenitic CrNi-steels are often used due to their good resistance to acetic anhydride.

In a CrNi-steel coil, ketene is formed from acetic acid at about 1,000 K (727°C) under reduced pressure in the presence of volatile organic phosphoric esters as catalysts.

Material	Acetic acid with		
	1 % Sulfuric Acid	5 % Sulfuric Acid	4 % Formic Acid
SAE 304 L	> 1.25 mm/a (> 49.2 mpy)	> 1.25 mm/a (> 49.2 mpy)	> 1.25 mm/a (> 49.2 mpy)
SAE 316	> 1.25 mm/a (> 49.2 mpy)	> 1.25 mm/a (> 49.2 mpy)	0.34 mm/a (13.4 mpy)
SAE 310	< 0.015 mm/a (< 0.59 mpy)	> 1.25 mm/a (> 49.2 mpy)	> 1.25 mm/a (> 49.2 mpy)
SAE 329 (Carpenter® 7 Mo)	< 0.015 mm/a (< 0.59 mpy)	> 1.25 mm/a (> 49.2 mpy)	> 1.25 mm/a (> 49.2 mpy)
Carpenter® 20 Cb	0.63 mm/a (24.8 mpy)	> 1.25 mm/a (> 49.2 mpy)	< 0.015 mm/a (< 0.59 mpy)

Table 18: Corrosion rates of various steels in acetic acid with 1 and 5 % sulfuric acid or 4 % formic acid (373 K (100°C)), test duration 96 h, specimen temperature 383 K (110°C) [241–243]

Reaction vessels made of austenitic CrNi-steels are also used for the production of acetyl cellulose and other esters for example polyacetoxy ester, using acetic anhydride [244].

When methacrylonitrile is prepared from acetone cyanohydrin and acetic anhydride in the presence of small amounts (1 %) of concentrated sulfuric acid at 823 to 853 K (550 to 580°C), reaction tubes made of CrNi-steel and containing ceramic packing materials are used [74].

The same materials which are used in the thermal cracking of acetic acid can also be employed in the thermal cracking of acetone at 873 to 973 K (600 to 700°C).

CrNiMo-steel 18 8 2-4 is used for the acetone evaporator. For the cracking pipes, in which the acetone is supposed to stay for only a short time pipes made of CrNi-steels 25 20 [74, 245, 246] and pipes of CrNi-steels which contain (%) 20 to 27 Cr- 10 to 12 Ni- < 0.25 C [74, 247, 248] and those containing (%) 18 to 40 Cr- 1.5 to 2 Ni- < 0.25 C [74, 249] are used in addition to Sicromal® 12 ((%) 23 to 25 Cr- 1.3 to 1.6 Al- 1.3 to 1.6 Si-max. 1Mn-max. 0.12C). A catalytic decomposition of acetone on the walls of the cracking tube is largely avoided by adding small amounts of carbon disulfide (0.1 %) directly to the acetone [250]. New cracking pipes require a certain time of operation until a carbon layer has been deposited on their inside surface areas. The cracking gases are immediately cooled down considerably in a mixing vessel made of CrNi-steel by means of the injected cold mixture of acetic anhydride plus acetic acid and passed into the reaction tower (CrNi-steel or copper), in which ketene is converted to acetic anhydride by means of the glacial acetic acid trickling down. The fractionation of acetic anhydride and acetic acid is carried out in copper columns, while the pure acetic anhydride – as in the thermal cracking of acetic acid – is condensed through a silver condenser and passed into a storage container made of pure aluminium. The pumps in this process consist of austenitic CrNiMo-steel or high-silicon cast iron [74].

For storage and transport of acetic anhydride, austenitic CrNi-steels, in most cases without molybdenum, are used [74, 251].

As for the behavior of stainless chromium-nickel steels towards acetic anhydride, see Table 19.

CrNi-steel	Acetic anhydride concentration %	Temperature K (°C)	Corrosion rate mm/a (mpy)	Remarks	Literature
SAE 304	crude	389 (116)	1.79 (70.5)		[252, 253]
	vapor	383 (110)	1.22 (48.0)		
SAE 347	crude	388 (115)	2.12 (83.5)		[252]
	vapor	383 (110)	1.85 (72.8)		
18 Cr, 8 Ni	no data given	293 (20)	< 0.1 (< 4)		[53]
		boiling	< 0.1 (< 4)		
	20	373 (100)	> 1.25 (> 50)		[54]
	30	352 (79)	> 1.25 (> 50)		
	50	373 (100)	< 0.5 (< 20)		
	90	373 (100)	> 1.25 (> 50)		
	100	373 (100)	< 0.5 (< 20)		
18-20 Cr, 8-11 Ni, stabilized	no data given	boiling		resistant, for tanks for storage and transport	[253]
18 Cr, 8 Ni	100	422 (149)	< 0.5 (< 20)	acetic anhydride in acetic acid	[54]

Table 19: Corrosion behavior of stainless chromium-nickel steels in acetic anhydride of various concentrations and temperatures

Table 19: Continued

CrNi-steel	Acetic anhydride concentration %	Temperature K (°C)	Corrosion rate mm/a (mpy)	Remarks	Literature
Remanit® steels corresponding to X5CrNi18-10 SAE 304 X6CrNiTi18-10 SAE 321	no data given	293 (20)	< 0.11 (< 4.33)	resistant	[238]
		boiling	< 0.11 (< 4.33)		
SAE 302, 304, 321, 347	10–70	373 (100)	< 0.5 (< 20)		[55]
	80	373 (100)	< 0.05 (< 2)		
	90	298 (25)	< 0.05 (< 2)	acetic anhydride in acetic acid	
	100	298–373 (25-100)	< 0.5 (< 20)		
	100	398 (125)	< 0.05 (< 2)		
	100	423 (150)	< 0.05 (< 2)		

Table 19: Corrosion behavior of stainless chromium-nickel steels in acetic anhydride of various concentrations and temperatures

At a sufficiently high oxygen concentration, the CrNiMo-steels 18 8 1 are very resistant in boiling crude acetic acid (70 % acetic acid + 0.5 % formic acid). Any destroyed passive layers are re-formed by the oxygen (Figure 1 and Figure 2) [52, 254]; see also Table 20. Oxidizing salts such as copper acetate also have a corrosion-inhibitoring effect: thus, the corrosion rate of 1.4571 (SAE 316 Ti) which is in conducting contact with copper in boiling technical grade glacial acetic acid which contains 0.25 % copper acetate in the liquid corresponds to a corrosion rate of 0.005 mm/a (0.20 mpy) after the first 144 h and only to a corrosion rate of 0.001 mm/a (0.039 mpy) after 1,200 h. Similar results are obtained with mercury salts [52]. As for the effect of oxygen dissolved in acetic acid, see Table 20 and [52].

Hot acetic acid vapors have only little effect on austenitic steels at 950 K (677°C) (Table 21), but the acid is decomposed catalytically during this process [52, 237, 255].

Austenitic chromium-nickel-molybdenum steels

– *Acetic acid* –

Stainless austenitic CrNiMo-steel ((%) 18Cr-10Ni-2.0 to 2.75 Mo) is the material of choice in the preparation of peracetic acid and acetaldehyde monoperacetate [256, 257].

In the oxidation of ethyl alcohol, which is particularly of importance for countries producing an excess of carbohydrates, mixtures of acetic with acetaldehyde are formed, the corrosion rates summarized in Table 22 being measured in a fractionating column [52].

The surface treatment of stainless steels greatly affects their chemical resistance in particular to stress corrosion cracking. In various chloride-containing aggressive media (acetic acid and the like) not only in operational but also in laboratory tests, transcrystalline stress corrosion cracking occurred on stainless steels in the surface zone which had been damaged by rough polishing, deformation by cutting, or scratches.

Thus, for example, in a column made of stainless CrNiMo-steel (DIN-Mat. No. 1.4571), which was exposed to a weak, chloride-containing acetic acid solution at 373 to 403 K (100 to 130°C), a branch in the bottom broke off after a short period of operation, because the entire zone affected by the welding had been damaged by the grinding of the scale and smoothing of the beads at the welds. It is recommended to remove this corrosion-susceptible deformation zone by vigorous pickling to a depth of about 3 mm; after this treatment, the steel 1.4571 shall be insentive to chloride contents [258].

Acid	Period of time	Corrosion rate, mm/a (mpy)			
		Nitrogen passed through		Air passed through	
		Liquid	Vapor	Liquid	Vapor
Technical grade glacial acetic acid*	first 144 h	0.032 (1.26)	0.019 (0.75)	0.016 (0.63)	0.019 (0.75)
	last 240 h	0.081 (3.19)	0.029 (1.14)	0.0 (0.0)	0.011 (0.43)
Crude pyroligneous acid**	first 360 h	0.108 (4.25)	0.090 (3.54)	0.027 (1.06)	0.057 (2.24)
	last 240 h	0.280 (11.0)	0.122 (4.80)	0.0 (0.0)	0.029 (1.14)

* overall test duration 1,200 hours
** overall test duration 600 hours

Table 20: Corrosion rate of 1.4571 (SAE 316 Ti) at the boiling temperature [52]

Experiences showed that 1.4571 is susceptible to pitting corrosion in acetic acid with small amounts of water in the presence of small chloride concentrations. In 99.5 % acetic acid with less than 1 ppm chloride, the steels 1.4439 (UNS S31726) and 1.4539 (UNS N08904) proved to be resistant to pitting corrosion at 391 K (118°C) and 423 K (150°C) as well as in 99.9 % acid with less than 1 ppm chloride up to 391 K (118°C). The special steel AVESTA® 254 SMO was resistant to acetic acid

Material	Test duration, h		
	450	1,769	1,384
	Corrosion rate, mm/a (mpy)		
Hastelloy® C	0.0025 (0.098) (increase)	0.0025 (0.098)	
SAE 317	0.01 (0.39)	0.01 (0.39)	0.03 (1.18)
SAE 316	0.01 (0.39)	0.015 (0.59)	0.0425 (1.67)
SAE 316 ELC		0.015 (0.59)	
Illium®	0.005 (0.197)		
Aloyco® 20		0.0025 (0.098)	
SAE 309		0.01 (0.39)	
Hastelloy® B	0.0075 (0.295)	0.0175 (0.689)	
Hastelloy® A	0.0825 (3.25)		
Inconel®	3.25 (128)	2.75 (108)	
Monel®	destroyed		
Nickel	destroyed		
SAE 410			0.0375 (1.48)
SAE 430			0.03 (1.18)
SAE 446			0.005 (0.197)

Table 21: Corrosion behavior in vapors of glacial acetic acid at 950 K (677°C). Vapor velocity 53 m/s. The specimens were located at the point of entry of the vapor into the reaction tube (operational test) [52, 237]

(99.5 %) up to chloride contents of 3 ppm and temperatures of 423 K (150°C). On access of air very high corrosion rates can be expected. The tests considered the corrosion behavior in dependence of the surface conditions (ground, pickled) of the specimens. For all specimens investigated (1.4541, 1.4571, 1.4439 and 1.4539) no susceptibility to stress corrosion cracking of looped specimens or ground weldings was observed [259].

The behavior of the SAE 316 is greatly affected by impurities in the acid. Thus, as little as 0.2 % acetic anhydride increases the corrosion considerably; likewise, the chloride ion content should not exceed certain limits, for example 18 ppm in the presence of traces of water in highly concentrated acid. At higher chloride concentrations, corrosion rates around 0.4 mm/a (15.7 mpy) have to be expected even in the presence of 1 % water. This can lead not only to uniform but also to pitting corrosion. This type of impurity probably explains the different behavior of two distillation columns in a plant: in one, the steel SAE 316 lost 4.3 mm/a (169 mpy), but

Materials	Corrosion rate, mm/a (mpy)			
	a	b	c	d
SAE 347	0.0025 (0.098)	0.0025 (0.098)	0.09 (3.54)	
SAE 316	0.0025 (0.098)	0.0025 (0.098)	0.0175 (0.689)	0.025 (0.98)
SAE 317	0.0025 (0.098)	0.0025 (0.098)	0.005 (0.185)	
Worthite®	0.0025 (0.098)	0.0025 (0.098)	0.0175 (6.89)	
Inconel®	0.0025 (0.098)	0.0025 (0.098)	destroyed	
Hastelloy® A	0.875 (34.4)	0.55 (21.7)	destroyed	
Monel®	1.075* (42.3)	0.475 (18.7)	destroyed	
Nickel	0.675 (26.6)	0.55 (21.7)	destroyed	
Copper	1.125 (44.3)	0.525 (20.7)	destroyed	0.675** (26.6)
Hastelloy® C	0.0025 (0.098)	0.0025 (0.098)	0.0175 (6.89)	

* minimum rate – The specimens were destroyed during the experiment.
** 10% Al-bronze with 5% Ni and 2.5 Fe
a) Operational test, duration 129 d, in 30% acetic acid and 0.5% acetaldehyde at 310.8 K (37.8°C) at the top section of the fractionating column
b) Operational test, duration 129 d, at the lower end of the same column in 30% acetic acid and 3% acetaldehyde at 316.3 K (43.3°C)
c) Operational test, duration 112 d, at the top section of a fractionating column in 20% acetic acid, 50% acetaldehyde, balance water, at 371.9 K (98.9°C)
d) Operational test, duration 255 d, at the top section of a column in 85% acetic acid, 8% acetaldehyde, 2% low-boiling components, 0.1% high-boiling components, balance water at 385.8 K (112.8°C)

Table 22: Operational tests on various alloys during acetic acid preparation [52]

Carpenter® 20 ((%) 20Cr-29Ni-3Cu-3Mo) only lost 0.7 mm/a (27.6 mpy); in the other, the steel SAE 316 lost 0.58 mm/a (22.8 mpy), but Carpenter® 20 3.86 mm/a (152 mpy) [260].

A water content of 1 to 5% in acetic acid shows no influence on the corrosion behavior of 1.4436, 1.4439 and 1.4571 at boiling temperature. Therefore, acid anhydride leads to a steep increase of the corrosion rates which amount at anhydride concentrations of 2 to 5% to more than 1 mm/a (39.4 mpy). Passivation does not take place. Test durations of up to 30 d were applied [261].

In the processing of pyroligneous acid according to Othmer, CrNiMo-steel 18 13 2 is used for the distillation vessel, the fractionating column and the condenser [52]. For the rectification of crude acetic acid from the dry distillation of wood, CrNiMo-steels 18 19 2 are sufficiently resistant (Figure 1 and Figure 2) [52, 53, 262].

However, the requirement for the favorable effect of molybdenum is a sufficiently low carbon content, since otherwise local chromium depletion is caused due to precipitation of chromium carbide, which leads to risk of inter-crystalline corrosion at these points. In this respect, a CrNiMo-steel 18 10 with 0.02% carbon is resistant to inter-crystalline corrosion in 20% boiling acetic acid at 375 K (102°C) [263].

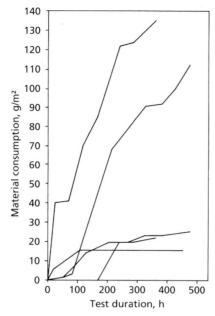

Figure 1: Corrosion behavior of austenitic CrNiMo-steels 18 8 2 (specimens from different melts) in boiling crude acetic acid (70 % acetic acid, 0.5 % formic acid) [52, 262]

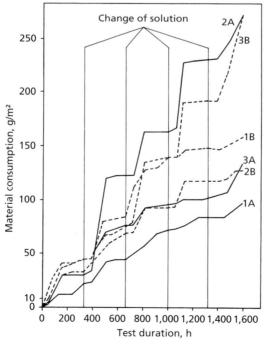

Figure 2: Corrosion behavior of austenitic CrNiMo-steel 18 8 2 (specimens from different parts of metal sheet) in crude acetic acid (70 % acetic acid, 0.5 % formic acid) [52, 262]

As for the effect of molybdenum on the corrosion resistance of the CrNi-steel 18 10 to the condensates formed when 80 % and 90 % acetic acid are boiled, see also Figure 3 [263].

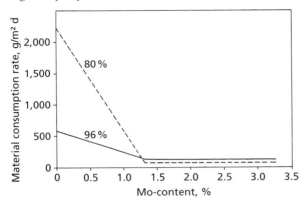

Figure 3: Effect of molybdenum on the resistance of CrNi-steel 18 10 in condensates of 80 and 90 % acetic acid [263]

In a manufacturing plant, an SAE 317 steel fractionating column for recovery corroded in acetic acid/acetic anhydride mixtures; by aerating the feed liquid, the corrosion was surprisingly increased [264].

CrNiMo-steels of the type 18 8 can be used in the preparation of acetic acid; a molybdenum content of 1.5 % is sufficient in these steels [265].

In [266] the use of 10Ch17N13M3T (1.4573, cf. SAE 316 Ti) for reactor bodies for the production of acetic acid from hydrocarbons is recommended.

In cellulose acetate plants, exclusively the CrNiMo-steel 18 8 2.5 is used for acetic acid and acetic anhydride [267]. In the preparation of vinyl acetate from ethene and acetic acid in the presence of palladium and alkali metal acetate, the corrosion of the apparatus can be reduced by passing steam into the acetic acid. The corrosion rates of the steel SAE 316 are 0.02 mm/a (0.79 mpy) at a water content of 2 %, while at a water content of 0.5 % 0.03 mm/a (1.18 mpy) are observed [268].

The presence of small amounts of chloride ions in the reaction mixture of (%) 98.97 acetic acid, 0.34 vinyl acetate, 0.43 crotonaldehyde, 0.04 acetaldehyde and 0.23 ethylidene diacetate at 386 K (113°C) in the production of vinyl acetate led to pitting corrosion of a distillation column made of stainless steel Ch17N13M2T (cf. 1.4571, SAE 316 Ti) [269].

The austenitic CrNiMo-steels (see also [270, 271]) SAE 316 and 317 are suitable for acetic acid of all concentrations up to the boiling point; as a rule, the steel SAE 317 has the best resistance to concentrated acetic acid at or near the boiling point [52, 237, 272–274]. The stainless particularly acid-resistant steel Weldshyne EWC® ((%) 18.25Cr-13.5Ni-3.75Mo-0.5Si-1.4Mn-0.03C) withstands hot acetic acid and acetic anhydride [275].

Mixtures of 30 % hydrogen peroxide and 10 % acetic acid do not attack stabilized CrNiMo-steels even at 363 K (90°C). Nor are these steels corroded by mixtures of glacial acetic acid and 50 % hydrogen peroxide [52, 276].

The material consumption rate of an austenitic steel ((%) 0.046 to 0.048 C, 17.34 to 17.39 Cr, 5.45 to 6.05 Ni, 0.32 to 1.012 Mo, 2.48 to 2.75 Mn, 1.144 to 1.53 Cu, 0.43 to 0.59 Si, 0.102 to 0.149 N), is said to be less than 0.048 g/m^2 h (0.053 mm/a (2.09 mpy)) in 5 % acetic acid [277].

As for the results of further operational tests, see Table 23 and 24 [52, 237].

Spray coatings made of austenitic stabilized CrNiMo-steels 19 9 2 did not show such a satisfactory corrosion behavior in 99.8 % acetic acid as the compact material [52, 278].

18 8 Steels with 2.5 % molybdenum and a maximum of 0.03 % carbon are resistant to acetic acid and in particular to pitting corrosion in this acid; for particularly aggressive media, the Ni-content is increased to 14 % and the Mo-content to 3.5 % [279].

Material	Corrosion rate, mm/a (mpy)		
	a	b	c
SAE 316	destroyed	1.375 (54.1)	0.0425 (1.67)
SAE 317	0.325 (12.7)	0.3 (11.8)	0.025 (0.98)
Copper	0.1 (3.94)	0.1 (3.94)	0.01 (0.39)
Aluminium bronze, 10 % Al	0.075 (2.95)	0.1 (3.94)	0.0175 (0.69)
Silicon bronze	0.075 (2.95)	0.1 (3.94)	0.0125 (0.49)
Copper-nickel 70/30	0.1 (3.94)	0.125 (4.92)	0.015 (0.59)
SAE 347	destroyed	destroyed	0.35 (13.8)
Monel®	0.075 (2.95)	0.225 (8.85)	0.0375 (1.47)
Nickel	0.375 (14.7)	destroyed	0.1625 (6.40)
Inconel®	0.525 (20.7)	destroyed	0.1 (3.94)
Hastelloy® C	0.0075 (0.30)	0.005 (0.19)	0.0025 (0.098)

a) Operational test in the middle of the fractionating column. 216 d in 75 % acetic acid, 20 % high-boiling substances, 5 % water, temperature 398.6 K (125.6°C).
b) Operational test in the same column and the same duration, but in the top section of the column 95 % acetic acid, 5 % high-boiling substances and water, temperature 488.6 K (215.6°C).
c) Operational test on the uppermost plate of the column. Duration 241 d in 99.7 % acetic acid at 390.8 K (117.8°C).

Table 23: Operational tests with different alloys under various conditions [52, 237]

The unexpected corrosion of the CrNiMo-steel (DIN-Mat. No. 1.4571) during the distillation of acetic acid was caused by by-products in the acid, which normally should not attack but can lead to a low redox potential. In this case, the passivity of

Material	Corrosion rate, mm/a (mpy)
Copper	0.0275 (1.08)
Silicon bronze	0.02 (0.79)
Hastelloy® B	0.185 (7.28)
Hastelloy® C	0.1 (3.94)
SAE 316	1.0* (39.4)

* The SAE 316 specimens were destroyed during the test. The rate was determined by means of the residual thickness of the mounting device for the specimen. Monel®, nickel, Inconel®, SAE 304, aluminium S 2 and soft steel were completely destroyed.

Table 24: Operational test during azeotropic distillation of ethyl acetate/acetic acid using a solution of 80% acetic acid, 2 to 3% formic acid, 3 to 5% propionic acid, ethyl acetate and a small amount of water at 363 K (90°C) and a liquid velocity of 22 cm/s (test duration: 1,126 d) [52, 237]

the steel could be stabilized by continuously passing air through. The state of passivity should be observed by continuous potential measurements in these cases [280].

As for the corrosion behavior of CrNiMo-steels in acetic acid, see Table 25.

In the mid-part of acetic acid recovery towers made of stainless steel 00Cr20Ni25-Mo4.4Cu, where mixtures of acetic and formic acid occur, corrosion damages have been observed. The titanium alloy Ti-0.8Ni0.3Mo was recommended for the renewal of the middle part of the column [281].

The corrosion mechanism of stainless steels with Mo- and with and without Cu-contents was investigated in mixtures of acetic acid with formic acid. It was found that the corrosion resistance can be improved by the alloying of Mo together with Cu in comparison to that of steels without Cu-content (00Cr18Ni13Mo2 and 00Cr20Ni25Mo5) [282].

In principle, CrNiMo-steels can be used for the processing of acetic acid under any conditions; in most cases, the standard 18 8 Mo-steels are resistant enough; only if high demands are made on the purity of the acid should CrNiMoCu steels 25 20 be recommended [283]. This type of steel containing 6.25% molybdenum (without copper) is said to have a corrosion rate of 0.03 mm/a (1.18 mpy) in boiling 20% acetic acid compared with 0.06 mm/a (2.36 mpy) for the steel SAE 316 [284].

Valves made of 18 10 Mo-steel are used for acetic acid [285]. A valve gate made of the steel SAE 316 was, however, corroded by a mixture of acetic acid (15 to 30%) and formic acid (1 to 2%), balance water, at 400 to 405 K (127 to 132°C) [286].

A container for the preparation of acetic acid was manufactured from CrNiMo-steel 18 14 2.5 Nb containing at most 1% ferrite, while using electrodes consisting of a steel with (%) 18 Cr, 10 Ni and 2 Mo and with 5 to 10% ferrite; the weld seams were covered with at least 3 mm thick CrNiMo-steel ((%) Fe-17Cr-14Ni-2.5Mo-0.03C) [287].

In 5 to 98% acetic acid, the Soviet steels Ch17T (1.4510, SAE 439), Ch18N10T (CrNiTi 18 10), Ch17N13M2T (CrNiMoTi 17 13 2), Ch17N13M3T (CrNiMoTi 17 13 3) and steels with a reduced Ni-content, such as 0Ch21N5T (CrNiTi 21 5) and

Steel Grade		Corrosion rate, mm/a (mpy)						
		Glacial acetic acid p. a. after		100 ml glacial acetic acid p. a. + 2 ml formic acid after			Crude acetic acid (70 % acetic acid, 0.5 % formic acid) after	
		24	144 h	24 h	168 h		24 h	144 h
Steel with 27 % Cr, 4.6 % Ni, 1.5 % Mo, annealed at 1,223 K (950°C) and quenched in water	vapor	0.01 (0.39)	0.02 (0.79)	0 (0)	0.04 (1.57)		0.05 (1.97)	0.01 (0.39)
	liquid	0.36 (14.2)	0.06 (2.36)	0 (0)	0 (0)		0.05 (1.97)	0.01 (0.39)
Steel with 27 % Cr, 4.6 % Ni, 1.5 % Mo, annealed at 1,223 K (950°C), quenched in water, then sensitized at 1,033 K (760°C) in a furnace for 6 hours	vapor	0.16 (6.30)	0.12 (4.72)	0.16 (6.30))	0.12 (4.72)		0.27 (10.6)	0.13 (5.12)
	liquid	0.50 (19.7)	0.09 (3.54)	0.40 (1.57)	0.06 (2.36)		0.57 (22.4)	0.10 (3.94)
Steel with 18 % Cr, 10 % Ni, 2 % Mo, annealed at 1,323 K (1,050°C) and quenched in water	vapor	0 (0)	0.03 (1.18)	not tested	not tested		0.33 (13.0)	0.13 (5.12)
	liquid	0 (0)	0 (0)				0 (0)	0 (0)

Table 25: Corrosion rates of CrNiMo-steels in acetic acid at the boiling point [52, 288]

0Ch21N6M2T (CrNiMoTi 21 6 2), are said to be completely resistant at temperatures of up to 373 K (100°C); in this range, the presence of 0.3 to 5 % formic acid is said to have no effect yet [289, 290]. Likewise, the resistance of the steel 10Ch17N13M2T (CrNiMoTi 17 13 2) is said to be virtually independent of the manufacturing method or heat treatment [291].

The stainless steel 0Ch20N6M2T (CrNiMoTi 20 6 2) having a Ti:C ratio of 5:1 has a corrosion rate of 0.37 mm/a (14.6 mpy) at 301 K (28°C) in acetic acid + methylene chloride after quenching down from 1,323 to 1,373 K (1,050 to 1,100°C) and tempering at 823 K (550°C) for 8 h [292]. According to a Soviet publication regarding phase stability and corrosion resistance of the steel 03Ch21N21M4B (CrNiMoB 21 21 4) (El-35) in various testing media – inter alia, 96 % acetic acid – the sigma phase does not impair the corrosion resistance; no case of inter-crystalline corrosion was detected [293].

As for the corrosion behavior of stainless CrNiMo-steels, see Table 26.

CrNiMo-steel	Acetic acid %	Temperature K (°C)	Corrosion rate mm/a (mpy)	Remarks	Literature
SAE 316 ((%) 19Cr, 12Ni, 3 Mo, 0.06 C)	24	308 (35)	< 0.0025 (< 0.098)	storage test/115 d	[52]
	98	308 (35)	< 0.0025 (< 0.098)	storage test/157 d	[52]
	99	boiling	0.015 (0.59)	operational test/11, 12, 21 d	[52]
CrNiMo 18 8	10 to conc.*	293 (20)	< 0.1 (< 4)		[53]
	10 to conc.	boiling	< 0.1 (< 4)		
	conc.	473 (200)	< 1.0 (< 39.4)	acetic acid conc., 473 K (200°C)/0.10 MPa	[53]
CrNiMo 18 10	50	375 (102)	< 0.025 (< 0.98)	without heat transfer	[294]
	50	383 (110)	< 0.025 (< 0.98)	with heat transfer	
	50	398 (125)	< 0.025 (< 0.98)		
	50	413 (140)	< 0.025 (< 0.98)		
	99.6	391 (118)	< 0.025 (< 0.98)	without heat transfer	
	99.6	383 (110)	< 0.025 (< 0.98)	with heat transfer	
	99.6	398 (125)	0.33 (13.0)		
	99.6	413 (140)	0.25 (9.84)		

* concentrated

Table 26: Corrosion behavior of stainless chromium-nickel-molybdenum steels in acetic acid of various concentrations and temperatures

Table 26: Continued

CrNiMo-steel	Acetic acid %	Temperature K (°C)	Corrosion rate mm/a (mpy)	Remarks	Literature
SAE 316, 317	10 to 50	373 (100)	< 0.05 (< 2)	air-free	[54, 55]
	60 to 100	373 (100)	< 0.5 (< 20)	air-free, intergranular corrosion	
	100	485 (212)	> 1.25 (> 50)	air-free, stress corrosion cracking	
	10 to 100	298 (25)	< 0.05 (< 2)		[54, 55]
	10 to 40	373 (100)	< 0.05 (< 2)		
	30	398 (125)	< 0.5 (< 20)		
	50 to 100	348 (75)	< 0.5 (< 20)	aerated, no HCl, H_2SO_4, NaCl	
	50	373 (100)	< 0.5 (< 20)		
	60 to 100	373 (100)	0.5–1.25 (20–50)		
	30	485 (212)	> 1.25 (> 50)		
	40	485 (212)	> 1.25 (> 50)		
	30	373 (100)	< 0.5 (< 20)	acetic acid vapor, intergranular corrosion	[54, 55]
	100	373 (100)	< 0.5 (< 20)		
	99	393 (120)	0.075 (2.95)	acetic acid vapors operational test/82 d	[52]
	30	323 (50)	< 0.05 (< 2)		[54, 55]
	100	373 (100)	< 0.05 (< 2)	90 % acetic acid with 7 % acetaldehyde + 2 % water	
	90	373 (100)	0.5–1.25 (20–50)		
	100	373 (100)	0.03 (1.18)	90 % acetic acid with 7 % acetaldehyde + 2 % water + 1 % formic acid	[295]

* concentrated

Table 26: Corrosion behavior of stainless chromium-nickel-molybdenum steels in acetic acid of various concentrations and temperatures

Table 26: Continued

CrNiMo-steel	Acetic acid %	Temperature K (°C)	Corrosion rate mm/a (mpy)	Remarks	Literature
Remanit® steel corresponding to X5CrNiMo17-12-2 X3CrNiMo17-13-3 (SAE 316) X6CrNi-MoNb17-12-2 (SAE 316 Cb)	10	293 (20)	< 0.11 (< 4.33)		[238]
	10	boiling	< 0.11 (< 4.33)		
	50	293 (20)	< 0.11 (< 4.33)		
	50	boiling	< 0.11 (< 4.33)		
	100	293 (20)	< 0.11 (< 4.33)		
	100	boiling	0.11–1.1 (4.33–43.3)		
	10	363 (90)	< 0.11 (< 4.33)	acetic acid + hydrogen peroxide	[238]
	50	363 (90)	< 0.11 (< 4.33)	acetic acid + hydrogen peroxide	[238]
SAE 316	100	373 (100)	< 0.05 (< 2)	acetic acid + acetic anhydride + peracetic acid	[54, 55]
	100	373 (100)	< 0.05 (< 2)	acetic acid + ether	[54]
	10	373 (100)	< 0.05 (< 2)	acetic acid + formic acid	[54, 55]
	60	373 (100)	0.5–1.25 (19.7–49.2)	acetic acid + formic acid	
	100	352	< 0.5 (< 20)	acetic acid + formic acid	
	100	373 (100)	0.5–1.25 (19.7–49.2)	acetic acid + formic acid	
	56	373 (100)	0.0025 (0.098)	condensate consisting of 6 parts of 56 % acetic acid + 1 part of 90 % formic acid/150 h	[52, 237]
	56	377.6 (104.6)	0.0025 (0.098)	as above, vapors/150 h operational tests	
	100	297 (24)	> 1.25 (> 50)	acetic acid + hydrobromic acid	[54, 55]
	100	297 (24)	> 1.25 (> 50)	acetic acid + hydrochloric acid	[54, 55]
	100	373 (100)	< 0.5 (< 20)	95 % acetic acid + 5 % propionic acid, stress corrosion cracking	[54]
	100	373 (100)	> 1.25 (> 50)	acetic acid + sulfuric acid	[54, 55]

* concentrated

Table 26: Corrosion behavior of stainless chromium-nickel-molybdenum steels in acetic acid of various concentrations and temperatures

Table 26: Continued

CrNiMo-steel	Acetic acid %	Temperature K (°C)	Corrosion rate mm/a (mpy)	Remarks	Literature
SAE 316	100	373 (100)	0.5–1.25 (20–50)	90 % acetic acid + 10 % salicylic acid	[54]
	100	352 (79)	0.5 (19.7)	acetic acid + salicylic acid vapors	[54, 55]
	86 to 91	399.7 (126.7)	0.85 (33.5)	86–91 % acetic acid + 8.4 % salicylic acid vapors, operational test/891 h	[52, 237]
	99.5	385.2 (112.2)	0.1025 (4.04)	99.5 % acetic acid + 0.05 % salicylic acid vapors, operational test/1,022 h	[52, 237]
SAE 316 ((%) 18 Cr, 10 Ni, 2.5 Mo)	25	377 (104)	0.025 (0.98)	25 % acetic acid + 1.5 % formic acid, operational test/129 d	[52, 296]
	25	377 (104)	0.0825 (3.25)	25 % acetic acid + 4 % formic acid, as above	
	30 to 50	377 (104)	0.5 (19.7)	30–50 % acetic acid + 2–10 % formic acid, as above	
SAE 316 ((%) 18 Cr, 13 Ni, 2 Mo, 0.07 C)	99.7	393 (120)	0.01 (0.39)		[52]
	25	383 (110)	0.08 (3.15)	25 % acetic acid + 4 % formic acid	[52]
	30	408 (135)	0.11 (4.33)	30 % acetic acid + 8 % formic acid	[52]
SAE 317 ((%) 17 Cr, 13 Ni, 5 Mo, 0.07 C)	24	308 (35)	< 0.0025 (< 0.1)	storage test/115 d	[52]
	98	308 (35)	< 0.0025 (< 0.1)	storage test/157 d	[52]
	99.7	393 (120)	0.013 (0.51)		
	50	391 (118)	0.008 (0.31)	50 % acetic acid + 2 % formic acid	[52]
	25	383 (110)	0.5 (19.7)	25 % acetic acid + 4 % formic acid	[52]
	30	408 (135)	0.05 (1.97)	30 % acetic acid + 8 % formic acid	[52]

* concentrated

Table 26: Corrosion behavior of stainless chromium-nickel-molybdenum steels in acetic acid of various concentrations and temperatures

Table 26: Continued

CrNiMo-steel	Acetic acid %	Temperature K (°C)	Corrosion rate mm/a (mpy)	Remarks	Literature
SAE 317	99.5	385.2 (112.2)	0.075 (2.95)	99.5 % acetic acid + 0.05 % salicylic acid vapors, operational test	[52, 237]
	86 to 91	399.7 (126.7)	0.275 (10.8)	86–91 % acetic acid + 8.4 % salicylic acid vapors, operational test/891 h	
SAE 317 ((%) 18 Cr, 10 Ni, 3.5 Mo)	25	377 (104)	0.0075 (0.30)	25 % acetic acid + 1.5 % formic acid, operational test	[52, 237]
	25	377 (104)	0.05 (1.97)	25 % acetic acid + 4 % formic acid, operational test/ 129 d	
	30 to 50	377 (104)	0.275 (10.8)	30–50 % acetic acid + 2–10 % formic acid, as above	
CrNiMoTi 17 13 2	5	438 (165)	0 (0)		[234]
	25	438 (165)	0 (0)		
	50	413 (140)	0.08 (3.15)		
	50	438 (165)	1.04 (40.95)		
	98	413 (140)	0.12 (4.72)		
	98	438 (165)	0.44 (17.3)		
CrNiMoTi 21 6 2	5	438 (165)	0 (0)		[234]
	25	438 (165)	0 (0)		
	50	413 (140)	0.04 (1.57)		
	50	438 (165)	0.53 (20.9)		
	98	413 (140)	0.15 (5.91)		
	98	438 (165)	0.53 (20.9)		
CrNiMo 20 10 2.5	30	450 (177)	0.03–0.5 (1.18–19.7)		[82]
	glacial acetic acid	383 (110)	0.03–0.5 (1.18–19.7)		
	glacial acetic acid	450 (177)	0.1 (3.94)		

* concentrated

Table 26: Corrosion behavior of stainless chromium-nickel-molybdenum steels in acetic acid of various concentrations and temperatures

Table 26: Continued

CrNiMo-steel	Acetic acid %	Temperature K (°C)	Corrosion rate mm/a (mpy)	Remarks	Literature
inksK>CrNi 20 25 Mo Cu	50	375 (102)	< 0.025 (< 0.98)	without heat transfer	[297]
	50	383 (110)	0.05 (1.97)	with heat transfer	
	50	398 (125)	0.08 (3.15)	with heat transfer	
	50	413 (140)	< 0.025 (< 0.98)	with heat transfer	
	99.6	391 (118)	0.18 (7.09)	without heat transfer	
	99.6	383 (110)	0.12 (4.72)	with heat transfer	
	99.6	398 (125)	0.05 (1.97)	with heat transfer	
	99.6	413 (140)	2.54 (100)	with heat transfer, pitting corrosion	
Carpenter® 20	99.5	385.2 (112.2)	0.055 (2.17)	99.5 % acetic acid + 0.05 % salicylic acid vapors, operational test/1,022 h	[52, 237]
	86 to 91	399.7 (126.7)	2.575 (101)	86–91 % acetic acid + 8.4 % salicylic acid vapors, operational test/891 h	

* concentrated

Table 26: Corrosion behavior of stainless chromium-nickel-molybdenum steels in acetic acid of various concentrations and temperatures

Ion-nitriding of the steel surface of SAE 316 does not improve the corrosion resistance in 0.1 mol/l acetic acid solution [298].

Heating coils from 1.4571 for the vaporization of acetic acid with excess of anhydride were destroyed after a short time of operation. In laboratory tests it was found that the access of oxygen reinforces the attack. By addition of water in an excess of 0.1 % the corrosion could be inhibited. Water acts as passivator and is necessary for the maintenance of the passive layer even in the presence of oxygen. Only in the presence of water oxygen shows its passivating effect [299, 300].

– Acetic acid vapor –

Under conditions under which the standard CrNiMo-steels 18 10 2 are no longer resistant, for example in the distillation of crude acetic acid (70 % acetic acid, 0.5 % formic acid), a steel containing (%) 24 Cr, 24.5 Ni and 1.15 Mo has proved to be extremely resistant in the liquid and in the vapor space. The steels having higher Ni-contents for example 10 and 21 %, which were tested for comparison, were not better. No weight loss could be detected after 144 h [301].

Under heat transfer conditions, the corrosion rate of the steel SAE 316 in glacial acetic acid increases considerably in the temperature range from 348 to 413 K (75 to 140°C) and goes through a maximum if the temperature of the metal of the heat transfer area is maintained at about the boiling point of the glacial acetic acid. Based on these results, it is recommended to maintain the temperature of the heating medium, for example vapor, distinctly above the temperature of the liquid and to make sure that the heating coils or heat transfer areas do not become dirty [302, 303]; see Figure 4.

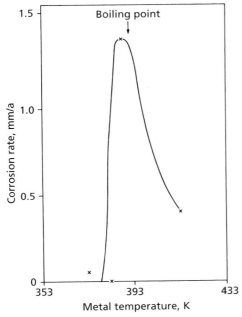

Figure 4: Effect of the temperature of the metal on the corrosion of SAE 316 in glacial acetic acid [302, 303].

However, the weld seams of CrNiMo-steels (SAE 316) are slightly attacked by hot acetic acid vapors which contain small amounts of phenylacetic acid at 588 K (315°C) [52].

– Acetic anhydride –

CrNiMo-steels (SAE 316 and 317) are used in the preparation and processing of acetic anhydride for distillation equipment, pumps, coolers, and in the preparation of cellulose acetate for acetylation equipment, pipelines and centrifuges for moist cellulose acetate (mixtures of acetic anhydride + acetic acid). They considered to be resistant materials up to the boiling point of acetic anhydride [114].

As for the corrosion behavior of stainless CrNiMo-steels towards acetic anhydride, see Table 27.

– Acetic anhydride vapor –

Acetic anhydride vapors are decomposed catalytically at 973 K (700°C) [114].

As for the behavior of various metallic materials towards mixtures of acetic anhydride + acetic acid at 950 K (677°C), see Table 28 [304].

– Vinegar –

The stainless austenitic steels (CrNi-, CrNiMo- and special steels) are not attacked by vinegar, which is the reason why in particular the austenitic CrNiMo-steels are increasingly used in the food and fermentation industry for tubes and other inserts of the fermenter, for cooling, but also for the entire plant [52, 305–308].

In tank cars made of stainless CrNi- and CrNiMo-steel, vinegar is transported without impairment of the container material or vinegar quality [309]. Gate valves made of CrNiMo-steel 18 10 are also used in the processing of vinegar-containing food products [310].

CrNiMo-steel	Acetic anhydride, %	Temperature K (°C)	Corrosion rate mm/a (mpy)	Remarks	Literature
SAE 316	crude	388 (115)	0.34 (13.4)		[311]
	vapor	383 (110)	0.44 (17.3)		
DIN-Mat. No. 1.4571	99	boiling	0.12–0.9 (4.72–35.4)		[312]
	99.5	373 (100)	0.12–0.9 (4.72–35.4)		
	no data given	943 (670)	to 0.12 (to 4.72)	acetic anhydride + acetic acid	
CrNiMo 18 8	no data given	293 (20)	< 0.1 (< 4)		[53]
		boiling	< 0.1 (< 4)		
SAE 316	30	297 (24)	< 0.05 (< 2)		[54]
	30	325–352 (52–79)	< 0.5 (< 20)		
	50	373 (100)	< 0.5 (< 20)		
	100	373 (100)	< 0.5 (< 20)		
	100	422 (49)	< 0.05 (< 2)	acetic anhydride + acetic acid	

Table 27: Corrosion behavior of stainless chromium-nickel-molybdenum steels in acetic anhydride of various concentrations and temperatures

Table 27: Continued

CrNiMo-steel	Acetic anhydride, %	Temperature K (°C)	Corrosion rate mm/a (mpy)	Remarks	Literature
Remanit® steels corresponding to SAE 316 SAE 316 Ti	no data given	293 (20)	< 0.11 (< 4.33)	resistant	[238]
		boiling	< 0.11 (< 4.33)		
Carpenter® 20 Cb-3	99	boiling	0.12–0.9 (4.72–35.4)		[313]
	99.5	373 (100)	0.12–0.9 (4.72–35.4)		
SAE 316, 317	10 to 70	373 (100)	< 0.5 (< 20)	acetic anhydride in acetic acid	[55]
	80	373 (100)	< 0.05 (< 2)		
	80	398 (125)	< 0.05 (< 2)		
	90	398 (125)	< 0.05 (< 2)		
	100	423 (150)	< 0.05 (< 2)		

Table 27: Corrosion behavior of stainless chromium-nickel-molybdenum steels in acetic anhydride of various concentrations and temperatures

Material	Corrosion rate, mm/a (mpy)		
	450 h	1,769 h	1,384 h
Hastelloy® C	0.0025 (0.098) (increase)	0.0025 (0.098)	
SAE 317	0.01 (0.39)	0.01 (0.39)	0.03 (1.18)
SAE 316	0.01 (0.39)	0.015 (0.59)	0.0425 (1.67)
SAE 316 ELC		0.015 (0.59)	
Illium®	0.005 (0.20)		
Aloyco® 20		0.0025 (0.098)	
SAE 309		0.01 (0.39)	
Hastelloy® B	0.0075 (0.30)	0.0175 (0.69)	

Table 28: Corrosion behavior in vapors of acetic anhydride + glacial acetic acid at 950 K (677°C), vapor velocity 53 m/s. The specimens were located at the point of entry of the vapor into the reaction tube (operational test) [314]

Table 28: Continued

Material	Corrosion rate, mm/a (mpy)		
	450 h	1,769 h	1,384 h
inksK>Hastelloy® A	0.0825 (3.25)		
Inconel®	3.25 (128)	2.75 (108)	
Monel®	destroyed		
Nickel	destroyed		
SAE 410			0.0375 (1.48)
SAE 430			0.03 (1.18)
SAE 446			0.005 (0.20)

Table 28: Corrosion behavior in vapors of acetic anhydride + glacial acetic acid at 950 K (677°C), vapor velocity 53 m/s. The specimens were located at the point of entry of the vapor into the reaction tube (operational test) [314]

In the food and canning industry, where vinegar is often used while hot and can contain chlorides, the susceptibility to pitting corrosion must be borne in mind. It is therefore advantageous to use molybdenum-alloyed steels of the type SAE 316. This also avoids the risk of contamination of the respective food products [315, 316].

Austenitic chromium-nickel steels with special alloying additions

– Acetic acid –

CrNiMo-steels form passive surface layers in acetic acid and phosphoric acid, which increase the corrosion resistance [317].

Pumps and valves made of Worthite® ((%) Fe, 24 Ni, 1.75 Cu, 20 Cr, 3 Mo, 3.25 Si, 0.6 Mn, max. 0.07 C) [318] or Durimet® 20 ((%) Fe, 29 Ni, min. 3.0 Cu, 20 Cr, min. 2.0 Mo, 1.0 Si, max. 0.07 C) [318] are resistant to 26% acetic acid + 5% sulfuric acid (balance water) at room temperature [52, 82, 319] and also to mixtures of crude acetic acid and formic acid at 366 K (93°C), in which low-alloyed steels led to failure after 3 weeks [52]. Durimet® 20 was not attacked by 5 to 100% acetic acid at the boiling temperature [53].

A corrosion rate of 0.036 mm/a (1.42 mpy) was given for an alloy ((%) Fe, 19 Cr, 10 Ni, 3 to 3.5 Mo, 2 Cu, 3 Si) in boiling glacial acetic acid [52, 320], while a different alloy ((%) Fe, 0.07 C, 24.96 Cr, 5.5 Ni, 2.3 Mo, 5.08 Co, 0.51 Si, 0.61Mn, 0.10 Cu) (heat treatment at 1,311 to 1,394 K (1,038 to 1,121°C), 1 h, cooled in water, tempered at 727 K (454°C) for 4 h) did not show any attack in boiling glacial acetic acid after 48 h [321].

As for the corrosion behavior of stainless CrNi-steels containing special alloying additions to acetic acid, see Table 29 and the references given there.

– *Acetic acid vapor* –

The NiCrMo-steel Sandvik® 2RK65 (1.4539, cf. SAE 904 L (%) Fe, max. 0.02 C, 0.45 Si, 1.8 Mn, max. 0.02 P, max. 0.02 S, 19.5 Cr, 25 Ni, 4.5 Mo, 1.5 Cu) has proved to be suitable in the production of acetic acid as material for waste gas condensers (inlet temperature 393 K (120°C), outlet temperature 365 K (92°C)); in contrast to tubes made of SAE 317 L ((%) Fe, 18 to 20 Cr, 11 to 15 Ni, 3 to 4 Mo, 2 Mn, 1 Si, 0.03 C), which has a service life of 18 months, tubes made of the Sandvik® steel remained virtually unattacked after 18 months [322, 323].

In the preparation of acetic acid from methanol or butanol and carbon monoxide, the Sandvik® steel 2RK65 has also proved to be suitable, especially in the condensers downstream from the reactor, where the acetic acid also contains formic acid [324].

– *Acetic anhydride* –

The high-alloy austenitic steels, for example Aloyco® 20 ((%) Fe, 20 Cr, 29 Ni, 4 Cu, 3 Mo) are completely resistant to acetic anhydride, even in a mixture with acetic acid, under most practical conditions of concentrations and temperatures and are therefore used for valves and pumps [74, 114].

As for the corrosion behavior of stainless CrNi-steels containing special alloying additions towards acetic anhydride, see Table 30.

Metal	Acetic acid %	Temperature K (°C)	Corrosion rate mm/a (mpy)	Remarks	Literature
Worthite® Durimet® 20	10–100	297 (24)	< 0.05 (< 2)	aerated	[54]
	30	422 (49)	< 0.5 (< 2)		
	60–100	325–373 (52–100)	< 0.5 (< 2)		
	100	422 (149)	> 1.25 (> 50)	aerated, pitting corrosion	
	10–100	325 (52)	< 0.05 (< 2)	air-free	

Table 29: Corrosion behavior of stainless chromium-nickel steels containing special alloying additions in acetic acid of various concentrations and temperatures

Table 29: Continued

Metal	Acetic acid %	Temperature K (°C)	Corrosion rate mm/a (mpy)	Remarks	Literature
Worthite® Durimet® 20	60–100	325–373 (52–100)	< 0.5 (< 2)	air-free	[54]
	80	422 (149)	> 1.25 (> 50)	air-free	
	100	477 (204)	> 1.25 (> 50)	air-free	
	30	352 (79)	< 0.5 (< 2)	acetic acid vapor	[54]
	100	325 (52)	< 0.5 (< 2)	acetic acid vapor	
	30	325 (52)	< 0.05 (< 2)	acetic acid + acetaldehyde	[54]
	100	373 (100)	< 0.05 (< 2)	acetic acid + acetic anhydride + peracetic acid	[54, 55]
	100	373 (100)	< 0.05 (< 2)	acetic acid + ether	[54, 55]
	100	373 (100)	< 0.05 (< 2)	acetic acid + formic acid	[54]
	100	422 (149)	< 0.5 (< 2)	acetic acid + formic acid	
	100	297 (24)	> 1.25 (> 50)	acetic acid + hydrobromic acid	[54]
	100	297 (24)	> 1.25 (> 50)	acetic acid + hydrochloric acid	[54]
	100	297 (24)	< 0.5 (< 2)	acetic acid + mercury salts	[54]
	100	422 (149)	> 1.25 (> 50)	90 % acetic acid + 10 % salicylic acid	[54]
	100	373 (100)	< 0.5 (< 2)	acetic acid + salicylic acid vapors	[54]

Table 29: Corrosion behavior of stainless chromium-nickel steels containing special alloying additions in acetic acid of various concentrations and temperatures

Table 29: Continued

Metal	Acetic acid %	Temperature K (°C)	Corrosion rate mm/a (mpy)	Remarks	Literature
Worthite® Durimet® 20	100	297 (24)	< 0.5 (< 2)	acetic acid + sulfuric acid	[54]
	100	373 (100)	0.5–1.25 (19.7–49.2)		
Worthite®	25	377 (104)	0.0325 (1.28)	25 % acetic acid + 1.5 % formic acid	[52, 237]
	25	377 (104)	0.075 (2.95)	25 % acetic acid + 4 % formic acid	[52, 237]
	30–50	377 (104)	0.5 (19.7)	30–50 % acetic acid + 2–10 % formic acid	[52, 237]
Durimet® 20 Fe-29Ni-20Cr-4Cu-2Mo-< 0.07C	10	no temperature given	< 0.0025 (< 0.098)	laboratory tests/ 168 h	[52]
	50	no temperature given	0.005 (0.20)		
	75	no temperature given	< 0.0025 (< 0.098)		[52]
	glacial acetic acid	no temperature given	0.025 (0.98)		
ACI CN 20 Fe-30Ni20Cr	10–50	373 (100)	< 0.05 (< 2)	aerated	[55]
	60–100	373 (100)	< 0.5 (< 2)		
	30	398 (125)	< 0.5 (< 2)		
	100	398 (125)	< 0.5 (< 2)		
	100	423 (150)	> 1.25 (> 50)	aerated, pitting corrosion	
	10–100	323 (50)	< 0.05 (< 2)	air-free	[55]
	10–50	373 (100)	< 0.05 (< 2)		

Table 29: Corrosion behavior of stainless chromium-nickel steels containing special alloying additions in acetic acid of various concentrations and temperatures

Table 29: Continued

Metal	Acetic acid %	Temperature K (°C)	Corrosion rate mm/a (mpy)	Remarks	Literature
ACI CN 20 Fe-30Ni20Cr	60–100	348–373 (75–100)	< 0.5 (< 2)	air-free	[55]
	80	398 (125)	> 1.25 (> 50)		
	100	485 (212)	> 1.25 (> 50)		
	30	298 (25)	< 0.05 (< 2)	90 % acetic acid + acetaldehyde + 7 % water	[55]
	10	373 (100)	< 0.05 (< 2)	acetic acid + formic acid	[55]
	100	373 (100)	< 0.05 (< 2)		
	100	398 (125)	< 0.5 (< 2)		
	100	298 (25)	> 1.25 (> 50)	acetic acid + hydrobromic acid	[55]
	100	298 (25)	> 1.25 (> 50)	acetic acid + hydrochloric acid	[55]
	30	348 (75)	< 0.5 (< 2)	acetic acid vapor	[55]
	100	323 (50)	< 0.5 (< 2)		
	100	298 (25)	< 0.5 (< 2)	acetic acid + mercury salts	[55]
	100	373 (100)	< 0.5 (< 2)	acetic acid + salicylic acid vapors	[55]
	100	298 (25)	< 0.5 (< 2)	acetic acid + sulfuric acid	[55]
	100	373 (100)	0.5–1.25 (19.7–49.2)		

Table 29: Corrosion behavior of stainless chromium-nickel steels containing special alloying additions in acetic acid of various concentrations and temperatures

Table 29: Continued

Metal	Acetic acid %	Temperature K (°C)	Corrosion rate mm/a (mpy)	Remarks	Literature
Aloyco® 31	40–50	377 (104)	0.18 (7.09)	40–50 % acetic acid + 4–6 % formic acid + 1 % methyl acetate, balance water	[82]
Remanit® steel corresponding to X1CrNi-MoN25-25-2 (SAE 310 MoLN)	glacial acetic acid	boiling	< 0.11 (< 4.33)		[238]

Table 29: Corrosion behavior of stainless chromium-nickel steels containing special alloying additions in acetic acid of various concentrations and temperatures

– Vinegar –

The Sandvik® 2RK65 (1.4539, cf. SAE 904 L) has proved to be suitable for the preparation of preserved cucumbers (vinegar containing sodium chloride), in which even the CrNiMo-steel SAE 316 is subject to pitting corrosion within 2 weeks [325].

Special iron-based alloys

– Acetic acid –

Austenitic cast iron grades are suitable for equipment in which products containing organic acids such as acetic acid, formic acid, oxalic acid and tar acids, which are formed in the distillation of wood products, are processed [326, 327].

Ni-Resist® type 1 (EN-JL3011, cf. UNS F41000 (%) Fe-15.5Ni-6.5Cu-2.5Cr-2.0Si-1.2Mn-2.8C) [328] loses at most 0.5 mm/a (19.7 mpy) in 25 % cold acetic acid when aerated; cast iron would corrode 40 to 50 times faster [329].

Corrosion rates of cast iron with Ni and Cr or Ni, Si and Cr are listed in Table 31.

These high-alloy cast irons are only used for unaerated acid and not above room temperature, due to their relatively low resistance. Only in the case of strongly dilute acids are higher temperatures also permissible [3].

The stainless semi-austenitic martensite-hardening steel AM 350 (1.5622, cf. UNS K13050 (%) Fe-0.07C-17Cr-4Ni-2.75Mo) can be compared with the stainless steel SAE 304 in its corrosion resistance to acetic acid [330].

Acetic acid/HNO$_3$/HCl mixtures (70:30:1) are used for the polishing of the Kovar® 29NK alloy (Fe-29Ni18Co) [331]. The layer removed during polishing was determined with 7.8×10^{-4} cm.

Metal	Acetic anhydride, %	Temperature K (°C)	Corrosion rate mm/a (mpy)	Remarks	Literature
Worthite® Durimet® 20	30	373 (100)	< 0.5 (< 2)		[54]
	50	373 (100)	< 0.5 (< 2)		
	100	297 (24)	< 0.05 (< 2)		
	100	325–373 (52–100)	< 0.5 (< 2)		
	100	422 (149)	< 0.05 (< 2)	acetic anhydride + acetic acid	[54]
ACI CN 20 Fe-30Ni20Cr	30	348 (75)	< 0.5 (< 2)		[55]
	50	373 (100)	< 0.5 (< 2)	acetic anhydride in acetic acid	
	100	348 (75)	< 0.05 (< 2)		
	100	373 (100)	< 0.5 (< 2)		

Table 30: Corrosion behavior of stainless chromium-nickel steels containing special alloying additions in acetic anhydride of various concentrations and temperatures [54, 55]

The stainless martensitic steel Custom 450® ((%) Fe, 14.5 to 16.5 Cr, 5.5 to 7 Ni, 1.25 to 1.75 Cu, Nb) is similar to the steel SAE 304 in its corrosion resistance in boiling 50 % acetic acid; it is also resistant to stress corrosion cracking and hydrogen embrittlement in 5 % acetic acid saturated with hydrogen sulfide [332].

The stainless martensitic-hardenable steel Custom 455® ((%) Fe, max. 0.03 C, max. 0.50 Mn, max. 0.50 Si, 11 to 13 Cr, 7 to 10 Ni, 0.90 to 1.40 Ti, 0.25 to 0.50 Nb/Ta, 1 to 3 Cu, max. 0.005 B) is superior to the Cr-steel 12 (SAE 410) in boiling 10 % acetic acid [333–335].

As for the corrosion behavior of special Fe-based alloys in acetic acid, see Table 32.

Material	Corrosion rate mm/a (mpy)	Literature
Ni-Resist® 2 18 to 22 Ni, 1 to 2.3 Cr UNS F41002	0.076 (2.99)	[335]
Ni-Resist® D-2 18 to 22 Ni, 1 to 2.5 Cr EN-JS3011	0.076 (2.99)	[3]
Ni-Resist® D-2B 18 to 22 Ni, 2.5 to 4 Cr GGG NiCr 20 3†	0.051 (2.01)	[3]
Nicrosilal® 18 to 22 Ni, 1.4 to 4.5 Cr	0.066 (2.60)	[3]
Nicrosilal® 18 to 22 Ni, 1 to 2.5 C	0.063 (2.48)	[3]

Table 31: Corrosion rates of Ni-Resist® and Nicrosilal® alloys in 2 % acetic acid

– *Acetic acid vapor* –

In the thermal cracking of acetic acid, strongly heated acetic acid vapors are cleaved into ketene and water in the presence of 0.2 to 0.3 % triethyl phosphate as catalyst under reduced pressure in coils made of CrAl-steel Sicromal® ((%) Fe, 23 Cr, 1 Si, 2.5 Al) in electrically heated furnaces at 973 to 1,023 K (700 to 750°C).

– *Acetic anhydride* –

In 100 % acetic anhydride, Ni-Resist® loses less than 0.5 mm/a (19.7 mpy) at 297 K (24°C), but is no longer resistant at 325 K (52°C) (corrosion rates above 1.25 mm/a (49.2 mpy)) [54].

– *Vinegar* –

Ni-Resist® ((%) Fe, 14 Ni, 6 Cu, 2 Cr, 2 Si or Fe, 20 Ni, 2 Cr, 2 Si) loses 0.46 g/m² d (0.021 mm/a (0.83 mpy)) in about 6 % vinegar at room temperature [52, 336].

Metal	Acetic acid %	Temperature K (°C)	Corrosion Rate mm/a (mpy)	Remarks	Literature
Ni-Resist®	20–100	297 (24)	> 1.25 (> 50)	aerated	[54, 55]
	10	297 (24)	< 0.5 (< 20)		
	10	325 (52)	> 1.25 (> 50)		
	20–100	297 (24)	> 1.25 (> 50)	air-free	
	10	297 (24)	< 0.5 (< 20)		
	10	325 (52)	> 1.25 (> 50)		
	100	373 (100)	> 1.25 (> 50)	acetic acid vapor	[54]
	100	297 (24)	> 1.25 (> 50)	acetic acid + formic acid	[54, 55]
	100	297 (24)	> 1.25 (> 50)	acetic acid + hydro-bromic acid	
	100	297 (24)	> 1.25 (> 50)	acetic acid + hydro-chloric acid	
	100	297 (24)	> 1.25 (> 50)	acetic acid + mercury salts	

Table 32: Corrosion behavior of special Fe-based alloys in acetic acid of various concentrations and temperatures

Bibliography

[1] Anonymous
Mater. Eng. 88 (1978) 6; Mater. Selector 1979, p. C2

[2] Wegst, C. W.
Stahlschlüssel (in German)
Verlag Stahlschlüssel Wegst KG., D-7142 Marbach, 1977

[3] Le Monnier, E.
in: Kirk-Othmer Encyclopedia of Chemical Technology
2nd ed., vol. 8, p. 386
John Wiley & Sons, Inc., New York-London-Sydney, 1965

[4] Dechema-Werkstoff-Tabelle "Essigsäure" (Dechema corrosion data sheets "Acetic acid") (in German)
DECHEMA, D-6000 Frankfurt am Main, 1957

[5] Nelson, G. A.
Corrosion Data Survey 1967 – Corrosion charts A-1/A-2 "Acetic acid"

[6] Hamner, N. E.
Corrosion Data Survey
5th ed., p. 2, p. 198
NACE, Houston (Texas/USA), March 1974

[7] Ritter, F.
Korrosionstabellen metallischer Werkstoffe (Corrosion charts of metallic materials) (in German)
4th ed., p. 88
Springer-Verlag, Wien, 1957

[8] Vdovenko, I. D.; Anikina, N. S.
Mechanism of iron dissolution in aqueous solution of acetic acid (in Russian)
Zashch. Met. 10 (1974) 2, p. 157

[9] Kuznetsova, S. P.; Zhuk, N. P.
Moscow Inst. of Steel and Alloys, Prot. Met. 11 (1975) 6, p. 726 (in Russian)

[10] Molodov, A. I. et al.
Mechanism of anodic dissolution of iron in acetic acid solution (in Hungarian)
Korroz. Figyelo 22 (1982) 4, p. 63

[11] Anikina, N. S.; Vdovenko, I. D.
Mechanism of armco iron dissolution in aqueous solutions of acetic acid (in Russian)
Zashch. Met. 11 (1975) 5, p. 607

[12] Anikina, N.S.; Vdovenko, I. D.
Hydrogen liberation during iron corrosion in the mixture CH_3COOH-H_2O (in Russian)
Ukr. Khim. Zh. 41 (1975) 11, p. 1151

[13] Versányi, M. L. et al.
Dissolution of iron in deaerated concentrated acetic acid solutions (in Hungarian)
Magy Kém. Foly. 90 (1984) 2, p. 89

[14] Markos'yan, G. N.; Siraki, L.; Molodov, A. I. et al.
Self-dissolution and passivation of iron in 95 % acetic acid in presence of oxygen (in Russian)
Elektrokhimiya 22 (1986) 2, p. 219

[15] Varshani, L. M. et al.
Self-dissolution and anodic dissolution of iron in glacial acetic acid deaerated solutions (in Russian)
Elektrokhimiya 20 (1984) 5, p. 625

[16] Markos'yan, G. N.; Siraki, L.; Molodov, A. I. et al.
Self-dissolution and passivation of iron in 95 % acetic acid in presence of oxygen (in Russian)
Elektrokhimiya 22 (1986) 2, p. 219

[17] Frenzel, F.
Über das korrosive Verhalten von Eisen in Essigsäure
(The corrosion behavior of iron in acetic acid) (in German)
Werkst. Korros. 27 (1976) p. 20

[18] Wiegand, H.; Piltz, H.-H.
Über das Zusammenwirken von Kavitation und Korrosion
(Interaction of cavitation and corrosion) (in German)
Werkst. Korros. 15 (1964) 3, p. 215

[19] Dechema-Werkstoff-Tabelle "Essigsäure" (Dechema corrosion data sheets "Acetic acid") (in German)
DECHEMA, D-6000 Frankfurt am Main, 1957

[20] Markovic, T.
Auflösung verschiedener Metalle im Kontakt mit rotierenden Pt-Elektroden
(Dissolution of various metals in contact with rotating Pt-electrodes) (in German)
Werkst. Korros. 5 (1954) p. 121

[21] Hackermann, N.; Glenn, E. E.
Corrosion of steel by air-free, dilute, weak acids
J. Electrochem. Soc. 100 (1953) p. 339

[22] Dechema-Werkstoff-Tabelle "Essigsäure"
(Dechema corrosion data sheets "Acetic acid") (in German)
DECHEMA, D-6000 Frankfurt am Main, 1957

[23] Ginding, A. G.; Kasakowa, W. A.
Metallkorrosion durch Fettsäuren
(Corrosion of metals caused by fatty acids) (in Russian)
Ber. Akad. Wiss. (1951) 80, p. 389

[24] Konradi, M. V.
Corrosion of copper, aluminium, and iron in solutions of acetic acid and ammonium sulfate (in Russian)
Izv. V.U.Z. Tsvetn. Metall. (1958) 4, p. 165
(C.A. 53 (1959) 7941e)

[25] Zavyalova, E. P.; Shchukurov, B. R.; Ionnidis, O. K.
Corrosion of metals in a hydrogen sulfide medium under the influence of organic acids (in Russian)
Korroz. Zashch. Neftedob. Prom. Nauch.-Tekh. Sb. (1967) 2, p. 5
(C.A. 69 (1968) 21409e)

[26] Stahl, R.; Kiefer, P.
Korrosion und Korrosionsschutz in der Lebensmittelindustrie
(Corrosion and corrosion protection in the food industry) (in German)
Werkst. Korros. 24 (1973) 6, p. 513

[27] Dechema-Werkstoff-Tabelle "Essigsäure"
(Dechema corrosion data sheets "Acetic acid") (in German)
DECHEMA, D-6000 Frankfurt am Main, 1957

[28] Sekine, I.; Senoo, K.
The corrosion behavior of SS 41 steel in formic and acetic acids
Corrosion Sci. 24 (1984) 5, p. 439

[29] Asatryan, V. G.; Balayan, A. M.
Electrochemical protection of carbon steel from corrosion in acetic acid (in Russian)
Khim. i Neft Mashinostr. (1965) 1, p. 28
(C.A. 63 (1965) 3885b)

[30] Farkhadov, A. A.; Balayan, A. M.; Asatryan, V. G.
Corrosion of carbon steel in acetic acid, and an electrochemical method for protecting the steel (in Russian)
Borba Korroz. Khim. Neft. Prom. (1967) 1, p. 86
(C.A. 69 (1968) 40660w)

[31] Süry, P.
Prüfmethoden der Spannungsrißkorrosion und Wasserstoffversprödung
(Testing methods for stress corrosion cracking and hydrogen embrittlement) (in German)
Mater. u. Tech. 4 (1976) 3, p. 140

[32] Anonymous
Corrosion in distillation plants
Chem. Process Eng. 42 (1961) 9, p. 385

[33] Schmidt, H. W.; Brouwer, A. A.
Three applications of cathodic protection for chemical equipment
Mater. Protection 1 (1962) 2, p. 26

[34] Legezin, N. E.; Zarembo, K. S.
Effect of carbon dioxide and organic acids on the corrosion of steel in gas-condensate well equipment (in Russian)
Mater. Nauch.-Tekh. Soveshch. Zashch. Korroz. Oborudovaniya Neft. Gazov. Skrazhin., Baku (1964) p. 89
(C. A. 68 (1968) 70968z)

[35] Kim, S. H.
Stress cracking of steel by hydrogen sulfide (in Korean)
Pusan Susan Teahak Yongu Pogo, Chayon Kwahak 13 (1973) 1, p. 57
(C.A. 80 (1974) 123643w)

[36] Nishimura, S.; Kurisu, A.; Otami, M.
Sulfide-corrosion cracking of welded high-strength steel
Zairyo 13 (1964) p. 527
(C.A. 64 (1966) 373f)

[37] Watanabe, M.; Mukai, Y.
Stress corrosion cracking of high-strength steels due to H_2S (in Japanese)
Technol. Repts. Osaka Univ. 14 (1964) p. 609
(C.A. 61 (1964) 14267g)

[38] Lang, C.; Magera, P.
Resistance of oil pipeline steels by hydrogen cracking (in Czech)
Hutn. Listy 28 (1973) 5, p. 330
(C.A. 79 (1973) 82343g)

[39] Nakauchi, H.; Togano, H.
Chemical studies on sulfide corrosion cracking of storage tank materials for liquid propane (LP) gas
II. Protection of high-strength steel immersed in LP gas containing hydrogen sulfide (in Japanese)
Tokyo Kogyo Shikensho Hokoku 62 (1967) 9, p. 319
(C.A. 68 (1968) 32414h)

[40] Dechema-Werkstoff-Tabelle "Essigsäure"
(Dechema corrosion data sheets "Acetic acid") (in German)
DECHEMA, D-6000 Frankfurt am Main, 1957

[41] Nelson, G. A.
Corrosion Data Survey 1967 – Corrosion charts A-1/A-2 "Acetic acid"
NACE, Houston (Texas/USA), 1968

[42] Hamner, N. E.
Corrosion Data Survey
5th ed., p. 2, p. 198
NACE, Houston (Texas/USA), March 1974

[43] Ritter, F.
Korrosionstabellen metallischer Werkstoffe
(Corrosion charts of metallic materials)
(in German)
4th ed., p. 88
Springer-Verlag, Wien, 1957

[44] Jahnke, H.; Schönborn, M.
Electrochemical corrosion measurements in motor fuels based on methanol and ethanol
Werkst. Korros. 36 (1985) 12; p. 561

[45] Heitz, E.
Vergleichende Korrosionsuntersuchungen in Ethanol-, Methanol- und wäßrigen Lösungen
(Comparative corrosion testings in ethanol-, methanol- und aqueous solutions)
(in German)
Werkst. Korros. 38 (1987) 12, p. 333

[46] Uller, L.; Bastos, S. M.; Wanderley, V. G.; de Miranda, T. R. V.
The detrimental effects on metal corrosion by some impurities present on hydrated ethanol
International Congress on Metallic Corrosion, Toronto (Canada), (Proc. Conf.) 3 (1984) p. 475
National Research Council of Canada, Ottawa

[47] Matsukura, K.; Sato, K.
Effect of metallurgical factors of low carbon steel sheets on stress corrosion cracking in methanol solution
Trans. Iron Steel Inst. Jpn. 18 (1978) 9, p. 554

[48] Krimcheeva, G. G.; Rozenfeld, I. L.; Vezirova, V. R.
The effect of hydrocarbons on the corrosion and hydrogen charging of a structural steel in aqueous solutions of sodium chloride and acetic acid (in Russian)
Korroz. Zashch. (1982) 12, p. 5

[49] Aal, M. S. A.; Wahdan, M. H.
Anodic behavior of mild steel in deaerated carboxylic acid solutions containing NO_2^-, Cl^- and NO_3^--ions
Br. Corros. J. 16 (1981) 4, p. 205

[50] Anikina, N. S.; Vdovenko, I. D.
Inst. Gen. Inorg. Chem. Acad. Sc., Proc. Met. 12 (1976) 3, p. 297 (in Russian)

[51] Wood, A. S.; Gensheimer, D. E.
Corrosion preventive composition
US Pat. 3 294 694 (General Aniline & Film Corp.) (27.12.1966)

[52] Dechema-Werkstoff-Tabelle "Essigsäure"
(Dechema corrosion data sheets "Acetic acid") (in German)
DECHEMA, D-6000 Frankfurt am Main, 1957

[53] Ritter, F.
Korrosionstabellen metallischer Werkstoffe
(Corrosion charts of metallic materials)
(in German)
4th ed., p. 88
Springer-Verlag, Wien, 1957

[54] Nelson, G. A.
Corrosion Data Survey 1967 – Corrosion charts A-1/A-2 "Acetic acid"
NACE, Houston (Texas/USA), 1968

[55] Hamner, N. E.
Corrosion Data Survey
5th ed., p. 2, p. 198
NACE, Houston (Texas/USA), March 1974

[56] Bartsch, E.; Danninger, H.
Über die Inhibitorwirkung der Sorbinsäure
(The inhibitory effect of sorbic acid)
(in German)
Werkst. Korros. 38 (1987) 2, p. 73

[57] Maltseva, V. P.; Maltseva, V. S.
The protective action of the rhodamine 6Zh KDM in acid media (in Russian)
Zashch. Met. 14 (1978) 6, p. 718

[58] Podobaev, N. I.; Kharkovskaya, N. L.; Korotkikh, E. V.; Ustinskii, E. N.
Dithiocarbamates as inhibitors of acidic corrosion (in Russian)
Zashch. Met. 16 (1980) 1, p. 73

[59] Kuznetsova, S. P.; Budnevskaya, G. A.; Zhuk, N. P.
Cathodic inhibitor protection of iron from corrosion in aqueous solutions of CH_3COOH (in Russian)
Zashch. Met. 15 (1979) 4, p. 465

[60] Kiriluk, S. S.; Titakova, I. K.; Korsunskaya, A. L.; Miskidzhyan, S. P.
Anticorrosion action of some urotropin derivatives (in Russian)
Zashch. Met. 16 (1980) 2, p. 180

[61] Schlerkmann, H.; Schwenk, W.
Untersuchung von Säureinhibitoren hinsichtlich ihrer Wirkung gegenüber H-induzierter Korrosion
(Investigation of inhibitors regarding weight loss corrosion and hydrogen absorption in acids) (in German)
Werkst. Korros. 35 (1984) 10, p. 449

[62] Konovalova, L. L.
Corrosion of steel in mixtures of acetic acid with aniline and pyridine (in Russian)
Uch. Zap., Perm. Gos. Iniv. (1966) p. 206
(C.A. 68 (1968) 89317h)

[63] Metsik, R. E.
Use of certain shale products as corrosion inhibitors (in Russian)
Tekhnol. Organ. Proiz. Nauch.-Proiz. Sb. 5 (1966) 41, p. 95
(C.A. 67 (1967) 24480r)

[64] Raudene, D.-E.S.; Persiantseva, V. P.
Corrosion behavior of steel St 3 in the methylene ethyl alcohol mixture
(in Russian)
Zashch. Met. 18 (1982) 2, p. 263

[65] Khachatryan, E. A.; Babayan, G. G.
The corrosion and passivation of the carbon steel St 3 in aqueous acetic acid solutions of sulphuric acid (in Russian)
Prom-st. Armenii (1985) 2, p. 36

[66] Cermakova, D.; Dohnalova, J.
Corrosion of metals by vapors of organic materials in microclimate (in Czech)
Koroze Ochr. Mater. (1963) 4, p. 5
(C.A. 60 (1964) 10336d)

[67] Rückert, J.
Korrosionsverhalten von Metallen in Verbindung mit Holz
(Corrosion behavior of metals in contact with wood) (in German)
Werkst. Korros. 37 (1986) 6, p. 336

[68] Schikorr, G.
Über die Korrosion von Metallen in hölzernen Gehäusen
(On the corrosion of metals in wooden casings) (in German)
Werkst. Korros. 12 (1961) 1, p

[69] Swales, G. L.
High-nickel alloys in the petroleum industry. Part 2
Corrosion Technology 6 (1959) 4, p. 119

[70] Hochmüller, K.; Maihöfer, A.; Braunstein, L.
Probleme in Dampferzeugern durch nichtionogene Inhaltsstoffe des Speisewassers
(Problems in steam-generators by non-ionogenic components of feedwater) (in German)
VGB Kraftwerkstech. 54 (1974) 3, p. 160

[71] Hill, G. V. G.
Corniform corrosion of mild steel exposed to acid and humid environments
Br. Corros. J. 8 (1973) 3, p. 128

[72] Donovan, P. D., Stringer, J.
Corrosion of metals and their protection in atmospheres containing organic acid vapours
Br. Corros. J. 6 (1971) p. 132

[73] Kulig, E.; Sokolowski, P.; Wojtas, R.
Aliphatic acids as corrosion inhibitors of steel in sulphuric acid solution (in Polish)
Ochr. Przed Koroz. 29 (1986) 1, p. 12

[74] Dechema-Werkstoff-Tabelle
"Essigsäureanhydrid"
(Dechema corrosion data sheets "Acetic anhydride") (in German)
DECHEMA, D-6000 Frankfurt am Main, 1957

[75] Holland-Merten, E. L.
Der Einfluß des Vakuums auf die Auswahl und die Korrosionsbeständigkeit der Werkstoffe
(The influence of vacuum on the choice and corrosion resistance of materials)
(in German)
DECHEMA Monographien 8 (1936) p. 200
Verlag Chemie, D-6940 Weinheim

[76] Rabald, E.
Corrosion guide, 2nd ed., p. 13
Elsevier Publishing Company, Amsterdam – London – New York, 1968

[77] Nelson, G. A.
Corrosion Data Survey 1967 – Corrosion charts A-1/A-2 "Acetic acid"
NACE, Houston (Texas/USA), 1968

[78] Hamner, N. E.
Corrosion Data Survey
5th ed., p. 2, p. 198
NACE, Houston (Texas/USA), March 1974

[79] Anonymous
Chemical safety data sheet SD-15 "Acetic anhydride"
Man. Chem. Ass., Washington, 1962

[80] Kühn-Birett
Merkblätter gefährliche Arbeitsstoffe, Blatt E 10 "Essigsäureanhydrid"
(Advisory sheets on hazardous materials handling, paper E 10 "acetic anhydride")
(in German)
Verlag Moderne Industrie Wolfgang Dummer & Co.
D-8000 München, 1977

[81] Eck, H.
in: Ullmanns Encyklopädie der technischen Chemie
(Ullmann's encyclopedia of industrial chemistry) (in German)
4th ed., vol. 11, p. 75
Verlag Chemie, D-6940 Weinheim, 1976

[82] Rabald, E.
Corrosion Guide, 2nd. ed., p. 3
Elsevier Publishing Company, Amsterdam-London-New York, 1968

[83] Gonzalez, O. D.; Josephic, P. H.; Oriani, R. A.
The undercutting of organic lacquers on steel
J. Electrochem. Soc. 121 (1974) 1, p. 29

[84] Levjanto, S. I.; Florianovich, G. M.
Electrochemical and corrosion behavior of chrome-plated tin in connection with its use in the canning industry (in Russian)

[85] Kolb, H.
Elektrochemische Untersuchungen zum Einfluß von Nitrat und Nitrit auf die Ausgangs-Ruhepotentiale und den Kurzschlußstrom des Elements Eisen/Zinn in Fruchtsäuren und -säften
(Electrochemical investigations into the influence of nitrate and nitrite in the starting-rest potentials and the short circuit current of the element iron/tin in fruit acids and juices) (in German)
Werkst. Korros. 27 (1976) p. 10

[86] Kopenhagen, W. I.
Metal Ind. 77 (1950) p. 137

[87] Anonymous
Mater. Eng. 88 (1978) 6; Mater. Selector 1979, p. C2

[88] Vaccari, J. A.
What's ahead in aircraft and aerospace materials
Mater. Eng. 79 (1974) 6, p. 45

[89] Le Monnier, E.
in: Kirk-Othmer Encyclopedia of Chemical Technology
2nd ed., vol. 8, p. 386
John Wiley & Sons, Inc., New York-London-Sydney, 1965

[90] Tsutsumi, N.
Corrosion of malleable iron
Rept. Coatings Research Lab., Waseda Univ., Tokyo (1956) 7, p. 19
(C.A. 51 (1957) 4247 g)

[91] Dechema-Werkstoff-Tabelle "Essigsäureanhydrid"
(Dechema corrosion data sheets "Acetic anhydride") (in German)
DECHEMA, D-6000 Frankfurt am Main, 1957

[92] Rabald, E.
Corrosion guide, 2nd ed., p. 13
Elsevier Publishing Company, Amsterdam – London – New York, 1968

[93] Dechema-Werkstoff-Tabelle "Essigsäureanhydrid"
(Dechema corrosion data sheets "Acetic anhydride") (in German)
DECHEMA, D-6000 Frankfurt am Main, 1957

[94] Ritter, F.
Korrosionstabellen metallischer Werkstoffe
(Corrosion charts of metallic materials) (in German)
4th ed., p. 88
Springer-Verlag, Wien, 1957

[95] Rabald, E.
Corrosion guide, 2nd ed., p. 13
Elsevier Publishing Company, Amsterdam – London – New York, 1968

[96] Rabald, E.
Corrosion guide, 2nd ed., p. 13
Elsevier Publishing Company, Amsterdam – London – New York, 1968

[97] Anonymous
Mater. Eng. 88 (1978) 6; Mater. Selector 1979, p. C2

[98] Vaccari, J. A.
What's ahead in aircraft and aerospace materials
Mater. Eng. 79 (1974) 6, p. 45

[99] Le Monnier, E.
in: Kirk-Othmer Encyclopedia of Chemical Technology
2nd ed., vol. 8, p. 386
John Wiley & Sons, Inc., New York-London-Sydney, 1965

[100] Timmerbeil, H.
Ein halbes Jahrhundert Gießereitechnik in Deutschland I. Entwicklung der Gußwerkstoffe. 7. MTC-beständiger Guß
(Half a century of foundry technology in Germany I. Development of cast materials. 7. MTC-resistant coatings) (in German)
Gießerei 46 (1959) p. 676

[101] Rabald, E.
Corrosion Guide, 2nd. ed., p. 3
Elsevier Publishing Company, Amsterdam-London-New York, 1968

[102] Rabald, E.
Corrosion Guide, 2nd. ed., p. 3
Elsevier Publishing Company, Amsterdam-London-New York, 1968

[103] Rabald, E.
Corrosion guide, 2nd. ed., p. 13
Elsevier Publishing Company, Amsterdam – London – New York, 1968

[104] Lefébure, R.
Métaux-Corros.-Ind. 25 (1950) p. 9

[105] Staley, W. D.
Chem. Eng. 53 (1946) 11, p. 256

[106] Rabald, E.
Corrosion Guide, 2nd. ed., p. 3
Elsevier Publishing Company, Amsterdam-London-New York, 1968

[107] Le Monnier, E.
in: Kirk-Othmer Encyclopedia of Chemical Technology
2nd ed., vol. 8, p. 386
John Wiley & Sons, Inc., New York-London-Sydney, 1965

[108] Rabald, E.
Corrosion Guide, 2nd. ed., p. 3
Elsevier Publishing Company, Amsterdam-London-New York, 1968

[109] Brockhaus, R.; Förster, G.
in: Ullmanns Encyklopädie der technischen Chemie
(Ullmann's encyclopedia of industrial chemistry) (in German)
4th ed., vol. 11, p. 57
Verlag Chemie, D-6940 Weinheim, 1976

[110] Ford, E.
Tantiron (in German)
Oberflächentechnik 50 (1973) 2, Supplement Metall-Journal, p. A3

[111] Ford, E.
Silizium macht Eisen säurefest
(Silicon makes iron acidproof) (in German)
Technica 21 (1972) 16, p. 1381

[112] Ford, E.
Korrosionsbeständige Metalle
(Corrosion-resistant metals) (in German)
Metall 28 (1974) 5, p. 459

[113] Udovitskii, V. I.
Diffusion silicon coatings on carbon steels (in Russian)
Metalloved. Term. Obrab. Met. (1971) 2, p. 59
(C.A. 74 (1971) 102290q)

[114] Rabald, E.
Corrosion guide, 2nd ed., p. 13
Elsevier Publishing Company, Amsterdam – London – New York, 1968

[115] Le Monnier, E.
in: Kirk-Othmer "Encyclopedia of chemical technology" 2nd ed., vol. 8, p. 405
John Wiley & Sons, Inc., New York-London-Sydney, 1965

[116] Wegst, C. W.
Stahlschlüssel (in German)
Verlag Stahlschlüssel Wegst KG., D-7142 Marbach, 1977

[117] Anonymous
Mater. Eng. 88 (1978) 6; Mater. Selector 1979, p. C16

[118] Medium-carbon chromium steels containing tungsten or molybdenum for razors or knives
Brit. Pat. 1 105 988 (Sandvikens Jernverks Aktiebolag) 13.3.1968)

[119] Ineson, E.
Materials for chemical plants: stainless steels
Chem. Process Eng. 41 (1960) 8, p. 357

[120] Bäumel, A.
Korrosionsverhalten nichtrostender Vergütungsstähle mit rund 13 % Chrom
(Corrosion behavior of stainless tempering steels with about 13 % chromium) (in German)
Werkst. Korros. 18 (1967) 4, p. 289

[121] Peter, W.: Gondolf, E. G.
Untersuchungen über die Korrosion von Stählen mit 13 % Cr
(Investigations on the corrosion of steels with 13 % Cr) (in German)
Arch. Eisenhüttenwes. 32 (1961) p. 337

[122] Anuchin, P. J.
Some aspects of the action of acetic acid on metals (in Russian)
Sb. Tr. Tsent. Nauch.-Issled. Proekt. Inst. Lesokhim. Prom. (1966) 17, p. 228
(C.A. 66 (1967) 97772b)

[123] Zotova, E. V.
Properties of chromium-molybdenum stainless steels (in Russian)
Sb. Tr. Tsentr. Nauchn.-Issled. Inst. Chernoir Met. (1965) 39, p. 94
(C.A. 64 (1966) 4698e)

[124] Charbonnier, J.-C.
Influence du molybdène sur la résistance des acier inoxydables à différents types de corrosion, en présence de solutions aqueuses minérales ou de milieux organiques (étude bibliographique)
(Influence of molybdenum on the resistance of stainless steels on different types of corrosion in the presence of aqueous solutions of minerals or organic media (bibliographical study)) (in French)
Métaux-Corros.-Ind. (1975) No. 598, p. 201

[125] Ueda, I.; Kawamura, M.
Some studies on the sulfide corrosion cracking of chromium molybdenum steels (II) (in Japanese)
Yosetsu Gakkaishi 36 (1967) 4, p. 429
(C.A. 68 (1968) 71519r)

[126] Fujii, T.; Horita, R.; Nomura, K.
The effect of temper embrittlement on hydrogen embrittlement of 2.25Cr 1Mo steel (in Japanese)
Tetsu-to-Hagané (J. Iron Steel Inst. Jpn.) 70 (1984) 2, p. 269

[127] Bui, N.; Dabosi, R.
Résistance à la corrosion notamment sous contrainte, des aciers à haute limite élastique du type maraging
(Resistance to the corrosion particularly under stress of high elastic limit steels of the maraging type) (in French)
Corr. Traitements, Protection, Finition 19 (1971) 1, p. 8

[128] Product Information
Chemische Beständigkeit der nichtrostenden Remanit® Stähle
(Chemical resistance of stainless Remanit® steels) (in German)
Thyssen Edelstahlwerke AG, D-4150 Krefeld, No. 1127/2, Nov. 1978

[129] Bünger, J.
Einige Beobachtungen über die Korrosionseigenschaften härtbarer Chromstähle
(Observations of the corrosion resistance of hardenable chromium steels) (in German)
Werkst. Korros. 12 (1961) 11, p. 681

[130] Volikova, I. G.
Corrosion resistance of steels Kh17T and Kh17N2 and their welded joints (in Russian)
Tr. Vses. Nauch.-Issled. i Konstrukt. Inst. Khim. Mashinostr. (1961) 37, p. 71
(C.A. 60 (1964) 3808b)

[131] Frolov, N. A.; Belinskii, A. L.; Fedorov, V. K.; Istrina, Z. F.
Properties of a cast stainless steel grade KH17M2TL and areas of its use in chemical machine construction (in Russian)
Tr. Vses. Nauchn.-Issled. i Konstrukt. Inst., Khim. Mashinostr. (1963) p. 84
(C. A. 61 (1964) 5281f)

[132] Konradi, M. V.
Corrosion of stainless steels in solutions containing acetic acid, ammonium sulfate, and phenols (in Russian)
Trudy Inst. Torfa, Akad. Nauk Beloruss. SSR (1959) 7, p. 174
(C.A. 55 (1961) 2447c)

[133] Sekine, I.; Hatakeyama, S.; Nakazawa, Y.
Corrosion behaviour of type 430 stainless steel in formic and acetic acids
Corros. Sci. 27 (1987) 3, p. 275

[134] Sekine, I.; Hatakeyama, S.; Nakazawa, Y.
Effect of water content on the corrosion behaviour of Type 430 stainless steel in formic and acetic acids
Electrochim Acta 32 (1987) 6, p. 915

[135] Sekine, I.; Ito, T.; Fujita, T.
Corrosion behavior of type 444 stainless steel in formic and acetic acids (in Japanese)
Corros. Eng. (Boshoku Gijutsu) 34 (1985) 9, p. 500

[136] De Benedetti, B.; Angelini, E.
Constitution and electrochemical properties of the surface layers on ion-nitrided stainless steels (in Italian)
Metall. Ital. (1986) 1, p. 31

[137] Zseitlin, K. L.; Korobkina, V. P.
Effect of sodium chloride on corrosion of stainless steels in organic acids (in Russian)
Khim. Prom. 42 (1966) 7, p. 543
(C.A. 65 (1966) 13314e)

[138] Pischik, L. M.; Tsinman, A. I.; Balvas, N. I.
Influence of iodides on corrosion of chromium-nickel steel in acetic acid solution
Prot. Met. (USSR) 18 (1982) 1, p. 72

[139] Kharyu, N. A.; Sokolov, M. M.; Garifzyanova, N. V.
Corrosion resistance of metallic materials under the conditions of the manufacture of optical sensibilizers (in Russian)
Khim. Prom. (1984) 11; p. 661

[140] Anonymous
Unique stainless steel vies for automotive, chemical industries
Mater. Eng. 71 (1970) 1, p. 30

[141] Anonymous
Nickel-free, austenitic special steels
Materials and Methods 45 (1957) 1, p. 104

[142] Shapovalov, E. T.; Kazakova, G. V.
The corrosion stability of the nickel-free chromium-manganese steels 10Kh14AG15 (D113) and 12Kh13G18D (D161) in organic acids (in Russian)
Zashch. Met. 20 (1984) 6, p. 929

[143] Yoshioka, K.; Suzuki, S.; Konishita, N. et al.
Ultra-low carbon and nitrogen high chromium ferritic stainless steels (in Japanese)
Kawasaki Steel Giho 17 (1985) 3, p. 240

[144] Product Information
Climax Molybdenum GmbH, Division of Amax Inc., D-4000 Düsseldorf, EUR 121 G (in German)

[145] Steigerwald, R. F.
New molybdenum stainless steels for corrosion resistance: A review of recent developments
Mater. Performance 13 (1974) 9, p. 9

[146] Product Information
Chemische Beständigkeit der nichtrostenden Remanit® Stähle
(Chemical resistance of stainless Remanit® steels) (in German)
Thyssen Edelstahlwerke AG, D-4150 Krefeld, No. 1127/2, Nov. 1978

[147] Bünger, J.
Einige Beobachtungen über die Korrosionseigenschaften härtbarer Chromstähle
(Observations of the corrosion resistance of hardenable chromium steels) (in German)
Werkst. Korros. 12 (1961) 11, p. 681

[148] Vaccari, J. A.
New ferritic stainless steels beat stress corrosion, ease fabrication
Mater. Eng. 82 (1975) 7, p. 24

[149] Lennartz, G.; Kiesheyer, H.
Korrosionsverhalten von hochchromhaltigen, ferritischen Stählen
(Corrosion behavior of ferritic steels with high chromium contents) (in German)
DEW-Tech. Ber. 11 (1971) 4, p. 230

[150] Desolneux, J. P.
Korrosionsprobleme in chlorhaltigen Medien: Lösungsmöglichkeiten durch einige nichtrostende Spezialstähle
(Corrosion problems in chlorine-containing media: some stainless special steels) (in German)
Werkst. Korros. 28 (1977) 5, p. 325

[151] Knoth, R. J.
High purity ferritic Cr-Mo stainless steel – Five years' successful fight against corrosion in the process industry
Werkst. Korros. 28 (1977) p. 409

[152] Dillon, C. P.
Use of low interstitial 26Cr-1Mo stainless steel in chemical plants
Mater. Performance 14 (1975) 8, p. 36

[153] Colombie, M. A.
Neue ferritische Stähle mit hohem Chromgehalt, die gegen Lochfraß-, Spalt- und Spannungsrißkorrosion beständig sind
(New ferritic steels with high chromium content, resistant to pitting, crevice and stress corrosion cracking) (in German)
Neue Hütte 18 (1973) 11, p. 693

[154] Desolneux, J. P.
Korrosionsprobleme in chlorhaltigen Medien: Lösungsmöglichkeiten durch einige nichtrostende Spezialstähle
(Corrosion problems in chlorine-containing media: resolving by some stainless special steels) (in German)
Werkst. Korros. 28 (1977) p. 325

[155] Oppenheim, R.; Lennartz, G.
Eigenschaften eines ferritischen 28-2 Chrom-Molybdän-Stahles in Superferrit-Güte
(Properties of a ferritic 28-2 chromium-nickel-molybdenum steel of super ferritic grade) (in German)
Chem. Ind. 23 (1971) 10, p. 705

[156] Tokareva, T. B.; Kolyada, A. A.; Smolin, V. V.; Medvedev, E. A.
Corrosion mechanical properties of high chromium content ferrite steels produced by vacuum smelting (in Russian)
Zashch. Met. 13 (1977) 5, p. 529

[157] Stricker, F.; Brauner, J.
Corrosion tests of chromium-molybdenum stainless steel, X 6 CrMo 17 and of microcracked chromium surface layers by electrochemical techniques
Electrodeposition Surface Treat. 1 (1973) 5, p. 395

[158] Colombier, L. (Product Information)
Molybdän in rost- und säurebeständigen Stählen und Legierungen
(Molybdenum contained in stainless and acid-resistant steels and alloys) (in German)
Climax Molybdenum GmbH (Amax/American Metal Climax, Inc.), D-4000 Düsseldorf

[159] Zseitlin, K. L.; Korobkina, V. P.
Effect of sodium chloride on corrosion of stainless steels in organic acids (in Russian)
Khim. Prom. 42 (1966) 7, p. 543
(C.A. 65 (1966) 13314e)

[160] Dechema-Werkstoff-Tabelle
"Essigsäureanhydrid"
(Dechema corrosion data sheets "Acetic anhydride") (in German)
DECHEMA, D-6000 Frankfurt am Main, 1957

[161] Sampaolo, A.; Rossi, L.; Esposito, G.; Delle Femmine, P.
Analytical research on the behavior of stainless steels in contact with foods (in Italian)
Rass. Chem. 23 (1971) 6, p. 226

[162] Product Information
Chemische Beständigkeit der nichtrostenden Remanit® Stähle
(Chemical resistance of stainless Remanit® steels) (in German)
Thyssen Edelstahlwerke AG, D-4150 Krefeld, No. 1127/2, Nov. 1978

[163] Anonymous
Materials: General survey
Chem. Process Eng. 45 (1964) 5, p. 224

[164] Sands, G. A.; Keady, M. B.
Comparison of the steels of the norm group 200 with 18/8-So-steel
Mater. Design Eng. 47 (1958) 4, p. 120

[165] Product Information
Chemische Beständigkeit der nichtrostenden Remanit® Stähle
(Chemical resistance of stainless Remanit® steels) (in German)
Thyssen Edelstahlwerke AG, D-4150 Krefeld, No. 1127/2, Nov. 1978

[166] Nijhawan, B. R.; Gupte, P. K.; Bhatnagar, S. S.; Guha, B. K.
Nitrogen-bearing stainless steels
Brit. Pat. 833 308 (Council of Scient. & Industr. Research) (21.4.1960)

[167] Wilson, H. J.
Stainless steels in the view of the consumer
Chem. Process. Eng. 38 (1957) 1, p. 5

[168] Shirley, H. P.; Truman, J. E.
J. Iron Steel Inst. 172 (1952) Part 4, p. 377

[169] Thomas, E. B.; Alcock, E. H.
E. Pat. 669 760 (British Celanese Ltd.) (1952)

[170] Reppe; von Kutepow; Titzenthaler
German Pat. 922 231 (BASF AG, D-6700 Ludwigshafen) (1952)

[171] Schierhold, P.
Säurebeständige Stähle in der Zellstoffindustrie unter besonderer Beachtung des Sulfitverfahrens (Acid-resistant steels in the cellulose industry with special respect to the sulfite-process) (in German)
Nickel-Bericht 19 (1961) 9/10, p. 257

[172] Johnson, P. R.
Corrosion in the production of acetic acid and its derivatives
Aust. Corr. Eng. 10 (1966) 7, p. 9

[173] Teeple, H. O.
Corrosion by some organic acids and related compounds
Corrosion 8 (1952) 1, p. 14

[174] Brink jr., J. A.; Shreve, R. N.
Nitration of 2-methylnaphthalene
Ind. Eng. Chem. 46 (1954) p. 694

[175] Perin, Y. I.; Valieva, R. A.; Burmistrova, L. K.; Kuznetsova, G. E.
Corrosion resistance of high-alloy steels and the alloy KhN65MV in certain media for the synthesis of terephthalic acid by the oxidation of p-xylene (in Russian)
Khim. Neft. Mashinostr. (1976) 10, p. 27 (C.A. 86 (1977) 94508x)

[176] Ludosan, E.; Cimpoeru, C.
Revista de chimie 31 (1980) 12, p. 1173 (in Rumanian)

[177] Leontaritis, L.; Horn, E.-M.
Zum Korrosionsverhalten nichtrostender austenitischer Chrom-Nickel-(Molybdän, Kupfer)-Stähle in nahezu wasserfreier Essigsäure
(Corrosion of stainless austenitic steels in almost anhydrous acetic acid) (in German)
Werkst. Korros. 39 (1988) 7, p. 313

[178] von Kutepow, N.; Himmele, W.; Hohenschutz, H.
Die Synthese von Essigsäure aus Methanol und Kohlenoxyd
(The synthesis of acetic acid from methanol and carbon monoxide) (in German)
Chem.-Ing.-Tech. 37 (1965) 4, p. 383

[179] Le Monnier, E.
in: Kirk-Othmer Encyclopedia of Chemical Technology
2nd ed., vol. 8, p. 386
John Wiley & Sons, Inc., New York-London-Sydney, 1965

[180] Wegst, C. W.
Stahlschlüssel (in German)
Verlag Stahlschlüssel Wegst KG., D-7142 Marbach, 1977

[181] Anonymous
Mater. Eng. 88 (1978) 6; Mater. Selector 1979, p. C16

[182] Le Monnier, E.
in: Kirk-Othmer Encyclopedia of Chemical Technology
2nd ed., vol. 8, p. 386
John Wiley & Sons, Inc., New York-London-Sydney, 1965

[183] Sekine, I.; Koeda, M.
Corrosion behavior of type 304 stainless steel in acetic acid solution (retroactive coverage) (in Japanese)
Corros. Eng. (Boshoku Gijutsu) 33 (1984) 9, p. 500

[184] De Benedetti, B.; Angelini, E.
Constitution and electrochemical properties of the surface layers on ion-nitrided stainless steels (in Italian)
Metall. Ital. (1986) 1, p. 31

[185] Bjorkroth, J.
Welding of low-carbon austenitic stainless steels (in Italian) Acciaio Inossidabile 33 (1966) 2, p. 39
(C.A. 67 (1967) 119542b)

[186] Fontana, M. G.
Ind. Eng. Chem. 39 (1947) 12, p. 91A

[187] Snair, G. L.
Chem. Eng. 56 (1949) 4, p. 217

[188] Colegate, G. T.
Metallurgia (Manchester) 41 (1959) p. 147, p. 259, p. 305, p. 362

[189] Poe, Ch. F.; van Vleet, E. M.
Action of organic acids on stainless steel
Ind. Eng. Chem. 41 (1949) p. 208

[190] Endo, H.; Itagaki, A.
Nippon Kinzoku Gakkai Shi 16 (1952) p. 352

[191] Mears, R. B.; Larrabee, C. P.; Fetner, C. J.
Spec. Tech. Publ. Amer. Soc. Test. Mat. Nr. 93 (1952) p. 183, p. 217

[192] Dillon, C. P.
Role of contaminants in acetic acid corrosion
Mater. Protection 4 (1965) 9, p. 20

[193] Takaaki, S.; Akira, T.; Shigetada, S.
Corrosion behavior of various metals and alloys in acetic acid environments
(in Japanese)
Corrosion Eng. (Boshoku Gijutsu) 15 (1966) 2, p. 49
(C.A. 64 (1966) 19086f)

[194] Stahl, R.; Kiefer, P.
Korrosion und Korrosionsschutz in der Lebensmittelindustrie
(Corrosion and corrosion protection in the food industry) (in German)
Werkst. Korros. 24 (1973) 6, p. 513

[195] Ineson, E.
Materials for chemical plants: stainless steels
Chem. Process Eng. 41 (1960) 8, p. 357

[196] Tajima, S.; Komatsu, S.; Momose, T.
The effect of ferric hydroxide on the oxygenation of ferrous ions in neutral solutions
Corrosion Sci. 16 (1976) 3, p. 191

[197] Heitz, E.
Vergleichende Korrosionsuntersuchungen in Ethanol-, Methanol- und wäßrigen Lösungen
(Comparative corrosion testings in ethanol-, methanol- und aqueous solutions)
(in German)
Werkst. Korros. 38 (1987) 12, p. 333

[198] Bourrat, J.
The corrosion resistance of stainless steel and alloys in organic acids
Corrosion et Anticorrosion 11 (1963) 4, p. 140

[199] Colombier, L. (Product Information)
Molybdän in rost- und säurebeständigen Stählen und Legierungen
(Molybdenum contained in stainless and acid-resistant steels and alloys) (in German)
Climax Molybdenum GmbH (Amax/American Metal Climax, Inc.), D-4000 Düsseldorf

[200] Pomey, J.; Voulet, P.
Chimie et Industrie (1930) March, p. 212

[201] Ford, E.
Silizium macht Eisen säurefest
(Silicon makes iron acidproof) (in German)
Technica 21 (1972) 16, p. 1381

[202] Bedea, T.; Atanasiu, I.
Electrochemical behaviour of 18Cr-10Ni stainless steel in organic acid solutions. II. Acetic acid
Rev. Roum. Chim. 20 (1975) 1, p. 45

[203] Valiyeva, R. A.; Perin, Yu. I.; Klochkova, M. R.; Tzinman, A. I.
Effect of oxygen on the corrosion of stainless steels and EP-567 alloy in acetic acid at high temperatures (in Russian)
Zashch. Met. 11 (1975) 4, p. 475

[204] Levin, I. A.; Kilchevskaya, T. E.
Corrosion of metals in the production of synthetic fatty acids (in Russian)
Bor'ba Korroz. Khim. Neftepererab. Prom. (1967) 1, p. 180
(C.A. 69 (1968) 38166w)

[205] Ekimov, Y. A.
Corrosion of stainless steels during nitric acid decomposition of phosphates in the manufacture of ammonium nitrate
(in Russian)
Tr. Tashk. Politekh. Inst. (1972) 90, p. 10
(C.A. 83 (1975) 138827v)

[206] Moskvicheva, A. F.; Zaretskii, E. M.; Klinov, I. Y.
Electrochemical and corrosion characteristics of stainless steel with lowered Ni-content in acetic acid solutions
(in Russian)
Tr. Mosk. Inst. Khim. Mashinostr. 28 (1964) p. 21
(C.A. 63 (1965) 9445b)

[207] Konradi, M. V.
Corrosion of stainless steels in solutions containing acetic acid, ammonium sulfate, and phenols (in Russian)
Trudy Inst. Torfa, Akad. Nauk Beloruss. SSR (1959) 7, p. 174
(C.A. 55 (1961) 2447c)

[208] Folek, K.
Corrosion resistance of selected stainless steels in nitric and acetic solutions
(in Polish)
Prace Inst. Met. (1979) 2, p. 85

[209] Tilman, M. M.
Effects of substituting cobalt for nickel on the corrosion resistance of two types of stainless steel
US Bur. Mines. Rept. Invest. No. 6591 (1965) 17 p.

[210] Anonymous
Maraging stainless steel – a new high-strength high-ductility material
Anticorros. Methods. Mater. 13 (1966) 2, p. 7

[211] Klinow, I. Y.; Levin, I. A.; Kochergina, D. G.
Intercrystalline corrosion of 21-5 type steels in formic and acetic acid solutions (in Russian)
Khim. Neft. Mashinostr. (1965) 6, p. 37
(C.A. 63 (1965) 9, 11088g)

[212] Vorobeva, G. Y.
Corrosion resistance of steels with a reduced nickel content in chemically active media (in Russian)
Bor'ba Korroz. Khim. Neftepererab. Prom. (1967) 1, p. 160
C.A. 69 (1968) 29572y)

[213] Anuchin, P. I.; Aleeva, T. K.
Corrosion of titanium in wood pulp chemical media (in Rumanian)
Khim. Pererabotka Drevesiny Sb. 26 (1963) p. 5
(C.A. 61 (1964) 14308f)

[214] Loginow, A. W.; Bates, J. F.; Mathay, W. L.
New alloy resists chloride stress corrosion cracking
Mater. Protection 11 (1972) 5, p. 35

[215] Geissler, G. N.; Smith, H. C.
Heat- and corrosion-resistant, iron-nickel-chromium alloy wire
US Pat. 3 753 817 (Driver, Wilbur B., Co.) (21.8.1973)

[216] Meckelburg, E.
Rostfreie Stahlbleche (Stainless steel sheets) (in German)
Blech 12 (1965) 11, p. 642

[217] Merrick, R. D.; Mantell, Ch. L.
Low-nickel stainless steels
Chem. Eng. 72 (1965) 18, p. 144

[218] Merrick, R. D.; Mantell, Ch. L.
Low-nickel stainless steels
Chem. Eng. 72 (1965) 18, p. 144

[219] Anonymous
Manganese versus nickel
Corrosion Technology 5 (1958) 9, p. 277

[220] Aniol, St.
Corrosion of 17 chromium-8 manganese-4 nickel steel in acetic acid (in Polish)
Ochr. Przed Koroz. 13 (1970) 6, p. 10
(C.A. 74 (1971) 102297x)

[221] Aniol, St.
Corrosion of 17 chromium-8 manganese-4 nickel steel in acetic acid (in Polish)
Ochr. Przed. Koroz. 13 (1970) 6, p. 10
(C.A. 74 (1971) 102297x)

[222] Kocich, J.; Tuleja, S.
A modified test of the resistance of austenitic stainless steel to point corrosion (in Slovakian)
(C. A. 68 (1968) 118768p)

[223] Class, J.; Gräfen, H.
Fortschritte und Erfahrungen in der Anwendung korrosionsbeständiger Stähle (Progress and experience in the use of corrosion-resistant steels) (in German)
Werkst. Korros. 11 (1960) 9, p. 529

[224] Bloom, F. K.
Stress corrosion cracking of hardenable stainless steels
Corrosion 11 (1955) p. 351t

[225] Brockhaus, R.; Förster, G.
in: Ullmanns Encyklopädie der technischen Chemie
(Ullmann's encyclopedia of industrial chemistry) (in German)
4th ed., vol. 11, p. 57
Verlag Chemie, D-6940 Weinheim, 1976

[226] Streicher, M. A.
Corrosion of stainless steels in boiling acids and its suppression by ferric salts
Corrosion 14 (1958) p. 59t

[227] Dorozhkin, N. N.; Zavistovskii, S. E.
Corrosion-mechanical wear of a powdered coating of PG-Sr3 nickel in acetic acid (in Russian)
Trenie Iznos. 4 (1983) 5, p. 822

[228] Le Monnier, E.
in: Kirk-Othmer Encyclopedia of Chemical Technology
2nd ed., vol. 8, p. 386
John Wiley & Sons, Inc., New York-London-Sydney, 1965

[229] Eisenbrown, Ch. M.; Barbis, P. R.
Corrosion of metals by acetic acid
Chem. Eng. 70 (1963) 9, p. 148

[230] Anonymous
Corrosion by acetic acid and acetic acid mixtures under heat transfer conditions
Mater. Protection 6 (1967) p. 77

[231] Groves, N. D.; Eisenbrown, C. M.; Scharfstein, L. R.
Corrosion of metals by weak acids under heat-transfer conditions
Corrosion 17 (1961) p. 173t

[232] Colombier, L. (Product Information)
Molybdän in rost- und säurebeständigen Stählen und Legierungen
(Molybdenum contained in stainless and acid-resistant steels and alloys) (in German)
Climax Molybdenum GmbH (Amax/American Metal Climax, Inc.), D-4000 Düsseldorf

[233] Bloom, F. K.
Stress corrosion cracking of hardenable stainless steels
Corrosion 11 (1955) p. 351t

[234] Janda; Morávek
Der erste internationale Kongreß des chemischen Ingenieurwesens, des Maschinenbaues und der Automation in Brno
(1rst International congress of chemical and mechanical engineering and automation in Brno) (in German)
Werkst. Corros. 14 (1973) 7, p. 631

[235] Product Information
Chemische Beständigkeit der nichtrostenden Remanit® Stähle
(Chemical resistance of stainless Remanit® steels) (in German)
Thyssen Edelstahlwerke AG, D-4150 Krefeld, No. 1127/2, Nov. 1978

[236] Product Information
Chemische Beständigkeit der nichtrostenden Remanit® Stähle
(Chemical resistance of stainless Remanit® steels) (in German)
Thyssen Edelstahlwerke AG, D-4150 Krefeld, No. 1127/2, Nov. 1978

[237] Teeple, H. O.
Corrosion by some organic acids and related compounds
Corrosion 8 (1952) 1, p. 14

[238] Product Information
Chemische Beständigkeit der nichtrostenden Remanit® Stähle
(Chemical resistance of stainless Remanit® steels) (in German)
Thyssen Edelstahlwerke AG, D-4150 Krefeld, No. 1127/2, Nov. 1978

[239] Product Information
Wiggin nickel alloys in the chemical process engineering Bulletin No. 3681 G, May 1976
Henry Wiggin & Company Ltd., Hereford (GB)

[240] Anonymous
Le leghe Wiggin di nickel negli impianti chimici ed industriali
(The Wiggin nickel alloys in chemical and industrial plants) (in Italian)
Ingegneria Chim. 27 (1978) 2, p. 12

[241] Eisenbrown, Ch. M.; Barbis, P. R.
Corrosion of metals by acetic acid
Chem. Eng. 70 (1963) 9, p. 148

[242] Anonymous
Corrosion by acetic acid and acetic acid mixtures under heat transfer conditions
Mater. Protection 6 (1967) p. 77

[243] Groves, N. D.; Eisenbrown, C. M.; Scharfstein, L. R.
Corrosion of metals by weak acids under heat-transfer conditions
Corrosion 17 (1961) p. 173t

[244] Anonymous
Hydroxylation: new unit process
Chem. Eng. 60 (1953) 8, p. 118

[245] Daroux, G. W.
Chem. Eng. 61 (1954) 2, p. 114

[246] Anonymous
J. Appl. Chem. 3 (1953) p. 241

[247] German Pat. 818 351 (Standard Oil Development Co.) (1960)

[248] US Pat. 2 393 778 (Tennessee Eastman Corp.) (1943)

[249] US Pat. 2 232 705 (Eastman Kodak) (1938)

[250] Tyler, D. L.
E.P. 629 211

[251] Anonymous
Chemical safety data sheet SD-15 "Acetic anhydride"
Man. Chem. Ass., Washington, 1962

[252] Anonymous
Materials: General survey
Chem. Process Eng. 45 (1964) 5, p. 224

[253] Sands, G. A.; Keady, M. B.
Comparison of the steels of the norm group 200 with 18/8-So-steel
Mater. Design Eng. 47 (1958) 4, p. 120

[254] Wetternik, L.; Zitter, H.
Säurebeständige Stähle bei der Holzessigfabrikation
(Acid-resisting steels in the production of wood vinegar) (in German)
Werkst. Korros. 6 (1955) p. 282

[255] Loveless, L. W. J.
Ind. Chemistry 25 (1949) p. 325

[256] Dechema-Werkstoff-Tabelle "Aliphatische Aldehyde"
(Dechema corrosion data sheets "Aliphatic aldehydes") (in German)
DECHEMA, D-6000 Frankfurt am Main, 1978

[257] Dechema-Werkstoff-Tabelle "Peressigsäure und Acetaldehyd-Monoperacetat"
(Dechema corrosion data sheets "Peracetic acid and acetaldehyde monoperacetate") (in German)
DECHEMA, D-6000 Frankfurt am Main, 1966

[258] Risch, K.
Einfluß der Oberflächenbehandlung nichtrostender Stähle auf deren chemische Beständigkeit insbesondere gegen Spannungsrißkorrosion
(The influence of surface treatment of stainless steel on their chemical resistance especially to stress corrosion cracking) (in German)
Werkstoff. Korros. 24 (1973) 2, p. 106

[259] Leontaritis, L.; Horn, E.-M.
Zum Korrosionsverhalten nichtrostender austenitischer Chrom-Nickel-(Molybdän, Kupfer)-Stähle in nahezu wasserfreier Essigsäure
(Corrosion of stainless austenitic steels in almost anhydrous acetic acid) (in German)
Werkst. Korros. 39 (1988) 7, p. 313

[260] Dillon, C. P.
Role of contaminants in acetic acid corrosion
Mater. Protection 4 (1965) 9, p. 20

[261] Vehlow, J.
Ermittlung der Korrosionsgeschwindigkeiten von hochlegierten Stählen und Hastelloy C in niederen Carbonsäuren mit Hilfe der Radionuklidtechnik
(The determination of corrosion rates of high-alloy steels and Hastelloy C in lower carbonic acids by radionuclid technology) (in German)
Werkst. Korros. 31 (1980) 2; 119

[262] Wetternik, L.; Zitter, H.
Säurebeständige Stähle bei der Holzessigfabrikation
(Acid-resisting steels in the production of wood vinegar) (in German)
Werkst. Korros. 6 (1955) p. 282

[263] Colombier, L. (Product Information)
Molybdän in rost- und säurebeständigen Stählen und Legierungen
(Molybdenum contained in stainless and acid-resistant steels and alloys) (in German)
Climax Molybdenum GmbH (Amax/American Metal Climax, Inc.), D-4000 Düsseldorf

[264] Bennion, D.
Typical works' corrosion problems
Chem. Eng. (London) (1970) No. 244, p. CE 419

[265] Ylijoki, P.
Selection of corrosion-resistant ferrous materials (in Finnish)
Kem. Teollisuus 24 (1967) p. 310
(C.A. 67 (1967) 93177j)

[266] Kilchevskaya, T. E.; Anokhika, T. P.
Corrosion resistance of construction materials used in media for the production of acetic acid from hydrocarbons (in Russian)
Khim. Neft. Mashinostr. (1979) 11, p. 24

[267] Evans, L. S.
Corrosion in the man-made fiber industry
Corrosion Technology 9 (1962) 2, p. 34

[268] Hirabayashi, M.; Kotani, Y.; Hakozaki, S.
Prevention of corrosion of the preparation apparatus for vinylacetate
Japan Pat. 7 601 411 (Nippon Synthetic Chemical Industry Co., Ltd.)

[269] Asatryan, V. G.; Farkhadov, A.; Orlovskaya, G. A.; Balayan, A. M.
Corrosion of an acetic acid distillation column in the production of vinyl acetate (in Russian)
Zashch. Met. 7 (1971) 3, p. 304
(C.A. 75 (1971) 79500t)

[270] Wegst, C. W.
Stahlschlüssel (in German)
Verlag Stahlschlüssel Wegst KG., D-7142 Marbach, 1977

[271] Anonymous
Mater. Eng. 88 (1978) 6; Mater. Selector 1979, p. C16

[272] Swandby, R. K.
Corrosion charts: Guides to materials selection
Chem. Eng. 69 (1962) 23, p. 186

[273] Charbonnier, J.-C.
Influence du molybdène sur la résistance des acier inoxydables à différents types de corrosion, en présence de solutions aqueuses minérales ou de milieux organiques (étude bibliographique)
(Influence of molybdenum on the resistance of stainless steels on different types of corrosion in the presence of aqueous solutions of minerals or organic media (bibliographical study)) (in French)
Métaux-Corros.-Ind. (1975) No. 598, p. 201

[274] Product Information
Jacob & Korves GmbH, D-4730 Ahlen

[275] Anonymous
New stainless steel fights corrosive acids
Mater. Eng. 66 (1967) 4, p. 54

[276] Anonymous
Hydroxylation: new unit process
Chem. Eng. 60 (1953) 8, p. 118

[277] Yokata, K.; Fukase, Y.; Chizawa, K.; Nemoto, C.
Low-nickel, austenitic stainless steel
Jap. Pat. 7 505, 646 (Japan Metallurgical Industries, Ltd.) (26.12.1969)

[278] von Rossum, O.
Metalloberfläche 5 (1951) p. A 113 (in German)

[279] Edström, J. O.
Rost- und säurebeständige Chrom-Nickel-Stähle mit max. 0,030 % als Konstruktionswerkstoffe für die chemische Industrie
(Corrosion and acid resistant chromium-nickel steels with a maximum carbon content of 0.030 % as a construction material for the chemical industry) (in German)
Werkst. Korros. 15 (1964) 9, p. 743

[280] Risch, K.
Korrosionsinhibitoren in der chemischen Industrie
(Corrosion inhibitors in the chemical industries) (in German)
Werkst. Korros. 25 (1974) p. 727

[281] Yu, C.; Lin, H.; Gao, J.
Material selection in renewal of mid-part for acetic acid recovery tower (D-230) – an analysis of material selection in the mixture of formic/acetic acid at high temperature (in Chinese)
Huagong Jixie (Chem. Eng. Mach.) 11 (1984) 6, p. 48

[282] Xin, O.-Y.; Wang, Z.
Combining effect of molybdenum and copper on the corrosion resistance of stainless steels – II
International Congress on Metallic Corrosion, Toronto (Canada) (Proc. Conf.) 2 (1984) p. 549
National Research Council of Canada, Ottawa (Canada)

[283] Colombier, L. (Product Information)
Molybdän in rost- und säurebeständigen Stählen und Legierungen
(Molybdenum contained in stainless and acid-resistant steels and alloys) (in German)
Climax Molybdenum GmbH (Amax/American Metal Climax, Inc.), D-4000 Düsseldorf

[284] Anonymous
New steel useful down to – 80 F
Mater. Eng. 84 (1976) 3, p. 20

[285] Kemplay, J.
Chem. Process Eng. 47 (1966) 7, p. 53

[286] Anonymous
Copper alloy No. 879 (silicon brass die-casting alloy)
Alloy Dig. (1977) June, Cu-335, 2 p.

[287] Saunders, E. A. D.; Thomas, S. N. G.; Horn, R.
Technology of the container manufacture
Chem. Process Eng. 49 (1968) 5, p. 75

[288] Levin, I. A.; Kilchevskaya, T. E.
Corrosion of metals in the production of synthetic fatty acids (in Russian)
Bor'ba Korroz. Khim. Neftepererab. Prom. (1967) 1, p. 180
(C.A. 69 (1968) 38166w)

[289] Vorobeva, M. A.; Klinov, I. Ya.
Corrosion of different alloys in fatty acids (in Russian)
Tr. Mosk. Inst. Khim. Mashinostr. 28 (1964) p. 55
(C.A. 63 (1965) 5309f)

[290] Moslov, V. A.; Ermolenko, Z. P.
The effect of high-frequency currents and quenching on the corrosion resistance of bubble caps of 18 : 10 type stainless steels in adsorption columns (in Russian)
Khim. Neft. Mashinostr. (1977) 1, p. 31

[291] Wetternik, L.; Zitter, H.
Säurebeständige Stähle bei der Holzessigfabrikation
(Acid-resisting steels in the production of wood vinegar) (in German)
Werkst. Korros. 6 (1955) p. 282

[292] Kristal, M. M.; Khalizova, V. N.; Belinkii, A. L.; Adugina, N. A.
Corrosion-resistant high-strength steels for chemical construction (in Russian)
Bor'ba Korroz. Khim. Neftepererab. Prom. (1971) 1, p. 130
(C.A. 69 (1968) 29571x)

[293] Khakhlova, N. V.; Sidorkina, Yu. S.; Yuganova, S. A.; Akshentzeva, A. P.; Sergeyeva, G. V.; Nesterova, M. D.
Phase composition and corrosion stability of steel 03Cr21Ni21M4B (Z 1.35) (in Russian)
Zashch. Met. 12 (1976) 1, p. 7

[294] Colombier, L. (Product Information)
Molybdän in rost- und säurebeständigen Stählen und Legierungen
(Molybdenum contained in stainless and acid-resistant steels and alloys) (in German)
Climax Molybdenum GmbH (Amax/American Metal Climax, Inc.), D-4000 Düsseldorf

[295] Rabald, E.
Corrosion Guide, 2nd. ed., p. 3
Elsevier Publishing Company, Amsterdam-London-New York, 1968

[296] Colombier, L. (Product Information)
Molybdän in rost- und säurebeständigen Stählen und Legierungen
(Molybdenum contained in stainless and acid-resistant steels and alloys) (in German)
Climax Molybdenum GmbH (Amax/American Metal Climax, Inc.), D-4000 Düsseldorf

[297] Colombier, L. (Product Information)
Molybdän in rost- und säurebeständigen Stählen und Legierungen
(Molybdenum contained in stainless and acid-resistant steels and alloys) (in German)
Climax Molybdenum GmbH (Amax/American Metal Climax, Inc.), D-4000 Düsseldorf

[298] De Benedetti, B.; Angelini, E.
Constitution and electrochemical properties of the surface layers on ion-nitrided stainless steels (in Italian)
Metall. Ital. (1986) 1, p. 31

[299] Risch, K.
Korrosionsinhibitoren in der chemischen Industrie
(Corrosion inhibitors in the chemical industries) (in German)
Werkst. Korros. 25 (1974) p. 727

[300] Risch, K.
Korrosionsschutz durch Inhibition
(Corrosion protection by inhibition) (in German)
VDI-Berichte 365 (1980) p. 115 (Proc. Conf.)
VDI-Verlag, D-4000 Düsseldorf 1

[301] Wetternik, L.; Zitter, H.
Säurebeständige Stähle bei der Holzessigfabrikation
(Acid-resisting steels in the production of wood vinegar) (in German)
Werkst. Korros. 6 (1955) p. 282

[302] Evans, L. S.; Morgan, P. E.
Effect of steam temperature on the corrosion of metals under heat-transfer conditions
Br. Corros. J. 2 (1967) p. 150

[303] Evans, L. S.; Morgan, P. E.
Effect of steam temperature on the corrosion of metals under heat-transfer conditions
Br. Corros. J. 2 (1967) p. 150

[304] Teeple, H. O.
Corrosion 8 (1952) 1, p. 20

[305] Hromatka, O.
Chemiker Ztg. 76 (1952) p. 776, 815
(in German)

[306] Anonymous
Food Process 18 (1949) p. 204

[307] Condylis, A.; Bayon, F.; Chateau, D.
The results of corrosion and contamination tests on austenitic stainless steels in the food industry (in French)
Rev. Métall. 73 (1976) 12, p. 787

[308] Bayon, F.; Chateau, D.
Results of corrosion and contamination tests on austenitic stainless steels in the food industry (in French)
Aciers Spéc. 40 (1977) p. 23

[309] Anonymous
All around the world tank trucks made of stainless steel
Int. Nickel Rev. (1965) 19, p. 8

[310] Kemplay, J.
Chem. Process Eng. 47 (1966) 7, p. 53

[311] Anonymous
Materials: General survey
Chem. Process Eng. 45 (1964) 5, p. 224

[312] Product Information
Jacob & Korves GmbH, D-4730 Ahlen

[313] Product Information
Jacob & Korves GmbH, D-4730 Ahlen

[314] Teeple, H. O.
Corrosion 8 (1952) 1, p. 20

[315] Colombier, L. (Product Information)
Molybdän in rost- und säurebeständigen Stählen und Legierungen
(Molybdenum contained in stainless and acid-resistant steels and alloys) (in German)
Climax Molybdenum GmbH (Amax/American Metal Climax, Inc.), D-4000 Düsseldorf

[316] Hale, E. C.
Welded stainless steel tube in food and drink processing
Anticorros. Methods Mater. 22 (1975) 8, p. 7

[317] Radovici, O.; Roman, E.
Rezistentia la coroziune a unui otel inoxidabil Cr-Ni-Mo-Cu in solutii apoase
(Corrosion resistance of a chromium-nickel-molybdenum-copper stainless steel in aqueous solutions) (in Rumanian)
Revista de Chimie 21 (1970) 9, p. 564

[318] Product Information
Bulletin 10 M 4-63, 3839, A 280
The International Nickel Company, Inc., New York (USA)

[319] Staley, W. D.
Chem. Eng. 53 (1946) 11, p. 256

[320] Khakhlova, N. V.; Sidorkina, Yu. S.; Yuganova, S. A.; Akshentzeva, A. P.; Sergeyeva, G. V.; Nesterova, M. D.
Phase composition and corrosion stabiliy of steel 03Cr21Ni21M4B (Z 1.35) (in Russian)
Zashch. Met. 12 (1976) 1, p. 7

[321] Johnson, T. E.
Corrosion-resistant alloys
Belg. Pat. 664 213 (Stainless Foundry & Engineering Inc.) (22.11.1965)

[322] Anonymous
Neuer Sandvik-Stahl für die chemische Industrie
(A new Sandvik steel for the chemical industry) (in Italian)
Ingegneria Chim. 21 (1972) 6, p. 13

[323] Katz, W.
Fortschritte auf dem Werkstoffgebiet – gezeigt auf der 17. ACHEMA 1973, Teil I: Metallische Werkstoffe
(Progress in the field of material – exhibited on the 17th ACHEMA 1973, part I: metallic materials) (in German)
Werkst. Korros. 24 (1973) 9, p. 791

[324] Edström, J. O.; Carlén, J. C.; Kämpinge, S.
Anforderungen an Stähle für die chemische Industrie
(Properties required for steels in the chemical industry) (in German)
Werkst. Korros. 21 (1970) 10, p. 812

[325] Brandberg, A.
Chloride-containing cooling water – A serious problem in the process industry
Chem. Age India 24 (1973) 11, p. 743

[326] von der Forst, P.
Gußeisen im Apparatebau der chemischen Industrie
(Cast iron for chemical plant construction) (in German)
Werkst. Korros. 10 (1959) 4, p. 216

[327] Swales, G. L.; Nickel, O.
Austenitische Gußeisenwerkstoffe in der chemischen und der Erdölindustrie
(Austenitic cast iron materials in the chemical and petroleum industry) (in German)
Werkst. Korros. 15 (1964) 9, p. 728

[328] Product Information
Bulletin 10 M 4-63, 3839, A 280
The International Nickel Company, Inc., New York (USA)

[329] Swales, G. L.; Nickel, O.
Austenitische Gußeisenwerkstoffe in der chemischen und der Erdölindustrie (Austenitic cast iron materials in the chemical and petroleum industry) (in German)
Werkst. Korros. 15 (1964) 9, p. 728

[330] Meckelburg, E.
Blech 12 (1965) 12, p. 749 (in German)

[331] Musina, A. S.; Lange, A. A.; Bukhman, S. P.; Semashko, T. S.
Changes in the state of the surface of Permalloy and Kovar during chemical treatment (in Russian)
Izv. Akad. Nauk Kazakh. SSR Khim. (1984) 3, p. 12

[332] Product Information
Bulletin 10 M 4-63, 3839, A 280
The International Nickel Company, Inc., New York (USA)

[333] Anonymous
Custom 450: a threat to 304 and 410 stainless steels
Mater. Eng. 72 (1970) 1, p. 19

[334] Anonymous
New stainless combines high yield strength with good ductility
Materials Design Eng. 61 (1965) 3, p. 11

[335] Anonymous
Maraging stainless steel – a new high-strength high-ductility material
Anticorros. Methods Mater. 13 (1966) 2, p. 7

[336] 5. Kongress der Europäischen Föderation Korrosion 24. – 28. Sept. 1973 in Paris; Proceedings p. 473

Alkanecarboxylic Acids

Unalloyed steels and cast steel
Unalloyed cast iron

The corrosion behavior of metallic materials in propionic acid and its solutions is comparable to that in acetic acid. These two acids have almost the same dissociation constant and no reducing double bond [1].

Unalloyed steel and cast iron as well as low-alloy steels are not resistant to propionic acid of any concentration at 293 K (20°C). The corrosion rates are above 3.0 mm/a (118 mpy) [2]. Corrosive attack causes discoloration of the solution, and this is undesirable in most cases. The corrosion behavior of unalloyed steel in acetic acid (comparable with propionic acid) is shown in Table 1. See [3] (Figure 1) for further details of corrosion in acetic acid.

The corrosion rate of unalloyed steel in static 1% butyric acid at 293 K (20°C) is 0.65 mm/a (25.6 mpy) [2, 4]. It increases with increasing concentration and temperature.

Alkanecarboxylic acids with 6 carbon atoms or more hardly attack metallic materials at all, having an almost neutral reaction. Unalloyed steel is attacked by caproic acid (hexanoic acid) dissolved in organic solvents only in the presence of air or oxygen. Small quantities of water have only little influence on corrosion. Cold caproic acid attacks unalloyed steel only slightly, the corrosion products discoloring the acid [5].

Metal	Acetic acid concentration %	Temperature K (°C)	Corrosion rate mm/a (mpy)	Remarks
Pure iron	33	boiling point	0.145 (5.71)	
Armco® iron	33	boiling point	0.16 (6.30)	
Steel	33	boiling point	2.33 (91.7)	
	5–20	room temperature	1.14–1.37 (44.9–53.9)	
	glacial acetic acid	room temperature	0.36 (14.2)	hydrogen bubbled through
	glacial acetic acid	room temperature	17.3 (681)	oxygen bubbled through
Steel (0,3% C)	33	boiling point	3.28 (129)	

Table 1: Corrosion behavior of unalloyed steel in acetic acid [3]

Table 1: Continued

Metal	Acetic acid concentration %	Temperature K (°C)	Corrosion rate mm/a (mpy)	Remarks
Mild steel	5	293 (20)	0.79 (31.1)	
	15	293 (20)	1.23 (48.4)	
	33	293 (20)	1.34 (52.8)	
Steel	5–100	297 (24)	> 1.25 (> 50)	aerated
	5–100	297 (24)	> 1.25 (> 50)	no air
	50	373 (100)	> 1.25 (> 50)	acetic acid vapor
	100	373 (100)	> 1.25 (> 50)	acetic acid vapor
	100	297 (24)	> 1.25 (> 50)	acetic acid + acetic acid anhydride + peracetic acid
	100	297 (24)	> 1.25 (> 50)	acetic acid + acetone + carbon tetrachloride
	100	297 (24)	> 1.25 (> 50)	acetic acid + ether
	100	297 (24)	> 1.25 (> 50)	acetic acid + ethyl alcohol
	100	297 (24)	> 1.25 (> 50)	acetic acid + formic acid
	100	297 (24)	> 1.25 (> 50)	acetic acid + hydrobromic acid
	100	297 (24)	> 1.25 (> 50)	acetic acid + hydrochloric acid
	100	297 (24)	> 1.25 (> 50)	acetic acid + mercury salts
	100	297 (24)	> 1.25 (> 50)	acetic acid + sulfuric acid
Cu-steel	3	311 (138)	1.21 (47.6)	
Mn-steel	33	boiling point	1.18 (46.6)	

Table 1: Corrosion behavior of unalloyed steel in acetic acid [3]

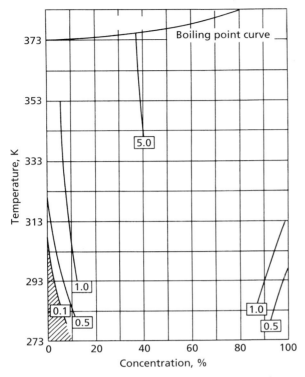

Figure 1: Corrosion curves (mm/a) for unalloyed steel and cast iron in acetic acid [6]

Unalloyed steel exhibits similar behavior in caprylic acid (octanoic acid). Corrosion at elevated temperatures is not very great, so that the material may be used, provided the reduced service life is sufficient and the discoloration of the acid is not an interference [7].

Capric acid (decanoic acid) slightly attacks unalloyed steel at 293 K (20°C) and discolors it in the presence of air and moisture. Crude acids may be distilled in columns made of this material [8].

Unalloyed steel and cast iron cannot be used in the manufacture of alkanecarboxylic acids containing more than 5 carbon atoms (Table 2), although these materials are suitable for storage of these acids.

When corrosion results from tests are given, the stated corrosion rate tends to be too high, since the specimens are corroded all round and in a practical operation the medium acts on only one side of the material as a rule.

Test 1: Specimens in vapor and spray of light fatty acid fractions having about 10 % water vapor above the top tray in the head of a deodorization column, temperature 423 to 448 K (150 to 175°C).

Test 2: Specimens in about 99 % fatty acid liquid mixture at the bottom of a collecting tank between the deodorizing stage and the distillation column. Temperature 423 to 448 K (150 to 175°C).

Material	Composition %	Form of material	Corrosion rate, mm/a (mpy) test			
			1	2	3	4
Inconel®	Ni + 14 Cr + 7 Fe	Plate	< 0.0025 (< 0.098)	< 0.0025 (< 0.098)	0.0025 (0.098)	< 0.0025 (< 0.098)
Inconel®		Plate, welded	< 0.0025 (< 0.098)	< 0.0025 (< 0.098)	< 0.0025 (< 0.098)	< 0.0025 (< 0.098)
Nickel	99.4 Ni	Plate	0.12 (4.72)	0.16 (6.30)	0.43 (16.9)	0.11 (4.33)
Monel®	Ni + 30 Cu	Plate	0.05 (1.97)	0.14 (5.51)	0.40 (15.7)	0.08 (3.15)
Steel SAE 302	Fe + 19 Cr + 9 Ni; ≤ 0.2 C	Plate	< 0.0025 (< 0.098)	< 0.0025 (< 0.098)	0.013 (0.52)	< 0.0025 (< 0.098)
Steel SAE 304	Fe + 19 Cr + 9 Ni; ≤ 0.08 C	Plate	< 0.0025 (< 0.098)	< 0.0025 (< 0.098)	0.008 (0.31)	0.0025 (0.098)
Steel SAE 304		Plate, welded but without post-treatment	< 0.0025 (< 0.098)	0.0025 (0.098)	0.068 (2.68)	0.010 (0.39)
Steel SAE 304		Plate, welded and annealed	< 0.0025 (< 0.098)	–	–	–
Steel SAE 347	Fe + 18 Cr + 10 Ni + 0.7 Nb; ≤ 0.008 C	Plate	< 0.0025 (< 0.098)	< 0.0025 (< 0.098)	< 0.0025 (< 0.098)	0.0025 (0.098)
Steel SAE 347		Plate, welded but without post-treatment	< 0.0025 (< 0.098)	< 0.0025 (< 0.098)	0.023 (0.91)	0.0025 (0.098)
Steel SAE 316	Fe + 18 Cr + 13 Ni + 2 Mo; ≤ 0.08 C	Plate	< 0.0025 (< 0.098)	< 0.0025 (< 0.098)	< 0.0025 (< 0.098)	< 0.0025 (< 0.098)
Steel SAE 316		Plate, welded but without post-treatment	< 0.0025 (< 0.098)	0.0025 (0.098)	< 0.0025 (< 0.098)	0.0025 (0.098)
Steel SAE 316		Plate, welded and annealed	< 0.0025 (< 0.098)	< 0.0025 (< 0.098)	< 0.0025 (< 0.098)	0.0025 (0.098)

Table 2: Results of operational corrosion tests in the continuous distillation of fatty acids from animal fats in vacuum columns (3,900 hours of operation) [9]

Table 2: Continued

Material	Composition %	Form of material	Corrosion rate, mm/a (mpy) test			
			1	2	3	4
Ni-Resist® 1	Fe + 15 Ni + 6 Cu + 2 Cr; 2.8 C		0.11 (4.33)	0.15 (5.91)	0.08 (3.15)	0.21 (8.27)
Cast iron	3.4 C; 1.8 Si		2.2 (86.6)	0.93 (36.6)	destroyed, original thickness 4.7 mm	destroyed, original thickness 4.7 mm

Table 2: Results of operational corrosion tests in the continuous distillation of fatty acids from animal fats in vacuum columns (3,900 hours of operation) [9]

Test 3: Specimens in the vapor and liquid phases of a mixture of fatty acid and about 20 % water vapor above the first tray of a column in the vicinity of the liquid inlet. Temperature 478 to 498 K (205 to 225°C).
Test 4: Specimens in a liquid consisting of 40 % free fatty acids and 60 % "pitch" at the bottom of a column near the discharge opening. Temperature 478 to 498 K (205 to 225°C).

The corrosion behavior of steel 20 (0.18 % C, 0.31 % Si, 0.51 % Mn) in butyric acid was investigated in [10] and the results given in Table 3.

The corrosion rate is only 0.05 mm/a (1.97 mpy) at 293 K (20°C) in the concentration range from 1 to 10 % in aqueous solution, and it increases in boiling 10 % butyric acid solution to 7.5 mm/a (295 mpy). The corrosion data in butanol as solvent, at a maximum of 1.2 mm/a (47.2 mpy), are substantially below those in aqueous solution [10].

Corrosion of carbon steel (St 3; cf. 1.0036, UNS K02502) in aqueous solutions of carboxylic acid concentrates can be inhibited in 20 to 40 % dilute solutions by thiourea or ammonium thiocyanate or mixtures of the two. A mixture of thiourea with ammonium thiocyanate or sodium thiosulfate is suitable for a concentrate of 70 %. The inhibitors are mixed in the ratio of 1:1 and have a concentration of 0.1 % in the solution. The concentrate of the 20 to 40 % test solutions consists of 14 % formic acid, 10 % acetic acid, 4 % propionic acid, 6 % butyric acid, 5 to 6 % esters, alcohols and other admixtures, that of the 70 % test solutions of 30 % formic acid, 24 % acetic acid, 6 % propionic acid, 8 % butyric acid and 10 to 15 % of other organic admixtures. The tests were carried out in a static medium at 293 K (20°C), the test durations being between 330 and 360 hours. The test results are shown in Table 4 and Table 5. The corrosion rates of 0.036 mm/a (1.42 mpy), equivalent to inhibiting efficiency of 96.7 or 97.7 % mean that steel offers very good resistance under the stated conditions (Table 4) [11].

Steel	Butyric acid concentration, %	Temperature					
		293 K (20°C)		323 K (50°C)		Boiling point	
		Aqueous solution	Butanol solution	Aqueous solution	Butanol solution	Aqueous solution	Butanol solution
		Corrosion rates, mm/a (mpy)					
Steel 20	1	0.04 (1.57)	< 0.001 (< 0.039)	0.1 (3.94)	0.01 (0.39)	1.4 (55.1)	0.1 (3.94)
	3	0.04 (1.57)	0.01 (0.39)	0.3 (11.8)	0.06 (2.36)	4.0 (157)	0.3 (11.8)
	5	0.05 (1.97)	0.01 (0.39)	0.3 (11.8)	0.08 (3.15)	5.0 (197)	1.0 (3.94)
	10	0.04 (1.57)	0.02 (0.79)	0.4 (15.7)	0.1 (3.94)	7.5 (295)	1.2 (47.2)
1Ch13 (cf. 1.4006, UNS S41000)	1	0.001 (0.039)	< 0.001 (< 0.039)	0.001 (0.039)	< 0.001 (< 0.039)	0.01 (0.39)	0.001 (0.039)
	3	0.001 (0.039)	< 0.001 (< 0.039)	0.001 (0.039)	< 0.001 (< 0.039)	0.01 (0.39)	0.001 (0.039)
	5	0.003 (0.12)	< 0.001 (< 0.039)	0.001 (0.039)	0.008 (0.31)	0.01 (0.39)	0.001 (0.039)
	10	0.05 (1.97)	0.001 (0.039)	0.001 (0.039)	0.01 (0.39)	0.05 (1.97)	0.006 (0.24)

Table 3: Corrosion rate of steel in butyric acid solutions (test duration 340 hours) [10]

Austenitic chromium-nickel-molybdenum steels/ Austenitic chromium-nickel steels with special alloying additions/ Special iron-based alloys

This steel also offers good resistance in a 70% carboxylic acid concentrate (corrosion rates of 0.131 mm/a (5.16 mpy)) where the inhibitor provides an inhibiting efficiency of 89.5% [11].

Inhibitor	Concentration of inhibitors, %	Corrosion rate mm/a (mpy)	Inhibiting efficiency, %
$CS(NH_2)_2$	0.10	0.036 (1.42)	96.7
	0.20	0.040 (1.57)	96.2
	0.30	0.073 (2.87)	93.2
	0.50	0.061 (2.40)	94.5

Table 4: Results of corrosion investigations (293 K (20°C), 330–360 hours in a static medium) of carbon steel St 3 (cf. 1.0036, UNS K02502) with various inhibitors in solutions of low-molecular weight carboxylic acids in concentrations of 20 to 40% [11]

Table 4: Continued

Inhibitor	Concentration of inhibitors, %	Corrosion rate mm/a (mpy)	Inhibiting efficiency, %
	1.00	0.091 (3.58)	90.1
	2.00	0.132 (5.20)	85.8
NH$_4$CNS	0.10	0.036 (1.42)	97.7
	0.20	0.043 (1.69)	96.2
	0.30	0.041 (1.61)	97.0
	0.50	0.065 (2.56)	93.6
CS(NH$_2$)$_2$ + NH$_4$CNS (1:1)	0.10	0.095 (3.74)	97.7
	0.20	0.116 (4.57)	84.5
Amines C$_{12}$C$_{15}$ + NH$_4$CNS (1:1)	0.5	0.056 (2.20)	95.5
Amines C$_{15}$ – C$_{18}$ + NH$_4$CNS (1:1)	0.5	0.031 (1.22)	98.0
Hydroxyl amine in organic compounds + NH$_4$CNS (1:1)	0.5	0.257 (10.1)	80.3

Table 4: Results of corrosion investigations (293 K (20°C), 330–360 hours in a static medium) of carbon steel St 3 (cf. 1.0036, UNS K02502) with various inhibitors in solutions of low-molecular weight carboxylic acids in concentrations of 20 to 40 % [11]

High-alloy cast iron, high-silicon cast iron

Cast iron containing no less than 14,5 % Si is resistant to propionic acid in all concentrations up to the boiling point.

The maximum corrosion rates are to be assumed to be 0.1 mm/a (3.94 mpy) [2, 4]. According to [12], they are in the range of up to 0.5 mm/a (19.7 mpy). Figure 2 shows the resistance to the rather more corrosive acetic acid in comparison with propionic acid [6].

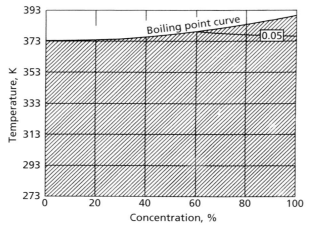

Figure 2: Corrosion curves (mm/a) for silicon cast iron 14–16 % Si in acetic acid [6]

Inhibitor	Concentration of inhibitor %	Corrosion rate mm/a (mpy)	Inhibiting efficiency, %
$CS(NH_2)_2 + NH_4CNS$ (1:1)	0.05	0.202 (7.95)	84.5
	0.10	0.179 (7.04)	86.0
	0.20	0.131 (5.16)	89.5
$CS(NH_2)_2 + Na_2S_2O_3$ (1:1)	0.10	0.197 (7.76)	85.0
	0.20	0.217 (8.54)	83.4
	0.50	0.371 (14.6)	71.7

Table 5: Results of corrosion investigations (293 K (20°C), 330–360 hours in a static medium) of carbon steel St 3 (cf. 1.0036, UNS K02502) with inhibitors and in solutions of low-molecular weight carboxylic acids, concentration 70 % [11]

Structural steels with up to 12 % chromium

The corrosion behavior of these steels in alkanecarboxylic acids and their solutions is dependent on their passivity under the relevant conditions. This passive condition mainly exists at 293 K (20°C). In 1 % butyric acid a steel containing 14 % Cr is still passive up to 333 K (60°C), the corrosion rate under these conditions being less than 0.05 mm/a (1.97 mpy) and increasing to more than 0.5 mm/a (19.7 mpy) in the active condition [4].

The steel 1Ch13 (cf. 1.4006, UNS S41000) containing 12 to 14 % Cr is attacked only slightly in aqueous butyric acid solutions (1 to 10 % butyric acid) [10]. In 10 % butyric acid solution the corrosion rates at 293 K (20°C) and at the boiling point are only 0.05 mm/a (1.97 mpy) (see Table 3). In butanol solutions of the same concentration they are even substantially less. In boiling 5, 10 and 98% butyric acid solution the corrosion rates can increase to 0.3 mm/a (11.8 mpy) and to 2.5 mm/a (98.4 mpy) in 25, 50 and 75% acid solution [10] (see also Table 13).

According to corrosion investigations in alkanecarboxylic acids with 7 to 9 carbon atoms at 373 K (100°C), the materials 1Ch13 and 2Ch13 (cf. 1.4021, UNS S42000) exhibited corrosion rates of 0.2 mm/a (7.9 mpy) each. They increase with increasing temperature and are about 1.2 mm/a (47.2 mpy) at 573 K (300°C) in the liquid for steel 2Ch13 [13].

Ferritic chromium steels with more than 12% chromium
Ferritic-austenitic steels with more than 12% chromium

These steels can be used in alkanecarboxylic acids mainly at 293 K (20°C), within the range in which they are passive. They are corroded by acetic acid (10 and 50%) at this temperature at the rate of up to 0.1 mm/a (3.94 mpy). In 50% acid the corrosion rates at the boiling point increase to 11 mm/a (433 mpy) [2]. These corrosion rates are also to be expected with propionic acid (Table 13). Chromium steels containing 17% Cr and 1.2% Mo are attacked less in acetic acid, the corrosion rate up to boiling point being less than 0.11 mm/a (4.33 mpy) [2].

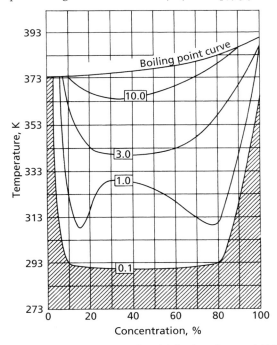

Figure 3: Corrosion curves (mm/a) for chromium steel (0.1% C, 17% Cr) in acetic acid [6]

Steel X 8 CrMoTi 17 (cf. 1.4523, X2CrMoTiS18-2) is used for handling propionic acid at low temperatures [1]. This reference roughly compares corrosion by propionic acid with that due to acetic acid. Figure 3 shows the corrosion of chromium steel (e.g. X 8 Cr 17) by acetic acid [6]. The molybdenum-alloyed steel is said to exhibit better corrosion behavior. Its range of application expands with increasing chain length of the alkanecarboxylic acids.

Cast chromium steel containing 27 % Cr is resistant to 50 % butyric acid at 293 K (20°C), the steel being passive. It is active at 373 K (100°C) and is then attacked (no corrosion rates available) [4].

At 378 K (105°C) 17 % chromium steel is very severely attacked by highly-concentrated butyric acid (no further details), whereas it is corroded at the rate of only 0.5 mm/a (19.7 mpy) in 50 % acid [4]. The chromium steels containing 12 to 20 % Cr are resistant to caproic acid (hexanoic acid) and higher alkanecarboxylic acids up to 373 K (100°C) [14–16].

Austenitic chromium-nickel steels

The corrosion behavior of the metallic materials on exposure to propionic acid is similar to that in acetic acid [1]. Corrosion tests found in the literature are therefore mainly for formic and acetic acid, whereas information on propionic and butyric acids is given in only very few cases [17]. Chromium-molybdenum and chromium-nickel-molybdenum steels are recommended for use in propionic acid at low temperatures (see Sections Ferritic-austenitic steels with more than 12 % chromium and Austenitic chromium-nickel-molybdenum steels – Table 10–15). From what has previously been said it follows that corrosion rates for chromium-nickel steels are similar to those in acetic acid.

Figure 4 shows corrosion of steel X5CrNi19-9 (cf. 1.4302, UNS S30888) in acetic acid [1].

The corrosion rates are substantially below those for chromium steels (see Figure 3). According to this reference, this steel can be used up to temperatures of 333 K (60°C) at maximum corrosion rates of 0.1 mm/a (3.94 mpy). The corrosion rate is stated to be 0.05 mm/a (1.97 mpy) in 20 to 95 % acetic acid at 300 K (27°C) [6]. The inadequate resistance in the range of the boiling point leads to selection of steels containing molybdenum, since the risk of pitting corrosion is greatly reduced in their case [1]. The corrosion rates as shown in Figure 4 should always be borne in mind when handling alkanecarboxylic acid solutions, since the latter are seldom present in the pure form (see Table 13).

Corrosion investigations with high-alloy steels in alkanecarboxylic acids confirm the inadequate resistance of chromium-nickel steels at elevated temperatures [18]. The corrosion rates are shown in Table 6, the test conditions in Table 7 and the alloy compositions of the materials investigated in Table 8.

The corrosion rates in Table 6 show that corrosion by alkanecarboxylic acids in the liquid phase is always more severe than in the vapor phase. The investigations further show that chromium-nickel steels may exhibit corrosion rates of about

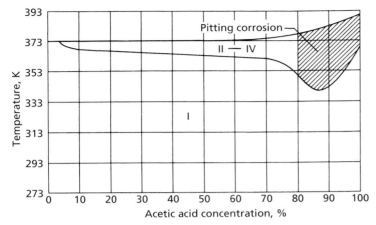

Figure 4: Corrosion of chromium nickel steel X5CrNi19-9 (cf. 1.4302, UNS S30888) in acetic acid (I < 0.11 mm/a (4.33 mpy), II 0.11–0.33 mm/a (4.33–13.0 mpy), III 0.33–1.1 mm/a (13.0–43.3 mpy), IV > 1.1 mm/a (> 43.3 mpy)) [1]

5 mm/a (197 mpy) in the region of the boiling point. Under these conditions steels containing more than 20 % Ni and more than 3 % Mo, like 000Ch21N21M4B and 0Ch23N28M3D3T (Russian types), are suitable. The steel Ch17N13M3T corrodes at the rate of 0.4 to 0.5 mm/a (15.7–19.7 mpy) [18].

The risk of pitting corrosion in the range between 90 and 100 % acetic acid (less in propionic acid) (see Figure 4) is greater in the case of the titanium-stabilized steel X6CrNiTi18-10 (cf. 1.4541, UNS S32100) than for the non-stabilized steel X5CrNi18-10 (cf. 1.4301, UNS S30400) [1]. The presence of chlorine ions substantially increases the corrosivity of the alkanecarboxylic acids.

Corrosion investigations with chromium-nickel steels in alkanecarboxylic acids confirm their good corrosion behavior in these acids. At 573 K (300°C) the corrosion rate was about 0.01 mm/a (0.39 mpy) [19]. Nevertheless, it is distinctly higher than that of the molybdenum-containing steels, whose corrosion rate was only about one-tenth of this figure.

Practical corrosion tests with heat exchangers in alkanecarboxylic acids $C_5 - C_{10}$ at 453–573 K (180–300°C) and 0.3–4 kPa resulted in corrosion rates of chromium-nickel steels of 0.25–1.29 mm/a (9.84–50.8 mpy) in the liquid and 0.12–0.67 mm/a (4.72–26.4 mpy) in the vapor phase after 120 hours [20]. The steels containing no molybdenum (12Ch18N10T, cf. 1.4541, UNS S32100) were susceptible to pitting corrosion. This was also observed with steels containing less than 3 % molybdenum (10Ch17N13M3T, cf. 1.4573, UNS S31635). The corrosion rate increased to 0.52–2.46 mm/a (20.5–96.9 mpy) on heating surfaces [20].

Material	Corrosion rates, mm/a (mpy)					
	$C_1 - C_4$		$C_5 - C_9$		$C_{10} - C_{17}$	
	Vapor phase	Liquid	Vapor phase	Liquid	Vapor phase	Liquid
000Ch21N21M4B	0.022(0.87)[1] / 0.000(0.0)	0.043(1.69) / 0.005(0.20)	0.086(3.39) / 0.008(0.32)	0.076(2.99) / 0.043(1.69)	0.184(7.24) / 0.016(0.63)	0.292(11.5) / 0.205(8.07)
0Ch23N28M3D3T	0.033(1.30) / 0.000(0.0)	0.049(1.93) / 0.011(0.43)	0.078(3.07) / 0.005(0.20)	0.113(4.45) / 0.049(1.93)	0.162(6.38) / 0.011(0.43)	0.356(14.0) / 0.329(13.0)
Ch17N13M3T	0.282(11.1) / 0.005(0.20)	0.303(11.9) / 0.033(1.30)	0.486(19.1) / 0.054(2.13)	0.882(34.7) / 0.151(5.94)	0.404(15.9) / 0.100(3.94)	1.372(54.0) / 0.551(21.7)
Ch17N13M2T	0.292(11.5) / 0.014(0.55)	0.369(14.5) / 0.035(1.38)	0.510(20.1) / 0.081(3.19)	1.085(42.7) / 0.270(10.6)	0.418(16.5) / 0.135(5.31)	1.304(51.3) / 0.540(21.3)
Ch17N5M3	0.070(2.76) / 0.005(0.20)	0.072(2.83) / 0.049(1.93)	0.940(37.0) / 0.113(4.45)	1.976(77.8) / 0.162(6.38)	0.734(28.9) / 0.713(28.1)	2.246(88.4) / 0.761(30.0)
Ch16N15M3	0.197(7.76) / 0.013(0.51)	0.313(12.3) / 0.027(1.06)	0.553(21.8) / 0.086(3.39)	0.888(35.0) / 0.162(6.38)	0.529(20.8) / 0.127(5.0)	1.480(58.3) / 0.531(20.9)
Ch18N10T	–	–	–	5.000 (197) / –	–	–
1Ch21N5T	–	–	6.448 (254) / –	7.330 (289) / –	–	–

[1] The corrosion rates in the numerator were determined at 0.1 MPa, those in the denominator at 1.33 kPa after a test duration of 24 hours on specimens measuring $40 \times 10 \times 1.8$ mm^3. See Table 7 and Table 8 for further data.

Table 6: Corrosion rates of chromium-nickel steels (Russian types) in boiling alkanecarboxylic acid fractions $C_1 - C_4$, $C_5 - C_9$ and $C_{10} - C_{17}$ [18]

Test conditions	Alkanecarboxylic acid fraction					
	$C_1 - C_4$		$C_5 - C_9$		$C_{10} - C_{17}$	
Pressure	0.1 MPa	1.33 kPa	0.1 MPa	1.33 kPa	0.1 MPa	1.33 kPa
Temperature, K (°C)	373 (100)	311–313 (138–140)	513 (240)	401 (128)	533 (260)	433 (160)
	374 (101)	333 (60)	536 (263)	438 (165)	613 (340)	553 (280)

Table 7: Test conditions in the determination of the corrosion rates in Table 6 [18]

Material	Alloying elements, %								
	C	Si	Mn	Cr	Ni	Mo	Ti	Cu	Nb
000Ch21N21M4B	0.03	0.6	1.9	20.8	21.2	3.7	–	–	0.7
0Ch23N28M3D3T	0.06	0.55	0.42	24.95	26.96	2.78	0.85	2.77	–
Ch17N13M3T	0.10	0.7	1.8	17.2	12.8	2.7	0.4	–	–
Ch17N13M2T	0.10	0.8	1.8	16.8	12.7	2.1	0.4	–	–
Ch17N5M3	0.11	0.3	0.4	17.1	5.1	3.1	–	–	–
Ch16N15M3	0.10	0.7	1.8	16.2	15.3	2.8	–	–	–
Ch18N10T	0.12	0.8	1.7	17.9	9.8	–	0.43	–	–
1Ch21N5T	0.12	0.71	0.52	20.9	5.6	–	0.29	–	–

Table 8: Compositions of the steels investigated as per Table 6 [18]

Austenitic chromium-nickel-molybdenum steels
Austenitic chromium-nickel steels with special alloying additions
Special iron-based alloys

The corrosivity of propionic acid is comparable to that of acetic acid [1, 21–23]. Molybdenum-alloyed chromium-nickel steels should mainly be used in propionic acid (see Ferritic-austenitic steels with more than 12 % chromium and Austenitic chromium-nickel steels), these specifically being X5CrNiMo17-12-2 (1.4401, UNS S31600), X6CrNiMoTi17-12-2 (1.4571, UNS S31635) and X6CrNiMoNb17-12-2 (1.4580, UNS S31640) [1].

Investigations were carried out to determine the corrosion rate in alkanecarboxylic acids ($C_1 - C_4$) with the aid of radionuclide technology [23]. The influence of water and the presence of acid anhydrides were also taken into consideration. The materials investigated were X2CrNiMoN22-5-3 (1.4462, UNS S32205), X2CrNiMo18-14-3 (1.4435, UNS S31603), X2CrNiMoN17-13-5 (1.4439, UNS S31726), X6CrNiMoTi17-12-2 (1.4571, UNS S31635) and the nickel-based alloys Incoloy® 825 and Hastelloy® C. To measure the corrosion rate, transfer of the corrosion products from a reactor-activated material specimen into the medium was determined from the increase in radioactivity.

This method is very sensitive and involves no disturbance of the corrosion system. It allows to record the chronological pattern of the corrosion reaction. The test durations were as long as 30 days. It was established that the steels behave approximately identically in all systems investigated. In all cases the alkanecarboxylic acids caused surface corrosion without formation of a surface coating, and the austenitic phase was preferentially attacked in the case of the semi-austenitic steel X2CrNiMoN22-5-3. As has also been stated in other references, the corrosivity of the alkanecarboxylic acids decreased with increasing chain length [1, 21, 24, 25]. In acetic acid the corrosion rates are below those in formic acid by a factor of 10, and by a factor of 100 in propionic and butyric acids. These results lead to the conclusion that corrosion by acetic acid causes ten times greater corrosion rates than propionic or butyric acid. Additions of 1–5 % water influence corrosion in acetic, propionic and butyric acids only insignificantly at the boiling point. Acid anhydrides leads to a marked acceleration in corrosion.

Corrosive attack of the CrNi-steels investigated is hardly detectable at all in pure propionic acid at the boiling point (414.5 K (141.5°C)) (corrosion rate less than 0.001 mm/a (0.04 mpy)) [26].

The steel X2CrNiMoN17-13-5 is resistant to propionic acid even when the water content is from 1 to 5 %. At a water content of 1 % the corrosion rates are 0.01 to 0.05 mm/a (0.39 to 1.97 mpy) for the steels X6CrNiMoTi17-12-2 and X2CrNiMoN22-5-3 and at a water content of 5 % 0.05 mm/a (1.97 mpy) for X2CrNiMoN22-5-3 and 0.01 mm/a (0.39 mpy) for X6CrNiMoTi17-12-2. After 5 days the corrosion rate for both steels decreases to 0.01 mm/a (0.39 mpy). The influence of propionic acid anhydride on the two steels was investigated at the boiling point. In this case, too, the two steels behaved roughly the same [26].

The addition of 2 % anhydride produces a corrosion rate of 0.05 to 0.1 mm/a (1.97 to 3.94 mpy). In 5 % solutions the corrosion rate increases to 1.5 mm/a (59.1 mpy) and in 10 % solutions to between 3 and 5 mm/a (118 and 197 mpy) [26]. In the case of propionic acid/propionic acid anhydride the conditions are the same as in the acetic acid/acetic acid anhydride system, although it must be borne in mind that the boiling point of propionic acid is 23 K (23°C) above that of acetic acid. The corrosion rates of the steels X2CrNiMo18-14-3, X2CrNiMoN17-13-5, X2CrNiMoN22-5-3 and X6CrNiMoTi17-12-2 in the acetic acid/acetic anhydride system at the boiling point, which rates also occur in the propionic acid/propionic anhydride system, are shown in Figure 5 [23, 26]. The corrosion rate increases steeply in the range up to 10 % anhydride.

No corrosive attack of the steels X2CrNiMoN22-5-3 and X6CrNiMoTi17-12-2 is observed in boiling butyric acid [26]. The same is also true of the steel X6CrNiMoTi17-12-2 with additions of 1 % water or 1 % to 5 % butyric anhydride.

Up to 5 mm/a (197 mpy) were lost from the steel X2CrNiMoN22-5-3 in butyric acid containing 1 % water. This high corrosion rate is attributed to the presence of chlorine ions. In butyric acid containing 5 % anhydride, the corrosion rate of this steel increased from about 0.1 mm/a (3.94 mpy) on the first day to 5 mm/a (197 mpy) after 10 days [26].

The corrosion behavior of chromium-nickel steels containing molybdenum is chiefly investigated in acetic acid with possible references to the less corrosive propionic and butyric acids. The experimental results obtained previously justify this procedure. Figure 6 shows the material consumption rate of chromium-nickel steels with increasing molybdenum content in boiling 80 and 95 % acetic acid; Figure 7 shows the corrosion curves for the same material with increasing acetic acid concentration at the boiling point [6].

The presence of molybdenum in the steel decreases the risk of pitting corrosion.

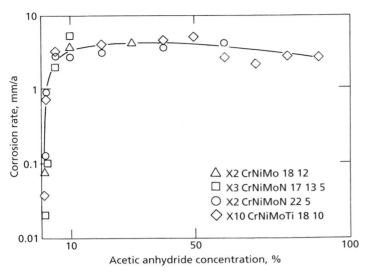

Figure 5: Corrosion rates of chromium-nickel-molybdenum steels in the boiling acetic acid system with various acetic anhydride contents (comparable behavior in the case of propionic acid) [23, 26]

If contamination of the medium is to be avoided in alkanecarboxylic acid solutions, which is important, for example, in the foodstuffs industry, a steel containing 20 % Cr, 25 % Ni and 3.5 % Mo should be used particularly when handling boiling lactic acid (which is duly mentioned here), even if alkanecarboxylic acids are contaminated by sulfuric acid or other corrosive media [27]. According to Table 6 and Table 8, chromium-nickel steels containing no less than 20 % Ni and 3 % Cr should be used in contact with carboxylic acid fractions $C_1 - C_4$ and $C_5 - C_9$ [18].

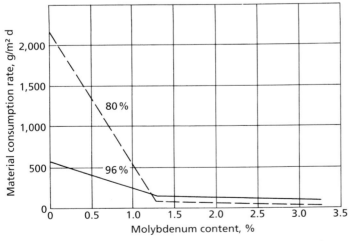

Figure 6: The influence of molybdenum on the corrosion resistance of a steel containing 18 % Cr and 10 % Ni in 80 and 96 % boiling acetic acid [27]

The corrosion rate in 99 to 100% butyric acid at 435 K (162°C) is stated to be 0.06 mm/a (2.36 mpy) for an 18 8 3 chromium-nickel-molybdenum steel [4]. According to the same reference, a steel containing 20% Cr and 7% Ni exhibits a corrosion rate of 0.11 mm/a (4.33 mpy) at 403 K (130°C) in concentrated butyric acid. With small quantities of water added, at 383 K (110°C) the corrosion rate of the steels SAE 316 (1.4401), SAE 317 (cf. 1.4449), SAE 310 (cf. 1.4841) and Carpenter® 20 (29 Ni, 20 Cr, 2 Mo, 3 Cu) is less than 0.025 mm/a (0.98 mpy) (Table 9).

Material	Alloying elements, %			Corrosion rates, mm/a (mpy)	
	Cr	Ni	Mo	in liquid	in vapor space
SAE 316	16 to 18	10 to 14	1.8 to 2.8	< 0.0025 (< 0.098)	< 0.0025 (< 0.098)
SAE 317	18 to 20	10 to 14	3 to 4	< 0.0025 (< 0.098)	< 0.0025 (< 0.098)
SAE 310	24 to 26	19 to 22	0	< 0.0025 (< 0.098)	< 0.0025 (< 0.098)
Carpenter®20	20	29	2 Mo 3 Cu	< 0.0025 (< 0.098)	< 0.0025 (< 0.098)

Table 9: Corrosion of chromium-nickel steels in butyric acid containing small quantities of water (383 K (110°C)) [4]

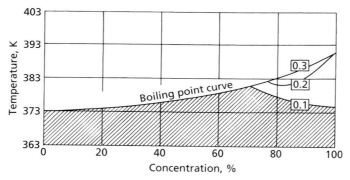

Figure 7: Corrosion curves (mm/a) for chromium-nickel-molybdenum steel (0.05% C, 18% Cr, 10% Ni, 2% Mo) [6]

The steel X3CrNiMo17-13-3 (1.4436, UNS S31600) is used in the manufacture of butyric acid, and steels containing 17 to 25% Cr, 10 to 30% Ni and 2 to 4% Mo as materials for pumps and valves.

Chromium-nickel-molybdenum steels are suitable materials for handling alkanecarboxylic acids having 3 carbon atoms or more, as the corrosion rates in Tables 10–15 show. These steels are resistant to these acids in all concentrations and in a wide temperature range.

Materials	Corrosion rate, mm/a (mpy)
EN AW-1100 (Al99 Cu)	0.36 (14.2)
EN AW-6061 (AlMgSi-alloy)	0.11 (4.33)
CuNi-alloy (70 30) (Alloy No. 715)	9.35 (368)
Hastelloy®B (cf. 2.4800)	0.12 (4.72)
Hastelloy®C (13–17Cr, 16–18Mo)	0.005 (0.20)
Hastelloy® D (NiSi-alloy)	0.14 (5.51)
Monel®400 (2.4360)	0.69 (27.2)
Nickel 200 (2.4060)	0.12 (4.72)
SAE 304 (X5CrNi18-10)	0.66 (26.0)
SAE 316 (X5CrNiMo17-12-2)	0.03 (1.18)
Unalloyed Steel	34.75 (1,368)

Table 10: Corrosion rates of metallic materials in boiling 100% propionic acid (unaerated, test duration 240 hours) [28]

Temp., K (°C)	Time, h	Corrosion rates, mm/a (mpy)			
		Unalloyed steel	SAE 304	SAE 316	EN AW-3003 (AlMn-alloy)
299 (26)	48	0.71 (28.0)	0.03 (1.18)	0.03 (1.18)	–
383 (110)	48	1.98 (78.0)	0.13 (5.12)	0.03 (1.18)	–
418 (145)	4,800	–	0.97 (38.2)	0.30 (1.18)	destroyed

Tabelle 11: The influence of temperature on the corrosion of metallic materials in 100% propionic acid [28]

Concentration, %						Temperature, K (°C)	Time, h	Corrosion rate, mm/a (mpy)				
Propionic acid	Acetic acid	Butyric acid	Nitric acid	Sulfuric acid	Other			SAE 304	SAE 316	SAE 317	Carpenter®20Cb-3	Titanium
99	–	–	–	1	–	411 (138)	240	0.51 (20.1)	0.11 (4.33)	0.13 (5.12)	–	–
99	–	–	–	1	–	412 (139)	288	1.85 (72.8)	0.28 (11.0)	0.36 (14.2)	–	–
98	2	–	–	–	–	415 (142)	5,900	–	0.02 (0.79)	0.003 (0.12)	–	–
97	3	–	–	–	–	415 (142)	15,120	–	0.02 (0.79)	0.005 (0.20)	–	–
95	5	–	–	–	–	413 (140)	4,800	–	0.03 (1.18)	–	–	–
94	5	0.5	–	–	0.5 solids	423 (150)	288	0.02 (0.79)	0.01 (0.39)	0.005 (0.20)	–	–
90	5	4	1	–	–	413 (140)	1,300	–	0.12 (4.72)	–	0.02 (0.79)	–
80	–	–	5	10	5 unknown	418 (145)	72	0.23 (9.1)	0.05 (1.97)	0.02 (0.79)	–	–
80	0.1	2.5	2	–	15.4 other higher acids	408 (135)	1,300	–	0.30 (11.8)	–	0.22 (8.66)	–
66	–	17	–	–	17 isobutyric acid	–	–	–	–	–	–	0.14 (5.51)
99	–	–	–	–	–	vapor	–	–	–	–	–	1)

1) destroyed

Tabelle 12: Corrosion rates (mm/a (mpy)) of metallic materials in high concentrated contaminated propionic acid [28]

Concentration, %						Corrosion rate, mm/a (mpy)				
Propionic acid	Acetic acid	Butyric acid	Other	Temperature, K (°C)	Time, h	Carbon steel	SAE 410	SAE 304	SAE 316	SAE 317
83	–	12	5H$_2$SO$_4$	411 (138)	240	–	–	1.14 (44.9)	0.08 (3.15)	0.09 (3.54)
70	1.5	12	16.5 esters	430 (157)	13,600	–	–	0.43 (16.9)	0.12 (4.72)	0.08 (3.15)
66	–	17	17 isobutyric acid	422 (149)	500	–	–	0.30 (11.8)	0.18 (7.09)	0.13 (5.12)
62	6	11	unknown	428 (155)	11,200	–	–	0.07 (2.76)	0.15 (5.91)	0.1 (3.94)
53	15	12	–	425 (152)	14,400	1)	1)	0.69 (27.2)	0.09 (3.54)	–
48	8	13	2HNO$_3$	383 (110)	6,000	1)	0.76 (29.9)	0.15 (5.91)	0.003 (0.12)	–
44	10	18	18 esters	413 (140)	8,592	1)	1)	0.58 (22.8)	0.08 (3.15)	–
25	15	30	acids	422 (149)	1,730	–	–	–	0.13 (5.12)	0.12 (4.72)

1) destroyed

Tabelle 13: Results of corrosion investigations in mixtures of alkanecarboxylic acids [28]

Concentration, %						Corrosion rate, mm/a (mpy)				
Propionic acid	Acetic acid	Copper, ppm	Iron, ppm	Temperature, K (°C)	Time, h	Carbon steel	SAE 304	SAE 316	EN AW-3003 (AlMn-alloy)	EN AW-5154 (AlMg-alloy)
99	–	300–60,000	< 10	417 (144)	1,128	8.9	0.04 (1.57)	0.08 (3.15)	3.3 (130)	2.5 (98.4)
96.2	3.4	5–30	35–2,500	417 (144)	816	–	0.03 (1.18)	0.02 (0.79)	1)	–
94	6	some	some	413 (140)	600	–	–	0.06 (2.36)	–	–
86.5/89	11/14	traces	traces	415 (142)	1,128	15.2	0.3 (11.8)	0.15 (5.91)	1)	1)

1) destroyed

Tabelle 14: Results of corrosion investigations on metallic materials in mixtures of propionic/acetic acid and propionic acid contaminated with copper and iron [28]

Concentration, %	Other	Temperature K (°C)	Time, h	Corrosion rate, mm/a (mpy)								
				Carbon steel	SAE 304	SAE 316	SAE 317	Nickel	Carpenter® 20Cb-3	Incoloy® 825	Monel® 400	Titanium
100	–	299 (26)	48	0.15 (5.91)	< 0.03 (< 1.18)	< 0.03 (< 1.18)	–	–	–	–	–	–
100	Sulfate addition	394 (121)	768	–	–	< 0.003 (< 0.12)	< 0.003 (< 0.12)	0.9 (35.4)	0.003 (0.12)	–	1.27 (50.0)	–
95	5 % acetic acid	401 (128)	1,300	–	–	0.04 (1.57)	–	–	0.03 (1.18)	0.03 (1.18)	–	–
80	10 % H₂SO₄ Vapor	439 (136)	240	–	6.6 (260)	0.61 (24.0)	0.3 (11.8)	–	–	–	0.05 (1.97)	–
100	Sulfate addition	383 (110)	769	–	–	< 0.003 (< 0.12)	< 0.003 (< 0.12)	0.23 (9.06)	< 0.003 (< 0.12)	–	0.30 (11.8)	< 0.05 (< 2)

Tabelle 15: Corrosion rates of metallic materials in butyric acid solutions [28]

Austenitic chromium-nickel steels containing 3% molybdenum are virtually not attacked by alkanecarboxylic acids $C_7 - C_9$. The corrosion rate in the temperature range from 293 to 573 K (20 to 300°C) does not exceed 0.001 mm/a (0.04 mpy) [29].

Investigations of corrosion in alkanecarboxylic acids $C_5 - C_{10}$ at 453–573 K (180–300°C) and 0.3–4 kPa (test duration 120 hours) showed corrosion rates of 0.25 to 1.29 mm/a (9.84 to 50.8 mpy) for steel 08Ch17N15M3T (comparable with 1.4571, SAE 316 Ti in the liquid and 0.12 to 0.67 mm/a (4.72 to 26.4 mpy) in the vapor. Pitting corrosion was not observed. The corrosion rate increases to between 0.52 and 2.46 mm/a (20.5 and 96.9 mpy) at heating surfaces. Under these conditions, the iron alloy 06ChN28MDT (26–29% Ni, 22–25% Cr, 2.5 to 3.0% Mo, 2.5–3.5% Cu, balance Fe) exhibits better corrosion resistance. It is corroded at the rate of 0.071 to 0.39 mm/a (2.80 to 15.4 mpy) in the liquid and 0.021 to 0.082 mm/a (0.83 to 3.23 mpy) in the vapor. The corrosion rate at heating surfaces is 0.27 to 0.78 mm/a (10.6 to 30.7 mpy) [30].)

Bibliography

[1] ABC der Stahlkorrosion
unter besonderer Berücksichtigung der
chemisch beständigen Stähle
(Steel corrosion) (in German)
Mannesmann AG, Düsseldorf, FRG,
Second Edition 1966

[2] Rabald, E.
Corrosion Guide, Second Edition
Elsevier Publishing Company, Amsterdam,
London, New York, 1968

[3] Rabald, E.
Dechema-Werkstoff-Tabelle, Essigsäure
(Dechema Corrosion Data Sheets, Acetic
Acid) (in German)
DECHEMA, Frankfurt am Main, 1957/1981

[4] Rabald, E.
Dechema-Werkstoff-Tabelle, Buttersäure
(Dechema Corrosion Data Sheets, Butyric
Acid) (in German)
DECHEMA, Frankfurt am Main, 1955

[5] Rabald, E.
Dechema-Werkstoff-Tabelle, Capronsäure
(Dechema Corrosion Data Sheets, Caproic
Acid) (in German)
DECHEMA, Frankfurt am Main, 1955

[6] Berg, F. F.
Korrosionsschaubilder
(Corrosion diagrams) (in German)
VDI-Verlag GmbH, Düsseldorf 1969,
Second Edition

[7] Rabald, E.
Dechema-Werkstoff-Tabelle, Caprylsäure
(Dechema Corrosion Data Sheets, Caprylic
Acid) (in German)
DECHEMA, Frankfurt am Main, 1955

[8] Rabald, E.
Dechema-Werkstoff-Tabelle, n-Caprinsäure
(Dechema Corrosion Data Sheets, n-Capric
Acid) (in German)
DECHEMA, Frankfurt am Main, 1955

[9] Rabald, E.
Dechema-Werkstoff-Tabelle, Fettsäuren
(höhere, ab C_6)
(Dechema Corrosion Data Sheets, Fatty
Acids) (in German)
DECHEMA, Frankfurt am Main, 1957

[10] Archakov, Ju. I.; Merkulova, O. P.
Influence of the addition of hydrochloric
acid and butanols on the corrosion of metals
in butyric acid (in Russian)
Zashch. Met. (Moscow) 10 (1974) 6, p. 714

[11] Sarnin, A. A.; Chanin, A. M.; Penzeva, T. A.
Corrosion inhibitors for carbon steel in
solutions containing low molecular
carboxylic acids (in Russian)
Zashch. Met. (Moscow) 7 (1971) 3, p. 340

[12] Hamner, N. E.
Corrosion Data Survey (Metals Section) 5th
Edition
National Association of Corrosion
Engineers, Houston, Tx. 77001

[13] Kartashova, K. M.; Sukhotin, A. M.;
Bobrova, M. M.; Shukan, N. Ya.
Corrosion of metallic construction materials
in fatty acids, higher amines and acetates of
higher amines (in Russian)
Tr. Gos. in-t prikl. khimii (1971) 67, p. 159

[14] Rabald, E.
Dechema-Werkstoff-Tabelle, Capronsäure
(Dechema Corrosion Data Sheets, Caproic
Acid) (in German)
DECHEMA, Frankfurt am Main, 1955

[15] Rabald, E.
Dechema-Werkstoff-Tabelle, Caprylsäure
(Dechema Corrosion Data Sheets, Caprylic
Acid) (in German)
DECHEMA, Frankfurt am Main, 1955

[16] Rabald, E.
Dechema-Werkstoff-Tabelle, n-Caprinsäure
(Dechema Corrosion Data Sheets, n-Capric
Acid) (in German)
DECHEMA, Frankfurt am Main, 1955

[17] Anonymous
NACE Publication 5A 180
Corrosion of metals by aliphatic organic
acids
Mater. Perform. 19 (1980) 9, p. 65

[18] Tveritinov, G. I.; Knyazev, V. V.; Chikov, V. A.
Corrosion resistance of stainless steels in
fractions of synthetic fatty acids (in Russian)
Khim. Neft. Mashinostr. (1974) No. 7, p. 23

[19] Kartashova, K. M.; Sukhotin, A. M.;
Bobrova, M. M.; Shukan, N. Ya.
Corrosion of metallic construction materials
in fatty acids, higher amines and acetates of
higher amines (in Russian)
Tr. Gos. in-t prikl. khimii (1971) 67, p. 159

[20] Kosmenko, Yu. L.; Baidin, I. I.
Corrosion resistance of materials in
synthetic acid production
(translated from Khimicheskoe i Neftyanoe
Mashinostroenie (1985) No. 10, p. 19)

[21] Heitz, E.
Grundvorgänge der Korrosion von Metallen in organischen Lösungsmitteln (I)
(Corrosion fundamental phenomena of metals in organic solvents (I)) (in German)
Werkstoffe u. Korrosion 21 (1970) 5, p. 360/7

[22] Heitz, E.
Korrosion von Metallen in organischen Lösungsmitteln
(Corrosion of metals in organic solvents) (in German)
Oberfläche-Surface 15 (1974) 11, p. 33

[23] Vehlow, J.
Ermittlung der Korrosionsgeschwindigkeiten von hochlegierten Stählen und Hastelloy® C in niederen Carbonsäuren mit Hilfe der Radionuklidtechnik
(Determination of the corrosion rates of high-alloy steels and Hastelloy® C in lower carboxylic acids by change of the radioactivity) (in German)
Werkstoffe u. Korrosion 31 (1980) 2, p. 120

[24] Ullmanns Encyclopädie der technischen Chemie – Fettsäuren
(Ullmann's Encyclopedia of Chemical Technology – Fatty Acids) (in German)
4th Ed., Vol. 11, p. 528 ff.
Verlag Chemie GmbH, Weinheim/Bergstraße, 1976

[25] Anonymous
NACE Publication 5A 180
Corrosion of metals by aliphatic organic acids
Mater. Perform. 19 (1980) 9, p. 65

[26] Vehlow, J.
Ermittlung der Korrosionsgeschwindigkeiten von hochlegierten Stählen und Hastelloy C in niederen Carbonsäuren mit Hilfe der Radionuklidtechnik
(Determination of the corrosion rates of high-alloy steels and Hastelloy C in lower carboxylic acids by change of the radioactivity) (in German) (unpublished manuscript)

[27] Product Information
Molybdän in rost- und säurebeständigen Stählen und Legierungen
(Molybdenum in stainless and acid-resistant steels and alloys) (in German)
Climax Molybdän Gesellschaft, Zürich

[28] Anonymous
NACE Publication 5A 180
Corrosion of metals by aliphatic organic acids
Mater. Perform. 19 (1980) 9, p. 65

[29] Kartashova, K. M.; Sukhotin, A. M.; Bobrova, M. M.; Shukan, N. Ya.
Corrosion of metallic construction materials in fatty acids, higher amines and acetates of higher amines (in Russian)

[30] Kosmenko, Yu. L.; Baidin, I. I.
Corrosion resistance of materials in synthetic acid production
(translated from Khimicheskoe i Neftyanoe Mashinostroenie (1985) No. 10, p. 19)

Carbonic Acid

Unalloyed and low-alloy steels/cast steel
Unalloyed cast iron and low-alloy cast iron

Behavior in gaseous carbon dioxide

The behavior of unalloyed and low-alloy steels as well as of temperature-resisting and heat-resistant steels in hot carbon dioxide gas is of interest in connection with their use in exhaust gases and in nuclear reactors cooled with CO_2 gas, and hence has been thoroughly investigated.

Carbon dioxide is the only medium known to be able to cause local scale break-offs and floret formation in unalloyed and low-alloy steels under pressure and at high temperatures. These scale break-offs may occur even after a long time. This needs to be observed when performing investigations and evaluating results. When plotting the mass increase against time for such tests, a curve is typically obtained with three different sections as schematically shown in Figure 1.

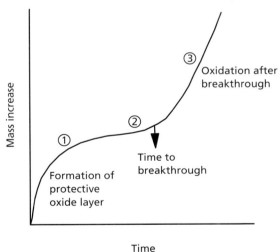

Figure 1: Schematic curve of mass change during the oxidation of low-alloy steels in carbon dioxide at high temperatures and pressures [1]

In section 1 protective layer of magnetite Fe_3O_4 is formed and the oxidation process follows an almost parabolic time law. In the transitional section 2 local break-offs occur in this magnetite layer, increasing in number and size in the course of time until the protective layer on the entire surface is finally destroyed. Thereafter, the oxidation process follows a linear time law in the third section.

This oxidation process has caused substantial damage to heat exchanger tubes and cladding tubes for fuel assemblies in nuclear reactors cooled with CO_2 gas.

The time until a breakthrough occurs decreases as

- the temperature increases,
- the pressure increases,
- the water vapor content of the carbon dioxide increases,
- the surface roughness increases.

In unalloyed and low-alloy steels the water vapor content in carbon dioxide exerts a clear effect on the oxidation behavior. It does not only shorten the time until a breakthrough occurs and hence supports the formation of non-protecting cover layers, but also influences the oxidation rate after a breakthrough. Lowering the water vapor content from 25 ppm to 1 ppm doubles the time till the breakthrough and at water vapor contents of less than 10 ppm scaling follows a parabolic time law. At higher water vapor contents the time dependence of the oxidation rate is almost a straight line with a gradient increasing as the water vapor content increases as shown in Figure 2 using the example of the temperature resisting steel 13CrMo4-5 (material No. 1.7335, cf. UNS K11547).

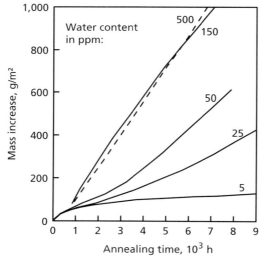

Figure 2: Oxidation behavior of the temperature resisting steel 13CrMo4-5 as a function of time and the water vapor content of carbon dioxide [2]

The formed scale always contains a considerable amount of carbon, the carbon content being especially high at the steel/scale phase boundary. The carbon content increases as the oxide layer thickness and the water vapor content in the carbon dioxide increase. The carbon content in the broken off layers (section 2 in Figure 1) is higher than in the initially formed protecting layers (section 1 in Figure 1).

Regarding the mechanism for breaking off the oxide layer it is assumed that the initially formed magnetite layer grows due to the migration of iron ions and electrons from the inside to the outside. This leads to the formation of voids at the steel/ Fe_3O_4 interface, penetrated by CO_2 along microcracks. Due to the low oxygen pressure in the pores a CO-CO_2 mixture is formed, which in redox reactions transports oxygen from the compact oxide layer to the steel surface so that the oxidation process will continue.

This mechanism also serves as an explanation for the formation of scale with two layers, which could always been found. Beneath an external compact Fe_3O_4 layer there is a porous Fe_3O_4 layer in contact with the steel surface, where alloying elements and minor ingredients of the steel accumulate. However, in equilibrium with Fe/Fe_3O_4 a CO-CO_2 mixture is thermodynamically instable below 873 K (600°C) and can deposit carbon according to Boudouard's equilibrium.

During scale tests with intermediate cooling higher mass increases compared to isothermal tests are only obtained in the beginning. However, after a longer test duration significant differences cannot be found since it is only the upper scale layers spalling during a temperature change, whereas the metal-near layers determining the rate are not removed.

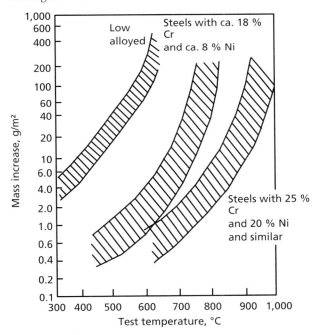

Figure 3: Mass increase of various steel groups in isothermal and cyclic scaling tests in pure dry carbon dioxide (test duration 1,000 h) [2]

Figure 3 summarizes the results of isothermal and cyclic scaling tests for several steel groups in pure dry carbon dioxide [2].

The scatter band of each steel group is rather narrow. The low-alloy, temperature resisting, ferritic steels containing molybdenum, chromium or molybdenum and chromium can be used up to temperatures of about 723 K (450°C). Unalloyed steels fall under the scatter band of these low-alloy steels up to a temperature of about

673 K (400°C), however their scaling rate is higher at higher temperatures and they exhibit breakthroughs of the scale layer at an early stage.

The influence of the flow rate of the gas on oxidation depends on the temperature. At flow rates up to 60 m/s below 723 K (450°C) an effect could not be established for the low-alloy chromium-molybdenum steels. From 763 K (490°C) oxidation increases as the flow rate increases, in particular due to the fact that the spalling of scale intensifies at rates from about 50 m/s.

For smelting steel in the converter with a higher input of scrap metal it is important to avoid oxidation of the iron with carbon dioxide, carbon monoxide, hydrogen and water vapor, which occur when fossil fuels are burnt. Studies in this connection involved investigations into the kinetics of the iron oxidation in carbon dioxide and in carbon dioxide/carbon monoxide mixtures in a temperature range from 1,573 to 1,723 K (1,300°C to 1,450°C) [3].

The oxidation tests were carried out in a Tamman furnace and the mass changes of the specimens were monitored by means of a thermogravity balance. The reaction gases were preheated to 1,373 K (1,100°C) in a preheating furnace. Since the formed iron oxide smelts at test temperatures above 1,650 K (1,377°C), the oxide dripping down was caught in a ceramic vessel. Figure 4 depicts the test apparatus used.

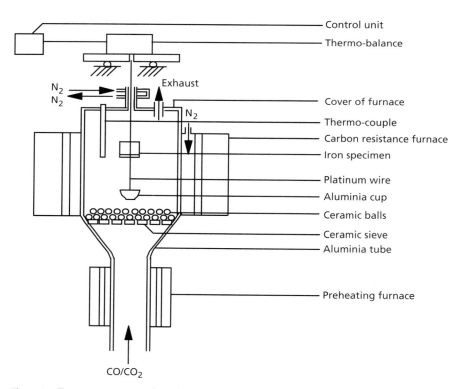

Figure 4: Test apparatus to perform the oxidation tests [3]

Figure 5 shows typical curves describing the time dependence of the mass increase of specimens for various carbon dioxide/carbon monoxide mixtures at 1,597 K (1,324°C) and a constant flow rate of 20 cm/s of the gases.

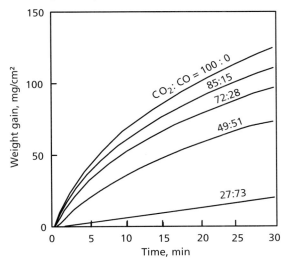

Figure 5: Dependence of mass increase of iron specimens in CO_2-CO mixtures at 1,597 K (1,324°C). Flow rate 20 cm/s [3]

If the carbon dioxide content is high, the curves initially follow a linear and subsequently a parabolic time law. If the carbon dioxide content is lower, the range of the linear time dependence covers a clearly longer period compared to high carbon dioxide contents. The mass increase is clearly reduced as the carbon dioxide concentration decreases.

In Figure 6 the oxidation curves of pure carbon dioxide are plotted against a constant temperature of 1,673 K (1,400°C) and various flow rates.

The time dependence of oxidation is a linear curve since at this temperature the formed iron oxides smelt and drop off and inhibition of the oxidation processes is not possible.

Figure 7 shows the oxidation curves obtained in pure carbon dioxide at temperatures between 1,373 and 1,673 K (1,100°C and 1,400°C).

At temperatures up to 1,633 K (1,360°C) a parabolic time dependence is observed after a short time and the reaction product mainly consists of wustite. At higher temperatures liquid iron oxides are formed and, therefore, the oxidation curve follows a linear time law as can be seen already in Figure 6.

Comprehensive investigations into the reaction mechanism and the kinetics of the oxidation of iron in carbon dioxide/carbon monoxide mixtures are also described in [4, 5]. Here, specimens with a thickness of 0.25 to 2 mm consisting of high-purity iron (99.99%) were examined in CO/CO_2 gas mixtures of various compositions at temperatures of 1,273 to 1,473 K (1,000°C to 1,200°C) at total pressures of 0.1 to 1.0 bar. The oxidation rates were thermogravimetrically determined. Figure 8 shows the

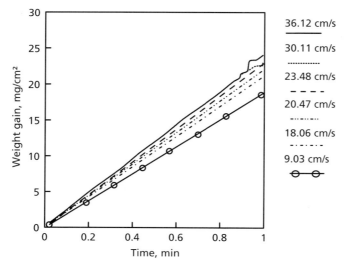

Figure 6: Mass increase of iron specimens in pure CO_2 at 1,673 K (1,400°C) and various flow rates [3]

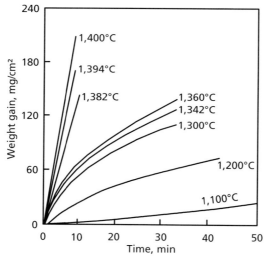

Figure 7: Mass increase of iron specimens in pure CO_2 at different temperatures [3]

influence of the flow rate of the test gases. At a test temperature of 1,473 K (1,200°C) there is practically no dependence of the oxidation rate on the flow rate if the flow rate is equal or higher than 1 cm/s. At a lower flow rate the oxidation rate decreases, which is attributable to insufficient access of the reaction gases to the surface. At the lower test temperature of 1,273 K (1,000°C) smaller flow rates are sufficient to reach a constant oxidation rate.

Figure 9 shows the connection between the mass increase of the iron specimens and the test temperature, the composition of the test gases and the test duration for the onset of the oxide formation and the first growth of the oxide layer.

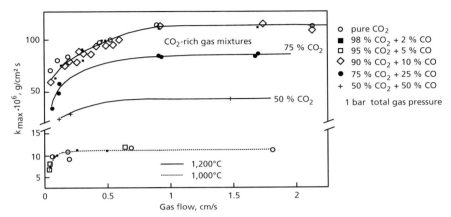

Figure 8: Maximum oxidation rate (k) of pure iron in different CO/CO_2 gas mixtures at a total pressure of 1 bar and temperatures of 1,273 and 1,473 K (1,000°C and 1,200°C) as a function of the gas flow rate [4]

The upper part of the figure shows that at a low total pressure of the gas the section of the linear rise of the mass increase over time is preceded by a period during which the reaction rate is lower. This induction phase attributable to the nucleation of oxides is shorter the higher the temperature and the CO_2 content in the gas are. At 1,473 K (1,200°C) the induction phase is so short that the regions of film growth and subsequent increasing oxidation rate overlap. This also holds true for the oxidation process at a higher total pressure as the lower part of the figure shows for a pressure of 1 bar.

If the test duration is longer, a linear rise of the mass increase of the specimens over time can be observed in practically all cases.

As can be seen in Figure 11, the maximum oxidation rate increases as the total pressure of the gas and the content of carbon dioxide in the test gas increase.

Behavior in aqueous carbon dioxide solutions

The behavior of the unalloyed and low-alloy steels in aqueous carbon dioxide solutions is of special technical importance in connection with the extraction, storage, transport and processing of crude oil and natural gas since the products to handle there almost always exhibit an aqueous CO_2-containing phase. CO_2 is already contained in the extracted products, in particular from deeper reservoirs, or it is injected into the reservoirs if the pressure drops. Correspondingly, there are many publications in the literature dealing with the mechanism of steel corrosion in aqueous carbon dioxide solutions [6–12].

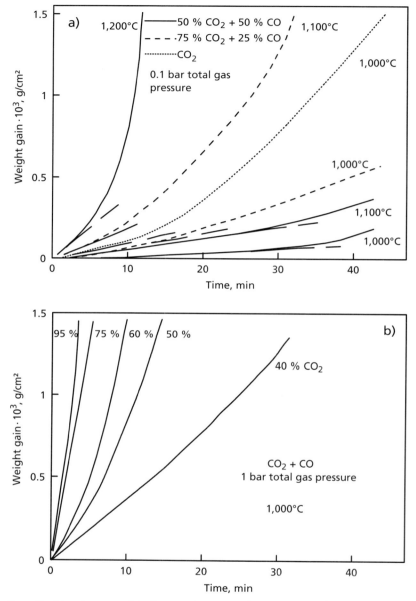

Figure 9: Mass increase of the iron specimens as a function of time for different compositions, temperatures and pressures [4]

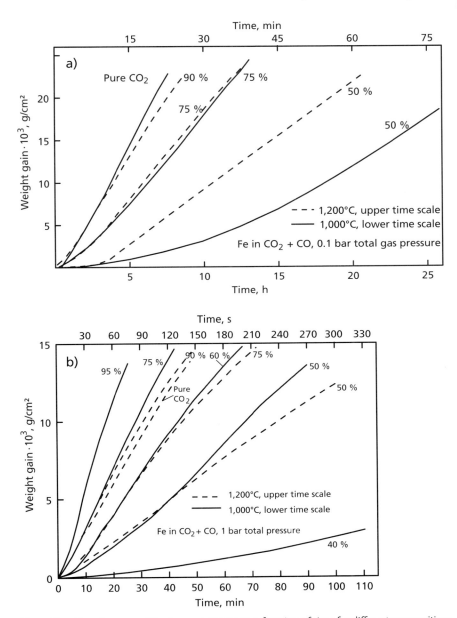

Figure 10: Mass increase of the iron specimens as a function of time for different compositions, temperatures and pressures as well as a longer test duration [4]

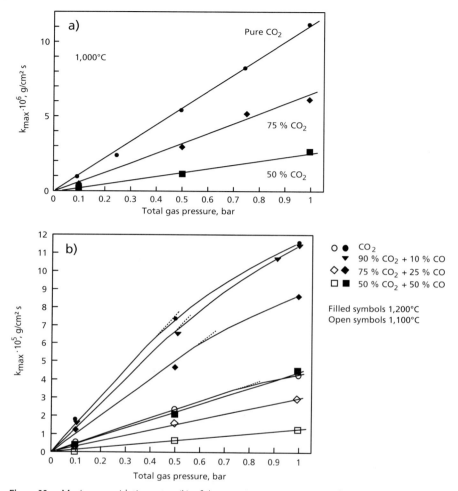

Figure 11: Maximum oxidation rates (k) of the pure iron specimens as a function of the total gas pressure at different temperatures and gas compositions [4]

CO_2 dissolved in aqueous fluids reacts according to Equation 1, forming the weak acid H_2CO_3. As can be seen from Figure 12 and Figure 13, the dissolved carbon dioxide causes the pH value of the solutions to decrease, and therefore the corrosion of steel in CO_2-containing oxygen-free waters occurs as an acid or water type corrosion. The entire reaction can be described with Equation 2.

Equation 1 $CO_2 + H_2O \rightarrow H_2CO_3$

Equation 2 $Fe + 2\,CO_2 + 2\,H_2O \rightarrow Fe^{2+} + 2H_2 + 2\,HCO_3^-$

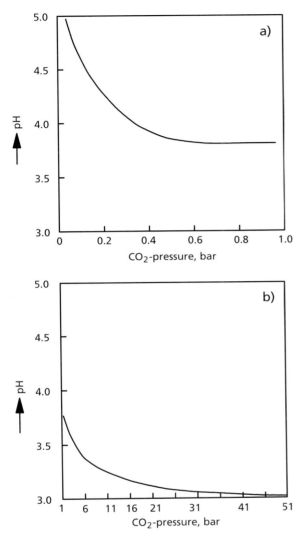

Figure 12: Dependence of the pH value of an aqueous solution on the partial pressure of CO_2 (T = 298 K (25°C)) [17]

The rather strong corrosive attack of CO_2 containing oxygen-free waters on steel is not solely attributed to the low pH value. It is known that the attack of carbon dioxide solutions on steels is stronger than that in diluted mineral acids with the same pH value [6, 7]. This is attributed to the hydrogen evolution intensified by carbon dioxide in the cathodic part process of the corrosion reaction. Hydrogen can be formed by the reduction of non-dissociated carbonic acid adsorbed on the metal surface by reduction [7]:

Equation 3 $H_2CO_3 \text{ (ad)} + e^- \rightarrow H \text{ (ad)} + HCO_3^-$

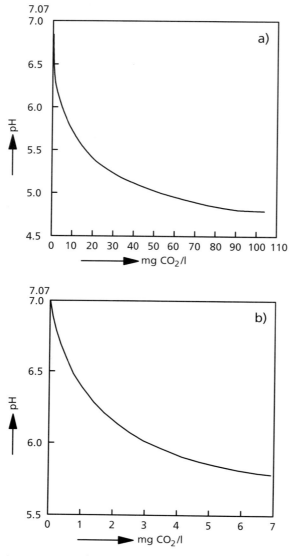

Figure 13: Dependence of an aqueous CO_2 solution on the dissolved amount of CO_2 (T = 298 K (25°C)) [28]

For details regarding the various mechanisms and individual steps reference is made to the cited literature.

The corrosion products formed during the reaction of the steel with the carbonic acid in oxygen-free solutions mainly consist of iron carbonate and offer a certain protection against further corrosion. The composition and structure of the layers depend on the steel grade and the surrounding conditions [13–16].

The corrosion of steel in oxygen-free CO_2-containing solutions is characterized in that the cathodic hydrogen evolution does not only occur in the form of a reduction of hydrogen ions, but also by direct reduction of adsorbed carbonic acid formed by hydration of CO_2. Here, the corrosiveness is a function of the partial pressure of CO_2 and the temperature. Basically, there is a risk of erosive corrosion above 0.5 bar. Temperature increases exert an especially noticeable influence. Freely corroding steel in CO_2-containing solutions absorbs almost double the amount of corrosion hydrogen compared to that in a sulfuric acid at identical pH values [17].

The investigations described in [8] determined the influence of the partial pressure of carbon dioxide on the corrosion of polished specimens of unalloyed steel in a 0.1 % NaCl solution at 298 and 333 K (25°C and 60°C) as shown in Figure 14. More results for different temperatures and partial pressures of carbon dioxide are compiled in Table 1.

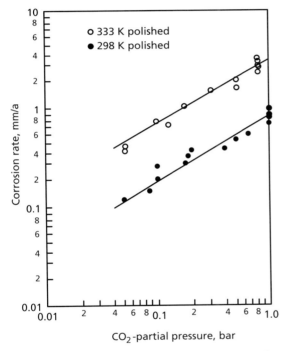

Figure 14: Influence of the partial pressure of CO_2 on the corrosion rates of unalloyed steel in 0.1 % NaCl solution at 298 and 333 K (25°C and 60°C) [8]

The high dependence of the corrosion rate of unalloyed steel in carbon dioxide-containing solutions on the content of dissolved CO_2 was also confirmed by electrochemical measurements in sodium sulfate solutions [9].

Carbonic Acid

Temperature K (°C)	Partial pressure of CO_2 bar	Corrosion rate mm/a (mpy)
278.5 (5.5)	1	0.4 (15.7)
285 (12)	1	0.6 (23.6)
288 (15)	1	0.7 (27.6)
	0.52	0.4 (15.7)
	0.37	0.3 (11.8)
	0.21	0.2 (7.87)
295 (22)	2	1.6 (63.0)
	1	0.9 (35.4)
303 (30)	1	1.3 (51.2)
313 (40)	0.92	1.7 (66.9)
323 (50)	0.88	2.3 (90.6)
333 (60)	0.80	3.9 (154)
343 (70)	0.69	4.3 (169)
353 (80)	0.53	5.7 (224)

Table 1: Influence of the partial pressure of CO_2 and the temperature on the corrosion rate of unalloyed steel in 0.1 % NaCl solution [8]

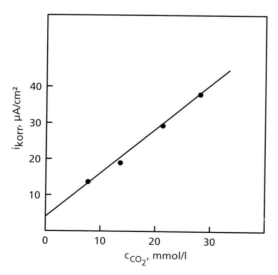

Figure 15: Dependence of the corrosion current density of unalloyed steel on the CO_2 content in an 0.5 M Na_2SO_4 solution at 298 K (25°C) [9]

Figure 15 shows the corrosion current densities measured at rotating disk electrodes from the unalloyed steel, material No. 1.1623 ((mass%: C 0,067, Si < 0,01, Mn 0,38, P 0,016, S 0,016, Al < 0,01, Cr 0,028, Ni 0,024, N 0,0015), in 0.5 M sodium sulfate solution as a function of the dissolved carbon dioxide at 298 K (25°C).

During the investigations described in [10] the influences were analyzed of lower alloy contents as well as of the purity degree of the carbon dioxide flushing gas on the corrosion of unalloyed steels in 0.5 M sodium sulfate solution at 298 K (25°C). The chemical analytical values of the examined steels are indicated in Table 2.

No.	Mat. No..	C	Si	Mn	P	S	Al	Cu	Cr	Ni	N
1	1.1623	0.067		0.380	0.016	0.016			0.028	0.024	0.0015
2	1.0375	0.040	0.020	0.210	0.016	0.020	0.035	0.020	0.010	0.030	0.0074
3	1.1003	0.020	0.010	0.010	0.002	0.009	< 0.001	< 0.01	0.010		0.0030
4	1.1104	0.004	0.010	0.190	0.010	0.023		0.030		0.080	
5	1.8961	0.090	0.290	0.580	0.016	0.025	0.057	0.520	0.640	0.270	
6	1.8962	0.090	0.430	0.270	0.080	0.023	0.037	0.400	0.560		0.0090
7	x)	0.011		< 0.01	0.006	0.005		0.006	0.145		0.012
8	x)	0.010		< 0.01	0.059	0.005		0.003	< 0.002		0.009
9	x)	0.008		< 0.01	0.010	0.005		0.175	< 0.002		0.010
10	x)	0.009		< 0.01	0.004	0.005		0.163	< 0.002		0.010

x) Special smelts

Table 2: Chemical analytical values of the examined steels [10]

The test solution was flushed with carbon dioxide until the desired CO_2 content was reached in the solution, while the maximum test duration was 500 hours. The material consumption of the rotating disk specimens was determined at intervals of 25, 50, 100, 250 and 500 h. To examine the influence of a low oxygen content in the carbon dioxide gas, apart from the standard CO_2 gas with 100 ppm O_2 also ultrapure carbon dioxide was used with an oxygen content of < 0.1 ppm O_2. The material removal values determined for steel 1 as a function of the test period in 0.5 M sodium sulfate solution with different standard CO_2 contents are plotted in Figure 16. The material consumption rates increase as the test period increases and, in particular, as the carbon dioxide content in the solution increases. But strikingly, lower material consumptions were measured for the highest carbon dioxide content of 27 mmol/l in the solution compared to the lower value of 21 mmol/l. This reproducible result is attributed to the fact that increasingly protective cover layers are formed at high carbon dioxide contents.

The corresponding measurement results for the tests with oxygen-free carbon dioxide are indicated in Figure 17.

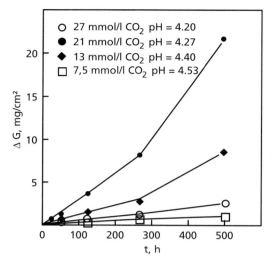

Figure 16: Time dependence of the material consumption rates of steel 1 in 0.5 M sodium sulfate solution with different standard CO_2 contents [10]

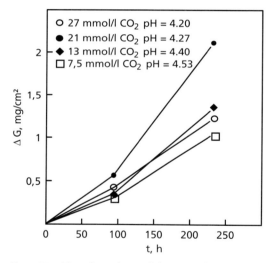

Figure 17: Time dependence of the material consumption rates of steel 1 in 0.5 M sodium sulfate solution with different oxygen-free CO_2 contents [10]

The material removal values are clearly lower than those of the oxygen-containing solution; but here again lower material consumptions are measured at the highest carbon dioxide content compared to lower values. Obviously, not only carbonates or hydrogen carbonates play a role in the cover layer formation responsible for that phenomenon, but also the other salts present in the solution, e.g. sulfates in this case, as shown by the tests in oxygen-free distilled water – Figure 18. The consumption values can be directly assigned to the increasing carbon dioxide contents.

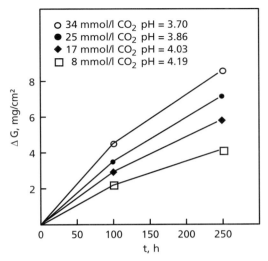

Figure 18: Time dependence of the material consumption rates of steel 1 in distilled water with different oxygen-free CO_2 contents [10]

Figure 19 shows the corrosion rates determined for all 10 steel grades in 0.5 M sodium sulfate solution saturated with carbon dioxide. The corrosion rates of the unalloyed steel grades No. 1 to 4 were about 0.05 mm/a (1.97 mpy), with a low oxygen content in the carbon dioxide gas practically remaining without influence. The attack on the lightly alloyed steels No. 5 and 6, in contrast, was much more considerable. Obviously, these small contents of alloying elements contributing to the formation of cover layers in the atmosphere, which provide a better protection, lead to the formation of less protecting layers under these conditions.

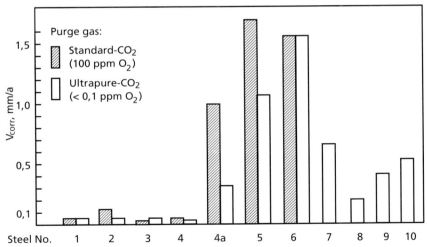

Figure 19: Corrosions rates of the 10 examined steels in 0.5 M sodium sulfate solution saturated with carbon dioxide at 298 K (25°C) [10]
(Steel 4a about 90% cold-formed)

To check which alloying elements in the steels No. 5 and 6 exert a negative influence on the corrosion behavior, the four special smelts No. 7 to 10 were examined, which were produced on the basis of pure iron with individual alloying elements, namely chromium, phosphorus and copper as well as copper plus phosphorus. It turned out that all three alloying elements reduced the corrosion resistance in carbonic solutions. Phosphorus exerts the rather lowest negative effect, yielding values for steel No. 8 which were only thrice as high as the values of the unalloyed steel grades No. 1 to 4. In contrast, the steels No. 7, 9 and 10 with increased copper and chromium contents exhibited corrosion rates which were up to ten times higher than those of the unalloyed steel grades No. 1 to 4. On these steels loose corrosion products were formed, in which considerable contents of copper were found.

The influence of cold forming was examined in steel No. 4, which was cold-formed by about 90 %. The corrosion rates determined for these specimens were five times higher in the oxygen-free test solution and even 30 times higher in the oxygen-containing solution compared to the non-formed specimens.

Compared to the test solution containing sodium sulfate, oxygen-free distilled water saturated with carbon dioxide exhibited a remarkably higher aggressiveness. Under conditions, which were identical otherwise, corrosion rates of 0.4 mm/a (15.7 mpy) were determined for unalloyed steel in this fluid. The reason is the lower pH value and the higher carbon dioxide content of these solutions. Only above pH 4.2 the presence of lower oxygen amounts resulting from the charge of carbon dioxide containing 100 ppm oxygen caused an acceleration of corrosion.

However, indicating linear material consumption rates is only of limited importance to the evaluation of the resistance of steels in the chosen fluid. All examined materials exhibited pitting corrosion after induction periods of several hundred hours, the occurrence of which was only attributed to the presence of carbon dioxide and/or carbonic acid. Regarding the pit diameter, pit depth and pit density, pitting corrosion was gradually weaker in low-alloy steels. In oxygen-free distilled water saturated with carbon dioxide shorter induction periods (about 250 hours) of pitting corrosion were observed compared to sulfate-containing solutions. The presence of oxygen traces caused an extension of pitting corrosion induction periods to a maximum of 1,000 hours, in particular for low-alloy steels. Local corrosion can also be explained by assuming active and passive regions on the metal surface.

Also newer investigations attempted to answer the question about the influence of chromium additions in steel on the corrosion in media containing carbon dioxide [18]. The test materials were pure iron (99.95 %) and three alloys of pure iron with 1 %, 3 % and 5 % chromium additions. The pure materials were chosen to avoid precipitations. All materials had a ferritic structure and the chromium was completely dissolved in the matrix even at 5 %. The test medium was a sodium chloride solution saturated with CO_2 and stirred with an NaCl concentration of 116 g/l and a pH value adjusted to 4.2, 5.0 or 6.0 by means of HCl or NaOH. The tests were performed at room temperature. Access of oxygen was excluded. After a test duration of 12 h, 24 h, 48 h and 72 h the material consumption of the specimens was determined.

The critical corrosion medium in the condensates for the internal corrosion of pipes in petrochemical plants are water, gases with acid reaction, such as carbon dioxide or hydrogen sulfide, and volatile organic acids occurring as byproducts of the hydrocarbons. For this reason 20 g/l acetate were added to the solution in several tests.

The material consumption measured for the specimens in the acetate-free NaCl solution with pH 4.2 is plotted against time in Figure 20. All materials tested show a linear time dependence of the material consumption. Although the corrosion values apparently decrease as the chromium content increases, the pure iron specimens and the specimens with 1 % Cr differ slightly only in their corrosion rates, whereas the behavior of the two higher alloyed materials is clearly better but again with slight differences only between them.

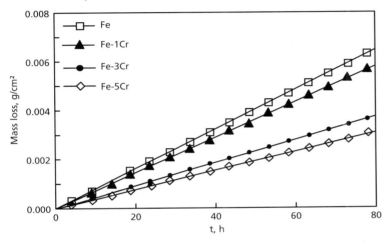

Figure 20: Material consumption of the four materials in acetate-free NaCl solution with pH 4.2 saturated with CO_2 [18]

In Figure 21 the material consumption measured for the pure iron specimens and the specimens alloyed with 3 % Cr is plotted at different pH values.

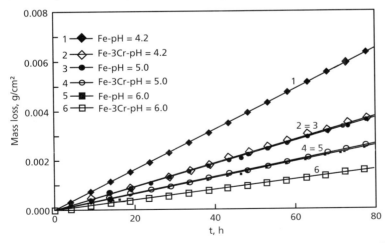

Figure 21: Material consumption of the pure iron specimens and the specimens alloyed with 3 % Cr in acetate-free NaCl solution saturated with CO_2 at different pH values [18]

Here again a linear time dependence of material consumption is found. The material consumption of both materials decreases as the pH value increases. A positive influence of the chromium content can only be detected at a low pH value.

Due to the linear dependence the corrosion rates indicated in Table 3 were extrapolated. According to the results indicated in [10] such linear time dependence can no longer be established for longer test durations (refer to Figure 16, Figure 17 and Figure 18).

Material		Fe	Fe-1Cr	Fe-3Cr	Fe-4Cr
	pH				
mm/a (mpy)	4.2	0.898 (35.4)	0.843 (33.2)	0.596 (23.5)	0.483 (19.0)
	5.0	0.52 (20.5)		0.395 (15.6)	
	6.0	0.34 (13,4)		0.342 (13.5)	

Table 3: Extrapolated corrosion rates of the examined materials in acetate-free NaCl solution saturated with CO_2 at different pH values [18]

The results of the tests during which acetate was added to the test solution as depicted in Figure 22 and Figure 23 suggest that there is no longer any linear time dependence of the material consumption under these test conditions.

The influence of acetic acid on the carbon dioxide corrosion of unalloyed steels in the horizontal flow of two-phase water-steam mixtures was investigated in connection with the problems in oil and gas pipelines with different corrosion manifestations occurring in 12 and 6 o'clock positions [19, 20]. To this end, from a circulation apparatus with horizontally arranged pipes specimens of the steels X 65 (material No. 1.8975, cf. API 5LX65) and C 1018 were taken, the composition of which is indicated in Table 4, were laid out in the 12 o'clock position (gaseous phase) and in the 6 o'clock position (liquid phase) and checked for their corrosion resistance.

	C	Si	Mn	P	S	Cu	Cr	Ni	Al	Nb	V
X 65	0.130	0.260	1.160	0.009	0.009	0.131	0.140	0.360	0.032	0.017	0.047
C 1018	0.160	0.250	0.790	0.008	0.029	0.250	0.063	0.078	0.001	0.006	0.004

Table 4: Composition of the investigated steels [19]

The liquid phase of the test medium mainly consisted of an aqueous 1% NaCl solution with different contents of acetic acid from 500 ppm to 5,000 ppm and a pH value of 5.5. The gaseous phase consisted of a gas-water steam mixture. The partial pressure of CO_2 amounted to 1.54 bar and the total pressure to 2 bar. The test temperature was 353 K (80°C) and the flow rates were 0.1 m/s in the liquid phase and 2 m/s in the gaseous phase.

As shown in Figure 24 using the example of C 1018 steel specimens in the 6 o'clock position, the corrosion rates increase as the content of acetic acid in the solution increases. Irrespective of the acid content, the values first increase as the test duration increases and then drop to a constant value below 1 mm/a (39.4 mpy).

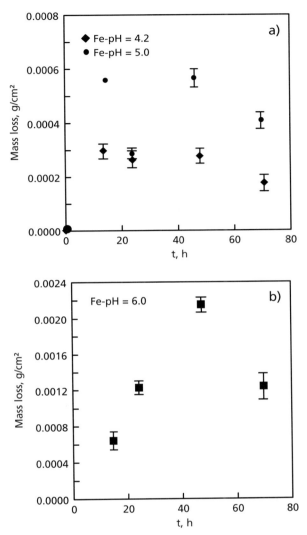

Figure 22: Material consumption of the pure iron specimens in acetate-containing NaCl solution saturated with CO_2 at different pH values [18]

The specimens in the 12 o'clock position (gaseous phase) and the specimens of the X 65 steel basically follow the same corrosion behavior with a clear increase of the corrosion rates at an increasing acetic acid content in the solution. This is attributed to the contribution of the hydrogen ions from free acetic acid to the cathodic partial reaction of the corrosion process. Both in the 12 o'clock position and in the 6 o'clock position local deposits of $FeCO_3$ were formed, causing a local corrosive attack.

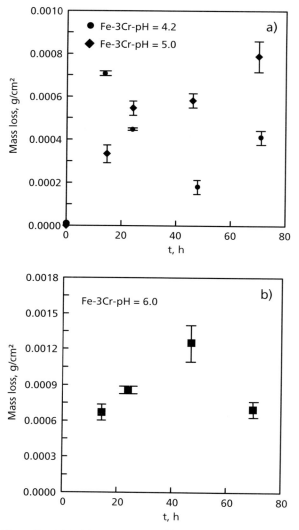

Figure 23: Material consumption of the specimens alloyed with 3 % Cr in acetate-containing NaCl solution saturated with with CO_2 at different pH values [18]

These local sites of attack can be found, in particular, in the upper pipe position, attributable to the enrichment of acetic acid in the condensate drops. This is also confirmed by other investigations in the circulation apparatus, which also present electrochemical examinations and model calculations proving the practical experience [21–24].

In this regard Figure 25 shows results of electrochemical investigations of pipe steel X 65 in 0.01 M NaCl solution saturated with 1 bar CO_2 and acetic acid additions of 0 to 1,000 ppm at room temperature [24]. Up to an acetic acid content of about

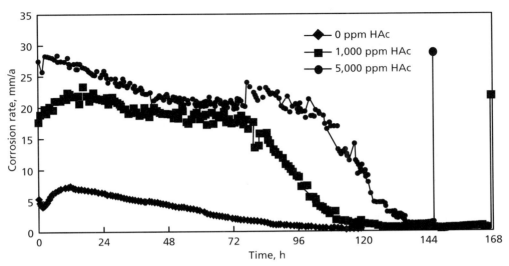

Figure 24: Influence of the acid content and the test duration on the corrosion of C 1018 steel in 1 % NaCl solution saturated with CO_2 with a pH value of 5.5 at 353 K (80°C) in the 6 o'clock position [19].

300 ppm the corrosion potential rapidly shifts into the negative direction. In contrast, higher contents have a less strong influence.

To avoid corrosion problems caused by aqueous phases containing carbon dioxide in the crude oil and natural gas production inhibitors on the basis of organic compounds with a higher molecular weight can be used.

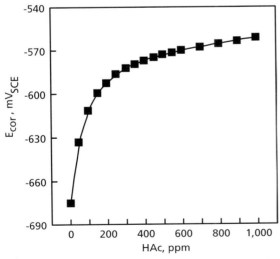

Figure 25: Corrosion potential of steel X 65 in 0.01 M NaCl solution saturated with 1 bar CO_2 and acetic acid additions at room temperature [24].

Hydrogen-induced stress corrosion cracking was not detected in the fairly soft steels examined in [10]. However, features of anodic stress corrosion cracking were found at higher partial pressures of CO_2 (equal/above 10 bar) and higher temperatures (> 313 K (> 40°C)) if the steels were subjected to static tensile stresses above yield strength.

Stress corrosion cracking of tempered steels in CO_2-containing attacking media has long been known for the $CO_2/H_2S/H_2O$ system. In this medium CO_2 plays a rather contributing instead of a primary role in the corrosion process. However, it is also known that higher strength steels, in particular tempered steels, may be affected by stress corrosion cracking in wet carbon dioxide and in general in the $CO_2/CO/H_2O$ system.

But also in the simple two-substance CO_2/H_2O system stress corrosion cracking can occur. This turned out in damage investigations of numerous cracked or ruptured CO_2 gas bottles or fire extinguishers attributable to the residual moisture resulting from a previous water pressure test. Here, the crack path was usually an intercrystalline one. From all experience it can be concluded that stress corrosion cracking depends on the presence of CO_2 and CO as well as water. No damage is known if only CO_2 or only CO were present. Partial pressures of 60 mbar of any gas type turned out to be sufficient to cause stress corrosion cracking [25].

In systematic investigations with tempered low-alloy steels of grade 36Mn7 (material No. 1.5069, cf. UNS H13400), 34CrMo4 (material No. 1.7220, cf. UNS G41350) and 30CrNiMo8 (material No. 1.6580, cf. UNS G43400), which were present in three strength levels, it was found out in experiments at a constant force of 50 to 90 % of the yield strength as well as at a constant strain rate between 10^{-5} and 10^{-7} 1/s in the temperature range from 273 to 333 K (0 to 60°C) and at a partial pressure of CO_2 from 1 to 60 bar that all examined materials may suffer stress corrosion cracking even at constant load [26]. The results are summarized in Figure 26.

Here, the main influencing factors are the strength of the steel, the level of loading, the partial pressure of CO_2 and the temperature. The cracking tendency rises as the strength of steel, the loading level and the temperature increase. At 333 K (60°C) and 60 bar partial pressure of CO_2, for instance, the region can be defined with high precision for all examined steels in which stress corrosion cracking occurs depending on the material strength and the level of loading. For the low-alloy Cr steels this is also possible at 313 K (40°C), the critical region being shifted towards higher strengths. The propensity to stress corrosion cracking is especially high at high partial pressures of CO_2. Above 10 bar CO_2 the lifetimes of constantly loaded tensile test specimen (0.9 Re) was generally below 1,000 h.

However, at this temperature the steel 36Mn7 exhibited a different behavior. Although at loads < 0.9 R_e cracks propagating from the surface were no longer found, crack initiation occurred in the V-notch shaped pitting corrosion sites, preferably developed at the breakthrough point of impact lines. The formation of crack initiation sites can be explained by the much higher stress intensity at the notch root. With reference to the real stress conditions, the susceptibility range for stress corrosion cracking indicated for the low-alloy Cr steels would also apply to the steel 36Mn7.

However, the transcrystalline damage caused by stress corrosion cracking on the outer side of gas transport lines with PE jacketings is indicative of the possibility

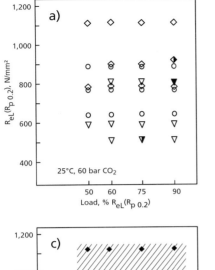
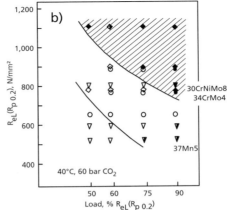
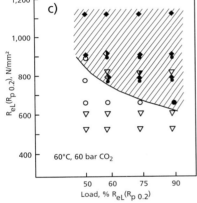

Figure 26: Cracking tendency of steels in the CO_2/H_2O system at 60 bar CO_2 pressure as a function of strength and the constant load level with reference to the yield strength (or 0.2 % elongation limit) at different temperatures. (Symbols: ◆ Steel 30CrNiMo8, ● Steel 34CrMo4, ▲ Steel 36Mn7; open symbols: small cracks, half filled symbols: crack initiation, fully filled symbols: cracks) [26]

that stress corrosion cracking in CO_2-containing aqueous media is also possible at low partial pressures of CO_2. The time until a crack is initiated may then cover a period of several years.

For all examined steel grades also mass loss measurements were performed in a temperature range from 273 to 353 K (0°C to 80°C) and at partial pressures of CO_2 from 1 bar to 60 bar. The results are summarized in Figure 27.

The corrosion rates increase with increasing temperature and carbon dioxide pressure. At 313 K (40°C) the steels 30CrNiMo8 (cf. UNS G43400) and 34CrMo4 (cf. UNS G41350) are less attacked than 36Mn7 (cf. UNS H13400) steel, whereas the attack is stronger than that at higher temperatures. At pressures up to 30 bar a maximum of uniform surface corrosion can be observed at about 333 K (60°C).

It can be seen also from these results that the protective action and stability of the carbonate cover layers are a function of temperature and CO_2 pressure. In general, the relative protective effects of the cover layers formed at temperatures above 313 K

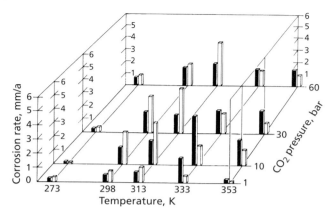

Figure 27: Influence of temperature and CO_2 pressure on the corrosion of tempered steels in the CO_2/H_2O system [26]

(40°C) was better the higher the CO_2 pressure was. The cover layers formed under these conditions also exhibited a much poorer dissolution behavior when being stripped compared to the corrosion products grown at lower temperatures and lower CO_2 pressures. A comparison of the results obtained from material consumption measurements with the results of the tensile tests shows that the highest susceptibility to stress corrosion cracking was found under conditions where also the highest pitting corrosion tendency was found. This is reasonable since local corrosion can only occur under the conditions of instable cover layers.

Regarding the question about the mechanism of the crack-forming corrosion in the CO_2/H_2O system, all observations are indicative of the fact that it is a classical anodic stress corrosion cracking, the formed hydrogen supporting crack propagation. The assumption of an anodic stress corrosion cracking is further supported by the fact that the cracking tendency increases as the temperature increases. In case of a pure hydrogen-induced stress corrosion cracking the highest susceptibility would rather occur at lower temperatures (around 298 K (25°C)) than at higher temperatures as found.

Scanning electron microscopic examinations of fracture surfaces and forced ruptures of tensile test specimens damaged by stress corrosion cracking did not reveal any impact fracture morphologies referring directly to hydrogen-induced crack formation. On the other hand, considerable hydrogen permeation rates were found in hydrogen permeation measurements, in particular at an elevated temperature as well as under the influence of tensile stress.

In the 60s and 70s crack damage caused by transcrystalline stress corrosion cracking was found on gas containers, bottles and separators of city and syngas plants of unalloyed and low-alloy steels if $CO/CO_2/H_2O$ mixtures were applied under operating conditions [27].

To date transcrystalline damage of a similar kind has been preferably found in the presence of a hydrogen-induced crack mechanism, e.g. in H_2S-containing sour gas condensate. Due to different signs and indications, the investigations performed did not permit a clear decision as to which mechanism of stress corrosion cracking was responsible here. Further investigations aimed at clarifying the damage mechanism and examine the influencing factors of media and materials.

For the investigations the fine-grain structural steels S355N (material No. 1.0545, cf. UNS K12709) and S500N (material No. 1.8907, cf. UNS K02001) were chosen as well as the steels S235JR (material No. 1.0037, cf. UNS K02502) and 13CrMo4-5 (material No. 1.7335, cf. UNS K11547). The test sheets were provided in a normalized condition. Fine-grain structural steel S355N is widely used for the manufacture of vessels and apparatuses in the chemical industry and was, therefore, used for these investigations as the basic material. The other steels mentioned served for comparative investigations.

Round test bars of the steels were tested in autoclaves in an electrochemically controlled process, with CO/CO_2-containing water (partial pressures up to 20 bar CO_2 and 10 bar CO) being applied with various potentials, temperatures and oxygen contents.

The investigation was mainly performed as slow strain rate tensile tests (CERT tests) during which the specimens were strained until fracture (in most cases at strain rates of 4.3×10^{-5} 1/s and 8.6×10^{-6} 1/s). Additional examinations at constant load (125 % R_{eL}) provided further information about the mechanism of crack propagation.

A transcrystalline crack formation occurred on all steel grades in $CO/CO_2/H_2O$ mixtures both under cathodic and anodic polarization. Regarding the free corrosion potential, the sensitivity to cracking was neglectably small. Crack formation is intensified in all materials examined on both increasing anodic and increasing cathodic polarization.

At a slower strain rate, the depth of cracks is clearly higher. Instead of polarization with direct current, the adjustment of the free corrosion potential in the range of stress corrosion cracking sensitivity can be also reached by the addition of oxygen (air).

Only anodic stress corrosion cracking can be the cause of the damage in practice since the cathocidally induced crack propagation is not typical for the $CO/CO_2/H_2O$ system but is observed in all weakly acidic hydrogen-supplying media and is of no relevance to practice as such conditions (cathodic current; rapid elongation) normally do not occur. The presence of atmospheric oxygen is responsible for the adjustment of the critical potentials.

In the potential range of anodic stress corrosion cracking the crack formation increases with increasing temperatures (up to about 333 K (60°C)), whereas surface corrosion increases at temperatures above that value and exclusively occurs at a temperature above 373 K (100°C).

Even under static load conditions (125 % R_{eL}) all steel specimens cracked in $CO/CO_2/H_2O$ mixtures on anodic polarization (lifetimes < 20 h).

The localization of the attack (crack nucleation) is also attributable to the formation of cover layers, which can be observed in the corrosion test in that the material consumption rates are considerably reduced (by about one order of magnitude).

Without polarization but with addition of oxygen (partial pressure of 2 bar) all specimens also failed, with the steels S355N and 13CrMo4-4 exhibiting especially short lifetimes. Only in air-flushed water the steel S355N did no longer exhibited any cracks.

S235JR steel reached a lifetime of up to 70 h, whereas the lifetime of the 13CrMo4-4 steel remained below 10 h. Due to its tendency to form cover layers, 13CrMo4-4 steel responds already to very small additions of oxygen.

The gas composition may vary over wide ranges without loosing effect. Only regarding the CO_2 content a lower limit concentration is required for crack nucleation. At CO_2 contents below 1 bar the crack nucleation is rather weak.

Therefore, anodic stress corrosion cracking can be considered the cause of the damage occurred in practice. An anodic polarization, which is finally the cause of a stress corrosion cracking risk, is easily reached by the presence of oxygen or other oxidants. At a moderately increased temperature (333 K (60°C)) the crack intensity of anodic stress corrosion cracking increases.

All tested materials are susceptible to cracking and exhibit only little difference in their sensitivity. However, the material 13CrMo4-4 is especially sensitive to small amounts of oxygen. Its alloying elements facilitate the formation of a cover layer with a strong protective effect resulting in a marked localization of the attack and a rapid crack growth.

Ferritic chrome steels with < 13 % Cr

In carbon dioxide at elevated pressures and temperatures the ferritic heat resisting chromium molybdenum steels with chromium contents of 9 % and 12 % exhibited an accelerated oxidation following a local floret development similar to the unalloyed and low-alloy steels. Below temperatures of 823 K (550°C) local breakthroughs do not occur. Above this temperature the oxidation process can be described again by means of the curve schematically depicted in Figure 1 comprising three phases. In the first phase, during which protecting oxide layers are formed, higher silicon contents up to 1 % in the steel exert a favorable effect. A decreasing water vapor content in carbon dioxide also reduces the oxidation rate. The first break-offs in the oxide layer occur at a mass increase of 20 mg/cm^2. The time until breakthroughs occur is reduced by increasing contents of carbon dioxide in water vapor and carbon monoxide. Higher silicon contents have practically no influence on the oxidation behavior in the third phase.

As in the low-alloy steels the protecting oxide layer has again a two-layer structure with a compact outer magnetite layer and a porous inner spinel layer enriched with chromium and silicon as oxides. And again the oxide layers contain higher amounts of carbon with a maximum amount at the steel/scale phase boundary. Also the base material is considerably carburized.

Behavior in aqueous carbon dioxide solutions

In aqueous solution of carbon dioxide with a partial pressure of 13.8 bar and a temperature of 328 K (55°C) the corrosion rate of chromium alloyed steels decreases if the chromium content increases as shown in Table 5 [28].

Chromium content, %	2.25	5.0	9.0	12.0
Corrosion rate, mm/a (mpy)	1.5 (59.1)	1.17 (46.1)	0.04 (1.57)	0.02 (0.79)

Table 5: Corrosion rate of chromium alloyed steels in CO_2 containing water after a test duration of 7 days [28]

With an increasing test duration the corrosion rates further decrease as a result of cover layer formation and in case of steel with 2.25 % Cr drop to

– 0.16 mm/a (6.30 mpy) after 28 days,
– 0.08 mm/a (3.15 mpy) after 70 days

and in case of steel with 12.0 % Cr drop to

– 0.005 mm/a (0.20 mpy) after 70 days.

Ferritic chromium steels with ≥ 13 % Cr
High-alloy multiphase steels
Ferritic/perlitic-martensitic steels

As shown in Table 5 the steels with higher chromium contents are largely resistant to material removing corrosion in carbon dioxide-containing waters. But in aqueous solutions containing also hydrogen sulfide apart from carbon dioxide, as for instance occurring in the aqueous phases during the production and the transport of natural gas and crude oil, the chromium alloyed steels may be subject to the risk of stress corrosion cracking.

By reducing the carbon content and adding the alloying elements nickel, molybdenum, copper and nitrogen, the corrosion behavior of a common 13 % chrome steel according to API in CO_2 and H_2S-containing media can be clearly improved [29]. Table 6 shows the chemical analytical values of a modified steel compared to 13 % chrome steel.

Steel	C	Si	Mn	P	S	Cr	Ni	Cu	Mo	N
CRSS	0.017	0.30	0.51	0.017	0.001	12.8	5.90	1.56	1.97	0.017
API	0.190	0.25	0.58	0.011	0.003	12.6	0.13			

Table 6: Chemical compositions (% by weight) of both 13 % chromium steels examined [29]

Tests were performed to investigate the material removing corrosion and stress corrosion cracking using four-point bending specimens and tensile test specimens. A load of 100 % of the yield strength was applied as a test stress to round test bars and a load of 90 % of the yield strength was applied to the bend test specimens. Following the NACE Specification TM0177-90A the test duration was 720 hours each. Solutions containing acidic chloride, hydrogen sulfide and carbon dioxide as occurring in plants processing natural gas and crude oil were chosen as the test solutions. The test conditions and results are compiled in Table 7 and Table 8.

The higher resistance to material removing corrosion compared to normal 13 % chrome steel was attributed to the additional alloy contents of nickel and copper, whereas the better resistance to stress corrosion cracking was attributed to the molybdenum content.

Test solution	20 % NaCl + 100 mg/l NaHCO$_3$ 40 bar CO$_2$/0.1 bar H$_2$S Temperature: 433 K (160°C); pH value: 4.1	
	CRSS	API
Bend test specimens	no cracks	no cracks
Corrosion rate	0.027 mm/a (1.06 mpy)	0.30 mm/a (11.8 mpy) *)

*) local attack

Table 7: Test conditions and results for material removing corrosion and stress corrosion cracking of bend test specimens [29]

Test solution					Steel	
Chlorides ppm	CO$_2$ bar	H$_2$S bar	pH		CRSS	API
7,200	50	0.02	2.9		no cracks	cracks
55,800	50	0.02	2.7		cracks	cracks
1,000	45	0.10	3.0		no cracks	cracks
68,000	45	0.10	3.3		no cracks	cracks
68,000	20	0.10	3.8		cracks	cracks

Table 8: Test solutions and results of stress corrosion cracking tests with tensile test specimens [29]

The good resistance to stress corrosion cracking of a steel containing 18 % chromium and 2 % molybdenum is also confirmed in [30].

Apart from the resistance to material removing corrosion, pitting corrosion and stress corrosion cracking, frequently materials used in the production of gas and oil are also required be resistant to erosion corrosion. In this connection tests were performed with X20Cr13 (1.4021, cf. UNS S42000) steel to establish the erosion corrosion behavior in simulated carbon dioxide-containing formation water at 333 K (60°C) [31, 32].

Composition of the test solution: 3.80 % NaCl
0.44 % $CaCl_2$
0.07 % $MgCl_2$
partial pressure of CO_2: 3 bar
pH value: 4.2.

As a solid body 0.1 % sand with a grain size of 0.4 mm was added. A segmented coiled column served as the test track as schematically shown in Figure 28. The figure contains the material removal obtained plotted against the various Reynolds numbers Re for the X20Cr13 steel. For comparison, the values obtained for the unalloyed steel C15 (1.0401, cf. UNS G10150) at Re = 3.5×10^5 are also indicated. At the beginning of the pipe construction the maximum material removal is reached by working off the edges. Following the narrow pipe section a maximum is found again between 2 and 3 $1/D_0$. Both maximum values depend on the flow rate. According to these investigation results a flow rate with a Reynolds number of 2.5×10^5 and a corrosion rate of about 0.5 mm/a (19.7 mpy) would still be acceptable for technical applications.

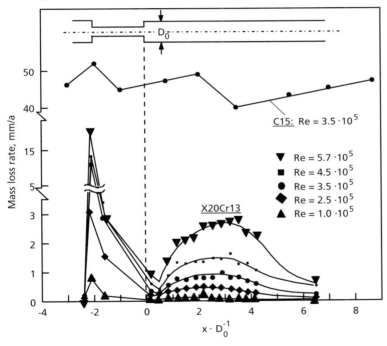

Figure 28: Results of erosion corrosion tests in carbonated water with various Reynolds numbers Re for X20Cr13 [33, 34]

Compared to an unalloyed steel C15 the erosion corrosion resistance of the X20Cr13 steel in carbonated aqueous solution is clearly better. The erosion corrosion behavior of steel X20Cr13 was examined in the same test medium with the Reynolds number 2.5×10^5, however with different solid contents. The results are shown in Figure 29.

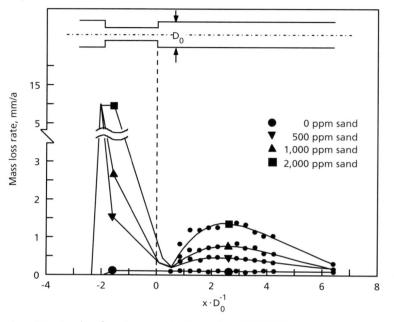

Figure 29: Results of erosion corrosion tests of steel X20Cr13 in carbonated waters with different solid additions [34]

An increase in the solid particle content of the corrosion medium leads to an increase in the material removal, resulting in an almost linear dependence between the sand content and the erosion corrosion rate in the carbonated water. Reduction of the chloride content in the test solution remained without any significant influence on the material removal. In the temperature range from 298 to 333 K (25 to 60°C) a change in the erosion corrosion loss cannot be detected. Increasing the temperature from 333 to 353 K (60 to 80°C) leads to an increase in the corrosion rate by 25 % [31, 32].

Using steel X2CrNiMoN22-5-3 (1.4462, cf. UNS S31803) instead of X20Cr13 may reduce corrosion by another 50 %.

The corrosion resistance of a 13 % Cr steel against carbon dioxide-containing waters may be increased by adding copper, nickel and molybdenum [35]. Starting from a common 13 % Cr steel six steel variants were investigated, the chemical composition of which is indicated in Table 9.

C	Si	Mn	P	S	Cr	Cu	Ni	Mo
0.025	0.25	0.45	0.015	0.002	13	0 to 2.0	4.0 to 5.0	1.0 to 2.0

Table 9: Chemical composition of the investigated steels [35]

The tests to investigate the behavior in carbon dioxide-containing solutions were performed by exposing the specimens in the autoclave to a 20 % NaCl solution saturated with 3 MPa CO_2 gas at 453 K (180°C). The corrosion rate was calculated from the material consumption of the specimens after a test duration of 7 days. The results are shown in Figure 30.

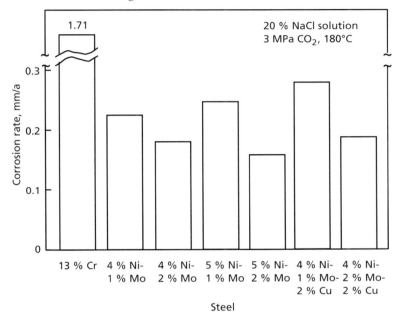

Figure 30: Influence on various alloying elements on the corrosion of a 13 % Cr steel against carbon dioxide-containing solution at 453 K (180°C) [35]

The addition of the alloying elements mentioned leads to a clear reduction of the corrosion rates compared to non-modified 13 % Cr steel. At the best, i.e. steel with 5 % Ni and 2 % Mo, the corrosion rate is reduced from 1.7 mm/a (66.9 mpy) to about 0.15 mm/a (5.91 mpy). Also the behavior with regard to stress corrosion cracking if carbon dioxide and hydrogen sulfide are present at the same time, is improved by the addition of alloying elements when comparing modified steels and conventional 13 % Cr steel, the major effect being attributed to molybdenum.

Austenitic chromium-nickel steels

Stainless austenitic steels are widely used in power plants as well as in chemical and petrochemical plants in gas atmospheres exerting an oxidizing or carburizing effect, such as air, steam, carbon dioxide or combustion gases in a temperature range from 573 to 973 K (300°C to 700°C). The resistance to aggressive gases at high temperatures is the result of the formation of a chromium-rich cover layer. Therefore, the steels used under these conditions contain not less than 20 % chromium.

The publications described in [36, 37] deal with the resistance of stainless steels with 20 % Cr and 25 % Ni in CO_2/CO gas mixtures. The chemical composition of the three investigated steels is shown in Table 10.

Steel	C	Si	Mn	P	p	Cr	Ni	Ti	Nb
20Cr25Ni/Nb	0.02	0.56	0.70		0.007	20.4	25.1		0.60
20Cr25Ni/Ti	0.015	0.66	0.81	0.005	0.001	19.7	24.1	0.23	
20Cr25Ni/TiNb	0.01	0.86	0.65			20.1	24.8	1.70	0.12

Table 10: Chemical composition of the three investigated steels [36]

The specimens with the dimensions of 20 mm × 5 mm × 0.4 mm from both steels 20Cr25Ni/Nb and 20Cr25Ni/Ti were subjected to recrystallization annealing in hydrogen at 1,203 K (930°C) for 1 h, yielding grain sizes of ~ 9 µm and ~ 2.5 µm in the surface area and of 13 µm and 20 µm in the core area. The third alloy 20Cr25Ni/TiNb with 1.7 % Ti was nitrided for 1 h at 1,423 K (1,150°C) in an atmosphere consisting of 95 % nitrogen and 5 % hydrogen, a finely dispersed TiN phase forming in the steel matrix. The mean grain size in the core was around ~ 100 µm and on the specimen surface about ~ 1 µm.

The tests were performed in carbon dioxide with different CO contents of 2 %, 10 %, 30 %, 75 % as well as in pure carbon monoxide at 923 K, 1,023 K and 1,123 K (650°C, 750°C and 850°C). The gas phase equilibria of both reactions according to

Equation 4 $2\,CO \leftrightarrow CO_2 + C$

Equation 5 $2\,CO + O_2 \leftrightarrow 2\,CO_2$

were used to calculate the partial pressures of oxygen p_{O2} and the carbon activities a_c for the various partial CO pressures and test temperatures as indicated in Table 11 according to Equation 6 and Equation 7.

Equation 6 $p_{O_2} = \dfrac{K_1 (p_{CO_2})^2}{p_{CO}}$

Equation 7 $a_C = \dfrac{K_2 (p_{CO})^2}{p_{CO_2}}$

The mass increase of the specimens as a function of the exposure duration and the CO content of the test gas is shown for the three steels in Figure 31, Figure 32 and Figure 33.

Following a short rise in the mass increase all steels reach an almost constant value in the gases with CO contents up to 30 %. At higher CO contents a further rise is observed over a longer test duration, which is highest in case of 20Cr25Ni/Nb steel and occurs in pure carbon monoxide only in case of 20Cr25Ni/TiNb steel.

Partial CO pressure bar	Temperature 923 K (650°C)		Temperature 1,023 K (750°C)		Temperature 1,123 K (850°C)	
	a_c	P_{O_2} / bar	a_c	P_{O_2} / bar	a_c	P_{O_2} / bar
0.02	1.29×10^{-3}	3.83×10^{-19}	1.46×10^{-4}	5.38×10^{-18}	2.45×10^{-5}	5.16×10^{-17}
0.10	3.51×10^{-2}	1.29×10^{-20}	3.97×10^{-3}	1.82×10^{-19}	6.67×10^{-4}	1.74×10^{-18}
0.30	4.07×10^{-1}	8.69×10^{-22}	4.59×10^{-2}	1.22×10^{-20}	7.73×10^{-3}	1.17×10^{-19}
0.75	7.12	1.77×10^{-23}	8.04×10^{-1}	2.49×10^{-22}	1.35×10^{-1}	2.34×10^{-21}
0.9998	1.58×10^{4}	6.39×10^{-30}	1.79×10^{3}	8.97×10^{-29}	3.00×10^{2}	8.60×10^{-28}

Table 11: Calculated carbon activities a_c and partial oxygen pressures P_{O_2} for various CO contents and test temperatures [36]

Figures 34 to 39 show the relevant curves of the mass increase as a function of the test duration and the gas composition for both test temperatures 1,023 K (750°C) and 1,123 K (850°C).

The curve at a test temperature of 1,023 K (750°C) is almost identical with that at 923 K (650°C), however the TiN-containing steel does not exhibiting any rapid rise of the mass increase in pure carbon monoxide at longer test durations either. A rise of the mass increase in pure carbon monoxide after lengthy test durations at 1,123 K (850°C) has not been observed for any of the three steels. However, the niobium-stabilized steel exhibited a clearly higher mass increase in 100 % CO compared to test gases with lower CO contents. Practically no influence of the CO content in the gas could be found in the nitrided steel at 1,123 K (850°C).

Examinations of the formed cover layers revealed that the mass increase is strongly influenced by the precipitation of carbon. In gases with a higher CO content the niobium-stabilized steel was particularly sensitive to carbon precipitations at all test temperatures, whereas the titanium stabilized steel and in particular the TiN-containing material exhibited a clearly better behavior. The differences in behavior can be explained by the different formation of chromium oxide layers on the surface. Steel 20Cr25Ni/TiN, for instance, forms a chromium oxide-rich cover layer under all test conditions, supported by the fine-grained structure in the surface area and inhibiting the precipitation of carbon. In contrast, more time and a higher temperature were necessary in case of the niobium-stabilized steel 20Cr25Ni/Nb to form a protective cover layer, attributed to its rather coarse-grained structure in the surface area. Regarding its behavior the steel 20Cr25Ni/Ti was ranking between the two other steels. The dependence of carbon precipitation on the CO content of the gas and the test temperature is interpreted as caused by the thermodynamics of the Boudouard reaction.

Due to damage to the cladding tubes of fuel assemblies in the nuclear reactors cooled with carbon dioxide gas, the steel with 20 % Cr, 25 % Ni and 1 % Nb used there was intensively examined. The time-related process of oxidation followed a logarithmic time law and, in general, the oxide layer exhibited again a two-layer

154 | *Carbonic Acid*

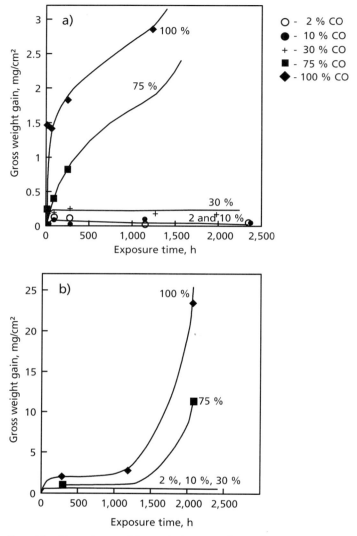

Figure 31: Mass change of the steel 20Cr25Ni/Nb as a function of the test duration and the CO concentration at 923 K (650°C) [36]

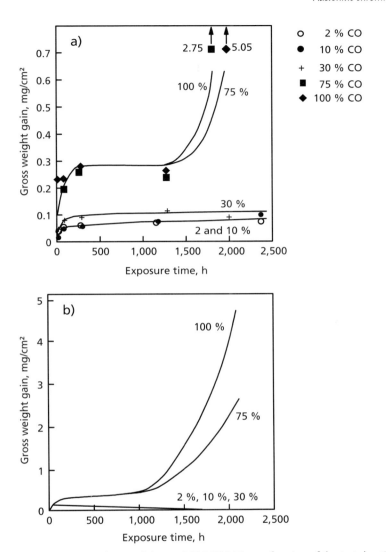

Figure 32: Mass change of the steel 20Cr25Ni/Ti as a function of the test duration and the CO concentration at 923 K (650°C) [36]

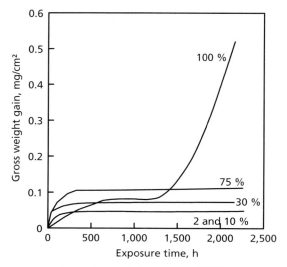

Figure 33: Mass change of the steel 20Cr25Ni/TiNb as a function of the test duration and the CO concentration at 923 K (650°C) [36]

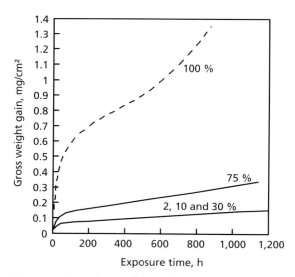

Figure 34: Mass change of the steel 20Cr25Ni/Nb as a function of the test duration and the CO concentration at 1,023 K (750°C) [36]

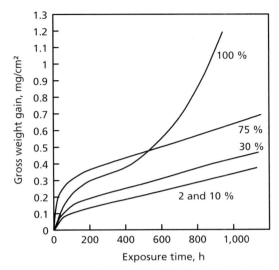

Figure 35: Mass change of 20Cr25Ni/Ti steel as a function of the test duration and the CO concentration at 1,023 K (750°C) [36]

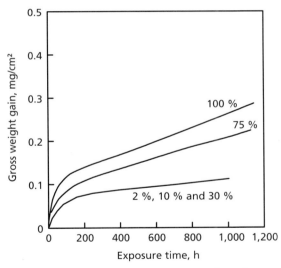

Figure 36: Mass change of 20Cr25Ni/TiN steel as a function of the test duration and the CO concentration at 1,023 K (750°C) [36]

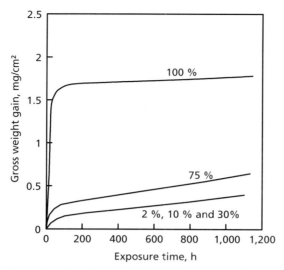

Figure 37: Mass change of 20Cr25Ni/Nb steel as a function of the test duration and the CO concentration at 1,123 K (850°C) [36]

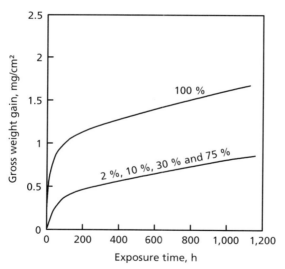

Figure 38: Mass change of 20Cr25Ni/Ti steel as a function of the test duration and the CO concentration at 1,123 K (850°C) [36]

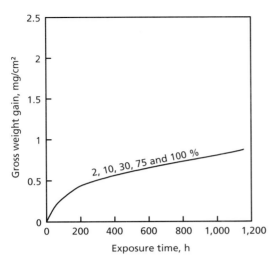

Figure 39: Mass change of 20Cr25Ni/TiN steel as a function of the test duration and the CO concentration at 1,123 K (850°C) [36]

structure. On the outer side there is an iron-rich spinel layer, whereas there is a Cr_2O_3 layer inside.

As shown in Figure 3 the austenitic 18-8 nickel chromium steels in pure carbon dioxide exhibit a good resistance to oxidation in a temperature range of up to about 803 K (530°C) – to a major extent irrespective of their chemical composition. However, at higher temperatures they are inferior to the steels with higher chromium and nickel contents.

Figure 40 and Figure 41 summarize the test results of the oxidation behavior of austenitic nickel chromium steels in CO-CO_2 mixtures [2].

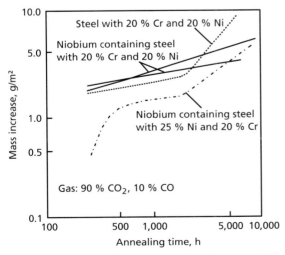

Figure 40: Time-related mass change of various austenitic nickel chromium steels in carbon dioxide with 10 % carbon monoxide at 1,023 K (750°C) [2]

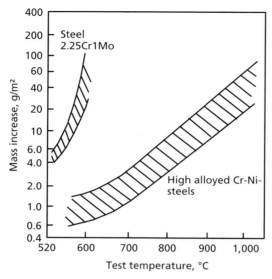

Figure 41: Mass change of various austenitic nickel chromium steels in carbon dioxide with 1 to 10 % carbon monoxide as a function of temperature. Test duration 1,000 hours [2]

Comparing Figure 40 and Figure 41 with Figure 3, the admixtures of carbon monoxide to carbon dioxide reduce the oxidation rates, in particular at lower temperatures. For 18-8 nickel chromium steels the time until the accelerated oxidation starts is clearly shifted to longer times due to the CO content of CO_2.

However, in case of steels with 20 % to 25 % Cr and 20 % to 30 % Ni only slightly lower mass changes were found in $CO-CO_2$ mixtures compared to pure carbon dioxide after a test duration of 10,000 h as shown in Table 12 [2].

Temperature K (°C)	Gas composition	Mass change g/m²
1,023 (750)	100 % CO_2	8 to 25
	90 % CO_2 + 10 % CO	3 to 20
1,173 (900)	100 % CO_2	30 to 100
	90 % CO_2 + 10 % CO	20 to 70

Table 12: Mass change of high-alloy austenitic nickel chromium steels in pure carbon dioxide and in carbon dioxide containing 10 % carbon monoxide at different temperatures [2]

In contrast to the low-alloy steels, austenitic nickel chromium steels show a clear influence of the flow rate of carbon dioxide even at lower rates at temperatures of 823 K (550°C) and 873 K (600°C) as shown in Figure 42 using the steel with the material No. 1.4541 (cf. UNS S32100) [2].

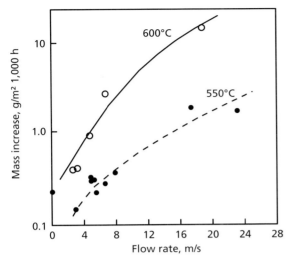

Figure 42: Dependence of the mass change of the steel with the material No. 1.4541 (cf. UNS S32100) in carbon dioxide on the flow rate of the gas [2]

[38] reports on investigations into the kinetics of high-temperature oxidation and the spalling of the cover layers of an Nb-stabilized 20% Cr-25% Ni steel in carbon dioxide at a pressure of 40 bar and temperatures of 1,123, 1,143, 1,183 and 1,223 K (850°C, 870°C, 910°C and 950°C). As the curves in Figure 43 show, the mass increase of the specimens follow a parabolic time law at all test temperatures. The activation energy is indicated to amount to 370 ± 36 kJ/mol.

The oxidation rate is determined by the diffusion of the internal chromium oxide layer of a two-layer oxide layer. An intercrystalline oxidation of silicon was observed beneath the oxide layer. The oxide layers spalled off at all temperatures and spalling increased linearly with the mass increase.

The high-temperature oxidation of a niobium-stabilized steel with 25% Ni, 20% Cr in carbon dioxide was examined gravimetrically and also with the help of thin layer activation [39]. During the thin layer activation a defined area of the steel surface was activated by exposure to deuteron radiation to generate radioisotopes of the elements chromium, manganese and cobalt. Measurement of the activity level of the radioisotopes before, during and after oxidation together with the gravimetrical determination of the spalled scale facilitates the assessment of the oxidation processes. The tests were performed under isothermal conditions with a test duration of up to 525 hours in a temperature range between 1,023 and 1,173 K (750°C and 900°C) and by subsequent cooling in the furnace. The flow rate of the carbon dioxide was 50 cm^3/min. The chemical composition of the examined steel is indicated in Table 13.

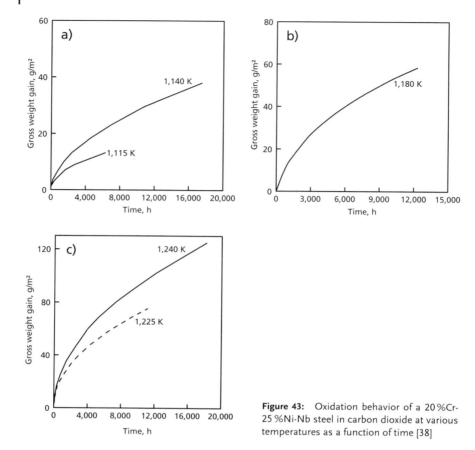

Figure 43: Oxidation behavior of a 20%Cr-25%Ni-Nb steel in carbon dioxide at various temperatures as a function of time [38]

C	Si	Mn	Cr	Ni	Nb	Co
0.054	0.62	0.74	19.6	24.7	0.61	0.007

Table 13: Chemical composition of the examined stainless steel [39]

These test results show that at 1,173 K (900°C) and 1,223 K (950°C) chromium and manganese contribute to the formation of a uniform, stable and protecting oxide layer in the same way. This layer consists of an external spinel layer of an Mn(FeNiCr)$_2$O$_3$ composition, a chromium oxide layer (Cr$_2$O$_3$) in the middle and an internal silicon oxide layer. During isothermal oxidation, adhesion of this protective layer is preserved. However, cracking and spalling of the oxide layer occurred after longer oxidation at 1,173 K (900°C).

Basically, oxidation processes may also impair the mechanical properties of a material. For instance, the load-bearing section may be reduced if it is uniformly attacked or the internal oxides may act as stress peaks. On the other hand, mechanical loads can also influence the oxidation process by causing defects in the protective oxide layers or causing such layers to spall off.

The investigations described in [40] analyzed the effect of oxidation in carbon dioxide at 1,273 K (1,000°C) on the grain growth and the long-term resistance of stainless 20Cr-25Ni-Nb steel at 1,023 K (750°C), on the one hand, and the effect of elongation under constant load on the oxidation of this steel under the same conditions on the other hand. Examinations were performed with two cold-rolled sheets of different thicknesses with the chemical compositions indicated in Table 14. Prior to the test start, the specimens were solution-annealed in a hydrogen atmosphere at 1,203 K (930°C) for 30 minutes and quenched in argon to room temperature.

Thickness	C	Si	Mn	P	S	Cr	Ni	B	N	Nb
0.38 mm	0.045	0.55	0.74	0.005	0.004	19.8	24.4	0.0004	0.005	0.61
0.60 mm	0.026	0.61	0.75	0.005	0.004	20.9	25.0	0.0004	0.005	0.59

Table 14: Chemical composition of the examined sheet specimens [40]

The specimens were oxidized in carbon dioxide at 1,273 K (1,000°C) and with an ambient pressure for an exposure duration up to 2,000 hours. The creep tests were performed with the oxidized specimens and, for comparison, with specimens subjected to heat treatment in air at 1,023 K (750°C) by applying a constant load of 46 N. Under the condition, that a cross section reduction did not occur during the oxidation, the initial load of the specimens was 69 MN/m^2. In another test series the specimens were oxidized under tensile stress in carbon dioxide at 1,273 K (1,000°C) for 100 hours. The initial load varied from 0 to 25.5 N according to the stresses of 0 to 11.0 MN/m^2. Table 15 contains the loads of the specimens and details regarding the oxide layers formed.

As results of the creep tests Figure 44 indicates the creep toughness and the time to fracture of the specimens.

Following the exposure to argon, the ductility (upper section of Figure 44) compared to the initial value of the reheated specimens up to 750 h slightly decreases and remains almost constant thereafter. The ductility of the specimens exposed to carbon dioxide, in contrast, clearly and continuously decreases as the test duration decreases. The lower section in Figure 44 shows that up to about 500 h the lifetime of the specimens stored in argon is slightly higher than the values of the specimens in delivery state, the values however slightly decreasing thereafter as the test duration increases. The specimens oxidized in carbon dioxide exhibited identical lifetimes compared to non-oxidized specimens up to 500 h. However, if the test duration is longer, a considerable drop of the lifetime can be observed.

If exposed to argon at 1,273 K (1,000°C) for 2,000 h, the grain size in the microstructure of the steel clearly grows, but obviously remains without any significant influence on the creep behavior. If exposed to carbon dioxide under the same conditions, the grain size does not grow as is attributable to the precipitation of carbides as a result of the oxidation reaction blocking the grain boundaries and thus preventing growth. But these carbide precipitations lead to embrittlement and therefore to a clear drop in the creep values.

Stress N	Stress MN/m²	Elongation %	Cover layer thickness µm	Depth of attack µm	Surface without cover layer %
0	0	0	15.5	30	0
0	0	0.5	15.5	43	0
10.5	4.1	3.9	16.6	55	27
12.1	4.9	5.1	14.2	54	50
8.9	3.8	5.2	15.5	55	0
14.5	5.6	9.4	16.6	69	0
22.8	9.9	13.5	16.6	66	60
20.8	8.1	19.7	14.2	70	0
19.2	7.5	19.8	14.2	70	0
24.2	10.5	27.5	14.2	66	8
25.5	11.0	42.5	14.7	75	19
25.5	11.0	51.2	14.7	74	14

Table 15: Load, initial stress and elongation, thickness of the protecting cover layer, depth of attack and share of the surface without protecting cover layer after 100 h oxidation of the specimens in carbon dioxide at 1,273 K (1,000°C) [40]

Since the specimens creep at a test temperature of 1,023 K (750°C), the specimen cross section changes during the test duration such that the effective stress increases. Table 16 depicts these changes in the test conditions.

Figure 45 shows the relationship between the effective stress and the time to rupture of the specimens starting from an initial stress of 46 N.

The elongation of specimens resulting from the combined effect of oxidation and loading was determined in the oxidation tests in carbon dioxide at 1,273 K (1,000°C) for 100 h and concurrent loading of the specimens. As shown in Figure 46 elongation clearly increases as the tensile stress increases.

Although the increasing elongation does not seem to impair the thickness of the uniform, chromium-rich and protecting outer cover layer, it increases the tendency to imperfections, leads to the formation of non-protecting cover layers and clearly increases the depth of the internal attack.

Chloride and sulfate-containing ash deposits resulting from corrosion by hot combustion gases in power plants can intensify the attack. Since biomass has higher contents of alkaline metals and chlorides, the components in contact with the exhaust gases of biomass-fueled power plants are endangered by ash deposits with high alkaline chloride contents. Therefore, the behavior of the ferritic and austenitic steels used in this area was investigated in simulated combustion gases with and without deposits within the framework of a research project [41].

Austenitic chromium-nickel steels | 165

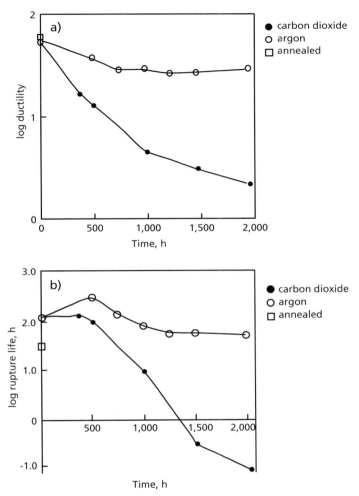

Figure 44: Results of the creep tests (load: 103 MN/m^2) at 1,023 K (750°C) in air performed on specimens following an exposure in carbon dioxide (●) or in argon (○) at 1,273 K (1,000°C) [40]

Oxidation period h	Effective specimen cross section mm^2	Effective stress MN/m^2
0	0.67	69
500	0.50	93
1,000	0.51	90
1,500	0.47	97
2,000	0.45	103

Table 16: Effective load of the specimens oxidized in carbon dioxide at 1,273 K (1,000°C) during the creep test at 1,023 K (750°C) [40]

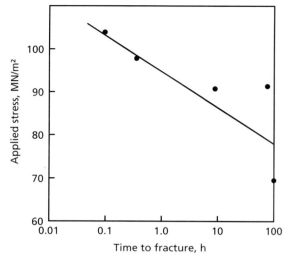

Figure 45: Relationship between the effective stress and the time to rupture of the specimens oxidized in carbon dioxide at 1,273 K (1,000°C) during the creep test with an initial stress of 46 N [40]

The composition of the investigated steels is shown in Table 17. The test atmosphere contained 22% H_2O + 5% O_2 + 0 to 25% CO_2 rest N_2. The exposure tests were performed at a temperature of 808 K (535°C) and the test duration was 360 hours.

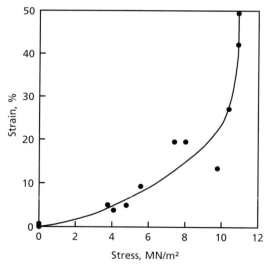

Figure 46: Influence of tensile stress on the elongation of specimens during oxidation in carbon dioxide at 1,273 K (1,000°C) for 100 h [40]

The results are summarized in Figure 47, Figure 48 and Figure 49. Figure 47 and Figure 49 show that ash deposits clearly increase the corrosive attack both in a CO_2-free and in a CO_2-containing atmosphere. Increasing CO_2 contents in the gas atmosphere intensify the corrosive attack irrespective of ash deposits. The behavior of the austenitic steels was clearly better than that of the ferritic steels in all tests.

Steel designation	Cr	Ni	Mo	V	C	Si	Mn	P	S	Nb
2.25Cr1Mo	2.29	0.44	0.96	0.01	0.09	0.23	0.59	0.02	0.02	0.02
X20	10.3	0.72	0.87	0.26	0.18	0.23	0.62	0.02	0.01	
X10	8.7	0.26	0.97	0.23	0.10	0.38	0.48	0.01	0.01	0.07
AC66	27.3	32.2			0.06	0.21	0.64	0.01	0.01	0.78
TP347H	17.6	10.7			0.05	0.29	1.84	0.03	0.01	0.56
Esshete®1250	15.0	9.65	0.94	0.22	0.08	0.58	6.25	0.02	0.01	0.86
Sanicro® 28	26.7	30.6	3.32		0.02	0.42	1.73	0.02	0.01	

Table 17: Chemical composition of the investigated steels [41]

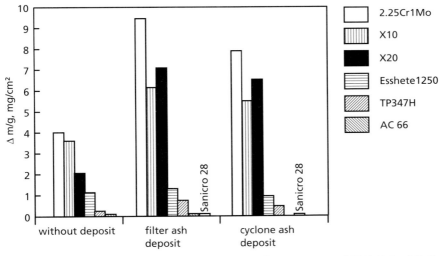

Figure 47: Mass change of the specimens in an atmosphere consisting of 22 % H_2O + 5 % O_2 + N_2 with and without ash deposits after 360 h [41]

Carbonic Acid

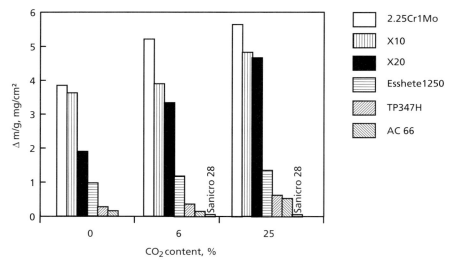

Figure 48: Mass change of the specimens in an atmosphere consisting of 22% H_2O +5% O_2 +x% CO_2 + N_2 without ash deposits after 360 h [41]

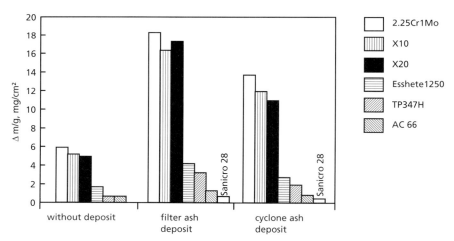

Figure 49: Mass change of the specimens in an atmosphere consisting of 22% H_2O +5% O_2 + 25% CO_2 + N_2 with and without ash deposits after 360 h [41]

Behavior in aqueous carbon dioxide-containing media

The austenitic stainless Cr-Ni steels exhibit a generally good resistance in aqueous carbon dioxide containing solutions with corrosion rates below 0.01 mm/a (0.40 mpy) [42, 43].

Results of corrosion tests with Cr-Ni steels of different alloy contents in carbonic aqueous solution with a partial pressure of CO_2 of 32.4 bar at 295 K (22°C) are indicated in Table 18 [44].

Steel	Cr, %	Ni, %	Si, %	Corrosion, mm/a (mpy)
1	16.8	7.9		0.0025 (0.10)
2	18.2	8.6	2.2	0.0018 (0.07)
3	17.0	23.7	2.9	0.0015 (0.06)

Table 18: Results of corrosion tests with Cr-Ni steels in carbon dioxide-containing solution [44]

Thus, the materials are resistant under the indicated corrosion conditions.

Austenitic CrNiMo(N) steels
Austenitic CrNiMoCu(N) steels

Also the molybdenum-containing austenitic CrNi steels of grade SAE 316 (material No. 1.4404) with about 18 % Cr, 12 % Ni and 2.5 % Mo form cover layers with a multilayer structure in carbon dioxide gas or in gas mixtures of CO_2 and CO at higher temperatures, consisting of Cr_2O_3, Fe_3O_4 and Fe-Cr-Ni spinels depending on the temperature and the gas composition [45].

Occasionally, damage caused by cracks is found in these steels in the temperature range of 773 to 973 K (500°C to 700°C), where oxidation processes play a role. To simulate this crack formation in the lab and examine the influence of the material as well as mechanical and chemical conditions, creep tests and slow strain rate tensile tests were performed with molybdenum-containing and molybdenum-free CrNi steels in different gas atmospheres at 883 K (610°C) [46].

The chemical composition of the investigated steels is shown in Table 19. Four commercial steels (304, 304 H and 316 L, 316 H) with different contents of molybdenum and carbon were used.

The steel 308 H was melted with an elevated manganese content to reach a stable austenitic condition.

To investigate the effect of a grain boundary sensitization by precipitations of chromium carbide ($Cr_{23}C_6$), 304 H steel was subjected to various heat treatment processes to vary the grain size and precipitations (Table 20). As the check for intercrystalline corrosion according to ASTM A 262 Practice B in boiling copper sulfate/sulfuric acid shows, all specimens were sensitized.

In addition, two welding materials were tested, i.e. 308 H steel on the one hand and a hard surfacing layer on the other hand, with 309 L steel as the base material and a cover layer of the unalloyed material A508C13, the contamination of the stainless steel 309 with A508C13 being adjusted to about 30 %.

Steel	C	Si	Mn	Cr	Ni	Mo
SAE 304	0.04	0.38	1.20	17.8	8.1	0.17
SAE 304 H	0.05	0.36	1.43	18.3	8.6	–
SAE 304 H	0.077	0.30	0.98	18.6	8.5	0.30
SAE 316 L	0.024	–	–	17.1	13.3	2.65
SAE 316 H	0.053	0.47	1.74	16.8	11.0	2.30
SAE 308 H	0.11	0.80	6.60	17.9	8.3	0.10
SAE 309 L	0.02	0.6	1.30	23.5	12.5	0.04
A508C13	0.16	0.2	1.30	0.2	0.7	0.5
X	0.06	0.4	1.3	16.0	9.0	0.15

X = mixed layer of SAE 309 L and A508C13

Table 19: Chemical composition of the investigated steels [46]

The tests were performed in vacuum (2×10^{-3} Pa), in air and in a gas mixture of air with 4 % carbon dioxide and 5.8 % or 8 % water vapor.

Creep tests were performed with round test bars of 6 mm in diameter consisting of 304 H steel on lever arm machines and slow strain rate tensile tests were performed with round test bars of 5 mm in diameter at strain rates of 3×10^{-8}, 10^{-7} and 10^{-6} 1/s for all steels.

Designation	Heat treatment	Grain size (ASTM)
G1	state as delivered sensitized at 973 K (700°C) for 30 min	5
G2	state as delivered annealed at 1,343 K (1,070°C) for 1 hour 5 % cold formed sensitized at 973 K (700°C) for 30 min	¾ to 2
G3	state as delivered sensitized at 883 K (610°C) for 15 h	5

Table 20: Heat treatment and grain sizes of steel 304 H [46]

The results of the creep tests are summarized in Table 21. Cracks did not occur in air during a test period of 10 h. The extension can be used to assess the creep strength. As expected, the material with the lower grain size is more sensitive than the coarse-grained material.

The microscopic test and the EDX analysis revealed that the cracks formed in the gas mixture during the test are covered by a black oxide layer with a thickness of around 5 μm mainly consisting of iron oxide, probably of magnetite Fe_3O_4. This oxide composition is unusual in a stainless austenitic steel where usually chromium-rich oxides are found.

Designation	Stress MPa	Test medium	Extension %	Crack length μm
G1	200	air	6.4	0
G2	200	air	0.9	0
G1	250	air	15.5	0
G2	250	air	5.4	0
G1	300	air	56	0
G2	300	air	46	0
G1	250	air/CO_2/H_2O	17.3	50
G3	250	air/CO_2/H_2O	23.0	70

Air with 4 % carbon dioxide and 5.8 % water vapor

Table 21: Results of creep tests [46]

The results of the slow strain rate tensile tests are summarized in Table 22. All cracks in the base materials were merely intercrystalline and interdendritic in the weld specimens. The sensitivity to cracking increases if the strain rate decreases from 10^{-6} 1/s to 10^{-7} 1/s.

Steel	Air + 4 % CO_2 + 8 % H_2O			Air	Vacuum
	Strain rate				
	10^{-6} 1/s	10^{-7} 1/s	3×10^{-8} 1/s	10^{-7} 1/s	10^{-7} 1/s
316 L – 0.024 %C	10	50			
316 H – 0.053 %C		20	50		
304 H – 0.040 %C	10	50	50		
304 H – 0.077 %C		300			
308 H – 0.11 %C	100	750		50	< 10
Hard surfacing layer 1		5,000			< 10
Hard surfacing layer 2		5,000			

Table 22: Results of slow strain rate tensile tests: Length of the main crack in μm [46]

The higher sensitivity of the hard surfacing layer is attributed to the higher carbon content and the dendritic structure with a higher grain size. The cracks exclusively occurred in the high-alloy part of the hard surfacing layer.

Bibliography

[1] Rahmel, A.; Schwenk, W.
Korrosion und Korrosionsschutz von Stählen, 1. Aufl.
Verlag Chemie GmbH, Weinheim, 1977, pp. 265–267

[2] Autorenkollektiv
Prüfung und Untersuchung der Korrosionsbeständigkeit von Stählen
Verlag Stahleisen, Düsseldorf, 1973

[3] Chengyu Yan; Oeters, F.
Kinetics of iron oxidation with CO_2 between 1300 and 1450°C
Steel Research 65 (1994) 9, pp. 355–361

[4] Bredesen, R.; Kofstad, P.
On the oxidation of iron in $CO_2 + CO$ mixtures: II.
Reaction Mechanisms During Initial Oxidation
Oxidation of Metals 35 (1991) 1–2, pp. 107–137

[5] Bredesen, R.; Kofstad, P.
On the oxidation of iron in $CO_2 + CO$ mixtures: III.
Coupled Linear Parabolic Kinetics
Oxidation of Metals 36 (1991) 1–2, pp. 27–56

[6] Schwenk, W.
Korrosion von unlegiertem Stahl in sauerstofffreier Kohlensäurelösung
Werkstoffe und Korrosion 25 (1974) 9, pp. 643–646

[7] De Waard, C.; Milliams, D. E.
Carbonic acid corrosion of steel
Corrosion (Houston) 31 (1975) 5, pp. 177–181

[8] Schmitt, G.; Rothmann, B.
Untersuchungen zum Korrosionsmechanismus von unlegiertem Stahl in sauerstofffreien Kohlensäurelösungen – Teil I. Kinetik der Wasserstoffabscheidung
Werkstoffe und Korrosion 28 (1977) 12, pp. 816–822

[9] Schmitt, G.; Rothmann, B.
Untersuchungen zum Korrosionsmechanismus von unlegiertem Stahl in sauerstofffreien Kohlensäurelösungen – Teil II. Kinetik der Eisenauflösung
Werkstoffe und Korrosion 29 (1978) 2, pp. 98–100

[10] Schmitt, G.; Rothmann, B.
Zum Korrosionsverhalten von unlegierten und niedriglegierten Stählen in Kohlensäurelösungen
Werkstoffe und Korrosion 29 (1978) 4, pp. 237–245

[11] Schmitt, G.
NACE Corrosion 83
Paper 43
Houston, TX, USA, 1983

[12] McIntrire, G.; Lippert, J.; Yudelson, J.
The effect of dissolved CO_2 and O_2 on the corrosion of iron
Corrosion 46 (1990) 2, pp. 91–95

[13] Palacios, C. A.; Shadley, J. R.
Characteristics of corrosion scales on steel in CO_2 saturated NaCl brine
Corrosion 47 (1991) 2, pp. 122–127

[14] Jasinski, R.
Corrosion of N80-type steel by CO_2/water mixtures
Corrosion 43 (1987) 4, pp. 214–218

[15] Oblonsky, L. J.; Devine, T. M.
Corrosion of carbon steels in CO_2 saturated brine. A surface enhanced Raman spectroscopy study
Journal of electrochemical society 144 (1997) 4, pp. 1252–1260

[16] Cheng, Y. F.
Corrosion of X-65 pipeline steel in carbon dioxide-containing solutions
Bulletin of Electrochemistry 21 (2005) 11, pp. 503–511

[17] Schmitt, G.
Zur Frage der Korrosivität kohlensaurer Kondensate in Gasverteilungssystemen
GWF, Gas- Wasserfach: Gas/Erdgas 122 (1981) 2, pp. 49–54

[18] Carvalho, D. S.; Joia, C. J. B.; Mattos, O. R.
Corrosion rate of iron and iron-chromium alloys in CO_2 medium
Corrosion Science 47 (2005), pp. 2974–2986

[19] Okafor, P. C.; Nesic, S.
Effect of acetic acid on CO_2 corrosion of carbon steel in vapor-water two-phase horizontal flow
Chem. Eng. Comm. 194 (2007), pp. 141–157

[20] George, K. S.; Nesic, S:
Investigation of carbon dioxide corrosion of mild steel in the presence of acidic acid
Corrosion 63 (2007) 2, pp. 178–186

[21] Nesic, S.; Lunde, L.
Carbon dioxide corrosion of carbon steel in two-phase flow
Corrosion (Houston) 50 (1994) 9, pp. 717–727

[22] Nesic, S.; Solvi, G.T.; Enerhaug, J.
Comparison of the rotating cylinder and pipe flow tests for flow-sensitive carbon dioxide corrosion
Corrosion 51 (1995) 10, pp. 773–787

[23] Nesic, S.; Postlethwaite, J.; Olsen, S.
An electrochemical model for prediction of corrosion of mild steel in aqueous carbon dioxide solutions
Corrosion (Houston) 52 (1996) 4, pp. 280–294

[24] Amri, J.; Gulbrandsen, E.; Nogueira, R. P.
The effect of acetic acid on the pit propagation in CO_2 corrosion of carbon steel
Electrochemistry Communications 10 (2008), pp. 200–203

[25] Schmitt, G; Kunze, E.
Spannungsrißkorrosion niedriglegierter Vergütungsstähle in CO_2-haltigen wäßrigen Angriffsmitteln
in: Gräfen, H.; Rahmel, A.
Korrosion verstehen – Korrosionsschäden vermeiden, 1. Aufl., Bd. 1
Verlag Irene Kuron, Bonn, 1994, pp. 86–92

[26] Schmitt, G.
Untersuchungen zur Gefährdung hochfester Stähle durch wasserstoffinduzierte Spannungsrißkorrosion in Kohlendioxid enthaltenden Kondensaten
Werkstoffe und Korrosion 34 (1983) 4, pp. 187–198

[27] Gräfen, H.; Schlecker, H.
Spannungsrißkorrosion unlegierter und niedriglegierter Stähle in $CO/CO_2/H_2O/O_2$-haltigen Wässern
in: Gräfen, H.; Rahmel, A.
Korrosion verstehen – Korrosionsschäden vermeiden, 1. Aufl., Bd. 1
Verlag Irene Kuron, Bonn, 1994, pp. 93–97

[28] Yanateva, O. K.
Bulletin of the Academy of Science of the USSR, Division of Chemical Science (English Translation) (1954) pp. 977–978

[29] Asahi, H.; Hara, T.; Sakamoto, S.
Corrosion properties and application limit of sour resistant 13 % Cr steel tubing with improved CO_2 corrosion resistance
Conference: EUROCORR 97, Vol. 1
Trondheim. Norway, 22–25 Sept. 1997

[30] Yamamoto, K.; Kagawa, N.
Ferritic stainless steels have improved resistance to SCC in chemical plant environments
Materials Performance 20 (1981) 6, pp. 32–37

[31] Lotz, U.; Schollmaier, M.; Heitz, E.
Flow-dependent corrosion. Ferrous materials in pure and particulate chloride solutions
Werkstoffe und Korrosion 36 (1985) 4, pp. 163–173

[32] Schollmaier, M.
Bericht zur Dechema-Jahrestagung 1984
4.3 Korrosions- und Verschleißerscheinungen in Zweiphasenströmungen
Werkstoffe und Korrosion 35 (1984), p. 429

[33] Lotz, U.; Schollmaier, M.; Heitz, E.
Flow-dependent corrosion – II. Ferrous materials in pure and particulate chloride solutions
Werkstoffe und Korrosion 36 (1985) 4, pp. 163–173

[34] Kohley, T.; Blatt, W.; Heitz, E.
Erosionskorrosion in Mehrphasensystemen der Offshore- und anderer Produktionstechniken – Teil C: Korrosionschemie und -schutzuntersuchungen unter Niederdruckbedingungen,
unveröffentlichtes Manuskript
Forschungsstelle: Dechema-Institut, Frankfurt/Main

[35] Kimura, M.; Miyata, Y.; Yamane, Y.; Toyooka, T.; Nakano, Y.; Murase, F.
Corrosion resistance of high-strength modified 13 % Cr steel
Corrosion (Houston) 55 (1999) 8, pp. 756–761

[36] Holm, R. A.; Evans, H. E.
The resistance of 20Cr/25Ni steels to carbon deposition. I. The role of surface grain size
Werkstoffe und Korrosion 38 (1987) 3, pp. 115–124

[37] Holm, R. A.; Evans, H. E.
The resistance of 20Cr/25Ni steels to carbon deposition. II. Internal oxidation and carburisation
Werkstoffe und Korrosion 38 (1987) 4, pp. 166–175

[38] Emsley, A. M.; Hill, M. P.
High-temperature oxidation of a 20/25 stainless steel in high-pressure carbon dioxide
Oxidation of Metals 33 (1990) 3/4, pp. 265–278

[39] Asher, J.; Sugden, S.; Benett, M. J.; Hawes, R. W. M.; Savage, D. J.; Price, J. B.
An investigation of the high temperature oxidation of a 20 % Cr/25 % Ni/Nb stainless steel in carbon dioxide using thin layer activation
Werkstoffe und Korrosion 38 (1987) 9, pp. 506–516

[40] Bennett, M. J.; Roberts, A. C.; Spindler, M. W.; Wells, D. H.
Interaction between oxidation and mechanical properties of 20Cr-25Ni-Nb stabilised stainless steel
Materials Science and Technology 6 (1990) 1, pp. 56–68

[41] Sroda, S.; Mäkipää, M.; Cha, S.; Spiegel, M.
The effect of ash deposition on corrosion behaviour of boiler steels in simulated combustion atmospheres containing carbon dioxide (CORBI PROJECT)
Materials and Corrosion 57 (2006) 2, pp. 176–181

[42] Montrone, E. D.; Long, W. P.
Choosing materials for CO_2 absorption systems
Chemical Engineering 78 (1971) 2, pp. 94, 96, 98–99

[43] Hamner, E.
Corrosion data survey, Fifth edition, 1974 NACE, Houston (Texas)

[44] Newton, L. E.; Hausler, R. H.
CO_2-Corrosion in oil and gas production
Selected papers, abstracts and references 1984 National Association of Corrosion Engineers, 1940 South Creek Drive, Houston, Texas, 77084

[45] Smith, A. F.
The duplex oxidation of vacuum annealed 316 stainless steel in CO_2/CO gas mixtures between 500 and 700°C
Corrosion Science 24 (1984) 7, pp. 629–643

[46] Le Calvar, M.; Scott, M. P.; Magnin, T.; Rieux , P.
Strain oxidation cracking of austenitic stainless steels at 610°C
Corrosion (Houston) 54 (1998) 2, pp. 101–105

Formic Acid

Unalloyed steels and cast steel
Unalloyed cast iron

Unalloyed steel dissolves in formic acid with evolution of hydrogen [1]. Corrosion increases with increasing temperature: The corrosion rate of carbon steel after exposure for 24 hours to 4.6 % formic acid at 443 K (170°C) in a sealed Pyrex® glass tube evacuated prior to the beginning of the test was about 30.5 mm/a (1,201 mpy) [2].

An exception to the use of non-alloy steel in contact with formic acid is in the manufacture of formic acid from formamide, water and sulfuric acid by the BASF process. In this process, hot mash already largely converted into formic acid and ammonium sulfate passes from a stirred vessel with protective lining, to which formamide and about 70 % sulfuric acid is continuously fed, into an externally heated iron rotary kiln. The formic acid formed in the stirred vessel is continuously drawn off from the rotary kiln together with the formic acid vapors produced there and is then fractionally distilled in a distillation column made of CrNiMo-steel. Hastelloy® B was used for oxygen-free zones of the column. Dry ammonium sulfate is discharged from the rotary kiln. Silver, glass, synthetic carbon or, in the case of highly-concentrated chloride-free acid, pure aluminium may be used as material of construction for the cooler.

Since formic acid remains in the rotary kiln for only a short time under this mode of operation, unalloyed steel can be used under these special conditions [3].

Gray cast iron is unsuitable as material for use in contact with formic acid. According to the results of an operational test, after exposure to a mixture of 50 % acetic acid and 90 % formic acid (6:1) at 373–377 K (100–104°C) for 180 hours the corrosion rate was 94.5 mm/a (3,720 mpy) [4].

High-alloy cast iron, high-silicon cast iron

The corrosion behavior of silicon castings containing 14 to 16 % silicon in formic acid is shown in Figure 1 [5].

The iso-corrosion curve, showing a corrosion rate 0.1 mm/a (3.94 mpy), on exposure to formic acid in the concentration range from zero to 100 % and temperatures up to 373 K (100°C) for silicon cast iron containing 14 to 16 % Si is in line with the data from a number of manufacturers. Addition of about 3 % molybdenum extends the range of resistance.

The resistance of the silicon castings to corrosion in acids is attributed to the presence of a Fe_3Si compound and to the formation of a protective coating of SiO_2 [6].

A corrosion rate of 7.4 mm/a (291 mpy) was determined after exposure to 4.6 % formic acid at 443 K (170°C) of silicon castings containing 14.5 % Si (Duriron®) in a sealed, evacuated Pyrex® glass tube [2].

Formic Acid

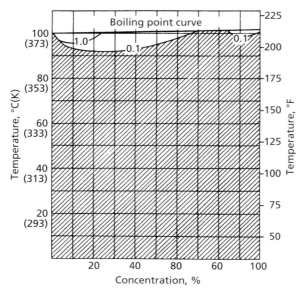

Figure 1: Iso-corrosion curves for silicon cast iron in formic acid [5]

The use of silicon cast iron in contact with formic acid above the relevant boiling points is not recommended.

Silicon castings containing about 14 to 16 % silicon without further alloying additions are known under the following trade names: Antacid®, Duracid®, Noricid®, Siferrid® (FRG), Duriron® (USA). Silicon contents of 15 to 18 %, e.g. in Thermisilid® (FRG), give improved resistance to acids at the expense of mechanical properties. Trade names for silicon castings containing about 3 % molybdenum include: Duracid®, Noricid® B (FRG) and Durichlor® (USA). Silicon castings are used for pumps, reaction vessels and valves. Corrosion resistance to acids, including the organic acids, almost reaches a maximum with a silicon content of 14 %. However, the mechanical-technological properties deteriorate with increasing percentage silicon content [6].

Structural steels with up to 12 % chromium
Ferritic chromium steels with more than 12 % chromium

Conventional chromium steels containing about 13 % and 17 % chromium, e.g. X12Cr13 (1.4006) and X6Cr17 (1.4016) corresponding to SAE 410 and SAE 430, are severely corroded by formic acid even at room temperature in almost the entire range of concentrations of the acid. Corrosion increases considerably with rising temperature [5]. Additions of molybdenum have a favorable effect on the corrosion behavior; a molybdenum content of 2 %, however, proved not to be adequate without other alloying elements [7].

Electrochemical investigations to determine the influence of increasing molybdenum contents, of nil, 2%, 3%, 4% Mo, on the corrosion behavior of 17% ferritic chromium steels in contact with 20% formic acid at 343 K (70°C) were carried out in an aerated and non-aerated medium. The tests were performed by recording polarization vs. time curves in combination with determinations of weight loss by analytical determination of the metal ions transferred into the medium. A 17% chromium steel without molybdenum and a 17% Cr steel with a molybdenum content of 2% (Fe-17 Cr-2 Mo) proved to be not resistant; the material consumption rate was 33,048 mg/dm^2 d for the 17% Cr-steel and 7,500 mg/dm^2 d for the chromium steel containing 2% Mo in the deaerated medium (0.1 ppm residual O_2) and 32,640 mg/dm^2 d and accordingly 10,800 mg/dm^2 d when saturated with oxygen (20 ppm O_2).

By activating the steel with a 2% molybdenum content, giving Fe-17Cr-2Mo before recording the polarization curves the existence of a metastable region was established which would not have been observed if the loss had been determined by weighing, and this would have been consonant with an overvaluation of the corrosion resistance of this type of chromium steel.

17% chromium steels containing 3% and 4% molybdenum exhibited resistance to corrosion in these investigations. The material consumption rates with a molybdenum content of 3% were 65 mg/dm^2 d and with 4% molybdenum content 57 mg/dm^2 d in the non-aerated medium and 0.8 mg/dm^2 d in the oxygen-saturated formic acid [7].

"Superferrites" with improved corrosion resistance were developed by increasing the chromium content to more than 25% and other metallurgical steps. Table 1 shows the behavior of the purely ferritic chromium steel X1CrNiMoNb28-4-2 (Remanit® 4575, (%) Fe-0.08C-0.18Si-0.14Mn-27.6Cr-3.49Ni-2.05Mo-0.45Nb; cf. UNS S32803) in boiling 20% formic acid in comparison with the ferritic chromium steels X8Cr18 (1.4015, cf. UNS S43080), X6CrMo17-1 (1.4113, cf. SAE 434) and the ferritic-austenitic steel X3CrNiMoN27-5-2 (1.4460, cf. SAE 329).

Chromium steel	Material consumption rate per unit area	Corrosion rate
	g/m^2 h	mm/a (mpy)
X1CrNiMoNb28-4-2 (Remanit® 4575)	0.1	0.12 (4.72)
X3CrNiMoN27-5-2 cf. SAE 329	0.1–1.0	0.12–1.23 (4.72–48.4)
X6CrMo17-1 cf. SAE 434	1.0–10.0	1.23–12.27 (48.4–483)
X8Cr18 cf. UNS S43080	> 10.0	> 12.27 (> 483)

Table 1: Corrosion behavior of chromium steels in 20% boiling formic acid [8]

Metallurgical measures led to a high-chromium ferritic steel (Fe-26Cr-1Mo, UNS S44627, trade name E-Brite® 26-1, see Table 2 for chemical composition) with improved weldability and toughness as well as increased corrosion resistance [9]. The corrosion resistance of E-Brite® 26-1 to boiling formic acid is shown by the following test example (Table 3) (comparison with the CrNiMo-steel SAE 316, corresponding to X 5 CrNiMo 18 10 with the DIN-Material No. 1.4401).

Chemical elements	Composition, %	ASTM Stand. (%) ASTM (Grade XM-217)
Carbon	0.001	0.01 max.
Nitrogen	0.010	0.015 max.
Chromium	26.0	25.0/27.5
Molybdenum	1.0	0.75/1.50
Silicon	0.25	0.40 max.
Manganese	0.01	0.40 max.
Sulfur	0.01	0.02 max.
Phosphorus	0.01	0.02 max.
Nickel + copper	0.12	0.50 max.
Copper	0.05	0.20 max.
Iron	balance	balance

Table 2: Composition of E-Brite® 26-1 [9]

Formic acid, %	E-Brite® 26-1	SAE 316
	Corrosion rates, mm/a (mpy)	
30	< 0.025 (< 0.98)	0.61 (24.0)
70	0.025 (0.98)	0.25 (9.84)

Table 3: Corrosion rates of E-Brite® 26-1 in comparison with SAE 316 in 30 and 70% boiling formic acid [9]

The improved ferritic chromium steels such as Remanit® 4575 and E-Brite® 26-1 are significantly superior in corrosion behavior against formic acid to the conventional ferritic chromium steels such as X6Cr13, X6Cr17 (with the DIN-Material Nos. 1.4000, cf. SAE 410 S and 1.4016, cf. SAE 430) [8, 9]. The available information on the concentrations of the formic acid and the test/operating temperatures do not cover the entire ranges of concentrations and temperature. Before using these steels, therefore, it is recommended that corrosion tests be carried out under the operating conditions [8, 9].

Ferritic-austenitic steels with more than 12% chromium

The development of chromium-nickel-molybdenum steels with mixed structure, varying from austenite containing only little ferrite to ferrites containing little austenite, was aimed at improved corrosion resistance to formic acid, increased resistance to stress corrosion cracking and to inter-granular corrosion. According to the results of a 7 day test, a steel in this group, Sandvik® 2 RN 65 (%) Fe-max. 0.02C-24Ni-17.5Cr-4.7Mo, exhibits better resistance to corrosion by formic acid in various concentrations (see Table 4) than the austenitic CrNiMo-steel SAE 316 L (X2CrNiMo17-12-2 or DIN Material No. 1.4404 with the following percentage composition: Fe-max. 0.03C-13.6Ni-17.6Cr-2.8Mo) [10].

Steel	Formic acid, %			
	5	10	25	90
	Corrosion rates, mm/a (mpy)			
Sandvik® 2 RN 65	0.05 (1.97)	0.1 (3.94)	0.25 (9.84)	0.3 (11.8)
SAE 316 L	0.15 (5.91)	0.3 (11.8)	0.57 (22.4)	1.9 (74.8)

Table 4: Corrosion rates of Sandvik® 2 RN 65 and SAE 316 L at 373 K (100°C) in formic acid of various concentrations (test duration 7 days) [10]

The corrosion behavior of a ferritic-austenitic steel, UHB® 44 LN, containing about 50% austenite in boiling formic acid in comparison with the austenitic chromium-nickel-molybdenum steel SAE 316 L (X2CrNiMo17-12-2, DIN Material No. 1.4404) is shown in Table 5 [11].

The chemical compositions of the steels UHB® 44 LN and SAE 316 L are shown in Table 6 [11].

Medium	Concentration, %	Corrosion rates, mm/a (mpy)	
		UHB 44 LN	SAE 316 L
Formic acid	30	0.2 (7.87)	0.5 (19.7)
Formic acid	100	0.1 (3.94)	0.9 (35.4)
Formic acid + acetic acid	50 / 50	0.2 (7.87)	0.1 (3.94)
Formic acid + acetic acid	70 / 30	0.2 (7.87)	0.1 (3.94)
Formic acid + acetic acid	90 / 10	0.1 (3.94)	0.8 (31.5)
Formic acid + acetic acid containing 0.1% Cl⁻	5 / 95	1.9 (74.8)	0.3 (11.8)

Table 5: Test of the corrosion behavior of UHB® 44 LN and SAE 316 L in boiling formic acid and in mixtures of formic and acetic acids [11]

Steel	Chemical composition, %						
	$C_{max.}$	Si	Mn	Cr	Ni	Mo	Other
UHB44LN®	0.030	0.4	1.7	25	6.2	1.7	N
SAE 316 L	0.03	1.0	2.0	17.6	13.6	2.8	–

Table 6: Compositions of UHB® 44 LN and SAE 316 L [11]

Austenitic chromium-nickel steels
Austenitic chromium-nickel-molybdenum steels
Austenitic chromium-nickel steels with special alloying additions

Chromium-nickel steels of the type 18 8 can be used in formic acid only at room temperature [3, 5]. The corrosion resistance of higher-alloyed CrNiMo-steels increases, although not for all concentration ranges (see Figure 2) [5].

Figure 2a shows the behavior of a CrNi-steel 18 8 in formic acid in the concentration range from 0 to 100% and temperatures from 273 K (0°C) to 373 K (100°C) or up to the boiling point, characterized by a broad range of inadequate corrosion resistance even at room temperature. The iso-corrosion curve of 10 mm/a (394 mpy) is exceeded in a formic acid concentration of about 50 to 70% at a temperature of 353 K (80°C) [5].

Figure 2b shows the iso-corrosion curve for the corrosion rate of 0.1 mm/a (3.94 mpy). From a formic acid concentration of 20% and higher, an increased corrosion rate with a resistance minimum in 60% formic acid between 303 K (30°C) and 313 K (40°C) can be discerned. The diagram is applicable to the higher-alloyed NiCrMoCu-steel containing Fe, 0.05% C, 20% Cr, 25% Ni, 3% Mo, 2% Cu [5].

Table 7 below contains the results of corrosion tests with CrNi- and CrNiMo-steels after exposure for 48 hours to boiling formic acid with concentrations of 10%, 50% and 90% with and without aeration [12].

The steels investigated are equivalent to the German steels listed in Table 8.

The appearance of the partially very much scattered values for the corrosion rate is probably due to variations in the test conditions as well as to differences in aeration and deaeration.

Table 9 shows the corrosion rates in mm/a (mpy) due to exposure to 50% boiling formic acid of various CrNiMoN-, NiCrMo-, CrNiMo- and CrNiMoCu-steels when air is excluded by gassing the test specimens with "special" high-purity nitrogen (99.996% N_2, test duration 14 days) [13].

Only the first three steels in this list (X1NiCrMoCu25-20-5, X4NiCrMoCuNb20-18-2 and X1NiCrMoCuN25-20-7) can be described as resistant and usable in the oxygen-free, reducing medium and in the formic acid concentration of 50% which is critical in respect of corrosive affect [13].

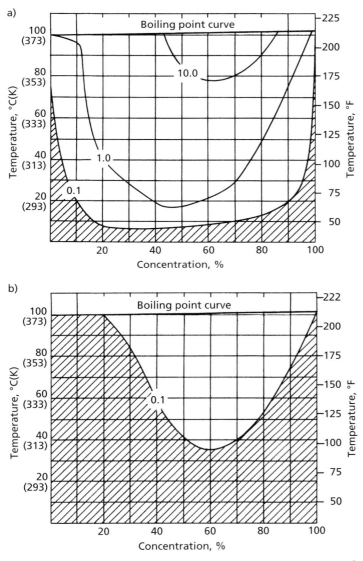

Figure 2: Iso-corrosion curves for CrNi-steel 18 8 (a) and NiCrMoCu-steel (b) in formic acid of various concentrations [5]

In corrosion tests in boiling 45 % formic acid the initially high corrosion rate of the CrNi-steel SAE 304 (1.4301) slowed down to a standstill after a test duration of 8 hours. When the steel specimen was left in the 45 % formic acid without changing the acid until it had been immersed for 52 hours, no corrosive attack was discernible. After removal of the test specimen without damaging the surface coating, the specimen was immersed in fresh 45 % formic acid; corrosion was not detectable.

HCOOH concentration, %	Steels	Corrosion rate, mm/a (mpy)			
		Not aerated		Aerated	
		Liquid	Vapor phase	Liquid	Vapor phase
10	SAE 316	0.025–0.48 (0.98–18.9)	0.20 (7.87)	–	–
50	SAE 316	0.025–0.25 (0.98–9.84)	0.76 (29.9)	0.025 (0.98)	0.025–1.45 (0.98–57.1)
90	SAE 304	12.7 (500)	1.78 (70.1)	–	–
90	SAE 310	10.2 (402)	1.88 (74.0)	–	–
90	SAE 316	0.025–1.40 (0.98–55.1)	0.13 (5.12)	0.025 (0.98)	1.02 (40.2)
90	SAE 317	0.08 (3.15)	0.13 (5.12)	–	–

Table 7: Corrosion rates of CrNi- and CrNiMo-steels in formic acid of various concentrations with and without aeration (test duration 48 hours) [12]

	DIN-designation	DIN-material No.
SAE 304	X5CrNi18-10	1.4301
UNS S31000	X15CrNiSi25-21	1.4841
SAE 316	X5CrNiMo17-12-2	1.4401
SAE 317	X3CrNiMo18-12-3	1.4449

Table 8: Equivalent German designations for the steels investigated in Table 7

DIN-designation	DIN-material No.		Corrosion rate
			mm/a (mpy)
X1NiCrMoCu25-20-5	1.4539	SAE 904 L	0.06 (2.36)
X4NiCrMoCuNb20-18-2	1.4505		0.1 (3.94)
X1NiCrMoCuN25-20-7	1.4529	UNS N08926	0.12 (4.72)
X6CrNiMoTi17-12-2	1.4571	SAE 316 Ti	0.3–0.5 (11.8–19.7)
X2CrNiMo18-14-3	1.4435	SAE 316 L	0.35 (13.8)

Table 9: Corrosion rates of various austenitic steels in 50 % boiling formic acid [13]

Following a total immersion time of 124 hours, the specimen was activated by pickling in acid and then immersed in 45 % formic acid. Severe corrosion, which started immediately, slowed down to a standstill even when the acid was changed every hour. During the repeated weighing, a protective film forms on the surface of the specimen due to the action of oxygen. The material consumption per unit area determined by weighing after exposure to the acid for 48 hours (without change of acid) was 10 g/dm^2, i.e. 1/10 lower than the material consumption rate of 2 g/dm^2 h after exposure to the acid for 1 hour [14]. This finding is of significance for the performance of corrosion tests with passivatable, metallic materials in accordance with DIN 50905 (boiling tests) and one of the causes of differences in the evaluation of tests with identical materials, identical corrosion media and identical temperature conditions. The marked decrease in the corrosion rate on exposure to boiling formic acid (373.5 K (100.5°C)); of austenitic steels (X2CrNiMo18-14-3, X2CrNiMoN17-13-5, X6CrNiMoTi17-12-2; DIN-Material Nos. 1.4435, 1.4439, 1.4571) to about 10 % of the initial corrosion rates within a test duration of 10 days has been confirmed in the literature. The corrosion rates of these high-alloy steels (and alloys) was determined with the aid of radionuclide technology [15].

– *Inhibitors* –

According to the results of a test report, corrosion of passivatable chromium and chromium-nickel steels by boiling formic acid in various concentrations is decreased or largely inhibited by additions of small quantities of soluble ferric and cupric salts [14, 16].

Example:

The effect of an addition of 0.15 g/l CuSO$_4$ · 5 H$_2$O (containing 0.038 g/l Cu^{2+}) on the corrosion of SAE 304 (1.4301) in boiling 90 % formic acid becomes clear on reference to the example in Table 10 [14].

The additional quantity of Fe^{3+} ions required for maximum inhibition is stated to be 0.024 g/l (equivalent to 0.17 g/l Fe(NO$_3$)$_3$ · 9 H$_2$O). The smallest added quantity below which the inhibiting effect is not exhibited was found to be 0.018 g Fe^{3+} or 0.13 g Fe(NO$_3$)$_3$ · 9 H$_2$O per liter [14].

Steel	Formic acid concentration, %	Corrosion rate, mm/a (mpy)	
		without inhibitor	with 0.15 g CuSO$_4$ × 5 H$_2$O*
SAE 304	90	26.8 (1,055)	0.30 (11.8))

* The addition of inhibitor corresponds to 0.038 g Cu^{2+}.

Table 10: The effect of inhibitor on the corrosion of SAE 304 in boiling 90 % formic acid [14]

Special iron-based alloys

According to a resistance table, Ferralium® alloy (cf. 1.4507) resists attack by formic acid at 293 K (20°C) with a corrosion rate of less than 0.15 mm/a (5.91 mpy). This data is based on the use of the alloys Ferralium® alloy and Langalloy® 20 V [17]. Corrosion tests were carried out in boiling formic acid with concentrations of 40 % and 88 % with a Ferralium® alloy (Fe-5.0Ni-27.5Cr-3.5Mo-1.7Cu-0.17N).

Apart from Ferralium® alloy, nickel-based alloys Hastelloy® C 276, the forerunner of Hastelloy® C 4 and Hastelloy® B 2, as well as the austenitic CrNiMo-steel SAE 316 L were investigated (see Table 11) [18].

Material	Formic acid concentration, %	
	40	88
	Corrosion rates, mm/a (mpy) (Average from 2 series of tests)	
Ferralium® alloy	0.20 (7.87)	0.97 (38.1)
Hastelloy® C 276	0.07 (2.76))	0.05 (1.97)
Hastelloy® B 2	0.01 (0.39)	0.02 (0.79)
SAE 316 L (1.4404)	1.07 (42.1)	0.48 (18.9)

Table 11: Results of laboratory tests on various alloys in boiling formic acid of various concentrations (test duration 24 hours) [18]

The austenitic casting materials Ni-Resist® (EN-GJLA-XNiCuCr15-6-2, DIN-Material No. EN-JL3011 old 0.6655 and GGL-NiCr 20 2, DIN-Material No. 0.6660) are not suitable for use in contact with formic acid at elevated temperatures. After exposure for 180 hours to a mixture of 56 % acetic acid with 90 % formic acid (6:1) at 373 to 377 K (100 to 104°C) the corrosion rate in this operational test was found to be 4.06 mm/a (160 mpy) [4].

Bibliography

[1] Iron, H.
Drogisten-Lexikon, Bd. 2 (1955) p. 69
Springer-Verlag Berlin, Göttingen, Heidelberg

[2] Miller, R. F.; Treseder, R. S.; Wachter, A.
Corrosion by acids at high temperatures
Corrosion (Houston) 10 (1954) p. 7

[3] Sachsze, W.
Ullmanns Encyklopädie der technischen Chemie
(Ullmann's Encyclopedia of Industrial Chemistry) (in German)
4th Ed., Vol. 7 (1974) p. 362
Verlag Chemie, 6940 Weinheim

[4] Anonymous
Die Ni-Resist-Gußeisenwerkstoffe
(The Ni-Resist® cast iron materials) (in German)
International Nickel, 3rd Ed. (1968) p. 102

[5] Berg, F. F.
Korrosionsschaubilder, Ameisensäure
(Corrosion Diagrams "Formic Acid")
(in German and English)
VDI-Verlag, 4000 Düsseldorf, 1965, p. 18

[6] Rabald, E.
Legierter Guß im chemischen Apparatebau
(Alloyed cast iron in chemical plant construction) (in German)
Werkstoffe und Korrosion 7 (1956) No. 8/9, p. 435

[7] Charbonnier, J.C.; Noual, P.
The influence of molybdenum on the behavior of 17Cr pure ferritic steels in a 20% HCOOH medium at 70°C
Corrosion Sci. 17 (1977) p. 1009

[8] Heimann, W.; Kiesheyer, H.
Ein weichmagnetischer Stahl für höchste korrosionschemische Beanspruchung: X 1 CrNiMoNb 28 4 2 (Remanit® 4575)
(A soft-magnetic steel to be used in high-corrosive media: X 1 CrNiMoNb 28 4 2 (Remanit 4575)) (in German)
Thyssen Edel-Stahlwerke AG, Technische Berichte 3 (1977) No. 1, p. 23

[9] Knoth, J.
High purity ferritic Cr-Mo stainless steel – Five years' successful fight against corrosion in the process industry
Werkstoffe und Korrosion 28 (1977) p. 409

[10] Steigerwald, R. F.
New molybdenum stainless steels for corrosion resistance: A review of recent developments
Mater. Performance 13 (1974) No. 9, p. 9

[11] Nordin, S.
Properties of a modified Type 329 weldable and SCC resistant stainless steel
Métaux Corrosion-Industrie 55 (1980) No. 659/660, p. 229

[12] Lackey, J. Q.; Degnan, T. F.
Formic acid corrosion of common materials of construction
Mater. Performance 13 (1974) No. 7; p. 13

[13] Schütze, K. G.
Korrosionsverhalten metallischer Werkstoffe in Ameisensäure
(Corrosion behavior of metallic materials in formic acid) (in German)
Werkstoffe und Korrosion 38 (1987) No. 1, p. 41

[14] Streicher, M. A.
Corrosion of stainless steels in boiling acids and its suppression by ferric salts
Corrosion (Houston) 14 (1958) No. 2, p. 59t

[15] Gassen, R.; Göhr, H.; Müller, N.; Vehlow, J.
Korrosion in Carbonsäuren, Teil II: Das Korrosionsverhalten hochlegierter Stähle, zweier Nickellegierungen und von Titan in Ameisensäure
(Corrosion in carboxylic acids, 2nd part: the Corrosion behavior of high-alloy steels, of two Ni-alloys and of titanium in formic acid) (in German)
Z. Werkstofftech. 17 (1986) Nr. 6, p. 218

[16] Brooke, J. M.
Corrosion inhibitor checklist
Hydrocarbon Processing 6 (1970) Aug., p. 107

[17] Product Information
Ferralium® Alloy – Hochfester säurebeständiger Edelstahl
(Ferralium® Alloy – High-strength acid resistant high grade steel) (in German)
Deutsche Langley Alloys GmbH, 6000 Frankfurt

[18] Asphahani, A. I.
Corrosion resistance of high performance alloys
Mater. Performance 19 (1980) No. 12, p. 33

Sulfonic Acids

Unalloyed steels and cast steel

The inhibiting action of Katapine (based on p-alkylbenzyl-pyridinium chloride, for example methyl-katapine [$CH_3(CH_2)_n$-CH_2-C_6H_4-CH_2N-C_6H_4]$^+$Cl$^-$) on the corrosion of Armco® iron in 0.5 mol/l sulfuric acid, with and without additives, is shown in Table 1. The inhibiting action of the Katapines is intensified both by sodium sulfide and by sulfosalicylic acid. The inhibiting effect attained in this way could be achieved with Katapine alone only in a 10 to 20 times higher concentration.

Additive mol/l	Inhibition value, γ^*		
	Methyl katapine	Ethyl katapine	Isopropyl katapine
–	110	440	35
0.0005 Na$_2$S	400	700	550
0.01 Sulfosalicylic acid	215	500	110

* $\gamma = k_o/k$, where k_o = corrosion rate without inhibitor and k = corrosion rate with inhibitor

Table 1: Inhibiting efficiency of Katapine (2.5 g/l) on the corrosion of Armco® iron in 0.5 mol/l sulfuric acid [1]

Benzenesulfonic acid and naphthalenedisulfonic acid have only slight inhibiting action on the corrosion of Armco® iron in 0.5 mol/l sulfuric acid; addition of 0.01 mol/l potassium bromide improves the efficiency only by a factor of 2 [2].

As a result of polarization of pure iron in 0.5 mol/l sulfuric acid containing 0.001 mol/l of sodium sulfide and 0.002 mol/l of the ammonium salt of tribenzyl-methylsulfonic acid, the corrosion current drops at –0.2 V$_{(NHE)}$ to about 10^{-7} A/cm^2, which corresponds to the negligible corrosion rate of only 0.001 mm/a (0.039 mpy) [3].

According to Figure 1, the inhibiting efficiency of sulfonic acid R-C_6H_4-SO_3H on the corrosion of steel (USSR designation 08-KP, cf. 1.0335, UNS G10060) in 3 mol/l sulfuric acid at 353 K (80°C) passes through a maximum for a given concentration. p-chloro-benzenesulfonic acid, at 0.003 mol/l KPI-3 (quaternary pyridinium salt), shows the greatest effect [4].

The results shown in Figure 1 are confirmed in a later paper [5]. It has been found, moreover, that the maximum of the inhibiting efficiency (as defined in Table 1) depends on temperature under these conditions and decreases from about 300 at 358 K (85°C) to 80 at 333 K (60°C) and 70 at 313 K (40°C).

Steel is corroded in a solution containing p-toluenesulfonic acid, polyvinylalcohol, sodium bichromate and N,N'-di-o-tolylthiourea (pH 1.8) at a rate of 0.131 g/m^2 h (0.14 mm/a), in the absence of thiourea at a rate of 3.075 g/m^2 h (3.4 mm/a) [6].

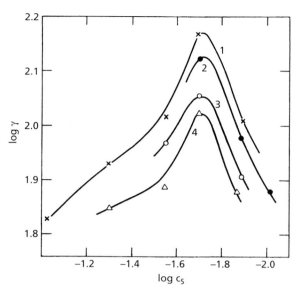

Figure 1: Relationship between the coefficient of the corrosion of steel 08-KP in 3 mol/l H_2SO_4 at 353 K (80°C) and the concentration of sulfonic acid $R(C_6H_4)SO_3H$ + 0.003 mol/l KPI-3, where R is H (1), Cl (2), CH_3 (3) and OH (4) [4]

As a result of the addition of dibenzyl sulfoxide and Resistin N® (= benzylpyridine isooctylxanthate), the polarization resistance during the dissolution of iron in 5 % aminosulfonic acid increases markedly and dissolution correspondingly decreases. Good protective action is achieved even at an inhibitor concentration of only 0.001 mol/l at 348 K (75°C).

Corrosion of low alloy steel in synthetic cooling water (ppm: 300 $CaCO_3$ + 100 $MgCO_3$ + 400 NaCl + 500 Na_2SO_4 + 10 alkalinity as $CaCO_3$) is inhibited by 100 ppm of sulfonic acid, $C_{13}H_{25}COCH_2CH_2SO_3H$, to the extent of 71 % and, if 25 ppm of zinc are present at the same time, to the extent of 82 %. At 66 %, the sulfonic acid, $C_{13}H_{27}CO\text{-}N(CH_3)\text{-}CH_2CH_2SO_3H$ is less effective [7].

Corrosion prevention in media containing hydrogen sulfide can also be achieved by utilizing the oxygen contained in the solution for oxidizing the hydrogen sulfide and removing it in this way. The recommended catalyst is a mixture of anthraquinonedisulfonic acid with Na_3VO_4 [8].

According to Table 2, the corrosion of carbon steel (St-3, cf. 1.0036, UNS K02502) in 5 % aminosulfonic acid at 363 K (90°C) is decreased by 3 g/l of an organic compound (USSR designaties PKU-M) and by 0.3 g/l of zinc or bismuth dithiocarbamate by more than 90 % [9]. In the absence of these additives, the corrosion rate of the steel in 5 % aminosulfonic acid was 161 g/m² h.

The corrosion of iron in 5 % hydrochloric acid is reduced by 0.2 % p-toluenesulfonic acid from 100 (without additive) to 78 and, if 0.5 % polyethoxylated alcohol (wetting agent) is added at the same time, to 14 % [10]. Additional incorporation of 0.5 % sulfuric acid has no influence.

Additives	Inhibition Value, %
PKU-M	77.3
PKU-M + DTC-Zn	97.7
PKU-M + DTC-Bi	97.8
PKU-M + DTC-Na	97.6
PKU-M + DTC-Ni	96.9

Table 2: Influence of dithiocarbamate (DTC) (0.3 g/l) on the inhibition of the corrosion of St-3 in 5 % aminosulfonic acid with 3 g/l of inhibitor (PKU-M) at 363 K (90°C) [9]

The corrosion of carbon steel (SAE 1020) in 0.1 mol/l hydrochloric acid at 298 K (25°C) is not inhibited, but accelerated, by addition of 0.1 mol/l benzenesulfonic acid [11].

According to Table 3, the corrosion of a low alloy steel (Fe-0.25C-0.01Si-0.76Mn-0.04Ni-0.09Cr-0.01Mo) in a mixture of water and petroleum (10:1), saturated with hydrogen sulfide and containing 3.5 % of NaCl, which is stirred at room temperature, is decreased synergistically by 200 ppm of a mixture of an aliphatic sulfonic acid and a sulfur-containing organic phosphorus compound (from 1.18 to 0.0023 mm/a (46.5 to 0.09 mpy)). Neither inhibitor has satisfactory action on its own [12].

To improve the protective action of oils, fats and waxes on the atmospheric corrosion of iron and steel, admixture of about 5 to 15 % of oil-soluble inhibitors, such as, for example, alkaline earth metal salts of long chain organic sulfonic acids, is recommended [13].

Carbon steels such as St-3 (cf. 1.0036, UNS K02502) are not suitable as materials for the preparation of sulfonated anthraquinone at temperatures between 340 and 420 K (67 and 147°C), owing to excessive corrosion of 98 and 1.8 g/m² h (108 and 2 mm/a) at 343 K (70°C) and 186 and 2.6 g/m² h (204 and 2.9 mm/a) at 383 K (80°C) [14]. The difference in the corrosion rate of the two steels cannot be explained so far.

Aqueous phase	Mean corrosion		Hydrogen	
	mm/a (mpy)	Protection, %	Current density, mA/cm²	Permeation, %
3.5 % NaCl	1.18 (46.5)	–	0.013	2.0
3.5 % NaCl + 200 ppm Inhibitor-A	0.065 (2.56)	95	0.0021	5.8
3.5 % NaCl + 200 ppm Inhibitor-B	0.55 (21.7)	56	0.0054	1.8
3.5 % NaCl + 200 ppm Inhibitor-C	0.0023 (0.09)	99	0.00069	56.0

Inhibitor-A: sulfur-containing organic phosphorus compound
Inhibitor-B: aliphatic sulfonic acid
Inhibitor-C: mixture of A and B

Table 3: Corrosion and hydrogen permeation through steel in a H_2S-saturated, stirred 10:1 water/petroleum mixture at room temperature [12]

According to Table 3a, the sulfonated residues from the phenol column inhibit, at 0.6 to 1.0 g/l, the corrosion of steel (St-3) in 0.1 mol/l hydrochloric acid between 293 and 353 K (20 and 80°C) quite noticeably, but still not sufficiently for application in the presence of hydrochloric acid [15].

Inhibitor g/l	Temperature K (°C)	Corrosion rate		Protective efficiency %
		g/m² h	mm/a	
–	293 (20)	1.65	1.8	–
	353 (80)	3.95	4.3	–
	373 (100)	7.65	8.4	
0.2	293 (20)	0.95	1.0	43
	353 (80)	2.65	2.9	33
	373 (100)	5.15	5.7	33
0.6	293 (20)	0.34	0.37	80
	353 (80)	1.05	1.2	74
	373 (100)	2.91	3.2	62
1.0	293 (20)	0.25	0.27	85
	353 (80)	0.77	0.85	81
	373 (100)	2.25	2.5	71

Table 3a: Corrosion rate of C-steel in 0.1 mol/l HCl with different sulfonated residues additives [15]

The inhibitor mixture has no influence on the corrosion of the steel in pit water (0.22 to 0.75 g/m² h) [15].

The anodic dissolution during the electro-polishing of low-carbon steel (St-15, 0.12 to 0.19 % C) in a solution of phosphoric and sulfuric acids (70:30 % by volume) at 0.8 $V_{(NHE)}$ is decreased with increasing concentration of sulfonic acid from 45 (without) to 25 mA/cm² (0.031 mm/h (10,691 mpy)) at 0.75 g/l and to about 10 mA/cm² (0.013 mm/h (4,483 mpy)) at 2.0 g/l of sulfonic acid [16].

Cleaning of iron surfaces by means of a solution of (g/l) 90 to 130 orthophosphoric acid (specific gravity = 1.6 g/ml) and 1 to 5 of a surface active agent is improved if the solution also contains 1 to 10 % sulfosalicylic acid, 2 to 20 % ammonium nitrate and 40 to 65 % acetone [17].

High-alloy cast iron, high-silicon cast iron

Having a corrosion rate of approx. 1 mm/a (39.4 mpy) in a reaction mixture formed during the photochemical sulfoxylation of hydrocarbons, and containing (%) 13 alkylsulfonic acid + 30 hydrocarbon + 4 sulfuric acid + 53 water at 393 K (120°C), high silicon cast iron can be regarded as not resistant to that mixture at room temperature [18]. It should be used only in exceptional cases.

Such an alloy (Fe-(0.3 to 0.5)C-(14 to 18)Si-(0.3 to 0.5)Mn-0.10P-0.07S) is not noticeably attacked by 20 % phenolsulfonic acid [19].

Ferritic chromium steels with more than 12 % chromium

Chromium steels (USSR designations) 2Ch13 (1.4021, cf. UNS S42000) and Ch15T (Fe-15Cr, Ti-stabilized) suffer considerable pitting in 0.05 mol/l sulfuric acid containing 0.05 and 0.15 mol/l sodium chloride at 0.35 and 0.55 $V_{(SCE)}$ respectively (Table 4); this attack can be greatly reduced by addition of 2,6-diphenylpyridylium sulfate (DPS). The protective efficiency, which is given at the same time, is calculated from the ration of the currents:

$$PE = 100 \, (i_o - i)/i,$$

where i_o is the anodic current density for the uninhibited solution and i the current density for the solution with the sulfonic acid.

Steel	DPS addition mol/l	V_P V	Protective efficiency, %*	Pits per cm^2	Mean pit diameter μm
2Ch13	–	0.35	–	3	500
	0.005		75	15	10
Ch15T	–	0.55	–	2	800
	0.002		8	2	500

* definition of protective efficiency see text

Table 4: Influence of the addition of 2,6-diphenylpyridylium sulfate (DPS) on the pitting of Cr-steels 0.15 mol/l NaCl at the pitting potential (V_P) and 298 K (25°C) [20]

Austenitic chromium-nickel steels

During the photochemical sulfoxidation of paraffin-type hydrocarbons (C_{10} to C_{20}) in the presence of water and oxygen, a mixture is formed consisting of (%) 18 sulfonic acid + 4.8 sulfuric acid + 34.8 hydrocarbons and 40 water + 2.4 sulfur trioxide. During a test period of 630 hours, steel Ch18N12M2T (Fe-18Cr-12Ni-2Mo, Ti-stabilized) corrodes under these conditions at 303 K (30°C) at a rate of 0.048 g/m^2 h (0.05 mm/a) [21].

For the same process, using a reaction mixture of (%) 13 alkylsulfonic acid + 30 hydrocarbon + 4 sulfuric acid + 53 water, the corrosion rate of steel Ch18N10T (Fe-18Cr-10Ni, Ti-stabilized, cf. 1.4541, UNS S32100) at 303 K (30°C), depends on the intensity of stirring and can be between 6.6 and about 0.9 mm/a (260 and about 35.4 mpy)) [18], so that this steel cannot be used in a stirred system.

Steel SAE 304 (Fe-0.043C-18.0Cr-9.12Ni-0.46Si-0.03P-0.003S) is susceptible to intercrystalline attack in 1 mol/l aminosulfonic acid (NH_2SO_3H) at 343 K (70°C); the attack is intensified by sensitization (14 hours at 873 K (600°C), see also Table 5) [22].

Benzotriazole, benzothiazole and dibenzyl sulfoxide have good inhibiting action in this case; the last two substances also prevent intercrystalline corrosion.

Petrov's catalyst (the Russian product equivalent to Twitchell's reagent), obtained by reaction of a kerosine distillate with sulfur trioxide, contains 50 to 55 % of sulfonic acids, in addition to small amounts of sulfuric acid. According to Table 6, chromium-nickel steels, with the exception of the chromium-nickel steel additionally alloyed with molybdenum and copper (see Austenitic chromium-nickel-molybenum steels), is attacked considerably with increasing temperature and catalyst concentration [23].

According to Table 7, pitting of steel 2Ch18N9 (Fe-0.2C-18Cr-9Ni, cf. 1.4310, UNS S30200) in a solution of 0.05 mol/l sulfuric acid + 0.25 mol/l sodium chloride at 298 K (25°C) and 0.55 $V_{(NHE)}$ can be reduced noticeably by addition of sulfonic acid [20].

Addition mol/l	Corrosion rate kg/m^2 h (mm/a)		Kind of attack
	unsensitized	sensitized	
–	0.06 (0.06)	1.1 (1.2)	IC
0.016 Acetylene derivatives	0.09 (0.10)	0.8 (0.9)	IC
0.010 Benzonitrile	0.12 (0.13)	0.9 (1.0)	IC
0.001 Benzimidazole	0.11 (0.12)	1.0 (1.1)	IC
0.001 Benzotriazole	0.002 (0.002)	0.3 (0.3)	IC
0.001 Dibenzylsulfoxide	0.03 (0.03)	0.04 (0.045)	–
0.001 Phenylthiourea	0.13 (0.14)	0.08 (0.09)	–
0.001 Benzothiazole	0.04 (0.04)	0.04 (0.045)	–
0.001 Mercaptobenzimidazole	0.20 (0.20)	0.24 (0.27)	–
0.001 2,5-Dihydroxy-1,4-dithiane	0.18 (0.19)	0.11 (0.12)	–

Table 5: Corrosion of SAE 304 in 1 mol/l aminosulfonic acid at 343 K (70°C) in the presence of various inhibitors (IC = intercrystalline corrosion); test duration 24 hours [22]

Steel	Temperature K (°C)	Catalyst content, %			
		10	20	50	100
		Corrosion rate, mm/a (mpy)			
12Kh18N10T*	293 (20)	0.20 (7.87)	0.24 (9.45)	0.22 (8.66)	0.25 (9.84)
	323 (50)	0.71 (28.0)	0.71 (28.0)	0.82 (32.3)	0.85 (33.5)
	353 (80)	0.71 (28.0))	0.95 (37.4)	2.5 (98.4)	12.2 (480)
	368 (95)	3.1 (122)	4.3 (169)	6.9 (272)	32.1 (1,264)
06Kh23N28M3D3T*	293 (20)	0.022 (0.87)	0.022 (0.87)	0.022 (0.87)	0.022 (0.87)
	323 (50)	0.10 (3.94)	0.10 (3.94)	0.066 (2.60)	0.077 (3.03)
	353 (80)	0.077 (3.03)	0.066 (2.60)	0.066 (2.60)	0.63 (24.8)
	368 (95)	0.055 (2.17)	0.055 (2.17)	0.066 (2.60)	2.9 (114)

* 12Ch18N10T = Fe-0.12C-18Cr-10Ni, Ti-stabilized (cf. 1.4541, UNS S32100)
06Kh23N28M3D3T = Fe-0.06C-23Cr-28Ni-(2 to 4)Mo-(2 to 3)Cu, Ti-stabilized

Table 6: Corrosion of chromium-nickel steels without and with Mo and Cu in Petrov's catalyst (a Twitchell's reagant) solution, 30 hour test [23]

Addition mol/l	PE %[1)]	Pits/cm^2	Mean diameter of pits, µm
–	–	4	300
0.1 Sulfonic acid	50	10	10
0.01 Aminosulfonic acid	76	14	10

1) PE: protective efficiency, %,
= 100 · $(i_o - i)/i$ where i_o is the anodic current density in the uninhibited, i the anodic current density in the inhibited solution

Table 7: Influence of sulfonic acids on the pitting corrosion of 2Kh18N9 (Fe18Cr9Ni-0,02C) in 0.05 mol/l H_2SO_4 + 0.25 mol/l NaCl at 0.55 V and 298 K (25°C) [20]

Steel El-417 (20Ch23N18 = Fe-0.20C-23Cr-18Ni; USSR grade, cf. 1.4845, UNS S31008) is unsuitable as a cathode material for the electrical reduction of 1-nitro-3,6,8-naphthalene-trisulfonic acid, since it is dissolved in the 0.4 mol/l sulfonic acid at a cathodic current density of 0.2 A/cm^2 at 323 K (50°C) at a rate of 37 mm/a (1,457 mpy). Here again (as in Table 6), steel El-943 (06Kh23N28M3D3T = Fe-28Ni-23Cr-3Mo-3CuTi, cf. 1.4503) is superior (0.11 mm/a (4.33 mpy)) [24].

Corrosion of reaction vessels of chromium-nickel steel by acidic catalysts, such as, for example, benzenesulfonic and p-toluenesulfonic acids, is said to be prevented to a large extent by hypophosphoric acid [25].

According to Table 7a, steel Ch18N10T (cf. 1.4541, UNS S32100) can be used as a material for a production unit for the sulfonation of anthraquinone [14]. The corrosion of the steel in the process solution at the phase boundary gas/solution is virtually identical with that in the liquid.

A potentiostatic etching method, using a solution of 5 % sulfosalicylic acid + 1 % tetramethylammonium chloride in methanol, is recommended for detecting carbide, nitride and boride precipitates in steel Fe-18Cr-25Ni-5Al-0.2Ti-0.25Zr [26].

Temperature, K (°C)			
343 (70)	353 (80)	363 (90)	373 (100)
Corosion rate, g/m^2 h (mm/a)			
0.12 (0.13)	0.09 (0.10)	0.21 (0.23)	0.07 (0.08)

Tabelle 7a: Corrosion rate of Ch18N10T (cf. 1.4541, UNS S32100) at the phase boundary gas/reaction solution [14]

Austenitic chromium-nickel-molybdenum steels

During the reduction of 0.4 mol/l 1-nitro-3,6,8-naphthalenetrisulfonic acid the cathode (El-943 = 06Ch23N28M3D3T = Fe28Ni23Cr3Cu3MoTi0.06C, cf. 1.4503), loses 0.099 g/m^2 h (0.11 mm/a) (298 K (25°C), 100 hour test) [24].

According to Table 8, the two steels, 0Ch23N28M3D3T and Ch18N12M2T (Fe-18Cr-12Ni-2Mo, Ti-stabilized) are suitable materials for a sulfoxidation unit containing the reaction solution (%): 18 sulfonic acid + 4.8 sulfuric acid + 34.8 hydrocarbons + 2.4 sulfur trioxide + 40 water [21].

Steel	Reactor, 303 K (30°C)		Extractor, 293 K (20°C)	
	Duration h	Corrosion mm/a (mpy)	Duration h	Corrosion mm/a (mpy)
Ch23N28M3D3T	630	0.044 (1.73)	1,320	0.00083 (0.033)
Ch18N12M2T	630	0.053 (2.09)	1,320	0.00084 (0.033)

Table 8: Corrosion of CrNiMo-steels in a sulfoxidation unit [21]

Steel 0Ch23N28M3D3T proved suitable in the production of alkylsulfonate and, even at 363 K (90°C), it was only slightly attacked (Table 9).

In the production of aromatic sulfonic acids from petroleum and kerosine, the corrosion of steels Ch18N10T (cf. 1.4541, UNS S32100) and Ch17N13M3T (cf. 1.4571, UNS S31635) is much too high (5.8 and 4.9 g/m^2 h resp., approx. (6.4 and 5.4 mm/a) at 323 K (50°C) and 20 and 12 g/m^2 h resp. (22 and 13 mm/a) at 343 K (70°C), whereas steel 0Ch23N28M3D3T is a very suitable material, losing only about 0.001 mm/a (0.039 mpy) between 323 and 363 K (50 and 90°C) [27].

Steel	Corrosion rate, mm/a (mpy)							
	Liquid phase		Gas phase		Liquid phase		Gas phase	
	T = 303 K (30°C)				T = 363 K (90°C)			
	max.	min.	max.	min.	max.	min.	max.	min.
Ch17N13M2T	0.11 (4.33)	0.01 (0.39)	0.02 (0.79)	0.01 (0.39)	32.8 (1,291)	0.08 (3.15)	0.25 (9.84)	0.01 (0.39)
0Ch23N28M3D3T	0.03 (1.18)	0.01 (0.39)	0.02 (0.79)	0.01 (0.39)	0.04 (1.58)	0.01 (0.39)	0.01 (0.39)	0.01 (0.39)

Table 9: Corrosion rate of CrNiMo-steels in an alkylsulfonate production unit [18]

According to Table 10, the influence of the atmosphere on the corrosion of chromium-nickel-molybdenum steels in the process solution ((%) 13.3 alkylsulfonic acid + 29.6 hydrocarbons + 3.7 sulfuric acid + 2.1 sulfur trioxide + 51.3 water) is considerable [18].

Steel	mm/a (mpy)			
	O_2	Air	Argon	SO_3
Ch17N13M2T (cf. 1.4571)	3.9 (154)	2.5 (98.4)	0.40 (15.7)	0.10 (3.94)
0Ch23N28M3D3T	0.05 (1.97)	0.01 (0.39)	0.09 (3.54)	0.07 (2.76)

Table 10: Corrosion rate of steels in a process solution ((%) 13.3 alkylsulfonic acid, 29.6 hydrocarbons, 3.7 sulfuric acid, 2.1 sulfur trioxide, balance water) at 303 K (30°C) in different atmospheres of 1 bar [18]

To supplement Table 6, Table 11 gives the corrosion rates of two chromium-nickel-molybdenum steels in the Petrov catalyst solution (= Twitchell's reagent) [23]. Here, again, the chromium-nickel-molybdenum-copper steel is the most resistant one.

According to a tank container design approved by the German Federal Institution for Testing Materials (BAM) the steels X5CrNiMo17-12-2 (Material No. 1.4401, SAE 316) and CrNiMo-steels 20 10 (Materials No. 1.4404 to 1.4410, cf. UNS S31603 and S32750) are approved for the storage and transport of chlorosulfonic acid, provided it is free from admixtures (in particular hydrochloric acid) and provided the tank container is filled under conditions of absolute absence of moisture and is then tightly closed, so as to exclude the possibility of penetration of moisture during transport. Moreover, the tank container has to be conveyed and stored in between in such a way as to prevent excessive heating up by climatic influences [22, 28]. The same is true for 1.4529 (cf. UNS N08926) and 1.4562 (cf. UNS N08031) [29].

Steel	Temperature K (°C)	Catalyst Content, %			
		10	20	50	100
		Corrosion rate, mm/a (mpy)			
10Ch17N13M3T (cf. 1.4571, UNS S31635)	293 (20)	0.022 (0.87)	0.011 (0.43)	0.08 (3.15)	0.09 (3.54)
	323 (50)	0.38 (15.0)	0.28 (11.0)	0.43 (16.9)	0.42 (16.5)
	353 (80)	0.51 (20.1)	0.51 (20.1)	0.40 (15.7)	2.2 (86.6)
	368 (95)	0.33 (13.0)	1.6 (63.0)	6.2 (244)	8.9 (350)
06Ch23N28M3D3T	293 (20)	0.022 (0.87)	0.022 (0.87)	0.022 (0.87)	0.022 (0.87)
	323 (50)	0.10 (3.94)	0.10 (3.94)	0.066 (2.60)	0.077 (3.03)
	353 (80)	0.077 (3.03)	0.066 (2.60)	0.066 (2.60)	0.63 (24.8)
	368 (95)	0.055 (2.17)	0.055 (2.17)	0.066 (2.60)	2.9 (114)

Table 11: Corrosion rates of CrNiMo-steels in Petrov's catalyst solution; 30 hour tests at different concentrations [23]

Austenitic chromium-nickel steels with special alloying additions

Steel Fe-18Cr-25Ni-5Al-0.2Ti-0.25Zr is not attacked below +0.8 $V_{(SCE)}$ by a solution of 5% sulfosalicylic acid +1% tetramethylammonium chloride in methanol. Above this potential the steel is attacked. This solution is therefore a suitable etching medium for the detection of carbide, nitride, boride, sulfide and oxide precipitations in the steel [26].

Bibliography

[1] Khasan, S. F.; Iofa, Z. A.
Increasing the inhibitory effect of Katapin on the corrosion of iron in sulfuric acid solutions (in Russian)
Zashchita Metallov 6 (1970) p. 231

[2] Iofa, Z. A.
Effect of synergism and antagonism on the adsorption and efficiency of surface-active agents on the electrochemical reactions and the corrosion of iron (in Russian)
Zashchita Metallov 8 (1972) p. 139

[3] Iofa, Z. A.
Mechanism of the action of hydrogen sulfide and inhibitors on the corrosion of iron in acids (in Russian)
Zashchita Metallov 16 (1980) p. 295

[4] Ledovskikh, V. M.; Sarycheva, I. V.
Investigation into the inhibiting properties of mixtures of surface-active cations with organic sulfonic and carboxylic acids during the corrosion of steels in acids (in Russian)
Zashchita Metallov 19 (1983) p. 895

[5] Ledovskikh, V. M.; Sarycheva, I. V.
Investigation into the inhibiting properties of mixtures of surface-active cations with organic sulfonic and carboxylic acids during the corrosion of steels in acids (in Russian)
Zashchita Metallov 19 (1983) p. 895

[6] Hatsutori, Ta.
Acid solution with thiourea derivative to prevent corrosion (in Japanese)
JP Patent 48-71328 (27 Sept. 1973)

[7] Zecher, D. C.
Corrosion inhibition by surface-active chelants
Mater. Performance 15 (1976) No 4; p. 33

[8] Redmore, D.
Oxygen scavenger
US Patent 3764548 (9 Oct. 1973)

[9] Podobaev, N. I.; Kharkovskaya, N. L.; Korotkikh, E. V.; Ustinskij, E. N.
Dithiocarbamate as inhibitors of the corrosion in acids (in Russian)
Zashchita Metallov 16 (1980) p. 73

[10] Constantinescu, A.; Buricatu, A.
Efectul unor substante tensioactive asupra inhibitorilor de coroziune
(Influence of some surface-active agents on corrosion inhibitors) (in Roumanian)
Rev. Chim. (Bucharest) 27 (1976) p. 631

[11] Foroulis. Z. A.
Molecular designing of organic corrosion inhibitors
Symp. Coupling of Basic & Appl. Corr. Res. 1956, publ. Houston (1969), p. 24

[12] Martin, R. L.
Hydrogen penetration and damage to oil field steels
Mater. Performance 13 (1974) No 7; p. 19

[13] Hoffmann, H.
Feasibility of combatting corrosion with corrosion inhibitors (in German)
Seifen-Öle-Fette-Wachse 110 (1975) p. 373

[14] Karuzin, V. N.; Oboyanskij, Yu. G.; Gobov, S. L.; Gushchina, Z. M.
Corrosion resistance of some materials used in equipment for the anthraquinone sulfonation (in Russian)
Tr. Buryatsk. Inst. Estestv. Nauk BF SO AN SSSR 14 (1977) p. 275

[15] Skachkov, E. A.; Papkov, B. M.
Investigation into the inhibiting properties of sulfonated residues from the phenol column (in Russian)
Koroz. Zashch. (1977) No 9; p. 12

[16] Tiranskaya, S. M.; Nemchinov, S. I.; Peretyatko, V. F.; Golovanova, S. K.
Influence of surface-active agents on the anodic dissolution of low carbon steel (in Russian)
Vopr. Khim. i. Khim. Tekhnol. Resp. Mezved. Nauchno-Tekhn. Sb. 47 (1977) p. 50

[17] Abrasimov, Yu. S.
Cleaning solution for ferrous metal surfaces (in Russian)
USSR Patent 594443 (25 Febr. 1978) (from Otkrytiya Izobret., Prom. Obraztsy, Tovarnye Znaki 55 (1978) No 32; p. 82

[18] Riskin, I. V.; Orlova, F. A.; Balakirev, E. S.
Corrosion resistance of metals under the conditions of alkylsulfonate manufacture (in Russian)
Khim. Prom. 47 (1972) p. 748

[19] Jordanov, D.; Statev, G.
High silicon cast iron resistant to phenolsulfonic acid (in Russian)
Mashinostroene 26 (1977) p. 304

[20] Ekilik, G. N.; Grigorev, V. P.; Ekilik, V. V.
Pitting corrosion of chromium and chromium-nickel steels in the presence of organic additives (in Russian)
Zashchita Metallov 14 (1978) p. 357

[21] Balakirev, E. S.; Ostroumova, V. V.; Gershenovich, A. I. et al.
Corrosion of materials under the condition of hydrocarbon sulfoxidation (in Russian)
Zashchita Metallov 6 (1970) p. 224

[22] Trabanelli, G.; Frignani, A.; Zucchi, F.; Zucchini, M.
The inhibition of the intergranular corrosion of a sensitized stainless steel in hot acidic solutions
Z. phys. Chemie Leipzig 264 (1983) p. 813

[23] Kopeliovich, D. Kh.; Trukhanova, I. A.; Verkhoglazov, Sh. B.
Corrosion resistance of some metallic materials in aqueous solutions containing Petrov's catalyst (in Russian)
Zashchita Metallov 14 (1978) p. 209

[24] Konarev, A. A.; Katunin, V. Kh.; Pomogaeva, L. S. et al.
Influence of the electrode material on the electrochemical reduction of 1-nitro-3,6,8-naphthalenetrisulfonic acid (in Russian)
Elektrokhimiya 20 (1984) p. 204

[25] Nukina, Sh.; Takeda, To.; Inoue, Ta.
Preventing corrosion during chemical reactions (in Japanese)
JP Patent 48-27583 (23 Aug. 1973)

[26] Kurosawa, Fu.; Taguchi, I.; Matsumoto, R.
Observation of precipitates and metallographic grain orientation in steel by a non-aqueous electrolyte potentiostatic etching method (in Japanese)
J. Japan Inst. Metals 43 (1979) p. 1068

[27] Allakhverdiev, G. A.; Mutallimov, M. D.; Mandzhgaladze, S. N. et al.
Corrosion by aromatic sulfonic acids (in Russian)
Vopr. Metalloved. Koroz. Metal., Tbilisi (1972) p. 346

[28] Anonymous
Approval, including supplements, of tank container design, covering containers for the transportation of dangerous goods (TC) Approval certificate for the design of a tank container design covered by Approval No D/ 70 200/TC (in German)
BAM (Bundesanstalt für Materialprüfung) Amts- und Mitteilungsblatt
12 (1982) p. 59

[29] Weltschev, M.; Bäßler, R.
Beständigkeit von hochlegierten Sonderedelstählen und Nickelbasislegierungen als Tankwandungswerkstoff für Behälter zum Transport von Gefahrgütern und wassergefährdenden Stoffen
(Resistance of high-alloy special stainless steels and nickel-based alloys as tank wall material for containers for transportation of hazardous goods and water pollutants)
Forschungsbericht 276
Bundesanstalt für Materialforschung und -prüfung, Berlin, 2006

Alkaline Earth Hydroxides

Unalloyed steels and cast steel

In the alkaline earth hydroxides corrosion system, in particular calcium hydroxide/steel, low-alloy but also low-alloy high-strength steels are of interest for the construction industry.

There is no need for application of high-alloy corrosion-resistant steels in contact with alkaline earth hydroxides [1].

Pitting corrosion, stress corrosion cracking and types of corrosion induced by foreign ions such as chloride and sulfate are of particular importance.

Iron and steel are passivated in a saturated air-free calcium hydroxide solution. In the presence of chloride and in the case of construction components which are under mechanical stress, severe stress corrosion cracking and even pitting corrosion can occur.

There is a minimum concentration of chloride ions which, if exceeded, will lead in all probability to corrosion (see Figure 1) [2].

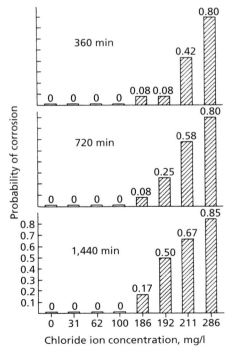

Figure 1: Probability of the occurrence of corrosion as a function of the chloride ion content; parameter: time of observation (360, 720 and 1,440 minutes) [2]

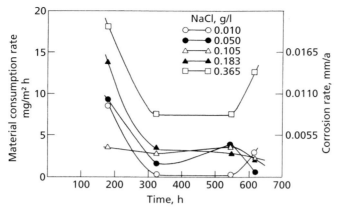

Figure 2: Change in the corrosion rates with time of carbon steel in 0.43 g/l Ca(OH)$_2$ solution as a function of the sodium chloride content [3]

This permissible value for the concentration of the chloride ions is about 100 mg/l. In construction practice it should be noted that due to rapid carbonatization of concrete, even at lower chloride concentrations, active-passive corrosion of the reinforcing steels can occur. The result is stress corrosion cracking, if the required mechanical limiting conditions for the particular steel are fulfilled [2].

In the calcium hydroxide system, the corrosion rates given in Figure 2–5 apply to carbon steels in the presence of sulfates or chlorides [3].

Not only complex iron-calcium hydroxo compounds but also iron(III) and iron(II–III) oxide are formed in the passive films.

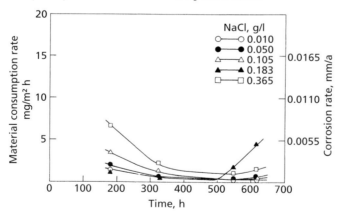

Figure 3: Change with time of the corrosion rates of carbon steel in 0.76 g/l Ca(OH)$_2$ solution as a function of the sodium chloride content [3]

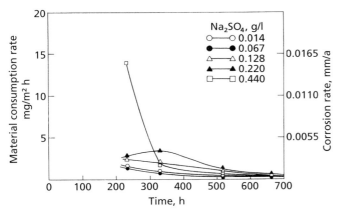

Figure 4: Change with time of the corrosion rates of carbon steel in 0.43 g/l Ca(OH)$_2$ solution as a function of the sodium sulfate content [3]

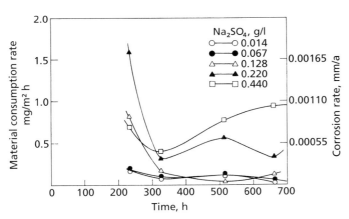

Figure 5: Change with time of the corrosion behavior of carbon steel in 0.76 g/l Ca(OH)$_2$ solution as a function of the sodium sulfate content [3]

The following conclusions can be drawn for practical application:

– 0.76 g/l calcium hydroxide protects steel against up to 0.366 g/l sodium chloride and 0.44 g/l sodium sulfate
– Low contents of 0.010 to 0.050 g/l sodium chloride in 0.76 g/l calcium hydroxide even increase the protective effect of lime water
– An increase in the sodium chloride concentration above 0.050 g/l increases the corrosion rate
– 0.014 to 0.067 g/l sodium sulfate in the presence of 0.76 g/l calcium hydroxide also increase the protective effect
– The protective effect of low salt concentrations is based on an initially increased corrosion and subsequent incorporation of the foreign ions into the protective film.

A comparable increase in the sulfate concentration is less dangerous [3].

Investigations of the road salts calcium chloride and sodium chloride did not result in any crucial differences in the corrosion behavior (Figure 6). Both salts cause pitting corrosion on steel, sodium chloride, however, activates passive steel more readily than calcium chloride [4].

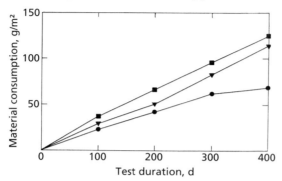

Figure 6: Material consumption of steel St 37 (cf. 1.0037, UNS G10150) in saturated Ca(OH)$_2$ solutions containing road salt [4]
● St 37/1 mol/l NaCl in saturated Ca(OH)$_2$ solution
■ St 37/1 mol/l CaCl$_2$ in saturated Ca(OH)$_2$ solution
▼ St 37/1 mol/l NaCl + 1 mol/l CaCl$_2$ (4:1) in saturated Ca(OH)$_2$ solution

The enormous damage caused by chloride corrosion on reinforcing steels led to a series of interesting measures [5, 6].

1. Protection against chloride by impermeable concrete
2. Protection by appropriate strength of concrete
3. Protection by impermeable covers
4. Wax beds
5. Membranes
6. Polymer impregnations
7. Sulfur infiltration
8. Prevention of the attack by salt on reinforcing steel
9. Zinc coatings
10. Epoxy resin coatings
11. Nickel coatings
12. Use of corrosion-resistant reinforcing materials
13. Alloyed steels
14. Fiber glass
15. Addition of inhibitors
16. Cathodic protection

Cathodic protection in solutions of alkaline earth chlorides depends on two criteria. The potential of steel must be less than the pitting potential to protect against the activation of a pore.

The pitting potential depends on the chloride concentration, the pH-value, the temperature and also the oxygen content of the solution.

Repassivation of active pits takes place below a threshold concentration of 0.03 mol/l chloride. A potential of 50 mV below the equilibrium potential of corrosion is necessary for cathodic protection [7].

Figure 7 shows good agreement of the pitting potentials determined from different groups and the effect of mechanical prestress.

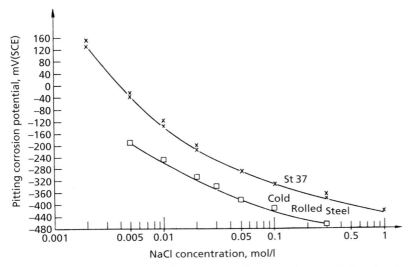

Figure 7: Pitting corrosion potential at different sodium chloride concentrations in saturated calcium hydroxide solution [7]

The dependence of the pitting potential on the temperature in the case of steel St 37 (cf. 1.0037, UNS G10150) can be seen from Figure 8 [7].

The oxygen content shows only a small effect on the pitting potential (see Figure 9).

Self-passivation takes place only with sodium chloride concentrations below the threshold value; the time depends on the chloride concentration (see Figure 10).

In all investigations pitting corrosion and not stress corrosion cracking was observed on the material in the low-alloy steel/calcium hydroxide/chloride system [2, 4, 7, 8].

For practical application in the construction industry, the content of water-soluble chlorides in the concrete mixture is the decisive factor. The primary supply of chlorides comes from the tricalcium aluminate "C_3A" and tetracalcium aluminate ferrite "C_4AF". The chloride binding capacity of Portland cements increases in accordance with their tricalcium content (Table 1) [8].

In cements having a C_3A content 12 %, 60 to 70 % of the chloride ions present become bound during the hydration phase of up to 7 days. In agreement with other investigations, a threshold value of 0.035 % of chloride is found for the chloride concentration in pitting corrosion [8].

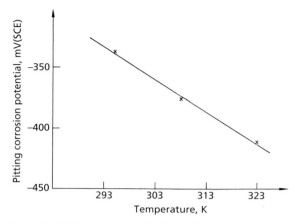

Figure 8: Pitting corrosion potential at different temperatures for St 37 (cf. 1.0037, UNS G10150) in 0.1 mol/l sodium chloride solution saturated with calcium hydroxide [7]

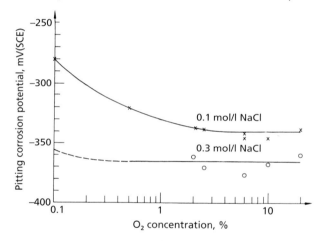

Figure 9: Pitting corrosion potential of steel St 37 (cf. 1.0037, UNS G10150) at different O_2 concentrations of an oxygen-nitrogen mixture in saturated calcium hydroxide solution [7]

Name of the cements	C_3A content, %
PZ 2/325	≤ 1.0
PZ 3/375	≤ 3.0
PZ 4/450	8.0
PZ 1/375	12.0
PZ 1/475	14.0
PZ 5/550	14.0

Table 1: C_3A contents of various Portland cements [8]

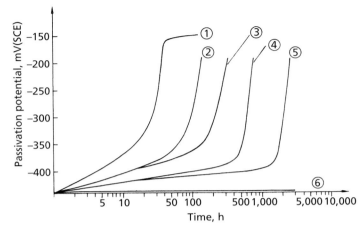

Figure 10: Duration of self-passivation of steel St 37 (cf. 1.0037, UNS G10150) in saturated calcium hydroxide solution at different NaCl concentrations [7]
① 0.01 mol/l NaCl
② 0.015 mol/l NaCl
③ 0.02 mol/l NaCl
④ 0.03 mol/l NaCl
⑤ 0.03 mol/l NaCl
⑥ 0.10 mol/l NaCl

The electrochemical reactions

$$Fe \rightarrow Fe^{2+} + 2e^-$$

and

$$1/2\, O_2 + H_2O + 2e^- \rightarrow 2\,(OH)^-$$

lead to an electric potential between the corrosion site and the passive metal surface (see Figure 11) [9].

As a result of the electric potential formed, the corrosion sites in concrete constructions can be recognized and localized. Empirical limiting values are used to evaluate the corrosion, see Table 2 [9].

This method has been used in the US since 1970 for the non-destructive testing of bridge constructions.

By means of a second electrotechnical method of measurement, the corrosion rate can be calculated from the polarization resistance, using a non-destructive method [9].

The corrosion-promoting effect of formate in the steel-calcium hydroxide system was determined in [10] by electrochemical and gravimetric methods and compared (Table 3).

Corrosion behavior	Potential	
	$mV_{(Cu/CuSO_4)}$	$mV_{(SCE)*}$
With 90 % certainty no corrosion	> −250	> −190
Medium range, to interpretation		
With 90 % certainty corrosion	< 350	< 290

* Saturated Calomel Electrode

Table 2: Empirical limiting values for measurements of the potential (ASTM C 876 – 80) on reinforced steel in concrete constructions under atmospheric conditions [9]

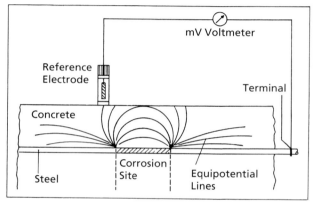

Figure 11: Measurement of the potential on the surface of concrete for finding corrosive zones (schematically) [9]

The results showed considerable deviations in the case of 0.05 mol/l formate. What is certain, however, is that increasing concentrations of formate ions, like chloride ions, depassivate steel in alkaline lime solution [10].

Tests for stress corrosion cracking, using Armco® iron and high-strength constructional steel in aqueous solution saturated with calcium hydroxide, calcium sulfide and calcium sulfate (pH 12.6), showed that after passivation of the surface and formation of a passive film no corrosion on the material occurs, as long as a limiting potential of 800 mV (E_H) is not exceeded.

Above E_H = 200 mV complete passivation takes place and above E_H = 300 mV initial corrosion occurs. Tensile stresses of up to 85 % of $R_{p0.2}$ do not cause SCC up to a potential of 500 mV, as long as the passive film is not damaged. The sulfate ion portion of the test solution was responsible for stress corrosion cracking, while the sulfide ion portion had no effect. The presence of oxygen favors the formation of stress corrosion cracks [11].

Barium hydroxide (3 %) shows low aggressivity towards various steels, such as, for example, St 3 (cf. 1.0036, UNS K02502), 93-1, 10Ch14AG15, 1381, 0Ch21N5T, Ch18N10T (cf. 1.4552, SAE 347) and Ch21N5T (Russian steels) in the system with formalin and alkylphenols. Except for St 3, rapid passivation takes place with all steels (see Figure 12 and Figure 13) [12].

| Test solution | Saturated Ca(OH)$_2$ solution containing mol/l | | | |
Corrosion	0.1 Cl⁻	0.72 Cl⁻	0.3 HCOO⁻	0.05 HCOO⁻
Weight loss, by experiment, mg	2.66	8.6	13.86	0.96
Weight loss, calculated, mg	2.28	11.9	12.00	0.10

Table 3: Comparison between experimentally determined and calculated corrosion rates in different media [10]

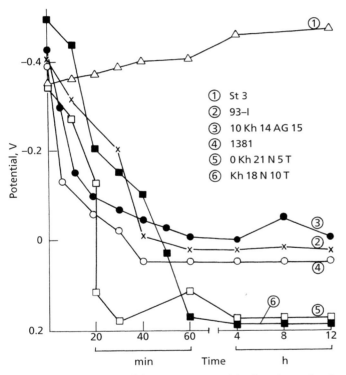

Figure 12: Measurement of the corrosion potentials of steels as a function of time in the system formalin + 3 % Ba(OH)$_2$ [12]

Organic lubricants based on condensation products in which barium hydroxide is incorporated combine lubricant and inhibitory properties for cylinders and pistons [13].

Barium hydroxide is also mentioned as a pigment in rust preventing agents [14].

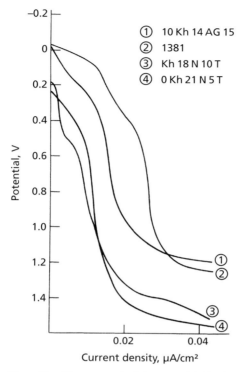

Figure 13: Measurement of the potential kinetic curves of steels in the system formalin + 3 % Ba(OH)$_2$ [12]

According to Table 4, which illustrates the significance of stress corrosion cracking in the field of corrosion, 23.4 % of the corrosion damages are cases of Stress corrosion cracking [15].

Eickemeyer has tried to coordinate some parameters which affect the occurrence of SCC (Figure 14); however, no statements which are generally valid can be made yet [15–18].

Proportion of cracks due to corrosion	56.9 %
Contribution of individual types of corrosion:	
Stress corrosion cracking	23.4 %
Uniform corrosion	31.5 %
Pitting corrosion	15.7 %
Intercrystalline corrosion	10.2 %
Erosion corrosion	9.0 %
Other types	10.2 %

Table 4: Analysis of damages due to corrosion [15]

The important factor for the occurrence of SCC is a critical threshold value of the stress intensity K_{ISCC} which, if exceeded, initiates the crack corrosion.

SCC occurs not only in iron alloys but also in high-strength titanium and aluminium alloys.

SCC is also considered to be hydrogen-induced brittle fracture [11–21].

In solutions of calcium hydroxide, chlorides are particularly damaging, followed by sulfates; nitrates has only a small effect on SCC [17].

It has already been mentioned that sulfate-free solutions have no tendency to SCC in the system steel – calcium hydroxide – sulfate – sulfide. Different statements about this system are found in [22] and [23]. The different results can be explained by the sensitivity to oxygen of aqueous alkaline sulfide solutions.

Investigations [23] confirm that no SCC occurs in the hydroxide – sulfide system. These investigations repeatedly confirm the effect of hydrogen on SCC. Thiocyanate ions, which prevent the recombination of hydrogen atoms to hydrogen molecules, promote stress corrosion cracking.

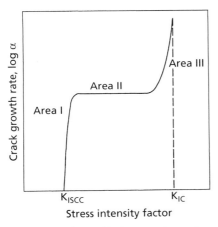

Figure 14: Relationship between the crack growth rate a and the stress intensity factor K_I in the environment-induced fracture of high-strength alloys (schematically) [16]

Figure 15 clearly shows the inhibiting effect of sulfide as well as the SCC inducing effect of chloride and thiocyanate [23].

Figure 16 shows that with increasing hydrogen charge the tendency to SCC also increases [23].

A comprehensive listing of media causing stress corrosion cracking on ferritic steels is given in [24]. A comprehensive theoretical explanation is not possible.

Known theories are discussed by Vasilenko, who confirms the effect of nitrate in solutions of calcium hydroxide, in which steel 40Ch (cf. 1.7035, UNS G51400) is less susceptible to SCC than steel 20ChGS2 (Figure 17) [25].

Steels 20ChGS2 and 20GS2 have only a low tendency to stress corrosion cracking in calcium hydroxide solution [26].

210 | Alkaline Earth Hydroxides

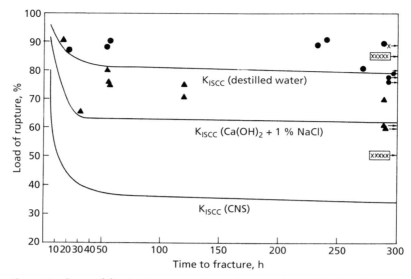

Figure 15: Susceptibility to stress corrosion cracking in calcium chloride solutions containing additives of sulfides and chlorides [23]
● Distilled Water
× Saturated Ca(OH)$_2$ + 0.5 % NaS
▲ Saturated Ca(OH)$_2$ + 1 % NaCl

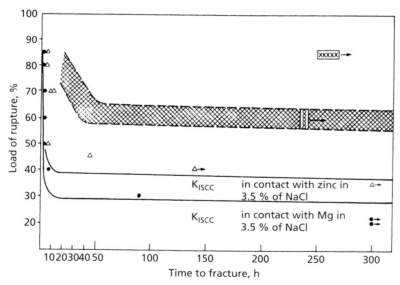

Figure 16: Effect of steel in contact with zinc on the fracture behavior [23]
○ in contact with zinc in cement containing 5 % NaCl
● in contact with Mg in 3.5 % NaCl
△ in contact with zinc in 3.5 % NaCl
× in contact with zinc in pure cement containing distilled water

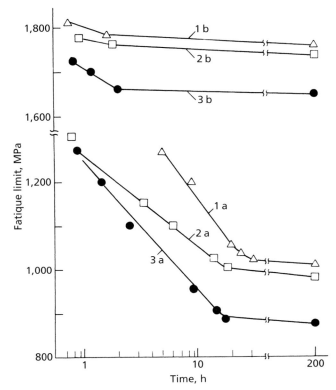

Figure 17: Fatigue limit of notched hardened specimens of steels 40Ch (a), 20ChGS2 (b) in pure Ca(OH$_2$) solution (1) and with additions of Cl$^-$ (2) and NO$_3^-$ (3) (T = 373 K (100°C), E = 700 mV) [25]

In the case of steel 20GS2, there is an increasing susceptibility to the media. Air < calcium hydroxide solution < distilled water < 3 % sodium chloride solution < moist hydrogen sulfide (see Figure 18) [27].

Stress corrosion cracking caused by fatigue is dependent on the frequency of the fatigue cycles and on the amount of additives, too (Figure 19) [28].

In pure calcium hydroxide solution, steels show a better behavior than in water.

Fatigue phenomena in the system reinforced steel-aqueous calcium hydroxide solution leads to a transition from general corrosion to stress corrosion cracking. The passive films are destroyed by increased cyclic stresses, in particular in the presence of chlorides, and corrosion potentials are reached (see Figure 20) [29].

These phenomena occur in particular on bridges and road constructions. These processes can be monitored by electrochemical methods [29].

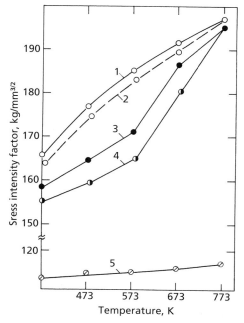

Figure 18: Temperature dependence of the stress intensity factor K_{ic} of the stress corrosion cracking limit on the temperature in various mediums [27]
1 Air
2 Calcium hydroxide solution
3 Distilled water
4 3 % solution of sodium chloride (pH 7.2)
5 Moist hydrogen sulfide

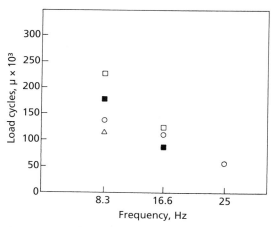

Figure 19: Number of fatigue cycles before reaching the fracture limit of steel specimens [28]
○ Water
□ Saturated Ca(OH)$_2$ solution
■ Saturated Ca(OH)$_2$ solution + 1,000 ppm Cl$^-$
△ Saturated Ca(OH)$_2$ solution + 2,000 ppm Cl$^-$

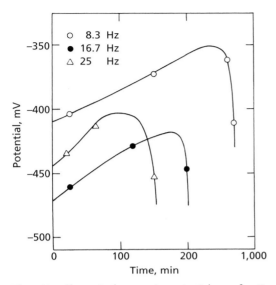

Figure 20: Change in the corrosion potentials as a function of the time of cyclic stress in saturated calcium hydroxide solution [29]

– Inhibitors –

The large number of corrosion mechanisms in the system steel-concrete, which is of such importance in industry, made it necessary to look for inhibitors which are capable of counteracting or even preventing the destruction of the reinforcing steels.

Depending on their effect in practice, the inhibitors are divided into primary and secondary inhibitors. Primary inhibitors do not react chemically with metals, whereas secondary inhibitors are modified at the metal boundary film by an electron reaction. It is desired to have systems which, even after the passive film has been damaged, are themselves capable of forming a new film.

Open systems can be protected by the action of external current, for example by sacrificial electrodes mode of aluminium, zinc or their alloys. In closed systems, inhibitors are preferred.

Type of cement	Composition, %									
	SiO_2	Al_2O_3	Fe_2O_3	CaO	MgO	SO_3	S	Loss on ignition	Free CaO	Insoluble residue
Portland cement	21.06	5.43	3.41	64.0	0.75	2.48	0	2.42	2.15	1.0
Slag cement	23.90	7.60	2.31	57.51	1.40	2.90	1.55	2.42	1.30	1.0

Table 5: Composition of various cement [30]

The mechanisms of primary inhibitors have been investigated by Horner [31].

In alkaline lime solution, hydrazine does not act as oxygen scavenger, unlike sodium sulfite, but blocks the anodic reaction in pores and areas which are endangered by corrosion [32].

At this point it should be mentioned once again that formate despite its reductive properties promotes the corrosion of steel in calcium hydroxide solutions to the same extent as chloride [10].

Investigations of the corrosion behavior of steel coated with Portland cement and slag cement (see Table 5) and containing inhibitor additives in the presence of sodium chloride show that a certain critical inhibitor concentration must be present for passivating the steel surface, in order to prevent corrosion (Table 6 and Table 7).

If the critical amount of inhibitor is exceeded, passivation occurs more easily and takes less time [33].

The measurement of the potential field distribution, which has already been discussed, gave unequivocal measured values in the corrosion system reinforcing steel-calcium hydroxide-calcium nitrite and clearly shows the inhibiting effect of nitrite (Figure 21) [34].

Inhibitor	Critical inhibitor concentration, %	
	Portland cement	Portland cement – blast furnace slag concrete
Sodium nitrite	0.75	1.5
Potassium chromate	3.00	3.5
Sodium benzoate	5.50	6.0
Disodium phosphate	5.50	6.0
Ammonium stearate	5.00	6.0

Table 6: Corrosion inhibition by addition of inhibitors in presence of 2% sodium chloride [33]

Coating	Critical concentration of admixed inhibitor	Maximum concentration of NaCl, %
Cement slurry	–	1.2
Cement slurry + 8% sodium nitrite	–	1.6
Cement slurry + 2% sodium nitrite	1% sodium nitrite	2.0
Cement slurry + 10% potassium chromate	–	1.5
Cement slurry + 3% potassium chromate	2.5% potassium chromate	2.0

Table 7: Effect of coating the reinforcing steel with Portland cement slurry, containing inhibitors, on corrosion behavior [33]

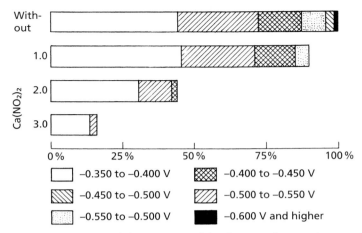

Figure 21: Experimental determination of the change in the corrosion rates in certain potential regions as a function of the inhibiting effect of the Ca(NO$_2$)$_2$ content in alkaline solutions [34]

For the inhibiting effect, the chemical reaction is:

$$2\,Fe^{2+} + 2\,OH^- + 2\,NO_2^- \rightarrow 2\,NO \uparrow + Fe_2O_3 + H_2O$$

90 % of the nitrogen oxide formed could be detected [34]. Nitrite is thus a typical secondary inhibitor. Nitrite has been shown to be a good inhibitor in the presence of chloride under other conditions as well (Figure 22, Table 8) [35].

In the same article good primary organic-based inhibitors are described.

The combination of nitrite with organic inhibitors does not necessarily show improved protection in these systems (Table 9) [36].

In the system cement-steel St 3 (cf. 1.0036, UNS K02502) and calcium chloride, it has been found that nitrite and organic inhibitor intensify each other in their effect. This effect was not found in the system without calcium chloride (Table 10) [37].

The investigations of Eickemeyer, who found that additions of nitrate, in particular in the acidified centers of stress corrosion cracking, slow down or prevent the growth of the cracks, are in agreement with the inhibiting effect of nitrite.

Via the equation:

$$NO_3^- + 3H^+ + 2e \leftrightarrow HNO_2 + H_2O \quad (13)$$

the transition is made to the system steel – calcium hydroxide – nitrite [38].

Nitrate and also perchlorate are said to have an inhibiting effect in the presence of halides (test solution: 0.05 mol/l H$_3$BO$_3$ + 0.004 mol/l KOH) and are added to the medium in suitable amounts to prevent pitting corrosion [39].

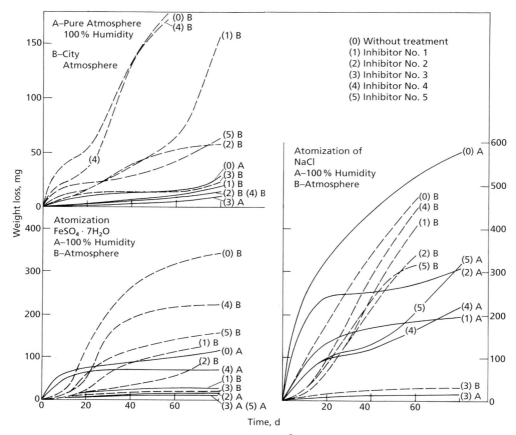

Figure 22: Weight loss of the sheet metal specimens of 80 cm² which have been treated with various inhibitors and exposed to the atmosphere (city air) or humidity [35] (Simulation of industrial atmosphere with iron sulfate and of sea atmosphere with sodium chloride)

No.	Type of medium	Composition	Trade name
1	organic	–	Shell Dromus B®
2	organic	–	Shell-Ensis-Rolling-Oil®
3	organic	–	Veto-Oxide 4-200
4	inorganic	$NaSiO_3$ (40° Be) + $NaNO_2$	–
5	inorganic	$Ca(OH)_2$ + casein	–

Table 8: Type and composition of various inhibitors [35]

Cl⁻, %	0.96 % NO₂	0.96 % NO₂ + 0.1 % SDB*
	% of Area attacked/Total area	
0.28	no corrosion	no corrosion
0.84	no corrosion	no corrosion
1.41	2.0	2.0
2.31	3.0	2.0
3.09	2.0	3.0
3.87	4.0	4.0

* Commercially available inhibitor

Table 9: Effect of nitrite and nitrite + additional organic inhibitor on the corrosion of steel at different Cl⁻ ion concentrations in cement [40]

Additive	Solution	
	0.2 % NaSO₄	3 % NaCl
	Corrosion rate, mm/a (mpy)	
	cement mixture	
Without additive	0.0023 (0.091)	0.00876 (0.345)
PAV*	0.0014 (0.055)	0.0692 (2.724)
CaCl₂	0.0051 (0.201)	0.0106 (0.417)
PAV + CaCl₂	0.0042 (0.165)	0.0092 (0.362)
NaNO₂	0.0009 (0.035)	0.0037 (0.146)
PAV + NaNO₂	0.0009 (0.035)	0.0009 (0.035)
PAV + NaNO₃ + CaCl₂	0.0014 (0.055)	0.0014 (0.055)
CaCl₂ + NaNO₃	0.0046 (0.181)	0.0046 (0.181)
	sat. Ca(OH)₂ solution	
Without additive	0.0028 (0.110)	0.0143 (0.563)
PAV*	0.0014 (0.055)	0.0023 (0.091)
CaCl₂	0.0115 (0.453)	0.0235 (0.925)
PAV + CaCl₂	0.0060 (0.236)	0.0160 (0.630)
NaNO₂	0.0014 (0.055)	0.0018 (0.071)
NaNO₂ + CaCl₂	0.0055 (0.217)	0.0148 (0.583)
PAV + NaNO₂	0.0009 (0.035)	0.0014 (0.055)
PAV + NaNO₂ + CaCl₂	0.0014 (0.055)	0.0018 (0.071)

* PAV = surface-active substance (surfactants)

Table 10: Corrosion rate of St 3 (cf. 1.0036, UNS K02502) in alkaline media with and without inhibitor (PAV) and inhibitor mixtures [41]

Nitrate acts simultaneously as primary and secondary inhibitor.

In hot solutions of calcium hydroxide, nitrate, chloride and sulfate in this order have an increasing effect on the susceptibility of high-strength steel 20ChGS2 to stress corrosion cracking [42].

Unalloyed cast iron

For iron, cast iron and steel vessels used in the paper and leather industry, the addition of 0.13 % calcium hydroxide to the water works as inhibitor [1].

Unalloyed and low-alloy cast iron are not attacked by alkaline earth hydroxides.

High-alloy cast iron, high-silicon cast iron

For the cast materials Ni-Resist® 1 and Ni-Resist® 2 (for chemical composition see Table 11), a material consumption rate of 0.10 g/m² d is reported in calcium hydroxide (lime water) at 289 K (16°C). Expressed in annual terms the corrosion rate is 0.005 mm/a (0.20 mpy).

Material UNS Designation	Trade Name	Chemical Composition, %					
		C max.	Si	Mn	Ni	Cr	Cu
EN-JL3011 UNS F41000	Ni-Resist® 1	3.0	1.0 to 2.8	1.0 to 1.5	13.5 to 17.5	1.0 to 2.5	5.5 to 7.5
GGL-NiCuCr 15 6 3 UNS F41001	Ni-Resist® 1b	3.0	1.0 to 2.8	1.0 to 1.5	13.5 to 17.5	2.5 to 3.5	5.5 to 7.5
0.6660 UNS F41002	Ni-Resist® 2	3.0	1.0 to 2.8	1.0 to 1.5	18 to 22	1.0 to 2.5	–
0.6661 UNS F41003	Ni-Resist® 2b	3.0	1.0 to 2.8	1.0 to 1.5	18 to 22	2.5 to 3.5	–

Table 11: Chemical composition of Ni-Resist® alloys [43]

Further laboratory experiments at 303 K (30°C) in saturated $Ca(OH)_2$ solution with O_2 aeration showed a slight weight increase of the specimens after 20 h [43].

Structural steels with up to 12 % chromium
Ferritic chromium steels with more than 12 % chromium
Ferritic-austenitic steels with more than 12 % chromium

The corresponding chromium-containing stainless structural steels are completely resistant to alkaline earth hydroxides within the range of normal application [1].

Austenitic chromium-nickel steels

Austenitic chromium-nickel steels are resistant under normal operating conditions.
At temperatures of 323 K (50°C) in calcium hydroxide solutions containing calcium carbonate, sodium hydroxide, sodium sulfide and with constant stirring and aeration, the corrosion rate is less than 0.0003 mm/a (0.012 mpy) [1].

Austenitic chromium-nickel-molybdenum steels

The same statements as in section austenitic chromium-nickel steels apply for austenitic chromium-nickel-molybdenum steels [1].

Austenitic chromium-nickel steels with special alloying additions

Steels of this group are resistant to calcium hydroxide under all conditions [1].

Bibliography

[1] Rabald, E.
Corrosion Guide, p. 133
Elsevier Publishing Company, Amsterdam-London-New York, 1968

[2] Grubitsch, H.; Binder, L.; Hilbert, F.
Über den Einfluß verschiedener Chloridionenkonzentrationen auf die Aktiv-Passiv-Korrosionsempfindlichkeit von Stahl in gesättigter Kalciumhydroxidlösung (The influence of various concentrations of chloride ions on the active-passive corrosion susceptibility of steel in saturated calcium hydroxide solution) (in German)
Werkstoffe und Korrosion 30 (1979) 4, p. 241

[3] Akolzin, A. P.; Ghosh, P.; Kharitonov, Yu. Ya.
Application and peculiarity of $Ca(OH)_2$ as inhibitor in presence of corrosion activators
Br. Corros. J. 20 (1985) 1, p. 32

[4] Anonymous
Aus dem Jahresbericht 1983 der BAM (From the BAM Annual Report (1983) (in German)
Werkstoffe und Korrosion 35 (1984) 11, p. 525

[5] Cook, A. R.
De-icing salts and the longevity of reinforced concrete
Corrosion '80, NACE, Houston (Texas), 1980, Paper No. 132, p. 117

[6] Clifton, J. R.; Beeghly, H. F.; Mathley, R. G.
Non-metallic coatings for concrete reinforcing bars
U.S. Dep. Commer. Nat. Bur. Stand. Build. Sci. Ser. (1975) 65, 31 p.

[7] Henriksen, J. F.
The corrosion and protection of steel in saturated $Ca(OH)_2$ contaminated with NaCl
Corros. Sci. 20 (1980) p. 1241

[8] Polster, H.
Über den Einfluß von Chloridionen auf das Korrosionsverhalten von Betonstahl in gesättigter Kalciumhydroxidlösung (The influence of chloride ions on the corrosion behavior of reinforcing steel in saturated calcium hydroxide solution) (in German)
Korrosion (Dresden) 11 (1980) 3, p. 126

[9] Elsener, B.; Böhni, H.
Elektrochemische Untersuchung der Korrosion von Armierungsstahl in Beton (Electrochemical investigation on the corrosion behavior of reinforcing steel in concrete) (in German)
Schweizer Ingenieur und Architekt 14 (1984) p. 264

[10] González, J. A.; Andrade, C.
Determinación electroquimica cuantitativa de la velocidad de corrosión de un acero de construcción. Evaluación de la susceptibilidad al ataque por picaduras (Quantitative electrochemical determination of the corrosion rate of a structural steel. Investigation in the pitting corrosion susceptibility) (in Spanish)
Mater. Construction (Madrid) (1977) 165, p. 69

[11] Janik-Czachor, M.; Riecke, E.
Untersuchungen zum Einfluß von Sulfid- und Sulfationen auf die Korrosion des Eisens in alkalischer Lösung (Investigations in the influence of sulfide and sulfate ions on the corrosion of iron in alkaline solution) (in German)
Werkstoffe und Korrosion 27 (1976) p. 5

[12] Yusifov, V. et al.
Corrosion resistance of structural steels during the concentration alkylphenol with formaldehyde in the presence of barium hydroxide (in Russian)
Deposited Document 1979, VINTI 339710 pp, Tiflis, UdSSR

[13] Rutkovskij, M. L.; Poznyak, T. A.
Corrosion of the cylinder and piston parts of an automobile engine and the mechanism of additive depletion in As-8 oil (in Russian)
Zashch. Met. 16 (1980) 4, p. 435

[14] Kühn, M.
Rostschutzanstrichmittel, gekennzeichnet durch die gleichzeitige Anwendung Anionen, insbesondere Sulfationen, bindender und die Bildung von Wasserstoffperoxid in Lokalelementen verhindernder Pigmentbestandteile (Rust-preventing coatings, characterized by the simultaneous application of pigment particles which can bind anions, especially sulfate ions, and prevent the formation of hydrogen peroxide in local elements)

(in German)
DOS 2 217 498 (25. October 1973)
[15] Eickemeyer, J.; Schlät, F.
Eine Methode zur Spannungsrißkorrosionsprüfung auf der Grundlage der Bruchmechanik
(A method to test stress corrosion cracking based on fracture mechanics) (in German)
Neue Hütte 19 (1974) 4, p. 232
[16] Eickemeyer, J.
Zur Kinetik des subkritischen Rißwachstums bei der Spannungsrißkorrosion hochfesten Stahls
(The kinetics of the subcritical crack growth during stress corrosion cracking of high-strength steel) (in German)
Neue Hütte 21 (1976) 3, p. 175
[17] Eickemeyer, J.
Subkritisches Rißwachstum in hochfestem Stahl bei statischer Belastung in gesättigten Ca(OH)$_2$-Lösungen
(Subcritical crack growth in high-strength steel with static load in saturated Ca(OH)$_2$ solutions) (in German)
Neue Hütte 20 (1975) 7, p. 422
[18] Eickemeyer, J.
Stress corrosion cracking of a high-strength steel in saturated Ca(OH)$_2$ solutions caused by Cl$^-$ and SO$_4^{2-}$ additions
Corros. Sci. 18 (1978) p. 397
[19] Isecke, B.; Stichel, W.
Untersuchungen zur Widerstandsfähigkeit von Spannstählen gegen wasserstoffinduzierte Spannungsrißkorrosion
(Investigations in the resistance of steel for prestressed concrete to stress corrosion cracking induced by hydrogen) (in German)
Amts- und Mitteilungsblatt der Bundesanstalt für Materialprüfung 13 (1983) 1, p. 12
[20] Riecke, E.
Über den wasserstoffinduzierten Sprödbruch hochfester Stähle
(On hydrogen-induced brittle fracture in high-tensile steels) (in German)
Arch. Eisenhüttenwes. 44 (1973) 9, p. 647

[21] Heiligenstaedt, P.; Bohnenkamp, K.
Untersuchungen zum wasserstoffinduzierten Sprödbruch verzinkter Spannstähle im Beton
(Investigation on the hydrogen-induced brittle fracture of galvanized prestressed steels in concrete) (in German)
Arch. Eisenhüttenwes. 47 (1976) 2, p. 107
[22] Gouda, V. K.; Abdul Azium, A. A.; Sayed, H. A. E.
The effect of stress on the corrosion behavior of mild steel in alkaline solutions
Corros. Sci. 9 (1979) p. 215
[23] McQuinn, K. F.; Elices, M.
Stress corrosion resistance of transverse precracked prestressing tendon in tension
Br. Corros. J. 16 (1981) 4, p. 187
[24] Parkins, R. N.
Environmental aspects of stress corrosion cracking in low-strength ferritic steels
Proc. Conf. NACE, Houston (Texas) (1977) p. 601
[25] Vasilenko, I. I.; Dykij, I. I.
Some aspects of steel corrosion cracking in the nitrate solution (in Russian)
Fiz. Khim. Mekh. Mater. 17 (1981) p. 82
[26] Petrivskij, R. I.; Krasovskaja, G. M.; Alekseev, S. N.; Dikij, I. I; Vasilenko, J. J.
Brittle fracture of high-strength steels in a calcium hydroxide solution (in Russian)
Fiz. Khim. Mekh. Mater. 14 (1978) 5, p. 38
[27] Petrivskij, R. I.; Krasovskaja, G. M.; Alekseev, S. N.; Koval, M. V.; Dikij, I. I.
Brittle fracture resistance of the 20GS2 reinforcing steel (in Russian)
Fiz. Khim. Mekh. Mater. 13 (1972) 2, p. 84
[28] Cerisola, G.; Busca, G.; De Anna, P. L.
Corrosion fatigue behavior of iron in different aqueous environments
Materials Chemistry and Physics 9 (1983) p. 387
[29] Bonora, P. L.; Cerisola, G.; De Anna, P. L.
Acil korróziós kifáradása vizes oldatokban – elektrokémiai vizsgálatok
(Corrosion fatigue of steel in aqueous solutions – electrochemical investigations) (in Hungarian)
Kemiai Közlemények 56 (1981) p. 429
[30] Gouda, V. K.; Halaka, W. Y.
Corrosion and corrosion inhibition of reinforcing steel
Br. Corros. J. 5 (1970) p. 198

[31] Horner, L.
Inhibitoren der Korrosion des Eisens – Wirkungsweise, elektronische und strukturelle Voraussetzungen
(Inhibitors against corrosion of iron – effect, electronic and structural conditions)
(in German)
Chemiker Zeitung 100 (1976) 6, p. 247

[32] Gouda, V. K.; Shater, M. A.
Corrosion inhibition of reinforcing steel by using hydrazine hydrate
Corros. Sci. 15 (1975) 3, p. 199

[33] Gouda, V. K.; Halaka, W. Y.
Corrosion and corrosion inhibition of reinforcing steel
Br. Corros. J. 5 (1970) p. 198

[34] Rosenberg, A. M.; Gaidis, J. M.
The mechanism of nitrite inhibition of chloride attack on reinforcing steel in alkaline aqueous environments
Mater. Performance 18 (1979) 11, p. 45

[35] Royuela, J. J.; Feliú, S.; Sánchez-Gálvez, V.; Elices, M.
Inhibition of atmospheric corrosion of prestressing steel reinforcement
(in Spanish)
Corros. Prot. 12 (1981) 2, p. 7

[36] Ostrovskij, A. V.; Novgorodskij, I. I.; Stantsel, E. E. et al.
Surfactants effect on corrosion of concrete-reinforcing steel (in Russian)
Zashch. Met. 16 (1980) p. 69

[37] Tupikin, E. I.; Chernov, M. S.; Platonova, E. E.
The corrosion resistance of steel St 3 reinforcements in alkaline media and in cement containing inorganic salts and additions of technical surfactants
(in Russian)
Korroz. Zashch. 5 (1981) p. 17

[38] Eickemeyer, J.
Inhibierende Wirkung von Nitrationen auf das Rißwachstum bei der Spannungsrißkorrosion eines hochfesten Stahls
(Inhibiting effect of nitrate ions on crack growth of stress corrosion cracking of a high-strength steel) (in German)
Neue Hütte 21 (1976) 5, p. 304

[39] Strehblow, H. H.; Titze, B.
Pitting potentials and inhibition potentials of iron and nickel for different aggressive and inhibiting anions
Corros. Sci. 17 (1977) 6, p. 461

[40] Ostrovskij, A. V.; Novgorodskij, I. I.; Stantsel, E. E. et al.
Surfactants effect on corrosion of concrete-reinforcing steel (in Russian)
Zashch. Met. 16 (1980) p. 69

[41] Tupikin, E. I.; Chernov, M. S.; Platonova, E. E.
The corrosion resistance of steel St 3 reinforcements in alkaline media and in cement containing inorganic salts and additions of technical surfactants
(in Russian)
Korroz. Zashch. 5 (1981) p. 17

[42] Dikij, I. I; Petrivski, J.; Alekseev, S. N.; Krasovskaja, G. M.
The effect of surface-active electrolyte ions on corrosion-mechanical strength of high-strength steel (in Russian)
Fiz. Khim. Mekh. Mater. 16 (1980) 1, p. 50

[43] Anonymous
Die Nickel-Resist-Gußeisenwerkstoffe
(Nickel-resist cast iron materials)
(in German)
International Nickel, 3rd ed. (1968) p. 92

Ammonia and Ammonium Hydroxide

Unalloyed and low-alloy steels/cast steel
Unalloyed and low-alloyed cast iron

The unalloyed and low-alloyed steels are practically resistant to dry gaseous ammonia. However, at higher temperatures nitrogen can be taken up in the surface zones. This reaction is exploited for nitride hardening of the steels. Thereby, as reversal of the reaction according to Equation 1, the primary atomic nitrogen produced by thermal dissociation diffuses at temperatures of 773 K to 803 K (500°C to 530°C) into the steel and produces nitrides giving high and uniformly distributed hardness values in the surface zones.

Equation 1: $N_2 + 3 H_2 \leftrightarrow 2 NH_3 + 92.44$ kJ

This procedure achieves hardness values of up to 1,000 HV 0.5. The hardness values decrease continuously with increasing distance from the surface, and at a distance of about 0.4 to 0.5 mm from the surface the initial values for the particular steel are reached again, as shown in Figure 1 for the example of a special constructional steel [1].

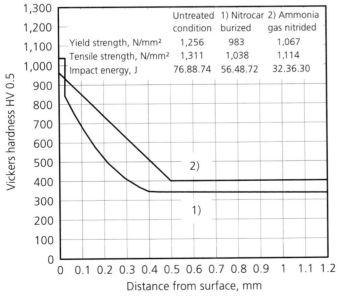

Figure 1:   Typical hardness profile for nitriding of a constructional steel [1]

In moist ammonia gas stress corrosion cracking can take place, as experience has shown in the gas space of spherical containers for liquid ammonia (see further below).

Liquid ammonia is a water-similar solvent, in which comparable electrochemical reactions can take place, and an active dissolution of iron is conceivable according to the following equations [2]:

Equation 2, anodic part reaction: $Fe \rightarrow Fe^{2+} + 2\,e^-$

Equation 3, cathodic part reaction: $2\,NH_4^+ + 2\,e^- \rightarrow 2\,NH_3 + 2\,H$

Equation 4, cathodic part reaction: $O_2 + 2\,NH_4^+ + 4\,e^- \rightarrow 2\,NH_3 + 2\,OH^-$

However, the uniform surface corrosion of iron and unalloyed steel in gaseous and water-free liquid ammonia is negligibly small at the free corrosion potential (see also Table 17), because due to the very small dissociation of the NH_3 according to

Equation 5 $\quad 2\,NH_3 \rightarrow NH_4^+ + NH_2^-$

the NH_4^+ ions required for the cathodic reaction are not available in sufficient number.

Various low-alloy steel grades are utilized in large quantities for tank vehicles and pipelines for transporting as well as for cylindrical pressurized and large spherical pressurized containers for storing liquid ammonia. Here stress corrosion cracking can take place in the liquid space as well as in the vapor space.

In 1956, the American specialist literature first mentioned cracking damage on containers made of water quenched and tempered fine grain constructional steels. The containers served for storage, transportation and bringing in of liquid ammonia for fertilizing the soil. The damage cause was recognized to be stress corrosion cracking [3]. The cracks appeared predominantly in the region of weld seams.

From the fact that these cases of damage appeared only on transportation or storage containers but not in production plants it was concluded that pure ammonia does not produce stress corrosion cracking, but that contaminations, that come into the ammonia through contact with air when filling and decanting the ammonia, are responsible for the damage.

Numerous investigations were carried out in connection with this damage, with the objective of clarifying the influence of contaminations of the ammonia, essentially by the gases air, oxygen, nitrogen and carbon dioxide as well as by water.

Already in 1962, the results of an extensive test program were reported, that was carried out as joint project by various industrial companies as a consequence of the damages observed in the USA [4]. Various steel grades utilized in the affected plant sections were examined in these investigations. The chemical composition and the chief mechanical key parameter values of these steels are specified in Table 1 and in Table 2.

For the investigations, welded as well as non-welded pre-stressed fork and ring specimens were exposed in containers of an ammonia distribution plant for agricultural utilization. The results of these exposure tests for the various investigated specimens of the steel grades T 1, ASTM A202 and A212, as well as ASME Case 1056 are collected in Table 3.

Table 1: Chemical composition of the steel grades tested in liquid ammonia [4]

Steel		C	Mn	P	S	Si	Cu	Ni	Cr	Mo	V	B
ASTM A 212 grade B		0.33	0.77	0.017	0.030	0.24	0.04	0.03	0.04			
ASTM A 285	cf. 1.0345	0.17	0.53	0.009	0.031	0.063	0.04					
ASTM E case 1056		0.30	1.13	0.017	0.030	0.03	0.04	0.11	0.07			
ASTM A 202 grade B	cf. 1.8812	0.19	1.46	0.016	0.020	0.77	0.04	0.05	0.45			
T 1	cf. 1.8719	0.16	0.88	0.017	0.019	0.22	0.34	0.84	0.53	0.48	0.05	0.002
SAE 4130	cf. 1.7218	0.33	0.54	0.009	0.018	0.28			0.98	0.19		
Spring steel		0.52	0.71	0.007	0.040	0.13	0.20	0.09	0.05	0.08	0.002	

Table 2: Mechanical key parameter values of the steel grades tested in liquid ammonia [4]

Steel	R_e N/mm^2	R_m N/mm^2	Elongation %	Hardness R_B	Hardness R_C
ASTM A 212 grade B	358	583	34	84	
ASTM A 285	260	415	45	66	
ASTM E case 1056	337	555	49	85	
ASTM A 202 grade B	400	655	41	94	
T 1	815	880	26		25
SAE 4130	836	942	19		29
Spring steel					55

The investigations of the influence of contaminations were carried out in special test containers or on pipes. Thereby the tests were carried out in one case in ammonia for fertilizing purposes, for which a lower purity is demanded, and in another case in ammonia for the refrigeration industry with higher purity. As examples, Table 4 and Table 5 show the influence of the additions of air and water on the behavior of the fork specimens made of the two higher strength steels T 1 and SAE 4130 in the two ammonia types.

The influence of carbon dioxide is shown in Table 6 based on the results for steel springs.

The results of these investigations confirm the experience in practice that higher strength tempered steels are significantly more sensitive to stress corrosion cracking in liquid ammonia than normalized annealed constructional steels, and that

Specimens	Number of specimens		Failure %	Number of cracked specimens after exposure duration of weeks				
	Checked	Cracked		0–50	51–100	101–150	151–200	201–250
A	61	37	60.6	20	2	9	1	5
B	22	12	54.5	10	0	1	1	0
C	78	33	42.5	14	9	3	4	3

A = cold formed, welded, prestressed, B = cold formed, prestressed, C = only prestressed

Table 3: Results for the specimens exposed in ammonia containers [4]

Additive		Number of specimens		
Air	Water, %	Checked	Cracked	Failures, %
No	0	14	3	21.4
Yes	0	44	10	22.8
Yes	0.1	4	0	0
Yes	0.25	4	0	0
	0.5	14	0	0
Yes	1.0	1	0	0
Yes	2.0	10	0	0
Yes	4.0	10	0	0

State: cold formed, welded, prestressed

Table 4: Influence of air and water on the stress corrosion cracking of T 1 and SAE 4130 in ammonia for the refrigeration industry (test duration: up to 160 weeks) [4]

Additive		Number of specimens		
Air	Water, %	Checked	Cracked	Failures, %
No	0	43	7	16.3
Yes	0	43	1	2.3
Yes	0.25	43	0	0.8

State: cold formed, welded, prestressed

Table 5: Influence of air and water on the stress corrosion cracking of T 1 and SAE 4130 in ammonia for fertilizer purposes (test duration: 48 weeks) [4]

Specimen	Additive	Number of specimens	
		Checked	Cracked
1	air + 0.25 % water + 1,000 ppm CO_2	18	0
2	air + 1,000 ppm CO_2	18	18

Table 6: Influence of carbon dioxide on the stress corrosion cracking of steel springs [4]

contaminations by air with normal contents of carbon dioxide are necessary to initiate stress corrosion cracking. Water fractions as from 0.1 % reduce the danger of stress corrosion cracking and can largely prevent crack formation in the softer steels. From the results the following recommendations were derived already in 1962 for the construction and operation of containers for liquid ammonia:

- Larger containers should be completely stress relief annealed after welding.
- Access of air must be avoided with great care. Before filling them, new containers or ones that have been opened should first of all be filled with nitrogen and then purged with ammonia to remove residual air.
- If permitted by the product, a water content of at least 0.2 % should be added.

In the United States of America a special stipulation of the Department of Transportation (DOT) makes this water addition obligatory for safe transport of liquid ammonia in containers made of tempered steel grades [5, 6].

In spite of these measures, cases of damage still occurred subsequently from time to time, whereby in particular transportation containers made of the tempered steel grades were involved. Accidents with road transport tanks, such as the bursting of a tank mounting unit made of steel T 1 with a load of 19 t of ammonia in 1968 in France, caused five fatalities [7], and the accident with a cylindrical storage container with 30 t of ammonia in 1973 in South Africa even caused 22 fatalities [8, 9]. In the latter case the explosion took place while the container was being filled from a rail tank truck and the crack was initiated at the weld seam of the dished floor. The pressure at the time of bursting was 6 bar and therewith well below the operating pressure rating of 19 bar, so that none of the safety valves responded. However, the involved floor had shown cracks in the region of the weld seam already two years before, that had been eliminated by grinding out and rewelding. After the welding a pressure test was made with 24 bar, but no stress relief annealing was carried out.

Also in the case of damage described in [10] the increased hardness in the weld seam region on the spherical floor of a road transport tank caused the cracking damage. At the time of the damage the internal pressure was 1.85 MPa, corresponding to only 40 % of the yield point of the container steel with a nominal container load of about 280 MPa. Further damages and their causes as well as preventive measures are described in [11–16].

The results of long term tensile tests confirm the negative influence of air and the positive effect of a water addition on the stress corrosion cracking resistance of the steel ASTM A 517 grade F in liquid ammonia [17]. The chemical composition of the investigated steel is shown in Table 7.

C	Mn	P	S	Si	Cu	Ni	Cr	Mo	V	Ti	B
0.18	0.86	0.015	0.018	0.23	0.36	0.74	0.52	0.48	0.036	0.005	0.0046

Table 7: Chemical composition of the investigated steel ASTM A 517 grade F [17]

The tensile tests were carried out with an elongation rate of 2×10^{-6} 1/s in pure ammonia for metallurgical application at room temperature. The influence of a contamination of the ammonia with air on the stress corrosion cracking behavior is shown by the specifications in Table 8.

Test medium	Necking at fracture %	Fractography
Air	50-54-56	ductile
Ammonia, pure	51-53	ductile
Ammonia, contaminated with air	20-23-28	weakly ductile subsidiary cracks

Table 8: Results of the slow tensile tests on steel ASTM A 517 grade F in liquid ammonia (several measurements) [17]

According to these results, stress corrosion cracking of the investigated material takes place only in ammonia contaminated with air. Also in other laboratory investigations no stress corrosion cracking took place in pure ammonia [18, 19].

The influence of a water content in the ammonia is illustrated by Figure 2. Already the addition of about 0.08 % of water to the ammonia that was contaminated with air was able to prevent stress corrosion cracking.

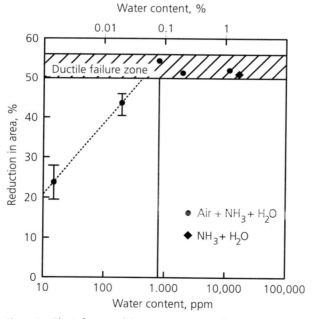

Figure 2: The influence of the water content in the ammonia on the sensitivity of the steel ASTM A 517 grade F [17]

In addition to the investigations at the free corrosion potential, tests were also carried out with anodic and cathodic polarization of the specimens in order to test the possibility of hydrogen induced cracking in this medium. Figure 3 shows correspondingly the necking values at fracture found on the specimens in ammonia contaminated with air, depending on the impressed current.

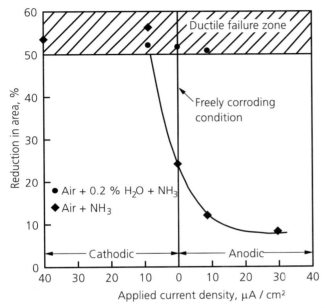

Figure 3: Influence of an impressed anodic or cathodic current on the stress corrosion cracking behavior of the steel ASTM A517 grade F in ammonia containing air [17]

The influence of the impressed current on the sensitivity for stress corrosion cracking in ammonia containing air can thus be explained only by an anodically controlled cracking mechanism that can, however, be inhibited by the addition of 0.2 % of water. According to these results, a hydrogen influence is unlikely, as also confirmed by other electrochemical investigations with regard to the problem of stress corrosion cracking in liquid ammonia [20–22].

Largely the same results were also obtained in the investigation of pipe steels [23]. For transporting ammonia in pipelines, electric resistance welded (ERP) pipe steels of low strength grade (API grades X-42 and X-46) in normal annealed state with maximum copper contents of 0.15 % were usually utilized. In order to assess the behavior of the higher strength ERP pipe steels, for which only the longitudinal seam is normalizing annealed after production, in liquid ammonia, the ERP pipe steels mentioned in Table 9 were investigated. To test the influence of the copper content, test melts with the basic composition of the X-46 but with graded copper contents were included in the investigation.

	C	Mn	P	S	Si	Cu	Ni	Cr	Mo
				Pipe steels					
X-42	0.25	0.64	0.009	0.022	0.011	0.14	0.064	0.046	0.014
X-46	0.24	0.84	0.012	0.028	0.011	0.026	0.060	0.054	0.015
X-52	0.23	1.25	0.007	0.022	0.024	0.007	0.016	0.032	0.011
X-60	0.32	1.14	0.010	0.017	0.17	0.012	0.015	0.014	0.010
				Test melts					
A	0.18	0.89	0.013	0.027	0.034	< 0.005	0.14	0.006	0.009
B	0.20	0.88	0.015	0.030	0.036	0.064	0.14	0.006	0.009
C	0.20	0.89	0.015	0.029	0.040	0.16	0.14	0.006	0.011
D	0.20	0.88	0.015	0.030	0.038	0.30	0.14	0.006	0.011
E	0.18	0.89	0.015	0.030	0.036	0.48	0.14	0.006	0.011

Table 9: Chemical composition of the investigated pipe steels and test melts [23]

The mechanical key parameter values of the materials are contained in Table 10. For the investigations tensile specimens of the pipes were taken from the base material transversally to the longitudinal seam and transversally with respect to a round seam. From the sheets of the test melts the tensile specimens were taken transversally to the rolling direction. The extension rate for the slow tensile tests was 1×10^{-6} 1/s. Three grades of ammonia that were taken from pipelines as well as three ammonia grades for metallurgical applications, were available for the investigations. The contents of water, oxygen and nitrogen of the six ammonia grades are specified in Table 11. On the one hand, the investigations were carried out on the six grades in the state as delivered and on the other hand in the metallurgical ammonia after contamination with air as well as after contamination and addition of 0.2 % of water.

The results can be summarized as follows:

- In the three ammonia grades from the pipelines (1–3) no stress corrosion cracking took place. Evidently the water content is adequate to prevent stress corrosion cracking even in the two grades that were already strongly contaminated with oxygen and nitrogen.
- The metallurgical ammonia grades (4–6) also did not lead to stress corrosion cracking in the state as delivered.
- After contamination with air, stress corrosion cracking took place on all specimens, i.e. in those of the pipe steels as well as those of the test melts. On the welded specimens of the pipe steels the cracks always appeared in the welds or in the heat-influenced zone.
- The sensitivity for stress corrosion cracking is independent of the strength grade, the heat treatment or the copper content of the materials.
- The addition of 0.2 % of water prevented the appearance of stress corrosion cracking in all cases.

	Yield point R_e MPa	Tensile strength R_m MPa	Fracture elongation A %	Necking at fracture Z %
X-42	362	516	27.8	58.1
X-46	422	517	22.9	51.7
X-52	422	560	26.6	54.6
X-60	505	609	21.1	48.2
A	261	446	37.3	66.7
B	251	457	35.0	63.9
C	276	470	34.3	64.8
D	286	474	33.7	64.2
E	301	487	34.0	65.1

Table 10: Mechanical key parameter values of the investigated materials [23]

	Water content	Oxygen concentration	Nitrogen content
Ammonia from pipeline			
1	2,200	282	446
2	1,800	282	297
3	4,300	9.4	0
Metallurgical ammonia			
4	6	9.4	71
5	1	51.5	127
6	-	141	463

Table 11: Contents (ppm) of water, oxygen and nitrogen of the ammonia kinds utilized for the investigations, in the state as delivered [23]

In the 1960 years spherical pressurized containers made of fine grain constructional steels with yield point grades of 355 N/mm² and 460 N/mm² were constructed in Germany and other European countries for storing liquid ammonia. At the end of the 1960 years and beginning of the 1970 years, damage appeared on such spherical pressurized containers due to stress corrosion cracking [24]. After a world wide survey of published data on the problem of stress corrosion cracking in spherical containers for storing liquid ammonia, carried out in 1982, it was found that of 121 checked containers 51 % showed more or less extensive damage [25].

In connection with the cracking damage that had appeared in spherical containers for storing liquid ammonia, investigations were carried out in a German chemical company to examine the susceptibility of various materials to stress corrosion cracking under the influence of liquid ammonia on the one hand, and to determine more exact conditions for the initiation of stress corrosion cracking and the inhibition by water on the other hand, taking into consideration the potential dependencies arising in connection therewith [2].

The investigations are divided into 3 groups:

- Exposure tests on welded prestressed steel specimens in a spherical pressurized container for liquid ammonia with the objective to obtain information regarding the susceptibility of welded joints between certain materials under conditions as encountered in practical operation of ammonia storage containers.
- Tests in the laboratory with slow tensile tests (CERT method) on tension rods made of the base material of the fine grain constructional steels P355N (Material No. 1.0562, cf. UNS K01600), P460N (Material No. 1.8905, cf. UNS K12004) and of the boiler sheet P235GH (Material No. 1.0345, cf. UNS K01700) at the free corrosion potential. The objective of these tests was to investigate the influence of individual contaminants in the liquid ammonia. As the test apparatus provided for these tests, the autoclave of a CERT testing machine and analyzer facility were integrated in a circuit for liquid ammonia. It was possible to add specific gaseous contaminants or water to the ammonia circulation. The analysis was made with a gas chromatograph.
- Electrochemically controlled CERT tests on tensile rods made of the fine grain constructional steel P460N.

For the exposure tests welded (UP and E-manual) bending specimens of the steel grades listed in Table 12 with the older designations, the material numbers and the newer designations were selected. The specimens were tested in the welded state as well as after subsequent stress relief annealing. The exposure took place in a spherical container with liquid ammonia at environmental temperature. The test duration was 197 days.

DIN 17 155	Mat. No.	DIN EN 10 028 T2	Comparable to
Boiler sheet H I[†]	1.0345	P235GH	UNS K01700
15Mo3[†]	1.5415	16Mo3	UNS K11820
13 CrMo 4 4[†]	1.7335	13CrMo4-5	UNS K11547
StE 355	1.0562	P355N	UNS K01600
StE 460	1.8905	P460N	UNS K12004
17 MnCrMo 3 3[†] N-A-XTRA 70	1.7279		

† old material designation

Table 12: Materials for the exposure tests in a spherical ammonia container [2]

With the visual and metallographic investigations of the welds and the base material of the 72 exposed specimens (12 specimens of each material) it was found that only the following specimens were without any cracks:

- Boiler sheet H I with E-manual and UP welding after stress relief annealing, and
- P355N with E-manual and UP welding in the state as delivered as well as after stress relief annealing.

Cracks were found in all other specimens, predominantly in the welds and in the weld transition zone. The materials 16Mo3 (Material No. 1.5415) and 13CrMo4-5 (Material No. 1.7335) showed cracking practically exclusively in the welds. Apart from occasional inter-crystalline crack components, the crack course in the weld material and in the base material was always trans-crystalline.

The steels P235GH, P355N and P460N were chosen for the tests in the laboratory. The slow tensile tests were carried out with an elongation rate of 1.25×10^{-4} mm/s. In these CERT tests oxygen was always added together with nitrogen, because only air can appear as contaminant in practice. Figure 4 shows the influence of the water content in the liquid ammonia on the fracture necking. Depending on the water content, a clear high and low with steep transition results for the values of the fracture necking.

Figure 5 shows the influence of the oxygen content on the fracture necking on the steels P460N (Material No. 1.8905), P355N (Material No. 1.0562) and P235GH (Material No. 1.0345) in the case of insufficient inhibition with water. In general, with increasing oxygen content the fracture necking decreases, i.e. the damage by stress corrosion cracking increases.

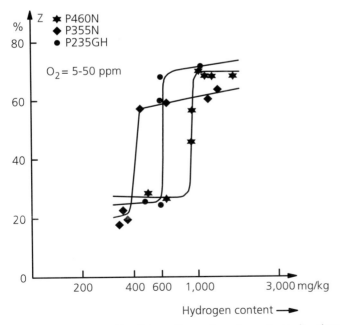

Figure 4: Fracture necking Z depending on the water content in liquid ammonia [2].

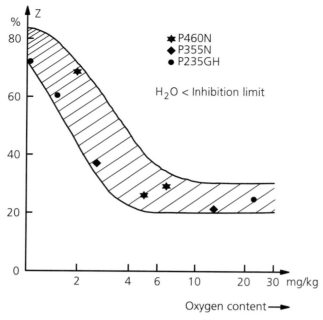

Figure 5: Fracture necking Z depending on the oxygen content in liquid ammonia (water content below the inhibition limit) [2].

The joint influence of the oxygen and the water content in liquid ammonia on the stress corrosion cracking is found to be as follows on the basis of the CERT tests:

- Ammonia in the state as delivered (purity higher than 99.9%) without addition of contaminants does not produce any stress corrosion cracking.
- Already small additions of oxygen (2–3 ppm) in the form of air can initiate stress corrosion cracking. Thereby no differences are found between the investigated materials (Figure 5).
- Water in liquid ammonia is able to inhibit the stress corrosion cracking caused by contaminations with air. For the individual materials steep transitions with different water contents are found as boundary between regions with and without stress corrosion cracking (Figure 4).
- Strongly increased oxygen contents require higher water contents for inhibition. With very small oxygen contents small water contents may suffice.

The results with regard to the damaging influence of oxygen and the inhibiting effect of water in the liquid phase can be summarized in the form shown in Figure 6.

Experience in practice had shown that in containers in which the ammonia was inhibited by the addition of water, stress corrosion cracking can take place in the vapor space. Therefore tests were carried out on pipe specimens stressed with internal pressure in an autoclave half filled with liquid ammonia. The specimens were only half immersed in the liquid ammonia. In the gas space the specimens could be slightly heated or cooled (see Figure 7). Thus with cooling condensation of a thin film of ammonia on the specimen surface was possible [25–28].

Unalloyed and low-alloy steels/cast steel | 235

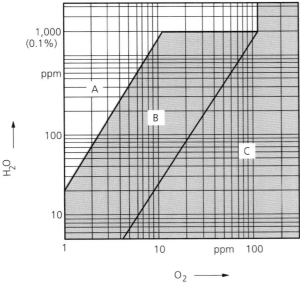

Figure 6: Hazard regions in liquid ammonia for steels up to yield point grade of 355 N/mm² [25]
Region A = no hazard
Region B = stress corrosion cracking possible
Region C = stress corrosion cracking very likely

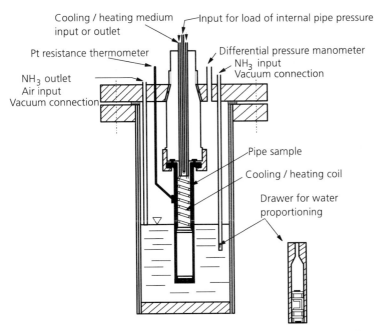

Figure 7: Test apparatus for testing pipe specimens (liquid phase and vapor phase) [25]

The tests on the steel grades specified in Table 13 were carried out at 291 K (18°C) in high purity ammonia that in the state as delivered contained max. 1 ppm of oxygen, max. 5 ppm of nitrogen and max. 5 ppm of water in the liquid phase.

Materials		Composition						Mechanical properties		
Designation	Steel grade	C	Mn	Si	P	S	Ni	R_e MPa	R_m MPa	Elongation %
St 52 1.0580	C-Mn steel normalized	0.18	1.42	0.31	0.014	0.018	0.04	390	540	36
NV4-4	C-Mn steel normalized	0.10	4.60	0.33	0.017	0.009	0.63	435	540	31
OX606 ASTM 537 CL. 2	C-Mn steel normalized	0.11	1.46	0.42	0.022	0.005		515	610	22
UHB 2N 50	5 % Ni steel tempered	0.08	0.57	0.27	0.004	0.007	4.92	475	570	28
UHB 2N 50 BS 1501-510	9 % Ni steel tempered	0.06	0.49	0.21	0.009	0.012	8.90	750	775	17

Table 13: Chief contents of alloying elements and mechanical properties of the investigated steel grades [28]

In the high purity ammonia in the state as delivered no stress corrosion cracking took place. However, additions of about 2 ppm of oxygen sufficed to initiate stress corrosion cracking. The cracking behavior of the materials was independent of whether the ammonia was contaminated with oxygen, with air or with oxygen/nitrogen mixtures between 5 ppm and 200 ppm. With only a few exceptions, stress corrosion cracking with the named contaminants took place only in the liquid phase.

Figure 8 shows the region endangered by stress corrosion cracking depending on the contents of water and oxygen in the case that the same temperature of 291 K (18°C) is present in the vapor space as well as in the liquid space. The results are partly identical to the limits of Figure 6 (broken lines). Under these conditions cracks appear only in the liquid phase.

The conditions do not change if the specimen surface in the gas space is slightly heated (1 or 3 K (1 or 3°C) higher temperature than in the liquid) (Figure 9).

If, in contrast thereto, condensation takes place on the specimen surface in the vapor space in response to slight cooling, stress corrosion cracking is observed, but only in the region of the vapor phase or the transition from the liquid phase to the vapor phase (Figure 10). Even water contents of more than 0.1 % in the liquid phase cannot prevent this.

On storage containers such temperature differences are possible in practice during the day and night transition, as measurements during the period from October to May show for an ammonia sphere without thermal insulation (Table 14) [25]. Water contents of 500 ppm or more in the ammonia prevented the appearance of cracks in the liquid phase as well as in the vapor region.

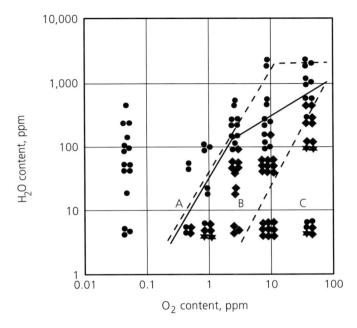

- ● Stress corrosion cracking in the liquid phase
- ◆ Stress corrosion cracking in the vapor phase
- ✱ No stress corrosion cracking

Figure 8: Hazard region for stress corrosion cracking on low-alloy steel depending on the contents of water and oxygen in liquid ammonia at 291 K (18°C) in the vapor space and liquid space [25]

Temperature measurement	Liquid region	Vapor region
Highest temperature, K (°C)	286.75 (13.75)	307.05 (34.05)
Lowest temperature, K (°C)	263.65 (−9.35)	267 (−6.0)
Maximum temperature difference on one day, K	6.3	21.4

Table 14: Temperature measurements on an ammonia sphere [25]

The migration of the damage from the liquid region to the vapor region is due to the different solubilities of oxygen and water in the two phases. Due to the distribution equilibrium, the solubility of oxygen in the vapor phase is higher by a factor of about 650 than that in the liquid phase, whereas the concentration of water in the vapor is lower by a factor of about 500 than that in the liquid ammonia. It is therefore necessary to monitor the contents of oxygen and water in storage containers for liquid ammonia. Appropriate methods of sampling and analysis are available for this [29].

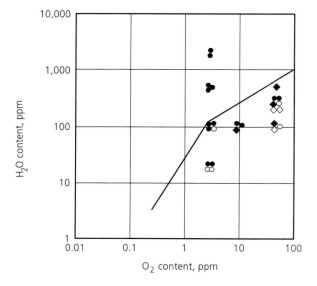

◇◆ 1 °C Temperature difference from NH$_3$(liquid) and NH$_3$-vapor
○● 3 °C
●◆ No stress corrosion cracking
○◇ Stress corrosion cracking

Figure 9: Hazard region for stress corrosion cracking on low-alloy steel depending on the contents of water and oxygen in liquid ammonia when heating the specimens in the vapor space [25]

Although in practice fewer cases of damage are known in plants that are operated practically without pressure at 240 K (−33°C), some tests were also carried out at 293 K (20°C) and at 240 K (−33°C). The critical stress limits are specified in Table 15 that are necessary at a temperature of 293 K (20°C) and of 240 K (−33°C) to initiate stress corrosion cracking in ammonia with an oxygen content of 50 ppm. Since there is practically no difference between the results obtained at the two test temperatures, only the values obtained at 240 K (−33°C) are specified.

Steel	Stress limit			Lowest extension at fracture	
	MPa		% R$_e$	%	%
	293 K (20°C)	240 K (−33°C)	293 K (20°C)	293 K (20°C)	240 K (−33°C)
St 52	425	520	112	0.8	1.9
NV4-4	450		103	1.7	
OX606	525	525	102	0.4	0.6
UHB 2N 50	485		102	1.7	
UHB 2N 90	585	770	75	0	0.1

Table 15: Stress limits for the appearance of stress corrosion cracking in liquid ammonia containing 50 ppm of oxygen at 293 K (20°C) and at 240 K (−33°C) [26]

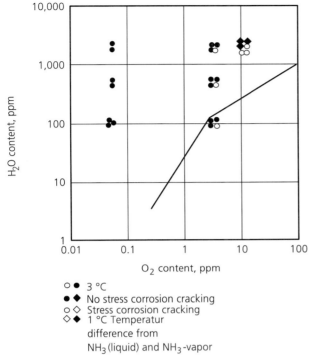

Figure 10: Hazard region for stress corrosion cracking on low-alloy steel depending on the contents of water and oxygen in liquid ammonia when cooling the specimens in the vapor space and condensation of ammonia takes place on the specimen surface [25]

For the low-alloy carbon-manganese steels and the 5 % Ni steel, stress corrosion cracking in ammonia containing oxygen takes place at room temperature only with loads above the yield point, i.e. slight plastic deformation is necessary for breaking protective cover layers. In contrast thereto, the 9 % Ni steel is susceptible to stress corrosion cracking already with a stress of 75 % of the yield point, and the ruptures take place practically without deformation. Significant differences were not found between the stress corrosion cracking behavior of the steels at 293 K (20°C) and at 240 K (−33°C). The fact that, in contrast thereto, the danger of stress corrosion cracking is significantly reduced with low temperature storage of ammonia can be attributed to the fact that in practice insufficient oxygen is dissolved in the ammonia at low temperatures.

Although for low strength steel grades stresses of the magnitude of the yield point are necessary to initiate stress corrosion cracking, existing cracks can grow even with smaller stress loads. Mechanical fracture investigations of CT specimens (compact tension) of the steel St 52-3 (S355J0, Material No. 1.0553, R_e = 380 MPa, R_m = 550 MPa, cf. UNS K12000) in liquid ammonia contaminated with air show that the crack growth rate increases with increasing stress intensity factor but decreases with time [30–33].

Since the crack growth is promoted by the oxygen content of the ammonia and thus determined by the rate of transport of this aggressive species to the crack tip, the decrease of the crack growth rate with time can be explained by a decrease of the oxygen content at the crack tip with increasing crack length and a resulting change of the electrochemical conditions in the crack [34]. For growth of the cracks significantly higher stress intensities are required at a temperature of 240 K (–33°C) of the ammonia than at a temperature of 291 K (18°C). Furthermore, the crack growth is considerably slower at the low temperature than at the higher temperature. This fact also contributes to the experience that low temperature storage tanks for ammonia are less endangered by stress corrosion cracking than ones in which the ammonia is stored at ambient temperature. By continuous monitoring of the crack length and crack depth in a gas sphere for the storage of liquid ammonia, and a subsequent fracture mechanical estimate of the crack behavior, it is thus possible to define the inspection intervals and the expected service life [35, 36].

For containers that are operated in the region A of Figure 6 normal inspection intervals suffice. Shorter inspection intervals are recommended for operation in the region B, and operation in the region C should not take place in any case [37–39].

Damage by stress corrosion cracking in ammonia containers takes place almost exclusively in the region of the weld seams. A comprehensive statistical evaluation of cases of cracking damage on storage and transportation containers made of fine grain constructional steels S355N (1.0545, StE 355) and P460N (1.8905, StE 460) showed that in particular weld seams that were welded with additives containing molybdenum were affected [40–42]. This led to the prohibition of additives containing molybdenum in the relevant recommendations and regulations [43, 44].

The experience gathered in recent decades in connection with the problems of stress corrosion cracking in liquid ammonia can be summarized with the following statements for the production and safe operation of containers for transportation and storage of liquid ammonia:

- Ammonia with high degree of purity (higher than 99.9 %) without contaminants does not lead to stress corrosion cracking.
- Oxygen taken up by access of air leads (as from 2 ppm) to a danger of stress corrosion cracking, already in very small concentrations, therefore all access of air to the ammonia must be avoided.
- Fine grain constructional steels with yield point grade of 355 MPa should be chosen.
- Welding additives of the same kind should be chosen, and the welding conditions must be chosen such that no hardening takes place in the weld region. Hardening alloying elements such as molybdenum or chromium must be avoided.
- Wherever possible the containers should be stress relief annealed.
- Grinding and shot blasting helps to eliminate internal stress in the weld region, whereby compressive stress is brought into this surface region that hinders the appearance of stress corrosion cracking.
- The ammonia should be inhibited with the addition of at least 0.2 % of water.

For further specifications and details, see the literature [43–47].

Carbon dioxide alone in liquid ammonia increases the general surface corrosion, but even in combination with oxygen or nitrogen does not produce any stress corrosion cracking [19], although in combination with air it significantly increases the susceptibility to stress corrosion cracking [4].

In connection with the investigations of stress corrosion cracking the behavior of various materials with regard to uniform surface corrosion and contact corrosion in liquid ammonia was also investigated [48]. The investigated materials are listed in Table 16.

Iron materials	Stainless steels	Nickel alloys	Others
Pure iron	SAE 304 1.4301 X5CrNi18-10	Monel® 400 2.4360 NiCu 30 Fe	pure aluminium EN-AW 3003
Unalloyed steel	SAE 316 1.4401 X4CrNiMo17-12-2	INCONEL® 600 2.4816 NiCr15Fe	AlMg2.5 EN-AW 5454
ASTM A 517 (F) (cf. 1.8850, 1.8907)	SAE 410 1.4006 X12Cr13	Hastelloy® C 4 2.4610 NiMo16Cr16Ti	zinc
	SAE 430 1.4016 X6Cr17		titanium

Table 16: Materials investigated with regard to uniform surface corrosion and contact corrosion in liquid ammonia [48]

The assessment of the general corrosive attack was made on the basis of the mass change of specimen sheets after exposure in an autoclave to liquid ammonia at room temperature and at a pressure of 8.75 bar. The exposure time was once for 1 month and once for 8 months. The results are summarized in Table 17.

Mass losses as well as mass increases due to thin surface films are observed. However, the mass losses are very small. Even for zinc, on which the highest mass losses were found, the highest value of 47 mg/dm^2 after 1 month corresponds to a corrosion rate of 0.076 mm/a (2.99 mpy). Taking the customary limit value of the corrosion rate of 0.1 mm/a (3.94 mpy) in practice as implying adequate durability, according to these results all tested materials can be classified as resistant. Local corrosive attack was not observed on the specimens.

In order to be able to assess a possible danger of contact corrosion in the case of electrically conducting contact between the different metals, the corrosion potentials of the materials in liquid ammonia were measured. The sequence of the potentials thereby found corresponds largely to the known sequence from the investigations in sea water (Figure 11). The passive or passivatable metals show positive potential values, whereas zinc takes on a negative potential. In ammonia containing air the potentials are shifted to significantly more positive values and all metals with the exception of zinc and to a slight extent aluminium are passivated (Figure 12).

	Mass change mg/dm²	
Material	Exposure duration: 1 month	Exposure duration: 8 months
Pure iron	−2.25	−1.5
Unalloyed steel	−3.76	−0.75
ASTM A 517 (F)	−3.0	−0.75
SAE 304	0	+3.0
SAE 316	0	+3.4
SAE 410	−7.52	+0.75
SAE 430	0	+3.8
Monel® 400	−1.5	−3.4
INCONEL® 600	−1.5	+2.6
Hastelloy® C 4	−1.13	−0.75
EN-AW 3003	+7.52	+1.9
EN-AW 5454	+12.8	+1.12
Zinc	−47.0	−21.0
Titanium	0	+3.0

Table 17: Mass change of the investigated materials after exposure in liquid ammonia (for the materials cf. Table 16) [48]

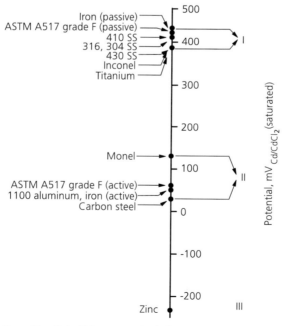

Figure 11: Potential sequence in air-free ammonia [48]

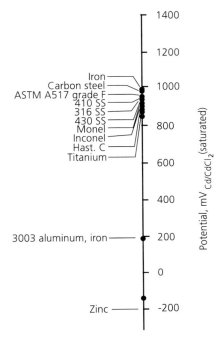

Figure 12: Potential sequence in ammonia containing air [48]

As shown by some supplementary stress corrosion cracking tests in this context, aluminium in contact with tensile specimens of the steel ASTM A 517 grade F (cf. 1.8850, 1.8907) can protect the tensile specimens cathodically, so that the sensitivity with regard to stress corrosion cracking is decreased significantly.

Since also the investigations of the potential dependence of the stress corrosion cracking in liquid ammonia have shown that the crack formation can be prevented by cathodic polarization, it is also possible to utilize cathodic corrosion protection as measure against stress corrosion cracking. However, the utilization of galvanically effective foreign anodes (sacrificial anodes) made of the metals zinc, aluminium or magnesium is problematic because due to the small electrical conductivity of the liquid ammonia adequately large current densities are not attained, and also because these anodes are not effective in the vapor space of a container. However, tests and experience in practice have shown that coatings of zinc and aluminium produced by flame sputtering can effectively prevent the stress corrosion cracking of steel in liquid ammonia [49]. For example, the successful utilization of zinc layers produced by flame sputtering, for protecting storage containers for liquid ammonia in Japan, is reported in [50]. After repairing the cracks that had occurred, zinc sputter layers of various thicknesses were produced in the critical region of the weld seams, and these layers were largely able to prevent the appearance of new cracks still after 8 years. With zinc layers less than 200 mm thick, new cracks under the zinc layer can be detected unambiguously by testing with magnetic powder. The corrosion rate of the zinc layers was about 10 μm/a (0.39 mpy), and blisters that had appeared in the zinc layer after 3 years did not impair the protective effect of the zinc layer.

As in most alkaline media, unalloyed and low-alloy steels are largely passive also in aqueous ammonia solutions. The attack is essentially determined by the dissolved oxygen. However, the resulting corrosion products – even in the case of only slight attack – can produce a yellow coloration of the ammonia solution.

Highly alloyed cast iron
Cast ferrosilicon
Austenitic cast iron (amongst others)
Ferritic chrome steels with < 13 % Cr
Ferritic chrome steels with ≥ 13 % Cr
Ferritic/perlitic-martensitic steels
Ferritic-austenitic steels/duplex steels
Austenitic chromium-nickel steels
Austenitic chromium-nickel-molybdenum(N) steels and CrNiMoCu(N) steels

The stainless ferritic and austenitic steels as well as the ferritic-austenitic steels are attacked by gaseous and liquid ammonia practically not at all [51, 52]. However, at higher temperatures take up of nitrogen in the surface regions of the steels can take place in ammonia gas [53].

Gas mixtures of ammonia and carbon dioxide, as present for example for the synthesis of urea, can, however, strongly attack the stainless steels at elevated temperatures. The reasons for this are the carbamic acid and carbamates that appear as intermediate products [54]. Although for the urea synthesis austenitic chromium-nickel-molybdenum steels, usually the steel Material No. 1.4435 (X2CrNiMo18-14-3), are utilized as inner lining of the pressurized containers, it is necessary to add oxidizing agents, e.g. atmospheric oxygen, to the reaction mixture in order to increase the passivity of the steels sufficiently for adequate service life of the plant units. Nevertheless, wall thickness decrease rates of 0.17 mm/a (6.69 mpy) to 0.25 mm/a (9.84 mpy) must be reckoned with, as measurements of the wall thickness made on a urea reactor in the course of 6 years have shown [55]. A mathematical model for estimating the corrosion behavior of the steel Material No. 1.4435 under the conditions of the urea synthesis is presented in [56].

In the stability lists of the manufacturers the stainless steels are reported as resistant also in pure aqueous ammonia solutions of all concentrations up to the boiling point [51, 52]. However, contaminations of the ammonia, particularly by chlorides, can lead to pitting corrosion depending on the magnitude of the chloride content and the actual temperature, and with the austenitic steels also to stress corrosion cracking.

At higher temperatures aqueous ammonia solutions also produce significant surface corrosion on stainless steels, as comparative tests on the materials specified in Table 18 show, that were carried out in connection with a damage caused by pitting corrosion [57]. The corrosion medium, that had produced damage on a heat exchanger pipe, was an aqueous ammonia solution containing about 300 ppm of chloride at operating temperatures between 473 and 573 K (200 and 300°C) and pressures between 40 and 100 bar.

Material No.	Abbreviation	cf. UNS	Cr	Ni	Mo	Mn	Si	Cu	Ti	W	Co	Zr
1.4435	X2CrNiMo18-14-3	S31603	18	14	3	2	1					
1.4462	X2CrNiMoN22-5-3	S31803	22	5	3	2	1					
1.4539	X1NiCrMoCu25-20-5	N08904	20	25	4	2	1	2				
1.4547	X1CrNiMoCuN20-18-7 254 SMO®	S31254	20	18	6	0.5	0.5	0.7				
1.4575	X1CrNiMoNb28-4-2	S32803	28	4	2	1	1					
1.4876	X10NiCrAlTi32-20	N08800	20	32	0	2	1		0.15			
2.4617	NiMo28	N10665	1	70	28	1	1	0.5				
2.4858	NiCr21Mo	N08825	22	42	3	1	3	0.5	1	1		
2.4602	NiCr21Mo14W	N06022	21	60	13	1	0.5			3	2	
2.4619	NiCr22Mo7Cu	N06985	22	40	7	1	1	2		1.5	5	
3.7055	Ti 3								100			
Zr (702)												100

Table 18: Chief alloying contents of the investigated materials [57]

The specimens were exposed in the plant in the medium stream between the damaged heat exchanger and the downstream reactor. After an exposure time of 3,000 h the corrosion rates specified in Table 19 were determined.

Titanium and zirconium are practically not attacked under these conditions, whereas the austenitic steels as well as the nickel-molybdenum alloy show significant corrosive attack. The ferritic-austenitic steel showed the most favorable behavior and was also chosen as replacement for the damaged heat exchanger.

Material No.	Corrosion rate mm/a (mpy)	Comment
1.4435	0.7 (27.6)	P
1.4462	0.02 (0.79)	
1.4539	0.5 (19.7)	P
1.4575	0.03 (1.18)	IC in the weld seam region
1.4876	0.9 (35.4)	P
1.4547	0.2 (7.87)	
2.4617	2.5 (98.4)	
2.4858	0.7 (27.6)	P
2.4602	0.1 (3.94)	
2.4619	0.05 (1.97)	
Ti 3	0.01 (0.39)	
Zr (702)	< 0.01 (< 0.39)	

P = pitting corrosion; IC = inter-crystalline attack

Table 19: Corrosion rates of the specimens after exposure for 3,000 h [57]

Bibliography

[1] Firmenschrift
Verarbeiten von hochverschleißfesten Thyssen-Sonderbaustählen
(Processing from high-wear-resistant Thyssen special construction steels) (in German),
TS/W/HD 29.08.94, 1994
ThyssenKrupp Schulte, Düsseldorf

[2] Gräfen, H.; Hennecken, H.; Horn, E.-M.; Kamphusmann, H.-D.; Kuron, D.
Spannungsrißkorrosion von unlegierten Stählen in flüssigem Ammoniak
(Stress corrosion cracking of unalloyed steel in liquid ammonia) (in German)
Werkst. Korros. 36 (1985), pp. 203–215

[3] Dawson, T. J.
Behaviour of welded pressure vessels in agricultural ammonia service
The Welding Journal 35 (1956) 6, pp. 568–574

[4] Loginow, A. W.; Phelps, E. H.
Stress corrosion cracking of steels in agricultural ammonia
Corrosion 18 (1962) 8, pp. 299–309

[5] Wilde, B. E.; Kim, C. D.
Galvanic factors in the corrosion of steel in liquid ammonia
Corrosion 38 (1982) 3, pp. 168–171

[6] Class, I; Gering, K.
Zur Frage der Spannungsrißkorrosion durch flüssiges Ammoniak
(To the question of the stress corrosion cracking by liquid ammonia) (in German)
Werkst. Korros. 25 (1974) 5, pp. 314–324

[7] Medard, L.
Rupture of an ammonia road tanker
Proceedings of the American Institute of Chemical Engineers Symposium on Safety in Ammonia Plant and Related Facilities
vol. 12, 1970, pp. 17–18

[8] Lonsdale, H.
Ammonia tank failure – South Africa
Proceedings of the American Institute of Chemical Engineers Symposium on Safety in Ammonia Plant and Related Facilities
vol. 17, 1975, pp. 126–131

[9] Lihou, D. A.
Failures of liquefied gas storage vessels
Proceedings of the institution of mechanical engineers
Journal of process mechanical engineering, 205, 1991, pp. 27–31

[10] Dahlberg, E. P.; Bradley, W. L.
Failure analysis of an ammonia pressure vessel
Conference: Fracture and Fracture Mechanics: Case Studies, Johannesburg, Republic of South Africa, 26–27 Nov. 1984
Publ.: Pergamon Press Ltd., Headington Hill Hall, Oxford OX3 0BW, UK
1985, pp. 327–337

[11] Baldock, P. J.
Accidental releases of ammonia: An analysis of reported incidents
Loss Prevention 13 (1980), pp. 35–42

[12] Pattnaik, K. C.; Gupta, M. P.
Stress corrosion cracking of ammonia receiver tank
British Corrosion Journal 30 (1995) 1, p. 80

[13] Kakas, D.; Sabo, B.; Bajic, V.
Failure analysis of liquid ammonia tanks produced of low alloy steel
Conference: Reliability and Structural Integrity of Advanced Materials,
Proceedings of the 9th Biennial European Conference on Fracture Vol. II Varna, Bulgaria, 21–25 Sept. 1992
Publ.: Engineering Materials Advisory Services Ltd., 339 Halesowen Rd. Cradley Heath, Warley, West Midlands B64 6PH, United Kingdom 1992, pp. 1221–1225

[14] Jackson, F.
The storage of liquid ammonia
VIGILANCE 4 (1983) 4, pp. 43–47

[15] Grover, S. K.; Parikh, K. N.
Stress corrosion cracking of ammonia storage sphere
Conference: 10th International Congress on Metallic Corrosion, Vol. I: Sessions 1–4, Madras, India, 7–11 Nov. 1987
Key Eng. Mater. 10–18 (1988) 1, pp. 835–842

[16] Tiner, N. A.
Brittle fractures in equipment failures
Mechanical Engineering 12 (1990) 6, pp. 60–63

[17] Deegan D. C.; Wilde, B. E.
Stress corrosion cracking behavior of ASTM A517 Grade F steel in liquid ammonia environments
Corrosion (NACE) 29 (1973) 8, pp. 310–315

[18] Hutchings, J.; Sanderson, G.
Stress corrosion of steels in anhydrous ammonia, 1975
Fulmer Research Institute

[19] Wilde, B. E.
Stress corrosion cracking of ASTM A517 steel in liquid ammonia: Environmental factors
Corrosion 37 (1981) 3, pp. 131–141

[20] Schmitt, W.; Heusler, K. E.
Spannungsrißkorrosion von Stählen in flüssigem Ammoniak
(Stress corrosion cracking of steel in liquid ammonia) (in German)
Werkst. Korros. 35 (1984) 7, pp. 329–337

[21] Klemm, M.; Heusler, K. E.
Anodische Spannungsrißkorrosion eines Stahls in flüssigem Ammoniak
(Anodic stress corrosion cracking of a steel in liquid ammonia) (in German)
Werkst. Korros. 39 (1988) 11, pp. 492–499

[22] Huerta, D.; Heusler, K. E.
Korrosion niedrig legierter Stähle in flüssigem Ammoniak
(Corrosion of low alloyed steel in liquid ammonia) (in German)
Werkst. Korros. 45 (1994) 9, pp. 489–497

[23] Kim, C. D.; Wilde, B. E.; Phelps, E. H.
Stress corrosion cracking of line-pipe steels in anhydrous ammonia
Conference: Mechanical working and steel processing XIII, Pittsburgh, Pennsylvania, USA, 22–23 Jan. 1975
Publ.: Iron and Steel Society/AIME, 410 Commonwealth Dr., Warrendale, Pennsylvania, 15056, 1975, pp. 315–337
also in
Corrosion (NACE) 31 (1975) 7, pp. 255–262

[24] Class, I; Gering, K.
Zur Frage der Spannungsrißkorrosion durch flüssiges Ammoniak
(To the question of the stress corrosion cracking by liquid ammonia) (in German)
Werkst. Korros. 25 (1974) 5, pp. 314–324

[25] Fäßler, K.
Spannungsrißkorrosion an Lagerbehältern für flüssiges Ammoniak
(Stress corrosion at storage vessels for liquid ammonia) (in German)
VGB Kraftwerkstechnik 67 (1987) 8, pp. 747–751
according to: Code of practice for the storage of anhydrous ammonia under pressure in the United Kingdom: Spherical and cylindrical vessels.
Publ. Chemical Industries Association Ltd. Jan. 1980

[26] Lunde, L.; Nyborg, R.
Stress corrosion cracking of some metallic materials in ammonia at ambient and low temperature
Conference: UK Corrosion '85: Coating and Preparation, Monitoring, Materials Selection, Harrogate, UK, 4–6 Nov. 1985
Institution of Corrosion Science and Technology, Exeter House, 48 Holloway Head, Birmingham B1 1NQ, UK, 1985, pp. 333–345
also in: Lunde, L.; Nyborg, R.
Stress corrosion cracking of some metallic materials in ammonia at ambient and low temperature
Industrial Corrosion 5 (1987) May, pp. 6–10

[27] Lunde, L; Nyborg, R.
Stress corrosion cracking of different steels in liquid and vaporous ammonia
CORROSION 87, 9–13, March, 1987, San Francisco, Calif. USA
Paper No. 174, pp. 1–14

[28] Lunde, L; Nyborg, R.
Stress corrosion cracking of different steels in liquid and vaporous ammonia
CORROSION NACE, 43, 1987, 11, pp. 680–686

[29] Ernst, J.
Qualitätssicherung beim Lagern von Flüssig-Ammoniak
(Quality assurance when storing liquid ammonia) (in German)
VGB Kraftwerkstechnik 72 (1992) 9, pp. 801–804

[30] Lunde, L.; Nyborg, R.
Stress corrosion crack growth rate of carbon-manganese steels in liquid ammonia
Conference: Corrosion Prevention in the Process Industries, Amsterdam, The Netherlands, 8–11 Nov. 1988
Publ.: NACE, 1440 South Creek Dr., Houston, Texas 77084 USA, 1990, pp. 211–214

[31] Nyborg, R.; Lunde, L.
Stress corrosion cracking of carbon steel in ammonia
Materials Performance 28 (1989) 12, pp. 29–32

[32] Nyborg, R.; Lunde, L. R.; Conley, M. J.
Integrity of ammonia storage vessels – Life prediction based on SCC experience
Materials Performance 30 (1991) 11, pp. 61–65

[33] Nyborg, R.; Lunde, L.
Life prediction of ammonia storage tanks based on laboratory stress corrosion crack data
Conference: Application of accelerated corrosion tests to service life prediction of materials, Miami, Florida, USA, 16–17 Nov. 1992
Publ.: ASTM, 1916 Race St., Philadelphia, PA 19103, USA, 1994, pp. 27–41

[34] Fäßler, K.; Spähn, H.
Grundlagen der Spannungsrißkorrosion unlegierter Stähle in flüssigem Ammoniak – Einflußgrößen und Gegenmaßnahmen
(Bases of the stress corrosion cracking of unalloyed steel in liquid ammonia – determining factors and counter measures) (in German)
VGB Kraftwerkstechnik 69 (1989) 11, pp. 1124–1131

[35] Games, G. A. C.; Geary, W.
An engineering critical assessment of a spherical liquid ammonia storage vessel
Conference: ECF 10-Structural integrity: Experiments-Models-Applications, Vol. II, Berlin , Germany, 20–23 Sept. 1994
Publ.: Engineering Materials Advisory Services Ltd. 339 Halesowen Rd. Cradley Heath, Warley, West Midlands, B64 6PH, UK, 1994, pp. 1337–1346

[36] Zeman, M.
Assessing cracks in ammonia vessels
Welding International 1 (1987) 3, pp. 236–239

[37] Corrosion by ammonia
in: Garverick, L.
Corrosion in the Petrochemical Industry
ASM International, Member/Customer Service Center, Materials Park, OH 44073-002, USA, 1994, pp. 212–214

[38] Towers, O. L.
SCC in welded ammonia vessels
Metal Construction 16 (1984) 8, pp. 479–485

[39] Gnirss, G.; Hrivnak, I.; Lukacevic, Z.
Recommendations for reliable operation of pressure vessels exposed to aggressive media like ammonia and sour gas
Welding in the World 35 (1995) 2, pp. 71–82

[40] Gayk, W.; Kautz, H. R.
Rißbildung in Schweißnähten von Ammoniak-Lagerbehältern aus StE355 – Maßnahmen zu ihrer Vermeidung (Cracking in welding seams of ammonia storage vessels from StE355 – measures to their avoidance) (in German)
VGB Konferenz "Werkstoffe und Schweißtechnik im Kraftwerk 1989" Essen 1989, pp. 186–204

[41] Gayk, W.
Empfehlungen zur Werkstoffauswahl und zu den Schweißdaten für NH_3-Lagerbehälter und NH_3-Kesselwagen (Recommendations regarding the material selection and regarding the welding data for NH_3-storage vessels and NH_3-tanker
Materialwissenschaft und Werkstofftechnik 19 (1988), pp. 126–133

[42] Kautz, H. R.; Zürn, H. E. D.
Modern welding techniques and component performance
Welding in the World 30 (1992) 11/12, pp. 337–341

[43] TRD 451 (12/1996)
Anlagen zur Lagerung von druckverflüssigtem Ammoniak für Dampfkesselanlagen-Druckbehälter – (Plants for the storage of pressure-liquefied ammonia for boiler plant pressure vessels) (in German)
Beuth Verlag GmbH, Berlin

[44] Gnirss, G.; Hrivnak, I.; Lukacevic, Z.
Recommendations for reliable operation of pressure vessels exposed to aggressive media like ammonia and sour gas
Welding in the World 35 (1995) 2, pp. 71–82

[45] Loginow, A. W.
A review of stress corrosion cracking of steel in liquefied ammonia service
Materials Performance 25 (1986) 12, pp. 18–22

[46] Krisher, A. S.
Material Requirements for Anhydrous Ammonia
in: Monitz, B.J.; Pollock, W.I.
Process Industries Corrosion – The Theory and Practice
NACE 1440 South Creek Dr., Houston, Texas 77084, USA, 1986, pp. 311–314

[47] Trettyak, I.-Yu.; Slipchenko, T. V.
Increasing the safety of vessels with liquid ammonia
Sov. Mater. Science 26 (1990) 2, pp. 243–245

[48] Jones, D. A.; Wilde, B. E.
Corrosion performance of some metals and alloys in liquid ammonia
Corrosion (Houston) 33 (1977) 2, pp. 46–50

[49] Lunde, L.; Nyborg, R.
SCC of carbon-manganese steel in ammonia efficiently prevented by metal flame spraying
Conference: Corrosion prevention in process industries, Amsterdam, The Netherlands, 8–11 Nov. 1988
Publ.: National Association of Corrosion Engineers, Houston, Texas, USA, 1990, pp. 253–263

[50] Imagawa, I.; Matuno, K.; Konishi, T.
Prevention of corrosion cracking in liquid ammonia tanks by zinc spraying
Corrosion Eng. (JPN) 38 (1989) 6, pp. 321–326

[51] Firmenschrift
Chemische Beständigkeit der nichtrostenden Remanit Stähle, Druckschrift 1127/2
(Chemical stability of the stainless Remanit of steel, print 1127/2) (in German), February 1992
Thyssen Edelstahlwerke AG, Krefeld

[52] Firmenschrift
Avesta Sheffield Corrosion Handbook, Information 18099GB, 8. ed, 1999
Avesta Sheffield AB, Avesta (Sweden)

[53] Halliger, H.; Wahren, R.
Hochtemperaturkorrosion austenitischer Stähle durch Ammoniak
(High-temperature corrosion of austenitic steel by ammonia) (in German)
Werkst. Korros. 44 (1993) 11–12, pp. 473–479

[54] Dillon, C. P.
Carbamates can cause corrosion problems
Materials Performance 38 (1999) 12, pp. 74–75

[55] Matza, G.; Gavrila, L.; Zichi, V.; Ivascan. S.
Corrosion behavior of W 1.4435.2 stainless steel in urea plants
International Congress Stainless Steels '96, 3–5 June 1966, Düsseldorf, Germany, pp. 171–176

[56] Gavrila, L.; Zichil, V.; Asaftei, S.
Mathematical model of the corrosion behavior of an industrial urea synthesis reactor
Conference: Stainless Steel '99, Science and Market, Sardinia, Italy, 6–9 June 1999
Publ.: Associazione Italiana di Metallurgia, Piazzale Rodolfo Morandi, 2, Milano, I-20121, Italy 1999

[57] Andreaus, A.; Cicciarelli, R.; Schönhofer, J.
Evaluation of duplex steel for a chemical production plant
Duplex Stainless Steels '91 Vol. 2, Beuane, Bourgogne, Frankreich, 28–30 Oct. 1991
Publ.: Les Editions, de Physique, Avenue du Hoggar, Zone Industrielle de Courtaboeuf, B.P. 112, F-91944 Les Ulis Cedex A, France 1992, pp. 1133–1139

Lithium Hydroxide

Unalloyed steels and cast steel

In steam boilers in power plants a pH between 10.0 and 10.5 is used in the water circuit with low oxygen contents of < 0.05 ppm in order to prevent corrosion of carbon steel and low-alloy steel. In the water circuits of nuclear power plants lithium hydroxide is used for this purpose because of the less severe radiation problems. Under these conditions at pH 10 the corrosion rate is held to < 0.03 mm/a (< 1.18 mpy) by the magnetite (Fe_3O_4) and hematite (Fe_2O_3) layer on steel produced between 470 and 550 K (197 and 277°C) [1]. The outer part of the oxide layer, produced in the water with a pH of 10.5 caused by the addition of lithium hydroxide, consists of α-Fe_2O_3.

In addition to FeOOH, small quantities of $LiFeO_2$ have also been detected [1]. In agreement with this, only α-$LiFeO_2$ has been detected in addition to Fe_3O_4 on steel in 5 mol/l lithium hydroxide solution, whereas with concentrations of lithium hydroxide of less than 1 mol/l, at 573 K (300°C) formation of the spinel $LiFe_5O_8$ predominates and ensures good protection against corrosion. On short term exposure Fe_3O_4 was mainly found, on exposure of over 16 hours, mainly α-$LiFeO_2$ [2].

If carbon steel (e.g. percent by weight): 0.3C-0.09Ni-0.1Cr-0.015Sn-0.1Mo-0.034Al-0.11Cu-0.75Mn-0.33Si) is brought into contact with platinum in 1 mol/l lithium hydroxide solution in a titanium autoclave, the number of $LiFeO_2$ crystallites per unit area formed and therefore the corrosion rate of the steel is reduced. Whereas $FeTiO_3$, – the titanium for which comes from the material of construction of the autoclave – is formed in addition to α-$LiFeO_2$ in sodium hydroxide solution, no such compound has been observed in the corrosion products in solutions containing lithium hydroxide [2].

Low-alloy steel (e.g. percent by weight): 0.3-0.5Cr, 0.3-0.5Ni, 0.2-0.3Co, 0.05-0.1V) exhibits a number of corrosion ranges in lithium hydroxide solution at 573 K (300°C) which depend on the concentration of the lithium hydroxide.

A minimum and a maximum of the corrosion rate appear within 100 hours in lithium hydroxide solutions with concentrations in excess of 1 mol/l. Below 0.7 mol/l lithium hydroxide the corrosion rate decreases uniformly and below 0.3 mol/l lithium hydroxide follows a parabolic law with a rate of < 0.05 mm/a (< 1.97 mpy) which is distinctly smaller than that in sodium and potassium hydroxide solutions [3]. Below 0.5 mol/l lithium hydroxide Fe_3O_4 and $LiFeO_2$ are the most important reaction products [4–7].

The magnetite (Fe_3O_4), the only solid corrosion product of steel, produced in the primary circuit at 573 K (300°C) is in the pressurized water reactor in the presence of water containing lithium hydroxide is resistant to lithium hydroxide in this dilution (up to 0.4 mol/l) [6].

The reduction of α-Fe_2O_3 by hydrogen, when α-iron is produced via Fe_3O_4, is accelerated in the presence of alkali hydroxides (LiOH, NaOH, KOH, RbOH, CsOH) [8].

The addition of less than 0.25 mol/l lithium hydroxide is recommended for corrosion protection of steel tubes in steam boilers operating at 570 K (297°C). The steels are attacked at this temperature at higher concentrations [9].

Small additions of lithium hydroxide to 5 mol/l sodium hydroxide solution reduce the corrosion rate of low-alloy steel at 573 K (300°C); this effect is said to be due to the reduction in the concentration of the iron ion vacant lattice sites in the dense inner oxide layer in which lithium ions are incorporated [10]. The extent to which this mechanism, which was experimentally detected in the NiO/Li_2O system [11], can be used here is a question which remains to be answered at the present time.

Oxidation tests with iron in air at 0.067 kPa at 1,133 K (860°C) containing Li_2O showed a negligibly small reduction in the oxidation rate [12].

According to a patent specification, the addition of 0.24 to 2,400 ppm lithium hydroxide to the water, thus bringing its pH to 9 to 13, is said to be able to prevent pitting corrosion in the steel tubes of a heating boiler operating at 590 K (317°C) and < 2.4 MPa in a steam generator. In contrast to sodium and potassium hydroxides, these small quantities of lithium hydroxide lead to the formation of a protective layer which is particularly effective provided that the spinel, $LiFe_5O_8$ is produced in accordance with the following reaction [4]:

$$5 Fe_3O_4 + 3 LiOH + H_2O \rightarrow 5/2 H_2 + 3 LiFe_5O_8$$

Confirmation of these findings is contained in a report from the US Marine Research Laboratory according to which a low-alloy steel for a pressurized water reactor was not significantly attacked in concentrated 28.5 % lithium hydroxide solutions at 589 K (316°C) after 8 days (0.07 g/m^2 h, corresponding to about 0.086 mm/a (3.39 mpy)), contrasting with severe corrosion in 40 % sodium hydroxide solution (75 g/m^2 h, equivalent to about 92 mm/a (3,622 mpy)) [13]. Figure 1 shows the chronological pattern of corrosion of a low-alloy steel in lithium and sodium hydroxide solutions. According to this diagram, corrosion is initially severe in lithium hydroxide but in the course of 7 days the material consumption rate decreases rapidly and subsequently more slowly to < 0.02 g/m^2 h, corresponding to a corrosion rate of about 0.024 mm/a (0.95 mpy) [13].

This proves that water containing lithium hydroxide at pH 10 can be recommended because of the inhibiting effect of lithium hydroxide for both steam generators and pressurized water reactors, since even low-alloy steel is protected not only against general corrosion but also against pitting corrosion.

On the basis of 7 day tests at 358 K (85°C), the inhibiting effect is at its maximum at concentrations between 100 (pH 10.5) and 1,000 ppm (pH 11.8), when the material consumption rate is 0.0037 g/m^2 h, corresponding to a corrosion rate of about 0.0045 mm/a (0.18 mpy), irrespective of the concentration of the lithium hydroxide [14].

In order to protect low-alloy steels against corrosion in the pressurized water reactor, according to a patent the steel is kept immersed in boiling lithium hydroxide solution at 422 to 505 K (149 to 232°C) and pH 10 for about 7 to 14 days before installation. This is said to form a dark oxidic protective coating which largely protects the steel against further corrosion [15].

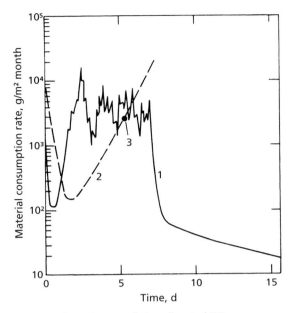

Figure 1: Corrosion rate of a low-alloy steel [13]
1 – in 28.5 % LiOH solution
2 – in 40 % NaOH solution
3 – test aborted, since the capsule was perforated

According to another patent [16], corrosion of iron and steel in an absorption cooler is reduced even in the presence of oxygen by the addition of lithium hydroxide to the absorption solution. The addition of antimony to the steel promotes the growth of a surface layer affording protection against corrosion; a suitable soluble antimony compound may also be added to the absorption solution for the same purpose.

Lithium chromate (Li_2CrO_4) is to be recommended as inhibitor for open cooling systems operated with 63 % lithium bromide solution as coolant. It is considerably more effective than lithium molybdate (Li_2MoO_4) containing lithium hydroxide (see Table 1). In closed systems the inhibiting effect of the lithium compounds at 455 K (182°C) on the corrosion of low-alloy steel in combination with Cu-10Ni and stainless steel SAE 304 is also good. The inhibiting effect of the lithium chromate is best on the addition of 0.005, and that of lithium molybdate on the addition of 0.02 % lithium hydroxide (pH 9 to 10.5) [17].

From the values given in Table 1, partial coating of the metals in the 63 % lithium bromide solution inhibited with lithium chromate/lithium hydroxide has no influence on the degree of corrosion, in contrast to the solution inhibited with lithium molybdate. The sole use of 0.02 % of lithium hydroxide as inhibitor under these conditions leads to the significantly higher corrosion rates of 0.058 mm/a (2.28 mpy) [17].

Inhibitor, %	Corrosion rate mm/a (mpy)		Current µA	Remarks
	Copper	Iron		
0.3 Li$_2$CrO$_4$ +0.005 LiOH	0.005 (0.2)	0.0 (0)	2–40	1)
0.1 Li$_2$MoO$_4$ + 0.2 LiOH	0.0075 (0.30)	0.01 (0.39)	25–320	1)
0.3 Li$_2$CrO$_4$ + 0.005 LiOH	0.005 (0.2)	0.00 (0)	2–40	2)
0.1 Li$_2$MoO$_4$ + 0.2 LiOH	0.0825 (3.25)	0.298 (11.7)	15–210	2)

1) completely immersed
2) completely immersed for 8 hours, only two-thirds immersed for 16 hours

Table 1: Galvanic corrosion of iron/copper couples in aerated 63 % LiBr solution at 433 K (160°C) with inhibitors; test duration 1 week [17]

In addition to K$_2$CrO$_4$, K$_2$Cr$_2$O$_7$, KNO$_2$, Na$_2$SiO$_3$ and Na$_3$PO$_4$, lithium hydroxide was also tested as inhibitor of corrosion of low-alloy steel in distilled water at 358 K (85°C). On the basis of the results obtained in this way, 200 ppm of lithium hydroxide or chromates are sufficient, whereas the remaining compounds are not sufficiently effective until the concentrations reach 1,000 ppm [14]. A good method for testing the effectiveness of inhibitors is based on the investigation of the anodic polarization in the solution containing the inhibitor. With a good inhibitor, low current densities of 7 µA/cm^2 are sufficient to shift the potential within 20 minutes from negative values to +0.7 to +0.9 V (Saturated Calomel Electrode (SCE)) whereas with inadequate inhibitor effect the potential remains significantly below +0.7 V.

Lithium hydroxide has also proved to be an effective inhibitor for the corrosion protection of low-alloy and unalloyed steels which are used for water circuits of large volume and also for parts of the pipe system, provided that there are no special requirements regarding temperature and pressure. As tests with steel St 37 (cf. S235JR, UNS K01702) in demineralized water saturated with air between 293 and 343 K (20 and 70°C) show, 25 ppm of lithium hydroxide is often sufficient to protect steel against general and pitting corrosion when chloride contents range up to 5 ppm [18].

As Figure 2 shows, the chloride content of 50 ppm of the lithium hydroxide solution saturated with air may be up to 70 ppm without the very low corrosion current of 0.2 µA/cm^2 increasing. At 333 K (60°C) the corrosion current density as a measure of the corrosion rate is about 0.1 µA/cm^2, even lower than at room temperature. In the higher temperature ranges the current density values scatter rather less. This behavior leads to the inference that a more stable and less porous protective coating is produced at higher temperatures.

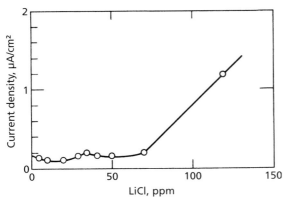

Figure 2: Corrosion current density of steel St-37 (cf. S235JR, UNS K01702) in 50 ppm LiOH solution saturated with air at room temperature as a function of the chloride concentration [18]

The results of cyclic voltametric measurements lead to the inference that Fe_3O_4 and $LiFeO_2$ are produced via the surface compound $HFeO_2$ [5]. It is noteworthy that the $LiFeO_2$ produced on the surface of the steel also inhibits stress corrosion cracking [7].

The susceptibility of steel St-37 (cf. S235JR, UNS K01702) to stress corrosion cracking is slight in water at 343 K (70°C) with lithium hydroxide concentrations up to 500 ppm, even if up to 100 ppm of chlorides are present. With higher concentrations of the hydroxide, however, intergranular incipient cracks occur whose number increases with the concentration of the hydroxide. Concentration of lithium hydroxide in such systems must therefore be prevented [18]. This is mainly likely to occur because of dead spaces resulting from poor design. If it is not possible to overcome these drawbacks steel X5CrNi18-10 (cf. 1.4301, SAE 304), which is virtually insensitive to stress corrosion cracking under such conditions, must be used [18, 19].

The inhibiting effect of lithium hydroxide on the corrosion of low-alloy steel in water at 633 K (360°C) and 18 MPa can reverse into an increase in corrosion in the steam boiler if dosage is too high [20].

Both in conventional and, more particularly, in nuclear power plants, the steam circuits must be operated with chemical conditioning in order to keep the corrosion of steel as low as possible and prevent the deposition of magnetite, which is the only solid corrosion product, on to, for example, the fuel rods of reactors operating at 570 K (297°C). The water is therefore conditioned with boric acid containing lithium hydroxide, when, for certain lithium hydroxide and boric acid concentrations, dissolubility minima for iron result. These concentrations should therefore be maintained within the range shown in Figure 3 [6].

In this context it should be noted that lithium hydroxide may be used for treating boiler feedwater only if the latter contains no phosphate, since otherwise precipitation of lithium phosphate occurs which jeopardizes the protective effects. Thus phosphate-free closed circuits are necessary, operated on the assumption that an intrusion of phosphate-containing cooling water is impossible [21]. If the water therefore contains no hardness-producing substances or iron oxides and the boiler

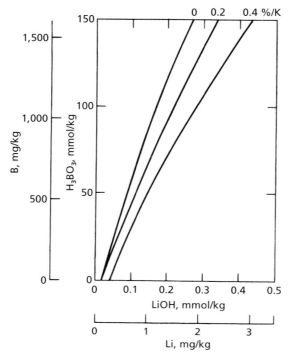

Figure 3: Lithium hydroxide and boric acid concentrations in water with different temperature coefficients of magnetite solubility at 573 K (300°C)] [6]

water contains no phosphate ions, the addition of lithium hydroxide to adjust the pH is advantageous, since it is less corrosive than sodium hydroxide in contact with carbon steel. As already mentioned, the protective coating on the steel consists of $LiFeO_2$ when lithium hydroxide concentrations are high, whereas it consists of $LiFe_5O_8$ when concentrations of lithium hydroxide are less than one percent.

The wear resistance, or erosion-corrosion of steel 45 (cf. 1.1191, SAE 1042) in hydroxides decreases as the hydroxide concentration increases from 0.001 to 1.0 mol/l and with decreasing radius of the alkali metal ion, i.e. erosion-corrosion is least in lithium hydroxide solutions [22].

Protective coatings produced by in-situ polymerization of thiophene and trichloroethylene in microwave plasma protects steel against hot water containing lithium hydroxide at a concentration of 10^{-4} mol/l [23].

For cleaning steels for mineral oil plants, specifically for removal of corrosion products, a solution consisting of 30% sodium + lithium hydroxides (3:1) with 1.5% sodium oleate, 3% triethanolamine, 0.1% sodium sulfosuccinate and 0.5% sodium metaphosphate is passed through the plants for 24 hours. Efficiency is said to be about 60% [24].

As shown in Figure 4 Armco® iron and copper in molten lithium hydroxide at 873 K (600°C) lose about 70 g/m² h corresponding to about 78 mm/a (3,071 mpy);

under these conditions a titanium-stabilized CrNi-steel X5CrNi18-10 (cf. 1.4301, SAE 304) is corroded substantially more slowly about 7.8 g/m² h, corresponding to a corrosion rate of about 8.6 mm/a (339 mpy) [25].

3 hour tests in the melts already mentioned (80 mol percent LiOH, 20 mol percent LiF, 1,073 K (800°C)) showed a corrosion rate for Armco® iron of 1,380 g/m² h (about 1,700 mm/a), i.e. rapid dissolution [26].

Unalloyed cast iron
High-alloy cast iron, high-silicon cast iron

No publications are known on the behavior of cast iron in solutions containing lithium hydroxide. Non-alloy cast iron, however, probably behaves in a similar manner to carbon steel in lithium hydroxide solutions.

Ferritic chromium steel with more than 12 % chromium

No literature references are available on the corrosion behavior of chromium steels in aqueous lithium hydroxide solutions. The chromium steel SAE 420 (16 % chromium) loses 820 g/m² h, equivalent to about 886 mm/a (34,094 mpy), in the melt already mentioned (80 mol percent LiOH, 20 mol percent LiF, 1,073 K (800°C)), i.e. it is rapidly dissolved [26]. No experimental results at lower temperatures are available.

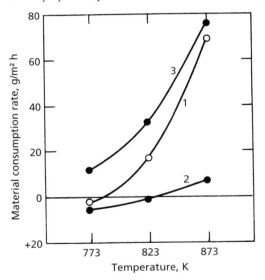

Figure 4: Material consumption rate per unit area as a function of temperature [25]
1 – Armco® iron;
2 – CrNi-steel Ch18N9T (cf. 1.4541, SAE 321);
3 – Copper in molten lithium hydroxide

Austenitic chromium-nickel steels
Austenitic chromium-nickel-molybdenum steels

The susceptibility of, for example, the niobium-stabilized steel SAE 347, to stress corrosion cracking in water containing alkali metal hydroxide increases in the sequence NaOH > KOH > LiOH, although it increases with increasing concentration of the alkali metal hydroxide [27, 28]. Table 2 lists two sets of results obtained with SAE 347 in comparison with those obtained with Inconel® 600. According to these results, resistance of the steel to stress corrosion cracking is better than that of the considerably more expensive nickel alloy.

Alloy	LiOH, mol/l	No. of cracks	Time to occurrence of cracks or test duration without cracking, d
SAE 347	0.01	0/6	180
	1.0	0/6	180
Inconel® 600	1.0	1/10	88

(0/6 = no cracks after 6 tests)

Table 2: Stress corrosion cracking test with steel SAE 347 and Inconel® 600 in high-pressure water containing LiOH at 608 K (335°C) [27]

Depending on the treatment of steel SAE 347, the material consumption rate in high-pressure water containing lithium hydroxide at 533 K (260°C) was between 0.0007 and 0.0013 g/m^2 h corresponding to a corrosion rate of 0.0009 to 0.0016 mm/a (0.035 to 0.063 mpy), i.e. it is negligible [27].

In demineralized water containing 0.05 ml/kg oxygen and 30 ml/kg hydrogen and 2 ppm lithium hydroxide, steel SAE 304 at 589 K (316°C) loses 0.0016 g/m^2 h corresponding to 0.002 mm/a (0.079 mpy), i.e. it is virtually resistant to corrosion [27].

In low-temperature water circuits 293 to 343 K (20 to 70°C) steel X5CrNi18-10 (cf. 1.4301, SAE 304) exhibited no stress corrosion cracking in water containing up to 500 ppm LiOH and up to 100 ppm chloride at 343 K (70°C), although such stress corrosion cracking cannot be excluded with long operating times such as are demanded today in plant construction, particularly when the chloride and lithium hydroxide become more concentrated in dead spaces due to bad design. For this reason corrosion tests with higher lithium hydroxide concentrations should also be done [18, 19]. Unfortunately, these papers do not contain any quantitative data on the corrosion rate.

The corrosion rate of steel SAE 304 in water containing lithium hydroxide at pH 9.5 can be very much decreased by polarization of −0.05 to −0.2 V (see Figure 5): between 20 and 220 mV (Saturated Calomel Electrode (SCE)) the current density (as a measure of the corrosion rate) falls from between 10 and 100 to 0.02 µA/cm^2 [29].

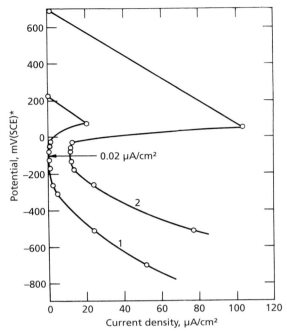

Figure 5: Anodic polarization of SAE 304 in water containing lithium hydroxide at pH 9.5, and 14 MPa [29]
1 – at 297 K (24°C); 2 – at 523 K (250°C)

Austenitic steel containing 17 % chromium and 13 % nickel under mechanical stress is less susceptible to intergranular stress corrosion cracking in desalinated and deaerated water containing lithium hydroxide at 633 K (360°C) than in water with sodium and potassium hydroxides contents of the same concentration [30, 31].

Molybdenum increases the susceptibility to stress corrosion cracking in water containing alkali metal hydroxide, whereas 4 % silicon improves the resistance of, for example, the CrNi-steel 17 14 only to corrosion by chlorides, but not by alkali metal hydroxides. A steel containing 17 % chromium and 30 to 45 % nickel is recommended for steam at 870 K (597°C) [31].

Table 3 shows the corrosion caused by the melt already mentioned (80 mol percent LiOH, 20 mol percent LiF) at 1,073 K (800°C) on various chromium-nickel steels. Although the values reported are only the results of 3 hour tests, they nevertheless clearly show that none of the chromium-nickel steels listed is suitable as material for such melts at 1,073 K (800°C) [26].

Figure 6 shows the corrosion rate of the Ti-stabilized steel Ch18N9T (cf. 1.4541, SAE 321) in molten lithium hydroxide between 773 and 873 K (500 and 600°C) as a function of temperature in comparison with Armco® iron and nickel [25]. According to this, the chromium-nickel steel at 823 K (550°C) is only slightly worse than nickel, with a material consumption of 1 g/m² h, corresponding to a corrosion rate of about 1.2 mm/a (47.2 mpy).

Steel	Composition, %					Corrosion rate mm/a (mpy)
	Fe	Ni	Cr	Mo	Cu	
SAE 304	70	10	19	–	–	100 (3,937)
SAE 310	53	21	25	–	–	1,000 (39,370)
SAE 316	67	14	17	–	–	187 (7,362)
SAE 321	72	10	18	–	–	133 (5,236)
Circle® L-34	44	27	21	3	3	1,506 (59,291)
Carpenter® 20	46	29	20	2	3	1,171 (46,103)

Table 3: Corrosion rate of some CrNi-steels in molten LiOH-LiF (80/20) at 1,073 K (800°C) (results from 3 hour tests) [26]

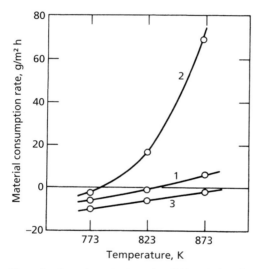

Figure 6: Corrosion rates in molten lithium hydroxide as a function of temperature [25]
1 – Ch18N9T (cf. 1.4541, SAE 321)
2 – Armco® iron
3 – copper

Bibliography

[1] Chauhan, P. K.; Sharma, S. K.; Gadiyar, H. S.
XPS study of oxides on carbon steel in LiOH at 250°C
Corrosion Sci. 21 (1981) p. 505

[2] Macdonald, D. D.; Rummery, T. E.
The growth of iron oxide films on carbon steel and platinum surfaces in LiOH and NaOH solutions at 285°C
Corrosion Sci. 15 (1975) p. 521

[3] Asai, O.; Kawashima, N.
Corrosion of steel in water at elevated temperatures and pressures. III. Corrosion of mild steel in high-temperature lithium hydroxide solutions
Denki Kagaku 34 (1966) p. 761

[4] Bloom, M. C.; Krulfeld, M.; Gothem, N.
Protection film formation in high-pressure steam generators
US-Patent 3 173 404 (16.3.1965)

[5] Macdonald, D. D.; Owen, D.
The electrochemistry of iron in 1 M lithium hydroxide solution at 22 and 200°C
J. Electrochem. Soc. 120 (1973) p. 317

[6] Bosselmann, W.; Kittel, H.; Schroeder, H. J.
Die Abhängigkeit zwischen pH-Wert, Lithiumhydroxid- und Borsäure-Konzentration im Primärkreislauf von Druckwasserreaktoren
(Relation between pH, lithium hydroxide and boric acid concentration in the primary circuit of pressurized water reactors)
(in German)
VGB Kraftwerkstechnik 60 (1980) p. 995

[7] Akolzin, P. A.; Mostovenko, L. N.
Corrosion of boiler steels in water treated with alkaline agents (in Russian)
Teploenergetika (1974) No 2; p. 79; (C.A. 80 (1974) 124 531v)

[8] Klisurski, D.; Mitov, I.
Moessbauer studies of the reduction of α-ferric oxide containing alkali metal hydroxides
J. Phys. Colloq., Orsay France (1979) No 2; p. 353

[9] Asai, O.; Kawashima, N.
Method for the corrosion protection of materials
Jap. Pat. 12 A 41, No 13 688 (19.9.1966/16.5.1970)

[10] Asai, O.; Kawashima, N.
Corrosion of steel in water of elevated temperature and pressure. V. Effect of LiOH on the corrosion of mild steel in high-temperature sodium hydroxide solution
Denki Kagaku 35 (1967) p. 702

[11] Pfeiffer, H.; Hauffe, K.
Über die Beeinflussung der Oxidationsgeschwindigkeit von Nickel und Titan durch Legierungszusätze und durch Behandlung von Metalloxiddampf
(Influencing the rate of oxidation of nickel and titanium by alloying additions and by treatment with metal oxide vapours)
(in German)
Z. Metallkunde 43 (1952) p. 364

[12] Brauns, E.; Rahmel, A.
Über die Beeinflussung der Oxidationsgeschwindigkeit des Eisens durch Behandlung mit Lithiumoxiddampf
(The influences of lithium oxide vapor treatments on the rate of oxidation of iron)
(in German)
Werkstoffe u. Korrosion 8 (1956) p. 448

[13] Bloom, M. C.; Frazer, W. A.; Krulfeld, M.
Corrosion of steel in concentrated lithium hydroxide solution at 316°C
Corrosion 18 (1962) p. 401t

[14] Gadiyar, H. S.; Balachandra, J.
Influence of inorganic inhibitors on the corrosion of mild steel in distilled water at 85°C
Trans. Indian Inst. Metals 19 (1966) p. 189; (C.A. 68 (1968) 65094)

[15] Patell, P. N. B.; Janusz, S. A.; Erwin, D.
Verfahren zur Herstellung oxidischer Schutzschichten
(Anticorrosive protective oxide coatings)
(in German)
DOS 2 035 690 (18.7.1970); (C.A. 74 (1971) 90241j)

[16] Greeley, E. M.; Stenerson, R. N.
Improvements in corrosion inhibition, particularly in a cooling plant
France Pat. 1 351 641 (7.1.1964); (C.A. 61 (1964) 9236e)

[17] Krueger, R. H.; Dockus, K. F.; Rush, W. F.
Lithium chromate: Corrosion inhibitor for lithium bromide refrigeration systems
ASHRAE (Am. Soc. Heating, Refrig., Air Cond. Eng.) J. 6 (1964) p. 40; (C.A. 60 (1964) 12952d)

[18] Böhm, H.; Carl, H.; Ebert, S.; Rao, D. K.
Inhibition von Niedertemperaturkreisläufen in Kraftwerken mit Lithiumhydroxid
(Inhibition of low temperature circulating systems with lithium hydroxide in power stations) (in German)
Werkstoffe u. Korrosion 30 (1979) p. 685

[19] Böhm, H.
Inhibition von Niedertemperatur-Kreisläufen in Kraftwerken mit Lithiumhydroxid
(Inhibition of low temperature circulating systems with lithium hydroxide in power stations) (in German)
Werkstoffe u. Korrosion 30 (1979) p. 728

[20] Mann, G. M. W.
History and causes of on-load waterside corrosion in power boilers
Brit. Corrosion J. 12 (1977) p. 6

[21] Gronskij, R. K.; Rychkova, V. I.
Some questions concerning the use of lithium hydroxide for the treatment of boiler feedwater (in Russian)
Teploenergetika (1980) No 8; p. 35

[22] Kolobov, Yu.; Melnikov, V. G.; Murzabekova, G. Ya.
Study of the influence of the properties of basic media on the wear rate of steel-45 (in Russian)
Ivanov. Khim.-Tekhnol. in-t Ivanovo (1960) 6 p., Bibliogr. 2 Nazv.

[23] Schreiber, H. P.; Tewari, Y. B.; Wertheimer, M. R.
Application of microwave plasmas for the passivation of metals
Ind. Eng. Chem., Prod. Res. Dev. 17 (1978) No 1; p. 27

[24] Trubac, K.; Hronec, M.; Sveutec, I.; Suchanek, V.; Revus, M.; Runa, A.; Krupova, O.; Floch, S.
Process for removing corrosion products and sulfidic deposits from plants (in Czech)
CSSR Patent Kl. S 23 G 1/16, No 181 998 (3.11.1975)

[25] Gurovich, E. I.
Reaktionen von geschmolzenem Lithium-, Natrium und Kaliumhydroxid mit Nickel, Kupfer, Eisen und Stahl
(Reactions of molten lithium, sodium and potassium hydroxide with nickel, copper, iron and steel) (in Russian)
Zhur. Priklad. Khim. 32 (1968) p. 817

[26] de Vries, G.; Grantham, L. F.
Corrosion of metals in the molten LiOH-LiF-H_2O system
Electrochem. Technol. 5 (1967) p. 335

[27] Berry, W. E.
The corrosion behavior of FeCrNi alloys in high-temperature water
Edit.: High temperature, high pressure electrochemistry in aqueous solutions
Stähle, R. W.; Jones, D.; Slater, L.; Houston (1977) p. 48

[28] Wanklyn, J. N.; Jones, D.
The intercrystalline corrosion of stainless steel in alkaline solution
J. Nuclear Mat. 2 (1959) p. 154

[29] Shock, D. A.; Riggs, O. L.; Sudbury, J. D.
Application of anodic protection in the chemical industry
Corrosion 16 (1960) p. 55t

[30] Coriou, H.; Grall, L.; Pelras, M.
Corrosion resistance of chromium-nickel-iron austenitic alloys used in nuclear technology
Tr. Mezhdunar. Kongr. Korroz. Metal. 3 (1966), publ. 4 (1968) p. 309; (C.A. 71 (1969) 118 705c)

[31] Coriou, H.; Grall, L.; Pelras, M.
Corrosion problems arising from the use of austenitic Cr-Ni-Fe alloys in nuclear applications
Energ. Nucl. 9 (1967) p. 303; (C.A. 68 (1968) 15 504v)

Potassium Hydroxide

Unalloyed steels and low-alloy steels/cast steel

As shown by Figure 1, the corrosion rate of unalloyed and low-alloy steels in aqueous media is generally reduced by higher pH values.

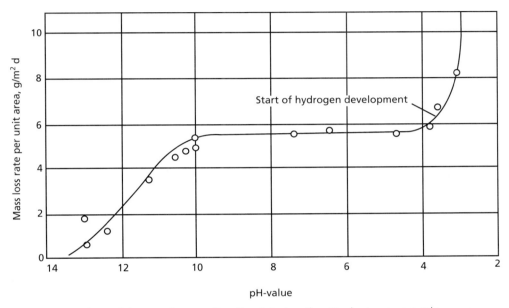

Figure 1: Dependence of the corrosion rate of unalloyed steel on the pH value in aqueous solutions [1]

The corrosion rate diminishes significantly as from pH 10 and reaches the very low value of $2\,g/m^2\,d = 0.08\,mm/a$ at pH 12. From Figure 2 it is evident that the corrosion rate of iron specimens in KOH solution drops to very low values already after short times [2].

With a shift of the potential to more negative values, longer times are required to form a protecting passivated layer, but the corrosion rate drops in any case to such low values that adequate resistance is achieved even at very negative potentials. Thus the following values are reported according to other investigations for unalloyed steel at room temperature, depending on the concentration of the potassium hydroxide solution and the potential [3]:

5 M KOH solution $U_H = -0.96\,V$ 0.09 mm/a (3.54 mpy)
1 M KOH solution $U_H = -0.98\,V$ 0.05 mm/a (1.97 mpy).

Elsewhere the values shown in Table 1 are reported [4].

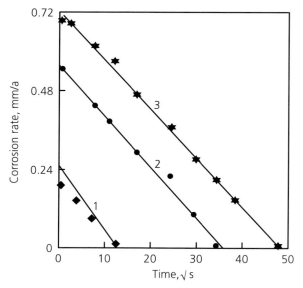

Figure 2: Time dependence of the corrosion rate of iron specimens in 4.25 M KOH solution at 298 K (25°C) and various potentials [2]
1: −600 to −500 mV
2: −700 to −600 mV
3: −800 to −700 mV

Steel	Corrosion rate mm/a (mpy)			U_H, V
	pH 12	pH 13	pH 14	
St 3 1.0333	0.09 (3.54)	0.075 (2.95)	0.055 (2.17)	−0.88 (−34.6)
St 45 1.0408 SAE 1020	0.14 (5.51)	0.12 (4.72)	0.06 (2.36)	−0.91 (−35.8)

Table 1: Corrosion rate of carbon steels in KOH solutions with various pH values at room temperature (test duration 900 h) [4]

In practice, low-alloy steels are the most frequently used materials in alkaline solutions up to concentrations of 50 % and temperatures up to 363 K (90°C). If the risk of stress corrosion cracking can be eliminated on consideration of the operating conditions, the utilization of low-alloy steels is ruled out essentially only in the case of demanded very low contamination of the product with iron.

Stress corrosion cracking

The uniform surface corrosion of unalloyed and low-alloy steels is therefore sufficiently low at temperatures that are not too high to permit the use of these steels in alkaline solutions. If necessary, surface corrosion can be countered by increasing the wall thickness to ensure sufficient service life of the component. However, at higher temperatures there is a risk of stress corrosion cracking also in KOH solution as in NaOH solution.

However, the system low-alloy steel/alkaline solution is a critical "material/corrosive agent" system with respect to stress corrosion cracking. Figure 3 and Figure 4 show the critical ranges of temperature and concentration of the NaOH solution. In complementation thereto, Figure 5 shows the range of stress corrosion cracking in a NaOH concentration–temperature diagram [5]. Although no correspondingly extensive information is available in the literature regarding the stress corrosion cracking behavior of the materials in potassium hydroxide solution, the behavior in KOH solution can be concluded with adequate reliability from these specifications [6].

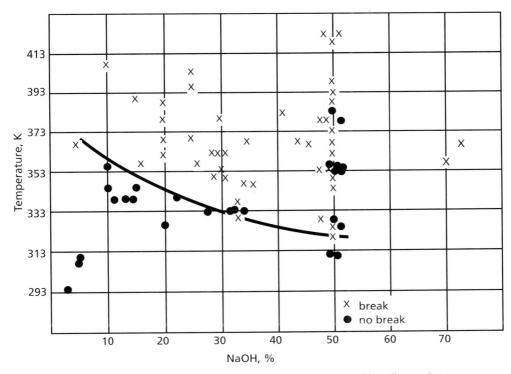

Figure 3: Influence of concentration and temperature on the SCC behavior of low-alloy steels in NaOH solutions [1]

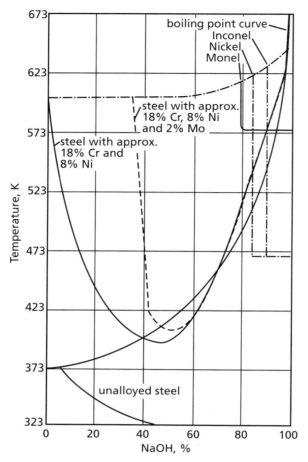

Figure 4: Regions in which there is a risk of stress corrosion cracking in NaOH solutions for various materials [1]

According to [7] carbon steels are suitable for transport and storage of alkaline solution. Depending on the concentration and temperature, welded components must be low stress annealed. Further details for this are given in Figure 5 in the section "Sodium hydroxide" in this book, and can be assumed to apply to potassium hydroxide solution too.

The critical potential range (critical) for the occurrence of stress corrosion cracking is shown in Figures 6 and Figure 7. Evidently the critical potential range for the occurrence of stress corrosion cracking is limited not only on the cathodic side but also on the anodic side, i.e. stress corrosion cracking can be prevented by potentials $U_H < -0.8$ or $U_H > -0.55$ V [8, 9].

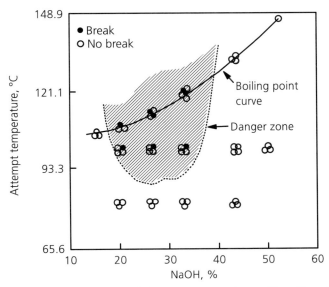

Figure 5: Range of susceptibility to stress corrosion cracking of low-alloy C-Mn steels in NaOH solutions of various concentrations at various temperatures [5]

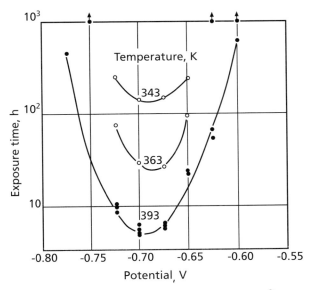

Figure 6: Dependence of the service life time of Armco® iron on the potential in 33 % NaOH solution [1]

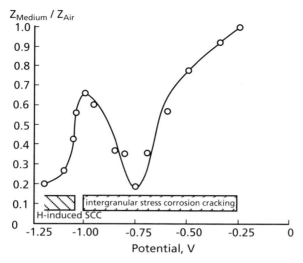

Figure 7: Dependence of fracture necking of tensile stress specimens made of S355N steel in boiling 35 % sodium hydroxide solution on the test potential (elongation rate $\dot{\varepsilon} = 1.4 \times 10^6$ 1/s) [10]

Extensive experience reports and investigation results are available in the literature with regard to the problem of stress corrosion cracking of unalloyed and low-alloy steels in alkaline solution, the influence of the chemical composition, of the microstructure, of the heat treatment, of the potential, of the operating conditions and of the kind of applied mechanical stress. Although most of these reports concern the behavior in NaOH solution, without doubt they are also extremely helpful for assessing the behavior in potassium hydroxide solution. Please consult the original references [11–22] and the chapter "Sodium hydroxide".

Corrosion fatigue cracking

Although only little is known about the corrosion fatigue cracking behavior in solutions containing alkali metal hydroxides, this type of corrosion must be contended with here for metallic materials, as in all aqueous media. Table 2 contains information based on results in NaOH solution concerning the potential range for corrosion fatigue cracking, that essentially corresponds to that for stress corrosion cracking [23].

U_H V	−0.55	−0.60	−0.65	−0.70	−0.75	−0.80	−0.825	−0.85
Δl mm/h	0.92	0.104	0.120	0.132	0.138	0.142	0.042	0.024

Table 2: Crack propagation rates Δl of a low-alloy steel with 0.1 % C in 10 M NaOH solutions at 393 K (120°C) as a function of the applied potential [23]

Unalloyed cast iron and low-alloy cast iron

Since cast iron shows similar corrosion behavior to that of the low-alloy steels in alkaline solutions, cast iron components have given satisfactory performance for certain applications in caustic solutions up to concentrations of 70 % and temperatures up to 363 K (90°C) [24]. Also in alkali electrolysis plants valves made of gray cast iron and pumps made of cast ferrosilicon are utilized [25].

Additions of 0.4 to 0.8 % chromium and 0.35 to 1 % nickel, which increase the strength of cast iron, have practically no influence on the corrosion resistance in alkaline solutions [26]. For a cast iron with a nickel content of 2 % a corrosion rate of 0.03 mm/a (1.18 mpy) was measured in 35 % potassium hydroxide solution at 298 K (25°C), but it increases with increasing temperature, to 0.12 mm/a (4.72 mpy) at 343 K (70°C) [27].

The cast iron grades Meehanite® with lamellar or spheroidal graphite are used as materials for evaporators in the processing of sodium hydroxide. The grade Meehanite-HD is recommended for vessels for NaOH melts at 813 K (540°C) [28]. It is to be expected that these grades can also be utilized in KOH solution under corresponding conditions.

High-alloy cast iron
Cast ferrosilicon
Austenitic cast iron (among other things)

As for most alloys, higher nickel contents improve the corrosion resistance in alkaline solution also for cast iron [29]. As Table 3 shows for the example of KOH solution and NaOH solution, the corrosion rate is reduced by a factor of ten when the nickel content is increased from 15 % to 20 % [30, 31].

Nickel content %	Corrosion rate in KOH mm/a (mpy)	Corrosion rate in NaOH mm/a (mpy)
0	0.53–0.75 (20.9–29.5)	1.9–2.3 (74.8–90.6)
3	0.075 (2.95)	
3.5		1.2 (47.2)
5		1.2 (47.2)
6.5	0.05 (1.97)	
12.4	0.01 (0.39)	
15		0.8 (31.5)
20		0.08 (3.15)
30		0.01 (0.39)

Table 3: Influence of the nickel content in cast iron on the corrosion in approx. 50 % KOH solution at 673 K (400°C) [31] and in boiling 50–65 % NaOH solution [32]

Additional chromium contents with a magnitude of 1 % to 6 % improve the resistance to concentrated KOH solution and NaOH solution of the austenitic cast iron grades with a nickel content of about 16 % and small amounts of copper and manganese [33].

The nickel alloyed cast iron materials listed in Table 4 have, as shown in Table 5, good resistance to potassium hydroxide solution, whereby the resistance here too increases with increasing nickel content [34].

Specimen	Composition, %						
	C	Si	Mn	P	V	Ni	Fe
FeNi-1	2.60	1.88	0.70	0.009	0.022	11.47	balance
FeNi-2	2.49	1.86	0.67	0.009	0.021	17.75	balance
FeNi-3	2.34	1.77	0.69	0.008	0.023	21.26	balance

Table 4: Chemical composition of iron-nickel cast alloys [34]

Alloy	KOH %	Corrosion rate	
		g/m² h	mm/a (mpy)
FeNi-1	30	0.016	0.019 (0.75)
FeNi-2		0.0096	0.012 (0.47)
FeNi-3		0.0042	0.005 (0.20)
FeNi-1	50	0.013	0.015 (0.59)
FeNi-2		0.0075	0.009 (0.35)
FeNi-3		0.0021	0.003 (0.12)

Table 5: Corrosion rate of three iron-nickel cast alloys in potassium hydroxide solution at 293 K (20°C) [34]

Ferritic chrome steels with < 13 % Cr

Hardly any information is available in the literature on the behavior of this group of materials in potassium hydroxide solution. However, it can be assumed that the extensive experience of the behavior in NaOH solution reported in the literature for the low-alloy CrMo steels and NiCrMoV steels frequently utilized in power plant construction can also be applied for assessing the behavior in potassium hydroxide solution [35–46] (see also the chapter "Sodium hydroxide").

Ferritic chrome steels with ≥ 13 % Cr

Chrome steels have adequate corrosion resistance (< 0.1 mm/a (< 3.94 mpy)) in hot concentrated alkaline solutions only with chromium contents greater than 18 %, whereby the addition of an inhibitor (0.1 to 0.2 % alkali metal chlorate) to the caustic solution is recommended [47].

Further information on the resistance of ferritic chrome steels in potassium hydroxide solution is contained in the Tables 8 and 10.

From the example of the results obtained in 48 % NaOH solution shown in Figure 8, the temperature dependence of the corrosion rate of steels with various chromium contents can be estimated also in potassium hydroxide solution [48]. According thereto, at 373 K (100°C) corrosion resistance (0.01 mm/a (0.39 mpy)) is already achieved with a Cr content of 24 %. However, at 473 K (200°C) only the steel with a Cr content of 30 % and a corrosion rate of approx. 0.1 mm/a (3.94 mpy) can be used.

The surface corrosion of the chrome steels alloyed with molybdenum is greater by a factor of 4 to 8 than that of the chrome nickel steels (SAE 304, corresponding to Material No. 1.4301, and SAE 310, corresponding to Material No. 1.4841) with comparable chromium contents. But in contrast to the austenitic stainless CrNi steels, they are largely resistant to stress corrosion cracking [49].

Ferritic/Perlitic-martensitic steels

The stainless martensitic heat-treatable steels with 12 or 17 % of chromium are utilized on account of their high strength values in steam turbine construction for parts subjected to high stress. Since alkaline additional media can come into contact with these components in the case of leaks in the steam or cooling water system, the stress corrosion cracking behavior of such steels was investigated in various alkaline media. There the investigated steels were found to be practically resistant to stress corrosion cracking, e.g. 17-4-PH (X5CrNiCuNb17-4-4, Material No. 1.4548) and SAE 410 (corresponding to Material No. 1.4006) [50–53].

Ferritic-austenitic steels/duplex steels

Since only very few informative literature reports on the behavior of the ferritic-austenitic steels in potassium hydroxide solutions exist, the extensively available reports of investigations in sodium hydroxide solutions must be resorted to for assessing the corrosion behavior in potassium hydroxide solution.

Further information on the resistance in potassium hydroxide solution is contained in the Tables 8 and 10.

The different corrosion resistance of austenite, ferrite and carbide in alkaline solutions can be exploited in metallography by etching in such solutions to distinguish between the microstructure components [54, 55]. In [55] for the duplex steel Al 2205® Alloy (UNS S31803, corresponding to Material No. 1.4462) with the

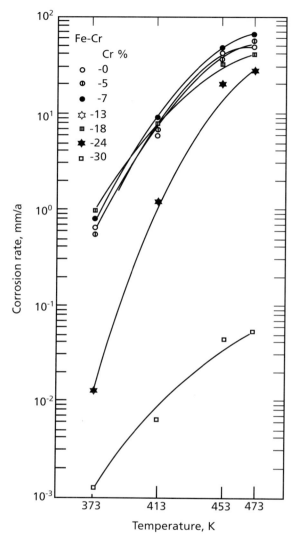

Figure 8: Temperature dependence of the corrosion rate of steels with chromium contents of up to 30 % in 48 % NaOH solution [48]

composition specified in Table 6, a corrosion rate of 0.61 mm/a (24.0 mpy) in boiling 50 % NaOH solution is reported, compared with 1.97 mm/a (77.6 mpy) for the austenitic steel grade SAE 316 L (corresponding to Material No. 1.4404; 1.4435).

C	Mn	P	V	Si	Cr	Ni	Mo	N	Fe
≥ 0.030	≥ 2.0	≥ 0.030	≥ 0.020	≥ 1.0	21.0–23.0	4.5–6.5	2.5–3.5	0.08–0.20	bal.

Table 6: Composition of the duplex steel Al 2205® Alloy (UNS S31803) [55]

As shown in Figure 9 for the example of boiling NaOH solution, for the customary duplex steels (composition as shown in Table 7) the corrosion rate increases with increasing concentration of the caustic solution, but with material consumption rates of $< 0.01\,g/m^2\,h$ in boiling caustic solution, adequate corrosion resistance is still achieved up to a concentration of 30 %. At higher concentrations, the corrosion rates increased noticeably and reached a maximum with 60 % NaOH. Under these conditions, the molybdenum-free steel 1.4362 is inferior to other steels. The tests were carried out on commercial sheets and welded materials in the solution-annealed state [56].

Steel	% C	% Si	% Mn	% P	% S	% Cr	% Mo	% Ni	% N	% Cu
1.4462	0.02	0.55	1.74	0.022	0.002	22.05	2.97	5.95	0.13	–
Uranus 50 UNS S32404	0.02	0.54	1.84	0.023	0.004	20.99	2.33	6.58	–	1.4
Ferralium® 255	0.03	0.24	1.52	0.023	0.002	25.93	3.15	5.6	0.18	1.71
1.4362 – 1	0.02	0.52	1.22	0.022	0.003	22.98	0.21	4.8	–	–
1.4362 – 2	0.02	0.52	1.84	0.027	0.001	23.05	–	4.2	–	–

The steel 1.4362 was present as 2 melts

Table 7: Chemical composition of the investigated duplex steels [56]

In investigations of the stress corrosion cracking behavior of these steels, that were carried out at the free corrosion potential as well as with potentiostatically controlled tests with specimens under constant deformation (U-bending specimens) and with slow tension stress tests (CERT tests), no stress corrosion cracking was observed even with caustic solutions of higher concentration up to 70 %.

According to these results, the use of ferritic-austenitic steels in hot alkaline solutions up to a concentration of approx. 30 % could be an alternative to nickel-based materials. In spite of their very good resistance to stress corrosion cracking, the use of these steels at higher concentrations of caustic solutions is not recommended due to the increasing material consumption rates.

Other investigations, too, confirm that the duplex steels are superior to the austenitic steels with regard to their resistance to stress corrosion cracking, and that for the ferritic-austenitic steels the general surface corrosion and not the stress corrosion cracking is the determining factor for the utilization limit in alkaline solutions. With regard to the stress corrosion cracking, increasing chromium contents have a positive influence, whereas higher fractions of nickel and molybdenum have an unfavorable influence [57–60].

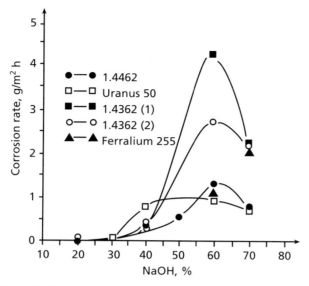

Figure 9: Corrosion rates of the ferritic-austenitic steels in boiling NaOH solutions [56]

Austenitic chromium-nickel steels

The resistance of the austenitic 18-8-CrNi standard steels to uniform surface attack in alkaline solutions strongly depends on the concentration and the temperature of the solution. The plain 18-8 steels, e.g. the chrome nickel steels frequently utilized in the foodstuffs industry of the grades SAE 304 L (X2CrNi19-11, Material No. 1.4306, UNS S30403), SAE 303 (Material No. 1.4305), SAE 316 L (corresponding to X2CrNiMo17-12-2, Material No. 1.4404; 1.4435), are adequately resistant to alkaline solutions only with small concentrations and temperatures up to 343 K (70°C). The good cleaning effect of dilute alkaline solutions can therefore be exploited in the foodstuffs industry for plants and containers made of chrome nickel steel [61].

Also for the austenitic CrNi steels the resistance in caustic solutions increases with increasing content of nickel [62]. In concentrated caustic solutions the following corrosion rates are reported for the CrNi steels:

- 8–12 % Ni (approx. 0.1 mm/a (3.94 mpy))
- 12–20 % Ni (approx. 0.5 mm/a (19.7 mpy))
- 33–36 % Ni (approx. 0.3 mm/a (11.8 mpy)).

Steels of type SAE 304 and 316 are susceptible to stress corrosion cracking at higher temperatures in alkaline solutions [59, 63, 64].

In investigations with the materials SAE 304 (UNS J92610, corresponding to Material No. 1.4301), Incoloy® 800 (X10NiCrAlTi32-21, Material No. 1.4876), Inconel® Alloy 690 (NiCr29Fe, Material No. 2.4642) and Inconel® 600 (NiCr15Fe, Material No. 2.4816) in solutions with each 25 % KOH and 25 % NaOH at high temperatures (approx. 598 K (325°C)) stress corrosion cracking took place only with the steel SAE 304 [65]. For the steel 304 the threshold temperatures for the appearance of stress corrosion cracking in 10 % NaOH solution lie at 363 K (90°C) and in 50 % NaOH solution at 323 K (50°C) [66].

In [6] a case of stress corrosion cracking damage of a bypass line made of steel SAE 304 L (UNS S30403, corresponding to Material No. 1.4306) is described. The line belonged to a steam/methane reformer plant for the production of hydrogen and was utilized only during the startup phase of the plant when no syngas was present as yet. Syngas, a mixture of hydrogen, carbon monoxide, carbon dioxide, methane and steam, is not a medium with which stress corrosion cracking of austenitic steels would be expected. The cracks showed an inter-crystalline, strongly branched course and were covered with black oxides. This is characteristic for stress corrosion cracking in alkaline media: Chlorides, that are a typical stress corrosion cracking initiating medium for austenitic steels, were not found in the cracks. But large amounts of potassium were found on the fracture surface. The investigations carried out in connection with this damage in the interior of the plant showed that potassium originating from a potassium doped catalyst utilized in the reformer migrated into the line and when undershooting the dew point produced potassium hydroxide with the condensed water. A subsequently carried out test program with regard to the stress corrosion cracking behavior of the steel 304 L in KOH solutions and with regard to the influence of the addition of CO, carbonate and hydrogen showed that at temperatures under 616 K (343°C) no stress corrosion cracking took place on the steel 304 L in pure KOH solution. However, in the presence of CO and/or hydrogen, stress corrosion cracking was observed already at temperatures above 461 K (188°C). In the presence of hydrogen cracks appeared at a temperature of 489 K (216°C) even with a KOH concentration of only 1 %. From the investigation results and from other literature reports the diagram shown in Figure 10 was derived, from which the critical temperature and concentration range for the appearance of stress corrosion cracking in KOH solution is seen for the austenitic CrNi steel 304 L. The corresponding range for NaOH solution is shown too (cf. also Figure 37 in the chapter "Sodium hydroxide" according to [67]). Evidently stress corrosion cracking can appear in NaOH solutions at considerably lower temperatures than in KOH solutions.

In investigations in NaOH solution, the high nickel content steel Incoloy® 800 (X10NiCrAlTi32-21, Material No. 1.4876) with low carbon content, that is utilized for steam boiler pipes, was found to be resistant to uniform surface corrosion as well as to stress corrosion cracking, even with high caustic solution concentrations, up to 458 K (185°C) [68]. Only at 623 K (350°C) cracks with up to more than 100 mm depth appeared in 10 % NaOH solution on Incoloy 800 in the course of up to 6,000 h [69].

Further information with regard to the stress corrosion cracking behavior of the steels Incoloy® 800, SAE 304 (corresponding to Material No. 1.4301) and SAE 316 (corresponding to Material No. 1.4401; 1.4436) as well as of the Ni alloy Inconel® 600 (Material No. 2.4816) in NaOH solution, that can serve for assessing the behavior of these materials in KOH solution, are contained in [70–75].

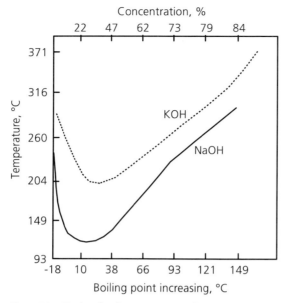

Figure 10: Region for the appearance of stress corrosion cracking (above the respective line) in NaOH and KOH for the austenitic CrNi steel 304 L depending on the caustic solution concentration and temperature [6]

The anodic dissolution of chromium-nickel steels decreases with increasing nickel and chromium contents. Already with nickel additions of 8% in chrome steels a significant improvement of the corrosion resistance in alkaline solutions at higher concentrations and temperatures is achieved [76].

The risk of stress corrosion cracking of chromium-nickel steels in alkaline solutions at high temperatures (543 to 603 K (270°C to 330°C)) is reduced if both the chromium and the nickel contents in the steel are increased [77].

In summary it can be said that the chrome nickel steels (e.g. SAE 304 and SAE 316) with higher chromium and nickel contents are also at higher temperatures resistant to uniform surface corrosion in diluted and concentrated alkaline solution, but they can be susceptible to stress corrosion cracking and in particular to corrosion fatigue cracking.

In alkali melts the austenitic CrNi steels, e.g. of the type 304 L (corresponding to Material No. 1.4306) and 310 can be attacked strongly. This is given as the reason for the frequently observed damage on the furnace side of steam boilers that is caused by alkaline-containing deposits from waste gases on components made of these steels [78].

Electrochemical investigations of carbon steel pipes cladded with steel 304 showed that, under molten alkali deposits the steel 304 has a less noble potential than that of the unalloyed steel. In the case of cracks or discontinuities in the cladding, the unalloyed steel thus has cathodic protection [78].

Austenitic CrNiMo(N) steels and CrNiMoCu(N) steels

Krupp Thyssen Nirosta [79] makes specifications of the resistance to potassium hydroxide solutions of various concentrations and at various temperatures for its entire range of stainless steel products; these specifications are shown in Table 8.

The resistance assessments given in Table 8 at higher temperatures and for the melt must be considered very critically, because the resistance specifications are too optimistic. In contrast thereto, the values specified by a second stainless steel manufacturer in Table 10 appear to be significantly more realistic [80]. It must furthermore be taken into consideration that the values in the two mentioned tables only make statements regarding the uniform surface corrosion and that all austenitic CrNi steels are endangered by stress corrosion cracking in potassium hydroxide solutions of higher concentration (> 20 %) and at higher temperatures (> 353 K (> 80°C)).

Attacking agent	Concentration %	Temperature K (°C)	Group 1	Group 2	Group 3	Group 4	Group 5	1.4465	1.4539	1.4565
Potassium hydroxide	20	293 (20)	0	0	0	0	0	0	0	0
	20	boiling	0	0	0	0	0	0	0	0
	50	293 (20)	0	0	0	0	0	0	0	0
	50	boiling	2	1	1	0	0	0	0	0
	saturated solution	boiling	2	1	1	0	0	0	0	0
	melts	633 (360)	3	3	3	3	3	0	0	0

0 = Material consumption rate < 0.1 g/m^2 h corresponding to a corrosion rate < 0.11 mm/a (< 4.33 mpy)
1 = Material consumption rate 0.1–1.0 g/m^2 h corresponding to a corrosion rate 0.11–1.1 mm/a (4.33–43.3 mpy)
2 = Material consumption rate 1.0–10.0 g/m^2 h corresponding to a corrosion rate 1.1–11 mm/a (43.3–433 mpy)
3 = Material consumption rate > 10.0 g/m^2 h corresponding to a corrosion rate > 11.0 mm/a (> 433 mpy)

Table 8: Stability of NIROSTA® steels in KOH solution [79]. For the group assignment see Table 9

Group 1 Mat. No.	Group 2 Mat. No.	Group 3 Mat. No.	Group 4 Mat. No.	Group 5 Mat. No.
1.4000			1.4301	
1.4002			1.4303	
1.4003	1.4016		1.4306	1.4401
1.4006	1.4057		1.4307	1.4404
1.4021	1.4120		1.4310	1.4429

Table 9: Classification in groups for the materials listed in Table 8 [79]

Table 9: Continued

Group 1 Mat. No.	Group 2 Mat. No.	Group 3 Mat. No.	Group 4 Mat. No.	Group 5 Mat. No.
1.4024	1.4305		1.4311	1.4435
1.4028	1.4427		1.4315	1.4436
1.4031	1.4509		1.4318	1.4438
1.4034	1.4510		1.4541	1.4439
1.4313	1.4511		1.4544	1.4462
1.4512	1.4520	1.4113	1.4546	1.4561
1.4589	1.4521	1.4568	1.4550	1.4571

Table 9: Classification in groups for the materials listed in Table 8 [79]

Concentration of the KOH solution, %	Temperature K (°C)	Material					
		1.4000, 1.4002, 1.4006, 1.4021, 1.4028, 1.4104, 1.4034, 1.4110, 1.4112, 1.4528	1.4016, 1.4057, 1.4122	1.4405, 1.4542, 1.4548	1.4301, 1.4304, 1.4306, 1.4361, 1.4544, 1.4546, 1.4550	1.4401, 1.4404, 1.4435, 1.4436, 1.4571	1.4439, 1.4460, 1.4462, 1.4465, 1.4466, 1.4467, 1.4539
20	293 (20)	1	1		1	1	0
20	boiling	1	1		1	1	1
50	293 (20)	1	1		1	1	0
50	boiling	2	2		2	2	1
Saturated solution	393 (120)	2	2		1	1	1
Melts	633 (360)	3	3	3	3	3	3

0 = Material consumption rate < 0.1 g/m^2 h corresponding to a corrosion rate < 0.11 mm/a (< 4.33 mpy)
1 = Material consumption rate 0.1–1.0 g/m^2 h corresponding to a corrosion rate 0.11–1.1 mm/a (4.33–43.3 mpy)
2 = Material consumption rate 1.0–10.0 g/m^2 h corresponding to a corrosion rate 1.1–11 mm/a (43.3–433 mpy)
3 = Material consumption rate > 10.0 g/m^2 h corresponding to a corrosion rate > 11.0 mm/a (> 433 mpy)

Table 10: Chemical stability of stainless steels in KOH solution [80]

The Figures 14–17 show the results of investigations of the influence of the concentration and the temperature of the hydroxide solutions LiOH, KOH and NaOH as well as of additions of chlorides and chlorates on the corrosion behavior of high-alloy Russian stainless steel grades [81]. The type of the materials tested, the testing media and testing conditions are given in Table 11.

No.	Materials	Testing medium	Concentration	Temperatures
1	ferritic steel 15Ch25T[1]	LiOH KOH NaOH	2.75–6.0 M 2.75–19.25 M 2.75–19.25 M	348–448 K (75–175°C)
2	austenitic steel 12Ch18N10T (cf. 1.4541)	LiOH KOH NaOH	2.75–6.0 M 2.75–19.25 M 2.75–19.25 M	348–448 K (75–175°C)
3	austenitic steel 10Ch17N13M2T (cf. 1.4571)	LiOH KOH NaOH	2.75–6.0 M 2.75–19.25 M 2.75–19.25 M	348–448 K (75–175°C)
4	ferritic-austenitic steel 08Ch22N6T[1]	LiOH KOH NaOH	2.75–6.0 M 2.75–19.25 M 2.75–19.25 M	348–448 K (75–175°C)
5	ferritic-austenitic steel 08Ch21N6M2T (corresponds to Material No. 1.4462)	LiOH KOH NaOH	2.75–6.0 M 2.75–19.25 M 2.75–19.25 M	348–448 K (75–175°C)
6	austenitic alloy 06ChN28MDT[1]	LiOH KOH NaOH	2.75–6.0 M 2.75 – 19.25 M 2.75–19.25 M	348–448 K (75–175°C)
7	Ni-Cr alloy ChN78T[1]	LiOH KOH NaOH	2.75–6.0 M 2.75–19.25 M 2.75–19.25 M	348–448 K (75–175°C)

1) For the composition of the materials cf. Table 12

Table 11: Materials, testing media and testing conditions for the investigations of the corrosion behavior of various metals in hydroxide solutions [81]

Material	Fe	Cr	Ni	Mn	Si	Ti	Mo	C	Cu	Other
15Ch25T	bal.	24–27	≤ 1.0	≤ 0.8	≤ 1.0	*	–	≤ 0.15	≤ 0.3	
08Ch22N6T	bal.	21–23	5.3–6.3	≤ 0.8	≤ 0.8	**	0.3	≤ 0.08	≤ 0.3 Cu	≤ 0.2 W
06ChN28MDT	bal.	22–25	26–29	≤ 0.8	≤ 0.8	0.5–0.9	2.5–3.0	≤ 0.06	2.5–3.5	
ChN78T	≤ 1.0	19–22	bal.	≤ 0.7	≤ 0.8	0.15–0.35	–	≤ 0.12	–	Al 0.15

* Ti ≈ 5 x %C; max. 0.9 %, ** Ti ≈ 5 x %C; max. 0.65 %

Table 12: Alloy components of the Russian materials in mass%

Tests were carried out to determine the corrosion rates in a test duration of up to 100 h. Stress corrosion cracking tests of welded samples were carried out for up to 255 h. The corrosion rates of the stainless steels (Nos. 1, 2, 3, 4 and 6) that were determined in the various testing media are shown in Figure 11 to Figure 15.

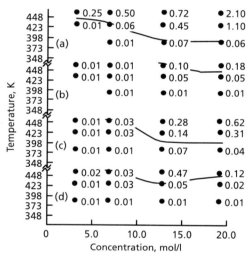

Figure 11: Corrosion rates (mm/a) and isocorrosion curve (0.1 mm/a (3.94 mpy)) for the material 15Ch25T of Table 11 in various testing solutions [81]
a = NaOH; b = KOH; c = NaOH + 5 g/l NaClO$_3$;
d = KOH + 5 g/l NaClO$_3$

Austenitic CrNiMo(N) steels and CrNiMoCu(N) steels | 281

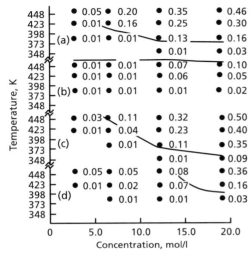

Figure 12: Corrosion rates (mm/a) and isocorrosion curve (0.1 mm/a (3.94 mpy)) for the material 12Ch18N10T of Table 11 in various testing solutions [81]
a = NaOH; b = KOH; c = NaOH + 5 g/l NaClO₃; d = KOH + 5 g/l NaClO₃

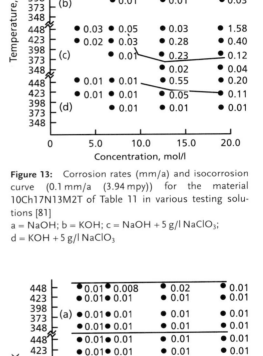

Figure 13: Corrosion rates (mm/a) and isocorrosion curve (0.1 mm/a (3.94 mpy)) for the material 10Ch17N13M2T of Table 11 in various testing solutions [81]
a = NaOH; b = KOH; c = NaOH + 5 g/l NaClO₃; d = KOH + 5 g/l NaClO₃

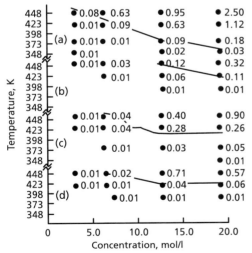

Figure 14: Corrosion rates (mm/a) and isocorrosion curve (0.1 mm/a (3.94 mpy)) for the material 08Ch22N6T of Table 11 in various testing solutions [81]
a = NaOH; b = KOH; c = NaOH + 5 g/l NaClO₃; d = KOH + 5 g/l NaClO₃

Figure 15: Corrosion rates (mm/a) and isocorrosion curve (0.1 mm/a (3.94 mpy)) for the material 06ChN28MDT of Table 11 in various testing solutions [81]
a = NaOH; b = KOH; c = NaOH + 5 g/l NaClO₃; d = KOH + 5 g/l NaClO₃

Independent of the concentration or the temperature, the corrosion rate of the nickel-based material No. 7 (ChN78T) was less than 0.1 mm/a (3.94 mpy). In the lithium hydroxide solutions the corrosion rates of all materials were less than 0.01 mm/a (0.39 mpy). Therefore, for stainless steels and nickel-based alloys, the corrosiveness of the hydroxide solutions increases in the following order:

LiOH < KOH < NaOH

As the concentration and temperature increase, the corrosion rate of the stainless steels increases from 0.01 mm/a (0.39 mpy) to less than 2.5 mm/a (98.4 mpy) in NaOH and to 0.3 mm/a (11.8 mpy) in KOH. In 2.75 M KOH, the investigated steels are practically resistant, even at temperatures up to 453 K (180°C), whereas at the same concentration in NaOH the corrosion rates were approx. 0.1 mm/a (3.94 mpy). With chlorides showing no effect, the oxidizing agent chlorate significantly improves the corrosion resistance in the positive sense. Thus the corrosion rate of some steels is lowered by a factor of ten in the presence of 5 g/l chlorate.

As a further result of the stress corrosion cracking tests, the limits for the susceptibility to stress corrosion cracking are given for three of the investigated materials in Figure 16 depending on the temperature and concentration of the caustic solution.

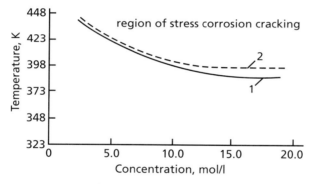

Figure 16: Range of stress corrosion cracking for the materials 12Ch18N10T and 10Ch17N13M2T (1) as well as 06ChN28MDT (2) in KOH solution and NaOH solution [81]

The ferritic steel and the two ferritic-austenitic steels did not exhibit stress corrosion cracking in the testing media. However, the ferritic-austenitic steels were susceptible to selective corrosion of the austenite phase. The welded nickel material showed a tendency for inter-crystalline corrosion (inter-crystalline) and cracking.

Extensive investigations of the stress corrosion cracking behavior on welded specimens of an austenitic CrNiMo steel of the type SAE 316 (Material No. 1.4401) and on mixed jointing weldings of this steel with the low-alloyed steel 10CrMo9-10 (Material No. 1.7380) or with the nickel-based material Inconel® 600 (Material No. 2.4816) are described in [82]. Further results of stress corrosion cracking investigations of three austenitic steels (ASTM F138, 1.4404 and 1.4539) compared with a ferritic-austenitic steel (1.4462) are reported in [83].

In connection with the development of an electrolysis plant for producing hydrogen, that operates at higher pressures and temperatures, the corrosion behavior of a number of stainless steels and higher alloyed nickel alloys was investigated in 30 % KOH solution at 403 K (130°C) and 423 K (150°C) under argon and oxygen gas at a pressure of 30 kg/cm² [84]. The approximate composition of the investigated materials and the obtained results of the corrosion tests are shown in Table 13.

The following conclusions can be derived from these results:

- The attack is stronger under oxygen atmosphere than under inert gas.
- 18-8-CrNi steels generally show poor corrosion resistance, even at 403 K (130°C), i.e. plants made of such steels, e.g. type 316/316 L, must be monitored diligently in long term operation to ensure that an operating temperature of 393 K (120°C) is not exceeded.
- Under oxygen atmosphere all kinds of corrosion are observed, such as uniform surface corrosion, pitting and crevice corrosion, stress corrosion cracking and inter-crystalline attack. Cracking took place even on some nickel alloys.
- No stress corrosion cracking of the ferritic-austenitic steels was found, but general attack and inter-crystalline corrosion took place.
- The austenitic steels with higher Cr and Ni contents showed on the whole better behavior, and among these steels the most favorable results were found for the ELC steel with 20 % Ni and 24 % Cr (type 316 L).

Contents of the main alloying elements			Temperature			
Ni %	Cr %	Others	403 K (130°C)		423 K (150°C)	
			Ar	O₂	Ar	O₂
9	19		–		–	–
9	17	Ti	–		–	–
10	18	LC	–		–	–
10	18	ELC		–	–	–
11	17	2 Mo	–		–	–
11	21	2 Mo – N – ELC		–	((+))	–
13	18	2 Mo – ELC	((+))		–	–
13	23		–		–	–
13	24	1 Mo – N		–	((+))	–
14	18	3 Mo – LC	((+))	–	–	–
15	12	ELC	–		–	–
16	16	2 Mo – 2 Cu	–		–	–

+ = resistant
(+) = conditionally resistant
((+)) = conditionally resistant for restricted utilization
– = not resistant

Table 13: Contents of the main alloying elements of the investigated materials and results of the corrosion tests in 30 % KOH solution [84]. LC = Low Carbon; ELC = Extra Low Carbon; ULC = Ultra Low Carbon

Table 13: Continued

Contents of the main alloying elements			Temperature			
Ni %	Cr %	Others	403 K (130°C)		423 K (150°C)	
			Ar	O₂	Ar	O₂
16	17	5 Mo – LC	–		–	–
20	24	ELC	(+)	+	+	+
20	25		+	+	+	–
23	18	2 Mo – 2 Cu	+	+	+	–
25	24	3 Mo – Nb – ELC	(+)	((+))	+	–
26	22	4 Mo – Ti – LC	(+)	((+))	+	–
29	20	2 Mo – 3 Cu – Nb	(+)	((+))	+	–
33	20	Ti – Al	(+)	+	+	+
41	22	3 Mo – 2 Cu – Ti	(+)	–	+	–
50	25	6 Mo – 1 Cu – Ti – ELC		+		+
59	16	16 Mo – 3 W – V – ULC		+		+
61	22	8 Mo – Ti – Nb – LC		+		++
62	27	Ti – Al – ELC		+		+
75	15	LC		+		
99				((+))		((+))
4	22	ELC		–	–	–
4	25	1 Mo – N – ELC		–	–	–
7	25	3 Mo – W – N –		–	–	–
	19	2 Mo – Nb – ULC	–	–	–	
	29	2 Mo – Nb – ULC	–	–	–	

+ = resistant
(+) = conditionally resistant
((+)) = conditionally resistant for restricted utilization
– = not resistant

Table 13: Contents of the main alloying elements of the investigated materials and results of the corrosion tests in 30% KOH solution [84]. LC = Low Carbon; ELC = Extra Low Carbon; ULC = Ultra Low Carbon

An addition of 2 to 5% molybdenum to austenitic CrNi steels can noticeably decrease the uniform surface corrosion in hot concentrated alkaline solutions. As example, that can also serve as information regarding the behavior in KOH solution, Figure 17 shows the favorable effect of molybdenum on the behavior of chrome nickel steels and nickel-chromium alloys in boiling 50% NaOH solution [85]. According thereto the steel Incoloy alloy 901 (42%Ni, 13%Cr, 6% Mo, corresponding to Material No. 2.4662) is more resistant, in spite of the lower Cr content, because of the higher molybdenum content, than the steel Incoloy® 825 (NiCr21Mo, Material No. 2.4858).

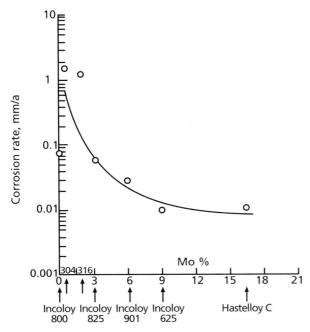

Figure 17: The influence of the molybdenum content on the corrosion of chrome nickel steels and nickel alloys in boiling 50 % NaOH solution [85]

Due to its high nickel content, the high-alloyed material X1CrNiMoCuN33-32-1 (Material No. 1.4591, Alloy 33, Nicrofer® 3033) with the composition specified in Table 14 is outstanding not only with regard to its good resistance in strongly oxidizing media, various acids and chloride solutions, but also with regard to its good resistance to alkaline solutions [86, 87]. In the material properties sheet for this high-alloy steel, utilization in the production and handling of KOH and NaOH solutions of up to 70 mass% concentration and maximum temperature of 443 K (170°C) is mentioned under the application fields [88].

% C	% Mn	% Si	% P	% S	% Cr	% Ni	% Mo	% Cu	% N	% Fe
≤ 0.015	≤ 2.0	≤ 0.5	≤ 0.020	≤ 0.010	31.0–35.0	30.0–33.0	0.5–2.0	0.3–1.2	0.35–0.60	bal.

Table 14: Chemical composition of Alloy 33 [86]

Bibliography

[1] Herbsleb, G.
Korrosionsschutz von Stahl
(Corrosion protection of steel) (in German)
p. 24
Verlag Stahleisen, Düsseldorf, 1997

[2] Hampson, N. A.; Latham, R. J.;
Marshall, A.; Giles, R. D.
Some aspects of the electrochemical behaviour of the iron electrode in alkaline solutions
Electrochimica Acta 19 (1974), p. 397

[3] Stepina, T. G.; Iofa, Z. A.; Kasatkin, E. V.; Shepelin, V. A.; Safonov, V. A.
Untersuchung von anodischen Oxidschichten auf Eisen in alkalischen Lösungen mit Hilfe der Ellipsometrie
Sow. Electrochem. 16 (1980), p. 1548

[4] Lesnikva, K. P.; Frjd, M. K.
The effect of boriding on corrosion resistance of carbon steels (in Russian)
Metallschutz Moskau 16 (1980), p. 163

[5] Berk, A. A.; Waldeck, W. F.
Caustic Danger Zone
Chemical Engineering 57 (1950), pp. 235–238

[6] Dean, S. W.
Caustic cracking from potassium hydroxide in syngas
Materials Performance 38 (1999) 1, pp. 73–76

[7] Ashbaugh, W. G.
Stress Corrosion Cracking of Process Equipment
Chem. Eng. Progr. 66 (1970) 10, p. 44

[8] Bohnenkamp, K.
Über die Spannungsrißkorrosion weicher Stähle in konzentrierter Natronlauge
(The stress corrosion of soft steel in concentrated caustic soda solution)
(in German)
Archiv Eisenhüttenwesen 39 (1968), pp. 361–368

[9] Mazille, H.; Uhlig, H. H.
Effect of Temperature and Some Inhibitors on Stress Corrosion Cracking of Carbon Steels in Nitrate and Alkaline Solutions
Corrosion NACE 26 (1972) 427

[10] Drodten, P.; Herbsleb, G.; Kuron, D.; Savakis, S.; Wendler-Kalsch, E.
Spannungsrißkorrosion Mo-freier und Mo-haltiger Stähle in Calciumnitratlösung und Natronlauge unter zeitlich konstanter Kraft und unter CERT-Bedingungen
Werkst. Korros. 42 (1991) 9, S. 473–489

[11] Jin, Z.; Zuo, J.; Lu, Y.
Mechanism of SCC for mild steel in 33 % NaOH solution at 115°C
Conference: Corrosion Control – 7th APCCC.
Vol. 1, China, 1991
Publ.: International Academic Publishers, Xizhimenwai Dajie, Beijing Exhibitio Centre, Beijing 100044, People's Republic of China, 1991

[12] Langenscheid, G.
Beitrag zur Korrosion weicher, unlegierter Stähle in Natronlauge
(Contribution for the corrosion of soft, unalloyed steels in caustic soda solution)
(in German)
Steel research 60 (1989) 6, pp. 281–287

[13] Vereshchagin, K. I.; Oskolkova, L. K.
Die Entstehung und Ausbreitung von Rissen in den Stählen St-3 und 16-GS bei der Untersuchung in alkalischen Medien
Phys.-chem. Mech. d. Werkstoffe, Kiew 11 (1975) 1, p. 106

[14] Vereshchagin, K. I.; Nechaev, V. A.; Podlipskij, L. A.; Rubenchik, Yu. I.; Karpenko, G. V.
Erhöhung der Beständigkeit von geschweißten Apparaten aus St-3 und 16-GS gegen Spannungsrißkorrosion
Phys.-chem. Mech. d. Werkstoffe, Kiew 8 (1972) 2, p. 3

[15] Pöpperling, R.; Schwenk, W.; Venkateswarlu, J.
Wasserstoffinduzierte Spannungsriß-korrosion von Stählen durch dynamisch-plastische Beanspruchung in Promotor-freien Elektrolytlösungen
(Hydrogen-induced stress corrosion cracking of steels by dynamic-ductile stress in activator-free electrolytic solutions)
(in German)
Werkst. Korros. 36 (1985) 9, pp. 389–400

[16] DIN 50915 (09/1993)
Prüfung von unlegierten und niedriglegierten Stählen auf Beständigkeit gegen interkristalline Spannungsrißkorrosion in nitrathaltigen Angriffsmitteln – Geschweißte und ungeschweißte Werkstoffe
(Examination of unalloyed and low-alloy steel on stability against intergranular stress corrosion in nitrate containing means of attack – welding and unwelded materials) (in German)
Beuth-Verlag GmbH, Berlin

[17] Diekmann, H.; Drodten, P.; Kuron, D.; Herbsleb, G.; Pfeiffer, B.; Wendler-Kalsch, E.
Zum Spannungsrißkorrosionsverhalten niedriglegierter Stähle in Nitrat- und Carbonatlösungen sowie in Natronlauge
(To the stress corrosion cracking behavior of low-alloy steel in nitrate and carbonate solutions as well as in caustic soda solution) (in German)
Stahl u. Eisen 103 (1983) 18, pp. 895–901

[18] Drodten, P.; Herbsleb, G.; Kuron, D.; Savakis, D.; Wendler-Kalsch, E.
Potentialabhängigkeit der Korrosion Mo-freier und Mo-haltiger Stähle in Calciumnitrat-Lösung und Natronlauge
(Potential dependence of the corrosion of Mo-free and Mo-containing steel in calcium nitrate solution and caustic soda solution) (in German)
Werkst. Korros. 42 (1991) 3, pp. 128–138

[19] Drodten, P.; Herbsleb, G.; Kuron, D.; Savakis, S.; Wendler-Kalsch, E.
Spannungsrißkorrosion Mo-freier und Mo-haltiger Stähle in Calciumnitrat-Lösung und Natronlauge unter zeitlich konstanter Kraft und unter CERT-Bedingungen
(Stress corrosion of Mo-free and Mo-containing steels in calcium nitrate solution and caustic soda solution under temporally constant force and on CERT conditions) (in German)
Werkst. Korros. 42 (1991) 9, pp. 473–489

[20] Drodten, P.; Herbsleb, G.; Kuron, D.; Savakis, S.; Wendler-Kalsch, E.
Der Einfluß mechanischer Wechselbelastungen auf die Beständigkeit Mo-freier und Mo-haltiger Stähle gegen Spannungsrißkorrosion in Calciumnitrat-Lösung und Natronlauge
(The influence of mechanical alternative loadings on the stability of Mo-free and Mo-containing steels against stress corrosion cracking in calcium nitrate solution and caustic soda solution) (in German)
Werkst. Korros. 42 (1991) 11, pp. 576–583

[21] Drodten, P.; Herbsleb, G.; Kuron, D.; Savakis, S.; Wendler-Kalsch, E.
Spannungsrißkorrosion von niedriglegierten Stählen in Calciumnitratlösung und Natronlauge
(Stress corrosion cracking of low-alloy steel in calcium nitrate solution and caustic soda solution) (in German)
Stahl und Eisen 111 (1991), pp. 117–124

[22] Drodten, P.; Schwenk, W.; Sussek, G.
Einflüsse auf die Spannungsrißkorrosion niedriglegierter Stähle
(Influences on the stress corrosion cracking of low-alloy steels) (in German)
Stahl und Eisen 120 (2000) 1, pp. 55–64

[23] Hoar, T. P.; Jones, R. W.
The Mechanism of Caustic Cracking of Carbon Steel – I. Influence of Electrode Potential and Film Formation
Corrosion Sci. 13 (1973), p. 725

[24] ASM International, Member/Customer Service Center, Materials Park, OH, 1994
Corrosion in the Chemical processing industry: corrosion by alkalis and hypochlorite
Corrosion in the petrochemical industry, pp. 204–210

[25] Isfort, D.
Moderne Alkalichlorid-Elektrolyse nach dem Hg-Verfahren – Ausrüstung und Konstruktionsmaterialien der Produktaufbereitungsanlagen
(Modern alkali chloride electrolysis in the Hg procedure – equipment and construction materials of the product reprocessing plants) (in German)
Chem.-Anl. Verfahren (1972) 9, p. 65

[26] Maslov, V. A. et al.
Korrosionsbeständigkeit niedriglegierter Gußeisen in alkalischen Medien bei erhöhter Temperatur
(Corrosion resistance of low-alloy cast iron in alkaline media at increased temperature) (in German)
Russ. Cast. Prod. (1968) 7, p. 317
Giesserei 58 (1971), p. 260

[27] Stypolova, B. et al.
Anodische Polarisation und das Problem des anodischen Schutzes von Gußeisen mit niedrigen Nickelgehalten in Kaliumhydroxid (in Polish)
Zesz. Nauk. Akad. Gorn.-Jutn. Cracow, Zesz. Spec. 45 (1973), p. 26

[28] Donaldson, E. G.
Der Einsatz von Meehanite-Gußeisen in Chemieanlagen
(The use of Meehanite cast iron in chemical plants) (in German)
Chem.-Techn. 4 (1975) 4, p. 127

[29] Nickel, O.
Eigenschaften der austenitischen Gußeisenwerkstoffe
(Properties of the austenitic cast iron materials) (in German)
Konstruieren & Giessen (1976) 3, p. 9

[30] ASM International, Member/Customer Service Center, Materials Park, OH, 1994
Corrosion in the Chemical processing industry: corrosion by alkalis and hypochlorite
Corrosion in the petrochemical industry, pp. 204–210

[31] Firmenschrift
Resistance of Nickel and its Alloys to Corrosion by Caustic Alkalis, Corrosion Engineering Bulletin CEB-2
The International Nickel Company, New York (NY/ZSA)

[32] ASM International, Member/Customer Service Center, Materials Park, OH, 1994
Corrosion in the Chemical processing industry: corrosion by alkalis and hypochlorite
Corrosion in the petrochemical industry, pp. 204–210

[33] Yukalov, I. N.
Korrosionsbeständigkeit austenitischer Nickel-Kupfer-Chrom-Gußeisen
(Corrosion resistance of austenitic nickel copper chrome cast irons) (in German)
Russ. Cast. Prod. (1968), p. 7
Giesserei 58 (1971), p. 260

[34] Cornea, I.; Perisanu, St.; Zaromb, S.
Sur la Résistance á la Corrosion des Fontes au Nickel
Rev. Roun. Chim. 16 (1971), p. 649

[35] Donati, J. R.; Grand, M.; Spiteri, P.
Corrosion procedures during the process of micro reactions between sodium and water within cracks, which take place in steam generator pipes of fast breeder reactors (in French)
J. Nucl. Mat. 54 (1974), p. 217

[36] Cowen, H. C.; Thorley, A. W.
Caustic cracking of 2¼ CrMo steel
Conference: Topical Conference on Ferritic Alloys for Use in Nuclear Energy Technologies, Snowbird, Utah, USA, 19.–23. June 1983
Publ.: The Metallurgical Society/AIME, Warrendale, PA, USA, 1984, pp. 51–56

[37] Wozadlo, G. P.; Roy, P.
Caustic stress corrosion behavior of 2.25 Cr-1Mo steels
Conference: Topical Conference on Ferritic Alloys for Use in Nuclear Energy Technologies, Snowbird, Utah, USA, 19.–23. June 1983
Publ.: The Metallurgical Society/AIME, 420 Commonwealth Dr., Warrendale, Pa. 15086, USA, 1984, pp. 57–64

[38] McIlree, A. R.; Pizzo, P. P.; Indig, M. E.
Corrosion Fatigue of 2¼Cr-1Mo Steel in Caustic at 316 C
Mat. Performance 19 (1980) 3, p. 30

[39] Indig, M. E.
Stress Corrosion Studies for 21/4 Chromium-1 Molybdenum Steel
Proc. Conf. Ferritic Steels for Fast Reactor Steam Generators
London 1977, 2 (1978), p. 408

[40] Lyle jr., F. F.; Burghard jr., H. C.
Cracking of Low Pressure Turbine Rotor Discs in US Nuclear Power Plants
Mat. Performance 21 (1982) 11, p. 35

[41] Lyle jr., F. F.
Stress corrosion cracking characterization of 3.5 NiCrMoV low pressure turbine rotor steels in NaOH and NaCl solutions
Corrosion 39 (1983) 4, pp. 120–131

[42] Shalvoy, R. S.; Duglin, S. K.; Lindinger, R. J. T.
The Effect of Turbine Steam Impurities on Caustic Stress Corrosion Cracking of NiCrMoV Steels
Corrosion NACE 37 (1981), p. 491

[43] Burstein, G. T.; Woodward, J.
Effects of Segregated Phosphorus on Stress Corrosion Cracking Susceptibility of 3Cr-0.5Mo Steel
Metal Sci. 17 (1983), p. 111

[44] Burstein, G. T.; Woodward, J.
Aspects of the Stress Corrosion Cracking of Cr-Mo Steels
8. Intern. Congress on Metallic Corrosion, Mainz (1981) Vol 1, p. 460

[45] Pyun, S.-I.; Kim, J.-T.; Lee, S.-M.
Effects of phosphorus segregated at grain boundaries on stress corrosion crack initiation and propagation of rotor steel in boiling NaOH solution
Materials Science and Engineering A 148 (1991), pp. 93–99

[46] Rechberger, J.; Tromans, D.; Mitchell, A.
Stress corrosion cracking of conventional and super-clean 3.5NiCrMoV rotor steels in simulated condensates
Corrosion 44 (1988) 2, pp. 79–87

[47] Sakaki, T.; Sakiyama, K.
Effects of $NaClO_3$ and NaCl on the Corrosion Behaviors of Iron-Chromium Alloys in Hot Concentrated Caustic Soda Solution
J. Japan Inst. Metals 43 (1979), p. 1186

[48] Sakaki, T.; Sakiyama, K.
Corrosion of Iron-Chromium Alloys in Hot Concentrated Caustic Soda Solution
J. Japan Inst. Metals 43 (1979), p. 527

[49] Davison, R. M.; Steigerwald, R. F.
The New Ferritic Stainless Steels
Metal Progress 115 (1979) 6, p. 40

[50] Wilson, I. L.; Pement, F. W.; Aspden, R. G.
Stress corrosion studies on some stainless steels in elevated temperature aqueous environments
Corrosion NACE 34 (1978) 9, pp. 311–320

[51] Shalaby, H. M.; Begley, J. A.; Macdonald, D. D.
Fatigue crack initiation in 403 stainless steel in simulated steam cycle environments: hydroxide and silicate solutions
British Corrosion Journal 29 (1994) 1, pp. 43–52

[52] Salinas-Bravo, V. M.; Gonzales-Rodriguez, J. G.
Stress corrosion cracking susceptibility of 17-4PH turbine steel in aqueous environments
British Corrosion Journal 30 (1995) 1, pp. 77–79

[53] Gonzales-Rodriguez, J. G.; Salinas-Bravo, V. M.; Martinez-Villafane, A.
Hydrogen embrittlement of type 410 stainless steel in sodium chloride, sodium sulfate and sodium hydroxide environments at 90°C
Corrosion 53 (1997) 6, pp. 499–504

[54] Naumann, F. K.; Spies, F.
Verzunderte und deformierte Brennerdüsen (Scaled and distorted burner nozzles)
(in German)
Prakt. Metallographie 17 (1980), p. 297

[55] Firmenschrift
Stainless Steel Al 2205 Alloy
Alleghenny Ludium Corporation, Pittsburgh (PA/USA)

[56] Horn, E. M.; Savakis, S.; Schmitt, G.; Lewandowski, I.
Performance of Duplex Steels in Caustic Solutions
Conference: Duplex Stainless Steels '91, Vol. 2, Bourgogne, France, 28.–30. Oct. 1991
Publ.: Les Editions de Physique, Les Ulis, France, 1922, pp. 1111–1119
also: Korrosionsverhalten nichtrostender ferritisch-austenitischer Stähle in Natronlauge
Werkst. Korros. 42 (1991) 10, pp. 511–519
also: Horn, E. M.; Schmitt, G.
Corrosion behaviour of stainless ferritic-austenitic steels in sodium hydroxide
Werkst. Korros. 41 (1990) 6, pp. 365–367

[57] Turnbull, A.; Griffiths, A.; Reid, T.
Corrosion and stress corrosion cracking of duplex stainless steels in caustic solutions at elevated temperatures
Stainless Steel World, Dec. (1999) pp. 63–67

[58] Standard practice for laboratory immersion corrosion testing of materials
G31-72, American Society for Testing of Materials, PA, USA

[59] Rondelli, G.; Vicentini, B.; Brunella, M. F.; Cigada, A.
Stress corrosion cracking of austenitic and duplex stainless steels in caustic environments
Conference: CASS 90: Corrosion – Air, Sea, Soil, Auckland, New Zealand, 19.–23. Nov. 1990 paper 37
Publ.: Australasian Corrosion Association Inc., New Zealand Branch, Auckland, New Zealand, 1990
also in: Werkst. Korros. 44 (1993) 2, pp. 57–61

[60] Ward, I.; Berlung, G.; Norberg, P.
Caustic and chloride stress corrosion cracking of austenitic and Duplex steels in the alumina and paper industries
5th Asian Pacific Corrosion Control Conference, Melbourne, Australia, 1987

[61] Condylis, A.; Bayon, F.; Chateau, D.
Resultats d'essais de corrosion et de polution d'aciers inoxydables austenitiques dans l'industrie alimentaire
Rev. Metallurgie (1976), p. 787

[62] Ahmad, Z.; Davami, P.
The Corrosion Resistance of 18Cr-37Ni Austenitic Steel in Alkaline, Acid and Neutral Solutions
Japan Soc. Corrosion Eng. 29 (1980), p. 595

[63] Wilson, I. L.; Pement, F. W.; Aspden, R. G.
Effect of Alloy Structure, Hydroxide Concentration and Temperature on the Caustic Stress Corrosion Cracking of Austenitic Stainless Steels
Corrosion NACE 30 (1974), p. 139

[64] Subrahmanyam, D. V.; Agrawal, A. K.; Staehle, R. W.
The Stress Corrosion Cracking Behavior of Type 304 Stainless Steel in Boiling 50 % NaOH Solution
Proc. 7. Intern. Congr. Metallic Corrosion, Rio de Janeiro (1978), p. 783

[65] Pement, F. W.; Wilson, I. L. W.; Aspden, R. G.
Stress corrosion cracking studies of high nickel austenitic alloys in several high-temperature aqueous solutions
Materials performance 19 (1980) 4, p. 43

[66] Anonymous
Anti-Corrosion Methods and Materials. 4: Forms of Corrosion
Anti-Corrosion 19 (1972) 12, p. 9

[67] Nicodemi, W.
Gli acciai inossidabili; Corrosion sotto tensione
Acciaio Inossidabile 55 (1988) 4, pp. 24–29

[68] Marsh, G. P.; Lawrence, P. F.; Carney, R. F. A.; Perkins, R.
The Corrosion Behaviour of Alloy 800 in Concentrated Sodium Hydroxide Solutions at Moderate Temperatures EUROCORR, 6. European Congr. Metallic Corrosion (1977), p. 505

[69] Blanchet, J.; Coriou, H.; Grall, L.; Mahieu, C.
The Stress Corrosion of Alloy 800 in Contaminated Water at High Temperature
Mém. Sci. Rev. Métallurgie 75 (1978), p. 237

[70] Berge, Ph.; Donati, J. R.; Prieux, S., Villard, D.
Caustic Stress Corrosion of Fe-Cr-Ni Austenitic Alloys
Corrosion NACE 33 (1977), p. 42

[71] Wilson, I. L. W.; Pement, F. W., Aspden, R. G., Begley, R. T.
Caustic Stress Corrosion Behavior of Fe-Ni-Cr Nuclear Steam Generator Tubing Alloys
Nuclear Technol. 31 (1976), p. 70

[72] Takano, M.; Teramoto, K.; Nakayama, T.; Yamaguchi, H.
Extremely Slow Strain Rate Stress-Corrosion Testing Machine and Some Experimental Results
J. Iron Steel Inst. Japan 65 (1979), p. 212

[73] Hickling, J.; Wieling, N.
Electrochemical Aspects of the Stress Corrosion Cracking of Fe-Cr-Ni Alloys in Caustic Solutions
Corrosion Sci. 20 (1980), p. 269

[74] Zheng, J. H.; Bogaerts, W. F.
Transpassive chromium dissolution – Interaction with stress corrosion cracking of austenitic stainless steel in caustic solution
in: Conference: Second International Conference on Corrosion-Deformation Interactions. CDI96', Nice, France, 24–26 Sept. 1996
Institute of Materials, London, UK, 1997, pp. 104–116

[75] Crowe, D. C.; Tromans, D.
Caustic cracking behavior and fractography of 316 and 310 stainless steels
Conference: Metallography and Corrosion, Calgary, Alberta, Canada, 25–26 July 1986
Publ.: National Association of Corrosion Engineers, Houston, Texas, USA, 1986

[76] Sakaki, T.; Simizu, Y.; Sakiyama, K.
Corrosion of Iron-Chromium-Nickel-Alloys in Hot Concentrated Caustic Soda Solution
J. Japan Inst. Metals 44 (1980), p. 582

[77] McIlree, A. R.; Michels, H. T.
Stress Corrosion Behavior of Fe-Cr-Ni and Other Alloys in High-Temperature Caustic Solutions
Corrosion NACE 33 (1977), pp. 60–67

[78] Tran, H.; Katiforis, N.A.; Utigard, T.A.; Barham, D.
Recovery boiler air-port corrosion, Part 3, Corrosion of composite tubes in molten sodium hydroxide
Tappi Journal, 78 (1995) 9, pp. 111–117

[79] Firmenschrift
Chemische Beständigkeit der NIROSTA-Stähle, 1 – 06/97
(Chemical resistance of NIROSTA-steels) (in German)
KRUPP THYSSEN NIROSTA – KRUPP, Düsseldorf

[80] Firmenschrift
Böhler – Chemische Beständigkeit nichtrostender Böhler-Edelstähle
(Böhler – chemical resistance of stainless Böhler high-grade steels) (in German)
AL 170 D – 1.89 – 3.000 Gi, January 1989
Böhler, Kapfenberg (Österreich)

[81] Levin, V. A.; Levina, E. E.
Corrosion resistance of metals in hot hydroxide solutions
Zashch. Met. 31 (1995) 3, pp. 262–268
in: Protection of Metals 31 (1995) 3, pp. 236–241

[82] Poulson, B.
Stress corrosion cracking of type 316 stainless steel in caustic solutions: Crack velocities and leak before break considerations
Corrosion Science 33 (1992), pp. 1541–1556

[83] Rondelli, G.; Vicentini, B.; Sivieri, E.
Stress corrosion cracking of stainless steels in high temperature caustic solutions
Corrosion Science 39 (1997) 6, pp. 1037–1049

[84] Abe, J.; Fujimaki, T.; Kajiwara, Y.; Yokoo, Y.
Hydrogen production by high temperature, high pressure water electrolysis I, Plant development
Proccedings of the World Hydrogen Energy Conference, Tokyo, Japan, June 1980, pp. 29–41

[85] Scarberry, R. C.; Graver, D. L., Stephens, C. D.
Alloying for Corrosion Control, Properties and Benefits of Alloy Materials
Mat. Protection 6 (1967) 6, p. 54

[86] Krupp VDM Company publication
Nicrofer 3033
Alloy Digest April 1996, Ni-508

[87] Köhler, M; Heubner, U.; Eichenhofer, K.-W.; Renner, M.
Alloy 33 A new corrosion-resistant austenitic material for many applications
Stainless Steel World 11 (1999) 4, pp. 38–49

[88] Firmenschrift
Nicrofer 3033 – alloy 33, material sheet No. 4142, October 1998
ThyssenKrupp VDM, Altena

Sodium Hydroxide

Unalloyed and low-alloy steels/cast steel

The addition of OH⁻ ions to aqueous media generally decreases the corrosion rates of unalloyed and low-alloy steels. Figure 1 shows the influence of the pH value on the corrosion rate of unalloyed steel.

Above pH 10, the corrosion rate noticeably decreases and reaches the very low value of $2\,g/(m^2 d) = 0.09\,mm/a$ (3.54 mpy) at pH 12, which corresponds to a NaOH content of approximately 0.5 g/l. In other investigations of low-alloy steels, samples were exposed to solutions containing sodium hydroxide for 48 h at room temperature. The initial pH of the solution was 9.5; this had decreased to 7.7 at the end of the test. The average corrosion rate was 0.004 mm/a (0.16 mpy) [1].

Depending on the concentration and temperature of the NaOH solution, the corrosion rates for unalloyed and low-alloy steels are 0.02 to 1.45 mm/a (0.79 to 57.1 mpy). Therefore, these steels can be passive in NaOH solutions over a wide range of concentrations and temperatures so that there is sufficient corrosion resistance. At low passivation potentials, magnetite layers are formed and at higher potentials, hematite layers.

In investigations on the passivation behaviour at a temperature of 373 K (100°C), the corrosion rate of a low-alloy steel in 33 % NaOH solution was 0.7 mm/a (27.6 mpy); after anodic passivation it was 0.4 mm/a (15.7 mpy) [2]. This slight difference indicates that unalloyed steel already has sufficient passivity even without polarisation.

In 0.01 M NaOH solutions at 573 K (300°C), a low-alloy steel showed a corrosion rate of approximately 0.006 mm/a (0.24 mpy). In 5 M NaOH solutions, this value increased to approximately 1 mm/a (39.4 mpy) [3].

In practice, low-alloy steels are the most frequently used materials in NaOH solutions at concentrations of 50 % and temperatures of 363 K (90°C). Their use is only limited by requirements of extremely low iron impurities in the product or by the risk of stress corrosion cracking.

Table 1 gives the results of experiments with rotating disk specimens of low-alloy steel in NaOH solutions with a concentration of 155 to 465 g/l NaOH or in aluminate solutions with up to 400 g/l NaOH. These tests showed that corrosion increased with temperature and movement [4].

Low-alloy steels are used in the pulp and paper industries in strongly alkaline solutions which may contain sulphur compounds. Sodium sulphide and sodium thiosulphate can increase the corrosion of these steels, while sodium sulphite and sulphate have practically no influence [5, 6].

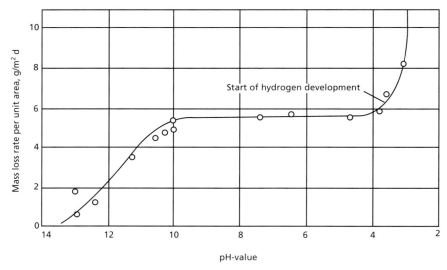

Figure 1: pH-Dependence of the corrosion rate of unalloyed steel in aerated NaOH and HCl solutions [8]

Temperature	Rotation speed	Corrosion rate	
K (°C)	U/min	g/(m² h)	mm/a (mpy)
298 (25)	880	0.151	0.17 (6.69)
353 (80)	100	0.727	0.81 (31.9)
	200	1.23	1.4 (55.1)
	1,000	1.79	2.0 (78.7)
388 (115)	80	3.19	3.5 (138)
	160	3.80	4.2 (165)
	800	6.70	7.4 (291)

Table 1: Corrosion rates from disk specimens of low-alloy steel in NaOH solution with a concentration of 300 g/l Na$_2$O at various temperatures and rotation speeds of the specimen [4]

Stress corrosion cracking

If the temperature or flow rates are not too high, the uniform surface corrosion of unalloyed and low-alloy steels is sufficiently low to allow the use of these steels in NaOH solutions. If necessary, surface corrosion can be countered by increasing the wall thickness to ensure sufficient service life of the component.

However, the system low-alloy steel/NaOH solution is a critical "material/corrosive agent" system with respect to stress corrosion cracking. Figure 2 and Figure 3 show the critical ranges of temperature and concentration of the NaOH solution. In addition, Figure 4 shows the range of stress corrosion cracking in a NaOH concentration – temperature diagram [7].

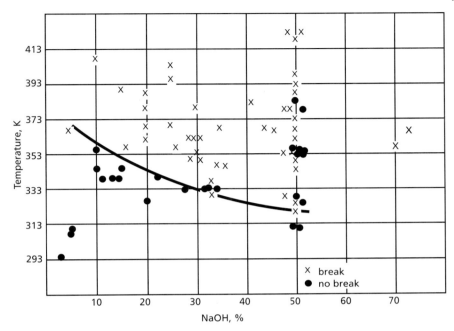

Figure 2: Influence of concentration and temperature on the SCC behavior of low-alloy steels in NaOH solutions [8]

According to [9], carbon steels are suitable for the transport and storage of NaOH solutions of medium and higher concentrations in the temperature range between 323 K and 373 K (50 and 100°C), if the conditions given in Figure 5 are observed. For concentrations and temperatures given in range B, welded components must be subjected to stress-free annealing. If the operating conditions lie within range C, then nickel alloys should be used instead of carbon steels.

It should be mentioned here that stress corrosion cracking damage was observed in a pipeline that was constructed of steels St35 (mat. no. 1.0308) and 16Mo3 (mat. no. 1.5415) which was used to carry 5 to 7% NaOH solution with a temperature of up to 323 K (50°C). During discontinuous flow, the caustic solution was heated to 373 K (100°C) during periods of standstill. Thus, the prerequisite for the occurrence of stress corrosion cracking was fulfilled by the medium. Damage occurred particularly in the areas of expansion compensation in the pipeline because the stress caused by deformation after bending of the elbow had not been relieved by stress-free annealing [10].

Figure 6 and Figure 7 show the critical potential range for the occurrence of stress corrosion cracking. Intergranular stress corrosion cracking of unalloyed steel occurred in boiling 33% NaOH solution only in a narrow potential range. The highest sensitivity is observed in the potential range $U_H = -0.65$ to -0.75 V. Therefore, the critical potential range for the occurrence of stress corrosion cracking is not only limited to the cathodic side but also to the anodic side, i.e. stress corrosion cracking can be prevented by potentials $U_H < -0.8$ or $U_H > -0.55$ V [11, 12].

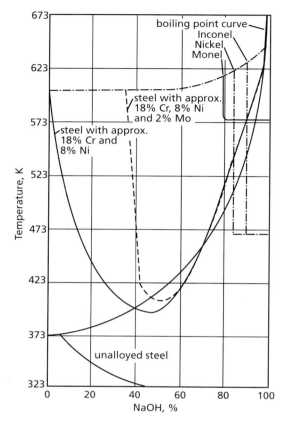

Figure 3: Regions in which there is a risk of stress corrosion cracking in NaOH solutions for various materials [8]

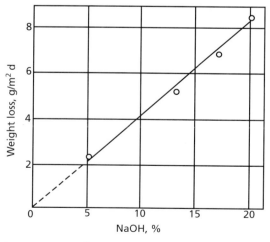

Figure 4: Range of susceptibility to stress corrosion cracking of low-alloy C-Mn steels in NaOH solutions of varying concentrations and temperatures [7]

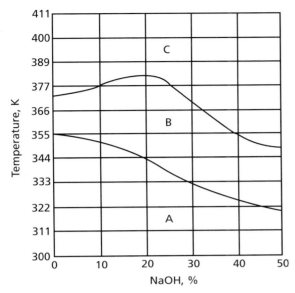

Figure 5: Temperature and corrosion ranges of the corrosion of carbon steels in NaOH solutions [9]
A: no stress-free annealing necessary
B: stress-release by heat treatment
C: use of C steels not permissible

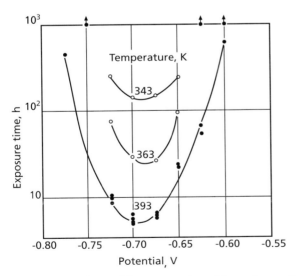

Figure 6: Dependence of the exposure time of Armco iron on the potential in 33 % NaOH solutions [11]

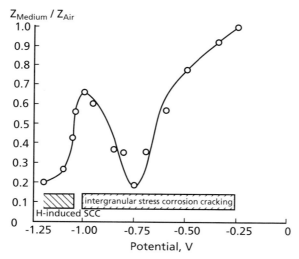

Figure 7: Dependence of the reduction of area of tensile specimens of steel S355N in boiling 35 % NaOH solution on the testing potential (extension rate: $\dot{\varepsilon} = 1.4 \times 10^{-6}$ 1/s, Z = fracture work) [11]

To determine the critical potential range for stress corrosion cracking, tensile specimens of 2 unalloyed steels (compositions given in Table 2) were investigated in 33 % NaOH solution at 388 K (115°C) in dependence on the potentiostatically defined potential [13].

Steel	% C	% Si	% Mn	% P	% S
1	0.130	0.24	0.68	0.016	0.019
2	0.076	0.24	0.34	0.029	0.036

Table 2: Analysis values of the investigated steels [13]

The experiments were carried out under constant a load of 0.8 Re as well as in slow strain-rate tests with an extension rate of 1.4×10^{-5} mm/sec. These experiments also showed a critical potential range at $U_H = -0.60$ to -0.85 V for the occurrence of stress corrosion cracking.

In order to explain this limited range of potentials for stress corrosion cracking of low-alloy steels, it is assumed that in this range the grain surfaces in the material are in the passive state while the grain boundaries are still in the active state. Passive grain surfaces and active grain boundaries are, however, an essential requirement for the occurrence of intergranular stress corrosion cracking. A passive grain surface is achieved by the formation of a passivating layer of Fe_3O_4, while the grain boundaries, at least at potentials where there is the greatest sensitivity to stress corrosion cracking, are not passivated or only passivated after a long delay. The current density – potential curves of steel samples containing varying amounts of carbon were recorded in saturated NaOH solution at 343 K (70°C). In addition to a current density maximum at approx. $U_H = -0.85$ V, which is attributed to the dissolution of ferrite, there was a second maximum at approx. -0.68 V that increased linearly with increasing content of cementite in the alloy and which is attributed to the dissolution of this cementite [14].

Steel samples with 0.06 % C were subjected to a suitable heat treatment so that the boundaries of the ferrite grains were strongly coated with cementite. These specimens were tested for stress corrosion cracking in the above NaOH solution with a tensile load of $0.85 \times R_m$ for up to 100 h. In the potential range of the first current-density maximum ($U_H = -0.87$ to $U_H = -0.82$ V), only the ferrite was attacked, but not the grain boundaries. At higher potentials (above $U_H = -0.75$ V), an increase in the formation of intergranular cracks was observed as a result of the dissolution of the cementite on the grain boundaries. After 60 h at $U_H = -0.68$ V the specimens fractured. Thus, the dissolution of cementite on the grain boundaries is the decisive process in the formation of intergranular cracks in NaOH solutions. Both the crack depth and the dissolution rate exhibited the highest values at a potential of $U_H = -0.68$ V.

At more positive potentials there was also an increasing passivation of the cementite so that the susceptibility to intergranular stress corrosion cracking was reduced or prevented. Markedly more negative potentials led to an activation, even of the ferrite, and surface corrosion became the dominant process.

In numerous investigations of the stress corrosion cracking of unalloyed steels in nitrate solutions, it was found that steels with higher phosphorus contents were much more sensitive than low-phosphorus steels. During such an investigation, comparative experiments were carried out in 33 % NaOH solutions at 393 K (120°C) [15]. The three investigated steels had comparable contents of carbon and manganese, but had differing phosphorus contents:

- steel 1: 0.15 % C   0.43 % Mn   0.002 % P
- steel 2: 0.15 % C   0.46 % Mn   0.029 % P
- steel 3: 0.14 % C   0.544 % Mn   0.045 % P

The steels were subjected to heat treatment at 1,213 K (940°C) for 1 h followed by 48 h at 773 K (500°C) to promote P segregation on the grain boundaries.

Figure 8 shows the results of CERT experiments on these steels in NaOH solutions. The measure used for the susceptibility to stress corrosion cracking in these slow extension tests was the relative fracture work expended until failure of the specimens, i.e. the ratio of fracture work for the tensile test in air to the fracture work for the tensile test in NaOH solution. The relative fracture work is plotted against the test potential for the three steels.

Intergranular stress corrosion cracking only occurred in the potential range between $U_H = -0.85$ and -0.50 V. The least resistance was shown by all samples at $U_H = -0.70$ V at the start of the passive range. The minimum value of the fracture work in this potential range was not influenced by the phosphorus content. The relative fracture work increased in both directions of potentials, i.e. the susceptibility to stress corrosion cracking decreased; the values of the low-phosphorus samples in the more negative direction were all slightly higher than those of the more P-rich

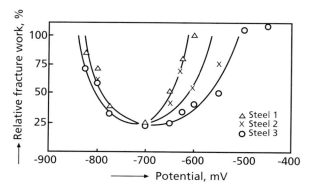

Figure 8: Relative fracture work of steels 1 to 3 in 33 % NaOH at 393 K (120°C) in dependence on the test potential [15]

samples. The differences in the anodic direction are more obvious: for example, steel 1 already reached 100 % of the fracture work in air at $U_H = -0.625$ V, whereas the value for steel 3 was only 40 %. Steel 3 is only resistant from $U_H = -0.50$ V. Thus, higher phosphorus contents in steel are also disadvantageous with respect to stress corrosion cracking in NaOH solutions.

In laboratory investigations, the critical potential range for stress corrosion cracking can not only be defined by a potentiostatically impressed current, but also by an additive which adjusts the redox potential of the solution into this region. An addition of 0.2 % PbO to the NaOH solution has been found to be suitable for this. Thus, investigations of welded samples of low-alloy steels in 25 % NaOH solution with 0.2 % PbO at 383 K (110°C) showed that the resistance can be increased by preheating during welding or by subsequent heat treatment at 623 to 723 K (350 to 450°C) [16, 17].

In plants using the Bayer process, in which aluminous clays are treated with a solution of sodium aluminate with a NaOH content of 130 to 150 g/l Na_2O at temperatures between 318 K and 333 K (45 and 60°C), stress corrosion cracking was observed on welded decomposers made of killed carbon steels with 0.14 to 0.22 % C, particularly on welded joints [18]. A passivating treatment of these steel vessels for 0.5 to 1 h at 333 K (60°C) in a solution of sodium hydroxide with 150 g/l Na_2O and hydrogen peroxide (0.010–0.075 M) prevented the occurrence of stress corrosion cracking, even after 120 days. During this time, the potential remained more positive than $U_H = -0.7$ V. If the potential were to decrease to this value, then the passivating treatment should be repeated to prevent further crack formation [18].

In the framework of efforts to increase the productivity of the Bayer digestion plants by increasing the concentration of the caustic solution and the operating temperatures, slow tensile tests were carried out on low-alloy steels of type A 285 and A 516 in NaOH solutions with concentrations of 135 to 240 g NaOH/l at temperatures up to 453 K (180°C) [19]. Necking of the tensile specimens after fracture was chosen as the measure of the susceptibility to stress corrosion cracking. Figure 9 shows that the lowest values for the reduction of area, and thus the highest susceptibility to stress corrosion cracking, were found at a test potential of approximately $U_H = -0.85$ V.

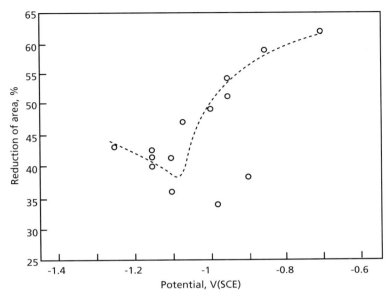

Figure 9: Influence of the potential on the reduction of area of the tensile specimens made of steel A 285 in a solution containing 143 g/l NaOH at 363 K (90°C) [19]

Figure 10 shows the influence of the temperature while Figure 11 shows the influence of the concentration of the caustic solution on the susceptibility to stress corrosion cracking using steel A 516 as an example.

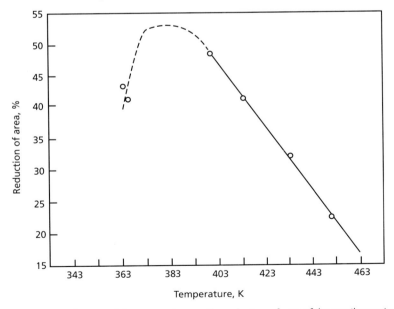

Figure 10: Influence of the temperature on the reduction of area of the tensile specimens of steel A 516 in a solution containing 173 g/l NaOH [19]

At temperatures above 403 K (130°C), the crack susceptibility increases linearly with increasing temperature (see Figure 11).

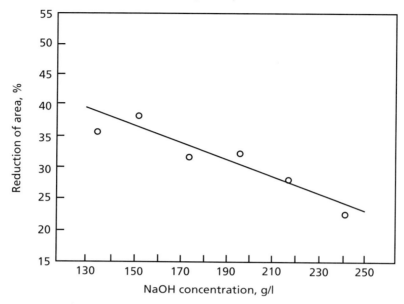

Figure 11: Influence of the concentration of caustic solution on the reduction of area of the tensile test specimens made of steel A 516 at 433 K (160°C) [19]

These results also show a linear relationship between the concentration and the sensitivity to stress corrosion cracking, which noticeably increases with increasing concentration.

The crack behaviour in artificial Bayer solutions at temperatures between 473 and 573 K (200 and 300°C) was investigated by slow tensile tests on three unalloyed steels used for pressure vessels (see Table 3) that contained differing amounts of sulphur [20].

The artificial Bayer solution had the composition given in Table 4. The extension rate for the slow tensile tests was 3.3×10^{-6} 1/s.

Steel	% C	% Si	% Mn	% P	% S	% Al	% Cu	% Mo	% Ni	% RE
16MnR	0.16	0.47	1.53	0.014	0.018	–	0.055	–	–	–
A48CPR	0.175	0.34	1.35	0.012	0.006	–	–	0.045	0.058	–
16MnRE	0.16	0.40	1.38	0.018	0.009	–	–	–	–	0.020

Table 3: Chemical composition of the investigated pressure vessel steels [20]

NaOH	Al$_2$O$_3$ × 3 H$_2$O	Na$_2$CO$_3$	Na$_2$SO$_4$	NaCl
7.42 M	1.32 M	0.3 M	0.14 M	0.14 M

Table 4: Composition of the artificial Bayer solution [20]

Figure 12 shows the influence of the temperature on the stress corrosion cracking behaviour of steel 16MnR. The determined reduction of area, extension and time-to-fracture determined for the specimens in the test solution are given. As a comparison, the values of the reduction of area in nitrogen are also given, which were approximately 80%, independent of the temperature. The susceptibility to stress corrosion cracking increased noticeably as the temperature increased. In these tests, the reduction of area was also the most sensitive characteristic to measure the susceptibility to stress corrosion cracking.

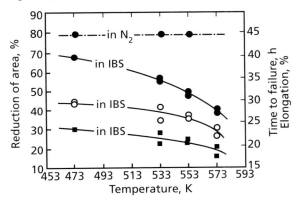

Figure 12: Reduction of area, extension and time-to-fracture of the tensile test specimens of steel 16MnR [20]
●: Reduction of area, ○: extension, ■: time-to-fracture; IBS: artificial Bayer solution

Figure 13 shows the behaviour of the three steels with regards to reduction of area of transverse and longitudinal specimens in the Bayer solution at 533 K (260°C). The higher sulphur content in steel 16MnR is particularly noticeable in the transverse specimens, while in low-sulphur steels the orientation of the specimens was only of very slight importance.

Slow strain-rate tests were used to investigate the influence of contaminants in the aluminate solution of the Bayer process on pressure vessel steels ASTM A 285 and ASTM A 516 grade 70 (compositions given in Table 5) [21–24]. The tests were carried out either at different testing potentials or at the critical potential for each individual testing medium that had already been determined in preliminary experiments. The testing temperature was 373 K (100°C).

Table 6 lists the test solutions and the results for steel A 516 carried out at the critical potential for each testing medium. The necking of the fractured samples was chosen as a measure of the susceptibility to stress corrosion cracking.

Sodium Hydroxide

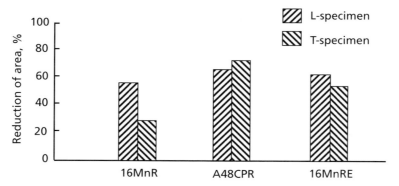

Figure 13: Values of the reduction of area for transverse and longitudinal specimens of steels 16MnR, 16MnRE and A48CPR in the Bayer solution at 533 K (260°C) [20]

Steel	% C	% Si	% Mn	% P	% S	% Cu	% Ni	% Cr	% Mo
A 285	0.19	0.01	0.71	0.020	0.016	0.007	0.004	0.033	0.004
A 516	0.24	0.25	1.15	0.008	0.007	0.044	0.025	0.033	0.004

Table 5: Chemical composition of steels A 285 and A 516 [22]

Test solution	Reduction of area %
pure 2.25 M NaOH	54
Bayer solution: 3.6 M NaOH + 0.67 M $Al_2O_3 \times 3H_2O$ + 0.46 M Na_2CO_3 + 0.02 M Na_2SO_4	42
aluminate solution: 3.6 M NaOH + 0.67 M $Al_2O_3 \times 3H_2O$	45
2.25 M NaOH + 0.46 M Na_2CO_3	55
2.25 M NaOH + 0.02 M Na_2SO_4	53
3.6 M NaOH + 0.67 M Al_2O_3 + 0.46 M Na_2CO_3 + 0.02 M Na_2SO_4	44
pure 7.73 M NaOH	40
7.73 M NaOH saturated with Al_2O_3	36

Table 6: Test solutions for the slow tensile tests and the resulting reduction of area of the samples [21]

The susceptibility to stress corrosion cracking increased with increasing concentration of the NaOH solution and the aluminate content. In comparison, the other contaminants present in the Bayer process play practically no role. The negative influence of the sodium aluminate content ($NaAlO_2$), or that of AlO_2^- ions was also found for the three solutions with steel A 516 (Table 7). The susceptibility to SCC was tested at two potentials: one at a potential which lies in the critical range for stress corrosion cracking in the active-passive transition ($U_H = -0.7$ V) and the other at an anodic protection potential ($U_H = -0.01$ V). The results were evaluated using the reduction of area Z on the fractured samples. The results are given as the ratio of reduction of area in the testing medium (Z_M) to that in oil (Z_{oil})

Test solution	Testing potential V_H	Z_M/Z_{oil}
2.25 M NaOH	−0.68	0.90
	−0.01	0.97
3.6 M NaOH + 0.67 M $Al_2O_3 \times 3H_2O$	−0.71	0.77
	−0.01	0.95
Bayer solution: 3.6 M NaOH + 0.67 M $Al_2O_3 \times 3H_2O$ + 0.46 M Na_2CO_3 + 0.02 M Na_2SO_4	−0.718	0.73
	−0.01	0.91

Table 7: Testing solutions and susceptibility to SCC of steel A 516 [22]

At a critical potential of approximately −0.7 V, a slight reduction of the reduction of area was also found in the pure NaOH solution. In solutions containing aluminate, the ratio Z_M/Z_{oil} was noticeably decreased. In these solutions, complete protection against stress corrosion cracking cannot even be attained by changing the potential in the anodic direction (anodic protection).

A negative influence of aluminate on the susceptibility to stress corrosion cracking was also found in the tests reported in [25] and [26]. In both slow strain-rate tests on round specimens as well as investigations on crack propagation of fracture mechanics specimens of steel A 516 grade 70 (in the solutions listed in Table 8 and at a test temperature of 365 K (92°C)), the addition of Al_2O_3 to the NaOH solution had an unfavourable effect on the cracking behaviour.

Test solution	Anion concentration mol/l		Testing potential V (U_H)
	$c\ (OH^-)$	$c\ (AlO_2^-)$	
2 M NaOH	2 M		−0.74
4 M NaOH	4 M		−0.83
8 M NaOH	8 M		−0.79
4 M NaOH + 1 M Al_2O_3	2 M	2 M	−0.72
*) 3.6 M NaOH + 0.67 M Al_2O_3	2.26 M	1.34 M	−0.78

*) Industrial Bayer solution that also contained 0.45 M Na_2CO_3, 0.019 M Na_2SO_4, 9 g/l organ. C

Table 8: Testing media and testing potentials for the crack corrosion behaviour of steel A 516 grade 70 [25]

Indeed, stress corrosion cracking was observed in all solutions in the potential range close to the active-passive transition; however, cracks also occurred in the passive range for solutions containing Al_2O_3 and at potentials in which no stress corrosion cracking was found in Al_2O_3-free solutions. In fracture mechanical investigations with increasing NaOH concentrations, a decrease of the critical stress intensity factor K_{ISCC} was observed. The addition of Al_2O_3 had the same effect.

The damage caused by stress corrosion cracking observed in plants for the treatment of bauxite in NaOH solutions at approximately 373 K (100°C), frequently occurs in the region of the welded joints. In fundamental investigations on the stress corrosion cracking behaviour in NaOH solutions of UP-welded joints made of the unalloyed carbon steels that are generally used for these plant components, it was found that the welded areas were not always more susceptible to stress corrosion cracking than the basis material [27, 28]. Three commercially available pressure vessel steels were used in this investigation. The chemical compositions and strengths are given in Table 9.

Steel	C %	Si %	Mn %	P %	S %	R_e MPa	R_m MPa	A %
ASTM A 285	0.19	0.01	0.71	0.020	0.010	263	424	32
ASTM A 516	0.24	0.25	1.15	0.008	0.007	357	541	38
SAE-SAE 1030	0.30	0.25	0.84	0.005	0.032	332	545	21

Table 9: Essential alloy contents and mechanical properties of the investigated steels [28]

Slow strain-rate tests (CERT tests) were carried out on round tensile specimens of the basis material as well as on longitudinal and transverse specimens taken from the welded joint. The tests were carried out with extension rates of 10^{-7} 1/s and 10^{-5} 1/s at 378 K (105°C) in 7.74 M NaOH solutions (approx. 310 g/l NaOH), which was deaerated by flushing with argon gas. The susceptibility to stress corrosion cracking was evaluated by the ratio of the necking of specimens fractured in NaOH solutions (Z_{NaOH}) to reduction of area of the specimens in oil (Z_{oil}), in dependence on the testing potential (Figure 14).

Again, the tests confirmed that the susceptibility to stress corrosion cracking had a maximum at potentials close to $U_H = -0.75$ V. Although the maximum hardness lay in the heat-affected zone of the welded joints, there was no great difference in the behaviour between the basis material and the welded area for the steels A 285 and A 516. The basis material, steel 1030, was not only noticeably more susceptible than the other two steels, but it was also much more susceptible than in the welded state.

Even carbon steels with a higher carbon content (0.80 % C; 0.56 % Mn, 0.20 % Si, 0.02 P, 0.02 % S) in the form of high-strength wires produced by cold deformation, undergo stress corrosion cracking in 33 % NaOH solutions at 393 K (120°C) if the potential lies in the active-passive transition range of approximately $U_H = -0.65$ V to $U_H = -0.90$ V. This potential range can be reached in the presence of oxidizing

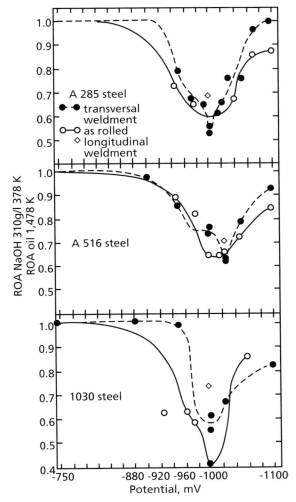

Figure 14: Dependence of the ratio of the reduction of area (ROA) Z_{NaOH}/Z_{oil} in NaOH solutions on the testing potential [28]

substances whose concentration is not sufficient to provide complete passivation of the steel [29].

Hydrogen-assisted stress corrosion cracking can occur at very negative potentials in NaOH solutions. Figure 15 shows the results of Constant Extension Rate Tensile (CERT-tests) of a low-alloy steel in 3 M NaOH solutions at 343 K (70°C) and at an extension rate of 2.2×10^6 1/s as a function of the applied potential. In this test procedure, the types of SCC are identified which require critical extension rates for crack initiation and propagation. Note the remarkable parallel course of the current density and reduction of area, which is used as a measure of the susceptibility to stress corrosion cracking (Figure 15) [30].

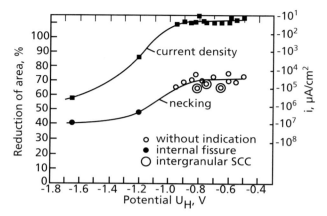

Figure 15: Potential dependence of the current density and reduction of area of a pipe steel in 3 M NaOH at 343 K (70°C) and an extension rate of $\varepsilon = 2.2 \times 10^{-6}$ 1/s (CERT test) [30]

The damage caused by intergranular stress corrosion cracking of pipelines made of unalloyed steel that are used for the transport of NaOH solutions is described in [31]. The damage is frequently caused by the incorrect choice of materials, processing faults or by unfavourable operating conditions that have led to increased concentrations in the solution or to increased temperatures.

Slow strain-rate tests and fracture mechanics tests were also used to test welded and unwelded samples of steel ASTM A 516 grade 70 (Table 10) in a NaOH solution containing Na$_2$S (3.35 M NaOH + 0.42 M Na$_2$S) at 365 K (92°C); these conditions simulate cellulose digestion [32].

Thickness	C %	Si %	Mn %	P %	S %	R_e MPa	R_m MPa
12.5	0.18	0.22	1.12	0.021	0.007	409	536
25	0.19	0.23	1.16	0.021	0.008	344	499

Table 10: Chemical composition and strength of the investigated steel ASTM A 516 grade 70 [33]

The results of the slow strain-rate tests are given in Table 11 and Table 12. They show the secondary cracking and the approximate crack propagation rates which were calculated from the length of the longest secondary crack and the time-to-fracture of the specimen at the individual testing potentials.

This data shows that the maximum susceptibility to stress corrosion cracking lies at approximately $U_H = -0.64$ V. The greatest number of secondary cracks and the greatest crack propagation rates are found in this potential range. At other testing potentials, secondary cracks only appeared in unwelded samples or in the basis material of welded samples. Stress corrosion cracking could not be excluded at more positive potentials in the passive range. In the case of cathodic polarisation to very negative potentials, the number of cracks increased again; this is probably caused by hydrogen-assisted crack formation.

Testing potential V (U_H)	Number of secondary cracks	Crack propagation rate m/s
−0.256	8	9.2×10^{-10}
−0.408	6	8.52×10^{-10}
−0.458	7	8.1×10^{-10}
−0.508	6	8.3×10^{-10}
−0.608	18	1.85×10^{-9}
−0.638	19	1.9×10^{-9}
−0.808	3	3.9×10^{-10}
−1.858	14	1.3×10^{-9}

Table 11: Results of the slow strain-rate tests on unwelded samples of steel ASTM A 516 grade 70 in 3.35 M NaOH + 0.42 M Na_2S at 365 K (92°C) [33]

Testing potential V (U_H)	Number of secondary cracks			Crack propagation rate m/s		
	ML	HAZ	BM	ML	HAZ	BM
−0.408	0	0	4			6.50×10^{-10}
−0.508	0	0	6			6.20×10^{-10}
−0.608	36	12	4	1.38×10^{-9}	7.4×10^{-10}	1.88×10^{-9}
−0.638	32	8	5	1.24×10^{-9}	7.1×10^{-10}	1.85×10^{-9}
−1.058	0	0	1			6.94×10^{-10}

ML = melt line, HAZ = heat-affected zone, BM = basis material

Table 12: Some results of the slow strain-rate tests on welded samples of steel ASTM A 516 grade 70 in 3.35 M NaOH + 0.42 M Na_2S at 365 K (92°C) [33]

The results of fracture mechanics tests are shown in Figure 16. The tests were carried out at only two potentials: firstly, close to the active-passive transition at $U_H = -0.638$ V, which corresponds to the most critical potential range according to the CERT tests, and secondly, at $U_H = -0.508$ V which is a sensible value for anodic protection because stronger anodic potentials would not provide any improvements according to the results of the CERT tests.

According to these results, there is a lower limiting value of the critical stress intensity factor of $K_{ISCC} \gg 20$ MPa$\sqrt{m}$, below which no further crack propagation was observed. Above this limiting value is a region in which crack propagation is dependent on the stress intensity factor. This is followed by a plateau between $K_I = 30$ to 50 MPa$\sqrt{m}$, in which crack growth is independent of K_I and is approximately 3×10^{-10} m/s. At higher K_I values, there is K_I-dependent crack growth.

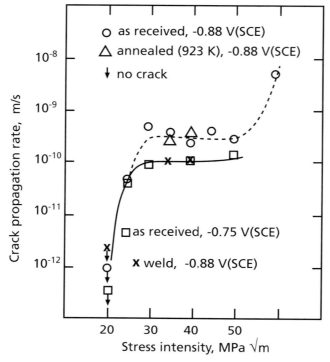

Figure 16: Influence of the stress intensity factor K_I and the testing potential on the growth of stress corrosion cracks in unwelded specimens of steel ASTM A 516 grade 70 in 3.35 M NaOH + 0.42 M Na$_2$S at 365 K (92°C) [33]

Since the stress corrosion cracking of low-alloy steels in NaOH solutions is limited to a relatively narrow potential range, this type of corrosion can be suppressed by polarisation into the anodic range. Investigations on the anodic protection of equipment used in the pulp industry made of steel ASTM A 516 grade 70 were carried out in the following solutions [34]:

– 40 g/l NaOH + 20 g/l Na$_2$S + 20 g/l Na$_2$CO$_3$ (383 K (110°C))
– 20 g/l NaOH + 20 g/l Na$_2$S + 20 g/l Na$_2$CO$_3$ (413 K (140°C))

The results of the slow strain-rate tests are given in Table 13. The susceptibility to stress corrosion cracking was evaluated by means of the maximum depth of secondary cracks in the fractured samples.

Thus, crack formation in this steel can be prevented in these solutions by anodic protection using more positive potentials than approximately $U_H = -0.6$ V.

Fracture mechanics tests were carried out on 2 melts of an unalloyed steel (composition given in Table 14) in hot 33 % NaOH solutions (12.5 mol/kg) by means of DCB (Double Cantilever Beam) specimens under conditions of free corrosion (potential $U_H \approx -1.0$ V) as well as under potentiostatically controlled conditions at $U_H = -0.76$ V [35].

Potential mV (U_H)	Crack depth mm	
	40 g/l NaOH 383 K (110°C)	20 g/l NaOH 413 K (140°C)
−752	0.00	0.05
−737	–	0.30
−722	0.25	–
−707	–	0.70
−692	0.90	0.30
−677	–	0.05
−662	0.10	0.00
−612	0.00	0.00
−562	0.00	0.00
−488	0.00	–
−358	0.00	–

– = not tested

Table 13: Penetration depth of secondary cracks in CERT tests of samples of steel A 516 in solutions containing 40 g/l NaOH and 20 g/l NaOH with each 20 g/l Na_2S + 20 g/l Na_2CO_3 at 383 and 413 K (110 and 140°C) [34]

	% C	% Si	% Mn	% P	% S	% Ni	% Cr	% Cu	% Mo
A	0.16	0.23	0.72	0.005	0.02	0.05	0.15	0.37	
B	0.20	0.25	0.58	0.014	0.02	0.04	0.15	0.12	0.04

Table 14: Chemical composition of the investigated melts of steel SAE C-1018 [35]

The measured crack propagation rates are plotted against the stress intensity factor in Figure 17 and against the temperature in Figure 18.

The results were similar to other systems: there are three regions with differing crack behaviour. In region 1 below a K_I value of approx. 25 MPa√m, K_I-dependent crack behaviour is observed followed by region 2 (approximately between K_I = 25 to 40 MPa√m), in which the crack propagation rate is independent of the stress intensity factor. In region 3, the crack propagation rate increases with higher stress intensity factor. A critical stress intensity factor K_{ISCC} for stress corrosion cracking, below which the crack propagation rate is negligible (e.g. < 10^{-10} m/sec), was not determined. However, it obviously lies below 18 MPa√m.

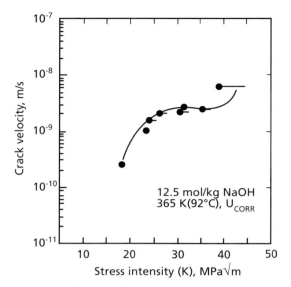

Figure 17: Influence of the stress intensity factor on the crack propagation rate in 33 % NaOH solution at 365 K (92°C) and $U_H \approx -1.0\,V$ [35]

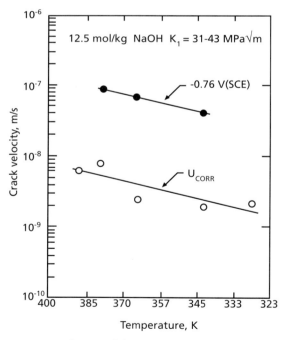

Figure 18: Influence of the temperature and potential on the crack propagation rate in 33 % NaOH solution [35]

The influence of the temperature was investigated at two potentials in region 2 for K_I values of approx. 35 MPa√m. At both potentials a linear relationship was observed between log(v) and 1/T. The crack propagation rates at potentials of $U_H = -0.76$ V were about one factor of ten higher than at $U_H \approx -1.0$ V. The potential $U_H = -0.76$ V lies in the transition range from the active to the passive state and has always proved to be especially critical with regards to stress corrosion cracking of unalloyed steels in NaOH solutions. To explain the strong negative potential of $U_H = -1.0$ V, it is assumed that there is involvement of hydrogen in the stress corrosion cracking process.

Although stress corrosion cracking is more common in caustic solutions than in nitrate solutions, the testing of unalloyed and low-alloy steels for their resistance to stress corrosion cracking is usually carried out in nitrate solutions according to DIN 50915 [36] for practical reasons. However, actual results show that the results obtained from the two media cannot always be compared, namely with respect to the steel composition and the type of mechanical loading. To elucidate these relationships, a comparative investigation of the stress corrosion cracking behaviour of eight steels, the compositions of which are given in Table 15, was carried out in nitrate solutions as well as in NaOH solutions.

Steel	C	Si	Mn	P	S	Cr	Mo	Al	N
P265GH (HII) (mat. no. 1.0425)	0.09	0.22	0.55	0.037	0.015		0.03	0.005	0.0098
S355N (mat. no. 1.0545)	0.18	0.430	1.38	0.015	0.003		0.01	0.040	0.0049
16Mo3 (mat. no. 1.5415)	0.14	0.26	0.68	0.011	0.008		0.30	0.055	0.0053
13CrMo4-5 (mat. no. 1.7335)	0.14	0.23	0.63	0.010	0.008	1.03	0.43	0.022	0.0064
10CrMo9-10 (mat. no. 1.7380)	0.11	0.23	0.51	0.013	0.003	2.07	0.95	0.026	0.0068
VS-A	0.08	0.04	0.80	0.005	0.005		0.98	0.080	0.0099
VS-B	0.08	0.01	0.78	0.007	0.003		0.01	0.009	0.0094
VS-C	0.18	0.31	1.04	0.007	0.004		1.10	0.031	0.0092

VS = test melts

Table 15: Essential values of the chemical analyses of investigated steels (proportional masses in %) [36]

In the evaluation of the resistance of steels to intergranular SCC, two assumptions were always made:
- the corrosion loading and the mechanical loading in the standard test in boiling calcium nitrate solution according to [36] are sufficiently high to allow a statement to be made on the relative resistance of the materials and

- the statement on the resistance in calcium nitrate solutions is valid in caustic alkaline solutions.

In the test according to [36], the bending specimens are loaded with a constant extension rate. Under operating conditions, other types of mechanical loading may be present:
- constant force
- alternating strain rates caused by fluctuating strains due to changes in the temperature and pressure.

The tests were carried out in the following media:
- boiling 60% calcium nitrate solution (b.p. 391–394 K (118–121°C)) according to [36] at a testing potential of $U_H = 0.15$ to 0.20 V, for which there is the greatest corrosion susceptibility. This potential was adjusted through the electrical contact of the samples with the test equipment made of CrNi steel.
- boiling 35% NaOH solution (b.p. 397 K (124°C)) at $U_H = -0.75$ V.

The following four types of test were used for the mechanical loading parameters:
- constant extension (ε = const.); tests with bending specimens corresponding to the loading given in [36]
- constant force (σ = const.); round tensile specimens under constant loading
- slow strain rate ($\dot{\varepsilon}$ = const.); round tensile specimens with unidirectional strain until failure; CERT test = Constant Extension Rate Tensile-Test. The extension rate in the tests was $\dot{\varepsilon} = 6 \times 10^{-6}$ 1/s
- slow strain rate ($\dot{\varepsilon}$ = const.); tensile specimens with cyclic loading (saw-tooth-shaped characteristics). In the tests, the average strain s_m was varied. The amplitude between s_{min} und s_{max} was chosen as:

$\sigma_{min} = \sigma_m - 0.1\, R_e$ $\quad$ $\sigma_{max} = \sigma_m + 0.1\, R_e$

The strain rates during loading were $\dot{\varepsilon} = 4 \times 10^{-6}$ 1/s and during unloading they were $\dot{\varepsilon} = 5 \times 10^{-3}$ 1/s. The frequency was 5×10^{-3} Hz.

The measured values determined were:
- for samples under constant load: the critical limiting stress
- for samples under slow extension: reduction of area Z as well as the maximum depth of attack in the necking area and shaft of the sample which was measured in the metallographic cross-section.

The test results of the steels are given in Table 16 for the individual testing media, testing potentials and types of loading.

In addition to detailed data on the mechanical loading, the summary also contains comments on the adjustment of the potential.

Tests according to DIN 50915 [36] only provide reliable data for CMn steels under elastic loading. These results are also applicable to NaOH solutions. In the plastic region, tensile tests under constant load can serve as a more sensitive test method for nitrate solutions. These results are, however, not applicable to materials containing Mo in NaOH solutions. Their resistance in NaOH solutions is noticeably lower than in nitrate solutions.

Steel	Type of loading	Calcium nitrate solution 60%; 393 K (120°C); $U_H = 0.2$ V	NaOH solution 35%; 397 K (124°C); $U_H = -0.75$ V
		O = resistant; X = susceptible limiting stress s_G/R_e,	
P265GH (HII)	constant deformation	X	X
S355N	bending sample	O	X
16Mo3	according to	O	X
13CrMo4-5	DIN 50 915	O	X
10CrMo9-10	ε = const.	O	X
VS-A		O	X
VS-B		O	X
VS-C		O	X
P265GH (HII)	constant load	0.96 (fracture)	0.33
S355N	tensile sample	> 1.4	0.72
16Mo3	σ = const.	> 1.3	0.29
13CrMo4-5		> 1.3	0.31
10CrMo9-10		> 1.1	0.28
VS-A		> 1.6 (fracture)	0.21
VS-B		1.4 (fracture)	0.57
VS-C		> 1.3	0.26
P265GH (HII)	constant extension	X	X
S355N	slow strain-rate test	X	X
16Mo3	unidirectional	X	X
13CrMo4-5	ε = const.	X	X
10CrMo9-10		X	X
VS-A		X	X
VS-B		X	X
VS-C		X	X
P265GH (HII)	cyclic loading	< 0.40	< 0.35
S355N	slow strain-rate test,	0.96	0.96
16Mo3	cyclic	0.90	< 0.30
13CrMo4-5	ε = const.	0.70	< 0.30
10CrMo9-10		0.70	< 0.30
VS-A		< 0.80	< 0.25
VS-B		< 0.80	0.70
VS-C		0.94	< 0.20

(VS = test melts)

Table 16: Summary of the test results of stress corrosion cracking of low-alloy steels in nitrate solution and in NaOH solution [36]

The amounts of alloying elements in the low-alloy steels have differing effects in the two corrosive media. In nitrate-containing solutions, aluminium-killed carbon-manganese, fine-grained construction steel of type S355N (mat. no. 1.0545) exhibited the best behaviour. The high-temperature construction steels alloyed with molybdenum or with chromium and molybdenum also showed very good resistance to stress corrosion cracking in nitrate solutions.

However, in NaOH solutions, the alloying element molybdenum on its own or in combination with alloying element chromium has a negative effect on the resistance. In this case, under critical conditions, only Al-killed, fine-grained C-Mn construction steels provide a satisfactory resistance. For all steels, the critical limit of the tensile stress is noticeably lower in NaOH solutions than in nitrate solutions.

The type of applied stress has a considerable influence on the corrosion behaviour. Under constant deformation, only those steels are susceptible to stress corrosion cracking whose critical limiting stress under constant loading lies in the range of the yield strength or below. For loading with slow strain-rates, all steels are susceptible to stress corrosion cracking [37–42].

Corrosion fatigue

Very little is known about this type of corrosion in solutions containing sodium hydroxide. If cyclic loading takes place, then it must be assumed that corrosion fatigue will also occur in NaOH solutions for all metallic materials in both the active as well as the passive state. The potential range close to $U_H = -0.7$ V also seems to be particularly critical for this type of corrosion. Table 17 gives the microscopically determined crack propagation rate Δl of specimens made of a steel containing 0.1 % carbon in 10 M NaOH solutions (corresponding to 44 % NaOH) for deformation rates between 1.5 and 436 % per minute. The potential range of the highest crack propagation rate lies between $U_H = -0.60$ and $U_H = -0.8$ V [43].

U_H V	−0.55	−0.60	−0.65	−0.70	−0.75	−0.80	−0.825	−0.85
Δl mm/h	0.92	0.104	0.120	0.132	0.138	0.142	0.042	0.024

Table 17: Crack propagation rates Δl of a low-alloy steel with 0.1 % C in 10 M NaOH solutions at 393 K (120 °C) as a function of the applied potential [43].

Unalloyed cast iron and low-alloy cast iron

Cast iron shows a similar corrosion behaviour in NaOH solutions as low-alloy steels. Cast iron components have proved successful for certain applications in NaOH solutions of concentrations of up to 70 % and temperatures up to 363 K (90 °C) [44].

In those parts of alkali-electrolysis plants which are in contact with NaOH solutions, valves of grey cast iron and pumps of cast ferrosilicon have been used because they are sufficiently resistant to 50 % NaOH solutions at temperatures up to 333 K (60 °C) [45].

Additions of 0.4 to 0.8 % chromium and 0.35 to 1 % nickel to the cast iron have practically no influence on the corrosion resistance in NaOH solutions at elevated temperatures; however they do increase the strength [46].

Slightly increased amounts of nickel and molybdenum can increase the resistance of ductile grey cast iron in NaOH solutions: the behaviour of an alloyed grey cast iron was compared to a standard nodular cast iron [47]. The composition of the two materials is given in Table 18.

Type of casting	% C	% Si	% S	% P	% Mn	% Mg	% Ni	% Mo
Unalloyed casting	3.12	2.21	0.02	0.11	0.44	0.02		
Alloyed casting	3.12	2.21	0.03	0.11	0.44	0.02	1.41	0.30

Table 18: Composition of the investigated types of cast iron [47]

The corrosion tests were carried out in 5 % NaOH solutions at 3 temperatures on samples that had had different pretreatments. The test conditions and results are given in Table 19. The corrosion rates were calculated from the current density–potential curves according to the Tafel method and from linear polarisation.

The alloyed cast iron showed in both the cast state as well as the tempered or cold-deformed state higher corrosion resistance than the correspondingly pretreated, unalloyed, grey cast iron. The resistance of both types was improved by tempering. Cold deformation of less than 15 % also had a favourable effect.

In most cases the corrosion resistance decreases in the following order:

slight cold deformation > tempered > cast state > strong cold deformation > austenitized/tempered.

Elevated temperatures led to increased attack on both types of cast iron.

State of sample	Testing temperature K (°C)	Corrosion rate mm/a (mpy)			
		unalloyed casting		alloyed casting	
		Tafel	linear polar.	Tafel	linear polar.
cast state	300 (27)	0.102 (4.0)	0.122 (4.8)	0.038 (1.5)	0.031 (1.2)
	325 (52)	0.13 (5.1)	0.165 (6.5)	0.083 (3.3)	0.099 (3.9)
	340 (67)	0.208 (8.2)	0.188 (7.4)	0.117 (4.6)	0.158 (6.2)
tempered 2 h at 873 K (600°C)	300 (27)	0.099 (3.9)	0.117 (4.6)	0.036 (1.4)	0.023 (0.9)
	325 (52)	0.127 (5.0)	0.163 (6.4)	0.081 (3.2)	0.081 (3.2)
	340 (67)	0.206 (8.1)	0.185 (7.3)	0.114 (4.5)	0.155 (6.1)
tempered cold rolled 7 %	300 (27)	0.084 (3.3)	0.091 (3.6)	0.033 (1.3)	0.02 (0.8)
	325 (52)	0.112 (4.4)	0.142 (5.6)	0.076 (3.0)	0.079 (3.1)
	340 (67)	0.191 (7.5)	0.165 (6.5)	0.112 (4.4)	0.15 (5.9)

Table 19: Corrosion rates in 5 % NaOH of cast irons listed in Table 18 [47]

Table 19: Continued

State of sample	Testing temperature K (°C)	Corrosion rate mm/a (mpy)			
		unalloyed casting		alloyed casting	
		Tafel	linear polar.	Tafel	linear polar.
tempered cold rolled 14 %	300 (27) 325 (52) 340 (67)	0.046 (1.8) 0.097 (3.8) 0.152 (6.0)	0.089 (3.5) 0.139 (5.5) 0.137 (5.4)	0.036 (1.4) 0.079 (3.1) 0.14 (5.5)	0.048 (1.9) 0.094 (3.7) 0.152 (6.0)
tempered cold rolled 21 %	300 (27) 325 (52) 340 (67)	0.051 (2.0) 0.079 (3.1) 0.158 (6.2)	0.094 (3.7) 0.142 (5.6) 0.16 (6.3)	0.041 (1.6) 0.081 (3.2) 0.145 (5.7)	0.076 (3.0) 0.109 (4.3) 0.155 (6.1)
tempered cold rolled 28 %	300 (27) 325 (52) 340 (67)	0.066 (2.6) 0.094 (3.7) 0.173 (6.8)	0.125 (4.9) 0.165 (6.5) 0.188 (7.4)	0.043 (1.7) 0.084 (3.3) 0.147 (5.8)	0.081 (3.2) 0.114 (4.5) 0.175 (6.9)
austen. 1,203 K (930°C) tempered 673 K (400°C)	300 (27) 325 (52) 340 (67)			0.091 (3.6) 0.14 (5.5) 0.15 (5.9)	0.137 (5.4) 0.152 (6.0) 0.305 (12)

Table 19: Corrosion rates in 5 % NaOH of cast irons listed in Table 18 [47]

The cast iron grades Meehanite®, with lamellar or spheroidal graphite, are used as materials for evaporators in the processing of solid sodium hydroxide. The grade Meehanite-HD is recommended for vessels for NaOH melts at 813 K (540°C) [48].

In the critical potential range close to $U_H = -0.65$ V, stress corrosion cracking also occurs in pure iron in NaOH solutions [49]. Table 20 gives the crack propagation rates of specimens of pure iron measured in slow strain-rate tests with a strain rate of $\dot{\varepsilon} = 10^{-6}$ 1/s in 35 % NaOH solutions at different testing potentials.

Testing potential V (U_H)	Crack propagation rate mm/s
−0.558	no cracking
−0.658	5.26×10^{-6}
−0.758	5.55×10^{-6}
−0.858	no cracking

Table 20: Crack propagation rate of pure iron in a 35 % NaOH solution at 373 K (100°C). CERT tests with a strain rate of $\dot{\varepsilon} = 10^{-6}$ 1/s [50]

High-alloy cast iron
Silicon cast iron
Austenitic cast iron (etc.)

Higher nickel contents in cast iron improve the corrosion resistance in NaOH solutions. As shown in Table 21, the corrosion rate is decreased by a factor of ten for a nickel content of 20 % [44].

Nickel content %	Corrosion rate mm/a (mpy)
0	1.9–2.3 (75–90)
3.5	1.2 (47)
5	1.2 (47)
15	0.8 (31.5)
20	0.08 (3.2)
30	0.01 (0.4)

Table 21: Influence of the nickel content in cast iron on the corrosion rate in boiling 50–65 % NaOH solutions [44]

In austenitic grades of cast iron with a nickel content of approx. 16 % and low contents of copper and manganese, chromium contents of 1 % to 6 % lead to an increased resistance to concentrated NaOH and KOH solutions [51].

In boiling 50 to 60 % NaOH solutions, the corrosion rate of cast iron is already reduced by additional alloying of approx. 15 % nickel from approximately 1.8 mm/a (70.9 mpy) to approximately 0.7 mm/a (27.6 mpy) [52]. At higher nickel contents, the corrosion rates decrease again, e.g. for

GGL-NiCr20-2 to 0.1 mm/a (3.94 mpy)
(mat. no. 0.6660)
(18–22 % Ni, 1–1.5 % Cr, 1–1.5 % Mn, 1–2.8 % Si, 3 % C)

GGL-NiCr30-3 to 0.04 mm/a (1.58 mpy)
(mat. no. 0.6676)
(28–32 % Ni, 2.5–3.55 Cr, 0.4–0.8 Mn, 1–2 % Si, 2.6 % C).

Ferritic chromium steels with < 13 % Cr

Low-alloy CrMo or NiCrMoV steels are frequently used in the construction of power stations. If there is an incursion of water, the sodium-cooled reactor can come into contact with a NaOH solution. This can lead to surface corrosion or to the formation of cracks. Tests on the unstabilized ferritic steel (2.25 % Cr-1 % Mo), which has been developed for sodium-cooled reactors, showed that incursion of water first led to surface attack and after 1,500 h there was increased crack formation [53].

In this context, investigations of stress corrosion cracking were carried out on a 2.2 % Cr steel (analysis values given in Table 22) in concentrated aqueous NaOH solutions, in anhydrous sodium hydroxide melts and in NaOH that contained additional sodium [54].

% C	% Cr	% Mo	% Si	% Mn	% S	% P	% Sn	% As	% Sb	% Fe
0.09	2.2	1.02	0.18	0.49	0.013	0.015	0.015	0.052	0.007	rest

Table 22: Composition of the investigated CrMo steel [54]

In the heat-treated state, the steel is susceptible to stress corrosion cracking in anhydrous NaOH melts and in concentrated aqueous solutions (75 and 92 % NaOH) at temperatures in the range 473 to 603 K (200 to 330°C). In NaOH + Na, SCC can occur at a temperature of 623 K (350°C).

In slow strain-rate tests on the same grade of steel in less concentrated NaOH solutions, the dependence of the SCC susceptibility on the temperature and concentration of the caustic solution is given in Figure 19 [55].

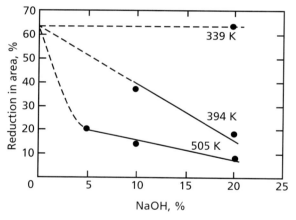

Figure 19: Influence of the temperature and concentration of the NaOH solution on the SCC behaviour of 2.25Cr-1Mo steel [55]

At temperatures up to 339 K (66°C) and NaOH concentrations up to 20 %, the steel is not susceptible to stress corrosion cracking. At temperatures of 394 and 505 K (121 and 232°C), stress corrosion cracking was observed above NaOH concentrations of 5 %, the intensity of which noticeably increased with increasing temperature and concentration.

In tests in 5 % NaOH solutions at 588 K (315°C) under slow cyclic mechanical loading of 1 Hz, the steel with 2.25 % Cr and 1 % Mo was found to be susceptible to corrosion fatigue after 4×10^3 to 4×10^4 load cycles [56, 57]. In 8 M NaOH solutions at 373 K (100°C), the crack propagation rates of the above steel were approx. 0.006 to 0.007 mm/h at potentials between $U_H = -0.9$ V and $U_H = -0.75$ V [56].

For steels 3CrMo and 3.5NiCrMoV, which are used for the rotors of low-pressure turbines in power stations, stress corrosion cracking can be expected to occur if sodium hydroxide and/or sodium chloride are able to leak into the water feed [58, 59]. The cracks are generally intergranular, are branched and are filled with corrosion products. The critical concentration of the NaOH solution that triggers stress corrosion cracking decreases with increasing temperature and amounts to

- at 341 K (68°C): 28 % NaOH
- at 430 K (157°C): 1 to 3 % NaOH
- at 477 K (204°C): 1 to 5 % NaOH

In contrast to the carbon steels, which are susceptible to stress corrosion cracking in 40 % NaOH solution at 373 K (100°C) only in the potential range between $U_H = -0.85$ V and $U_H = -0.6$ V, NiCrMoV steels are susceptible to stress corrosion cracking in a wide range of potentials above $U_H = -0.8$ V [60]. Precipitates of phosphorus, chromium or molybdenum compounds on the grain boundaries increase the susceptibility of these steels to stress corrosion cracking [61, 62].

Within the framework of fundamental investigations on the corrosion-assisted fracture behaviour of low-alloy CrNiMo steel, 3 different grades of heat-treated steels that are frequently used for steam turbine rotors in power stations, were tested in 30 % NaOH at temperatures between 298 and 353 K (25°C and 80°C). The composition of the alloyed steels is given in Table 23 [63]. The heat treatment and the mechanical properties determined for the steels are given in Table 24.

Steel	% C	% Mn	% Si	% Cr	% Ni	% V	% Mo
32CrNi3MoV	0.31	0.41	0.10	1.27	3.25	0.07	0.40
33CrNi3MoV	0.33	0.64	0.26	1.38	3.28	0.16	0.40
34CrNi3Mo	0.33	0.49	0.15	0.73	2.90	–	0.35

Table 23: Chemical composition of the investigated steels [63]

Steel	Heat treatment	R_e MPa	R_m MPa	A %	Z %	K_{IC} MPa $\sqrt{m}$
32CrNi3MoV	1,133 K (860°C), 1 h, 878 K (605°C), 6 h	1,042	1,103	15.0	64.0	130
33CrNi3MoV	1,143 K (870°C), 1 h, 878 K (605°C), 6 h	930	1,085	14.8	48.8	133
34CrNi3Mo	1,133 K (860°C), 1h, 883 K (610°C), 1 h	921	1,000	19.0	65.0	165

Table 24: Heat treatment and mechanical properties of the steels [63]

Investigations of the stress corrosion cracking and the corrosion fatigue behaviour were carried out on four-point bending samples with incipient cracks. Stress corrosion cracking tests of steel 34CrNi3Mo showed that SCC only occurred at temperatures $T \geq 323$ K ($\geq 50°C$) and at potentials $U_H = -0.65$ to -0.80 V. The tests were carried out at $T = 353$ K ($80°C$). The fracture mechanics properties obtained in the FCC tests, namely the stress intensity factor K_{ISCC} and the crack propagation rate da/dt, are given in Table 25.

Steel	K_{Iscc} MPa $\sqrt{m}$	da/dt mm/h	K_{Iscc}/K_{Ic}
32CrNi3MoV	67	4.3×10^{-5}	0.51
33CrNi3MoV	91	6.5×10^{-5}	0.68
34CrNi3Mo	83	6.4×10^{-5}	0.56

Table 25: Stress intensity factor K_{ISCC} and crack propagation rate da/dt of the investigated steels in 30 % NaOH at $T = 353$ K ($80°C$) [63].

The FCC tests were carried out with an R value of 0.75 at a frequency of 0.3 Hz. The results are given in Figure 20 for steel 33CrNi3MoV, although they are representative for the 3 steels, as a plot of da/dn against ΔK.

The curve can be divided into 3 regions. Up to a value of $\Delta K = 6.5$ MPa$\sqrt{m}$, the crack propagation rate da/dn noticeably increases with increasing ΔK to reach a plateau up to $\Delta K = 16$ MPa$\sqrt{m}$ which is then followed by another increasing gradient.

These results show that in the system low-alloy CrNiMo steels/NaOH solution, corrosion cracking occurs if the temperature reaches 323 K (50°C) and the potential lies in the region of the active/passive transition. The cracks are intergranular and are strongly branched.

Steels with elevated chromium contents of 9 % to 12 % are also susceptible to stress corrosion cracking in NaOH solutions. Even for the correct heat treatment, 9 % Cr-1 % Mo steel is susceptible to stress corrosion cracking at 573 K (300°C) even at low NaOH concentrations if it is tested under slowly increasing deformation. The cracking rate in 1 M NaOH solutions is approximately 0.01 mm/h. Above a concentration of the NaOH solution of more than 2.5 mol/l, crack formation becomes more obvious and the cracking rate was found to be 0.1 mm/h at 8 mol/l [64]. These results were obtained from 10 h tests and are in agreement with those obtained from long-term tests over 5,000 h at 623 K (350°C) in 1 to 5 M NaOH solutions [65].

Even the steel SAE 410 with approx. 12 % Cr is susceptible to stress corrosion cracking in boiling 70 % NaOH, and essentially independent of the applied potential. Therefore, it is not suitable for use in such concentrated NaOH solutions [66].

The susceptibility of 19 chromium steels, some of which are used as materials in steam turbines, to stress corrosion cracking was investigated in long-term tests in 28 % NaOH with 3.5 % sodium chloride at 339 K (66°C) [67]. The important test parameters and results are given in Table 26. After 24,000 h (= 2.74 years), all steels showed corrosion cracks at comparatively low tensile stresses.

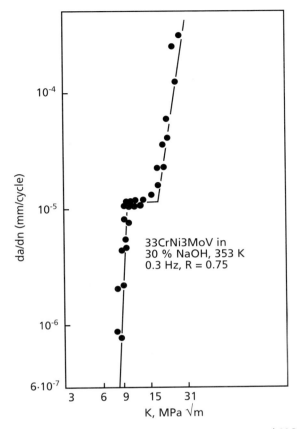

Figure 20: Results of the corrosion fatigue tests on steel 33CrNi3MoV in 30% NaOH at 353 K (80°C), 0.3 Hz and R = 0.75 [63]

Since the damage to turbine rotors at the Hinckly Point power station were shown to be caused by intergranular crack formation due to the caustic solution and Auger analyses showed phosphorus segregations on the grain boundaries, several other investigations have proved that segregations on the grain boundaries in low-alloy steels can promote intergranular corrosion as well as the generation and growth of cracks in NaOH solutions.

The unfavourable influence of higher contents of phosphorus in steel was proved in stress corrosion cracking tests of a NiCr steel with Ni and Cr contents at the usual levels used in turbine steels but with very low amounts of C, S and P in one grade and another grade with a markedly increased P content [68]. The composition of these two steels is given in Table 27. The contents of the elements not listed were less than 50 ppm. The samples were annealed in argon for 100 h at 948 K (675°C) to promote recrystallisation. The strengths of these steels are also given in Table 27. Some specimens were additionally aged in argon for 100 h at 753 K (480°C) to promote P segregations on the grain boundaries.

Cr content of the steel %	Tensile load ksi	Testing time h	Crack formation
1.75	100	100	yes
	70	3,600	yes
	47	24,000	yes
0.14	79	1,400	no
	38	24,000	yes
1.24	87	1,400	no
	37	4,000	no
	25	24,000	yes
1.27	97	1,400	no
	38	24,000	yes
1.67	78	3,600	no
	52	24,000	yes
0.95	123	1,500	yes
	67	4,000	yes
	45	24,000	yes
11.96	78	1,400	no
	33	24,000	yes
10.73	76	4,000	yes
	51	24,000	yes
13.96	40	1,500	yes
	30	4,000	no
	20	24,000	yes

Table 26: Chromium content, load, testing time and crack formation of the investigated steels [67]

Steel	% Ni	% Cr	% C	% S	% P	R_e N/mm^2	R_m N/mm^2
pure	3.5	1.7	0.0064	0.003	0.005	209	324
P doped	3.5	1.7	0.0064	0.003	0.033	217	365

Table 27: Contents of the major elements and strengths of the investigated steels [68]

The stress corrosion cracking tests were carried out in 9 M NaOH solutions at 371 K (98°C) under potentiostatic control at a potential of $U_H = -0.53$ V. The tensile specimens were tested with different initial loadings and the time-to-fracture recorded. The results are given in Figure 21.

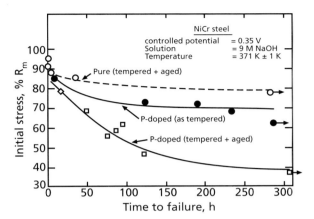

Figure 21: Exposure times to failure in dependence of the initial loading for the pure and the P-doped NiCr steel in 9 M NaOH solutions at 371 K (98°C) and $U_H = -0.53$ V [68]

The susceptibility to stress corrosion cracking noticeably increases with a higher phosphorus content and drastically increases again with increased segregations on the grain boundaries.

These results were confirmed by investigations of a NiCrMoV commercial rotor steel and a corresponding steel doped with phosphorus [69]. The composition of these two steels is given in Table 28.

Steel	% C	% Si	% Mn	% P	% S	% Ni	% Cr	% Mo	% V
NiCrMoV	0.21	0.05	0.27	0.007	0.007	3.50	1.61	0.36	0.09
P doped	0.25	0.11	0.32	0.063	0.015	3.42	1.57	0.38	0.10

Table 28: Composition of the two steels [69]

The samples were heat-treated (1,193 K (920°C)/4 h; quenched in oil; 823 K (550°C)/4 h) and subsequently aged for 100 h at 773 K (500°C) to achieve increased phosphorus precipitates on the grain boundaries. The tests were carried out on tensile specimens under constant load in a 40 % NaOH solution rinsed with nitrogen at the boiling temperature of 397 K (124°C) and a testing potential of $U_H = -0.53$ V. The results are presented in Figure 22. The load is referenced to the yield strength and given in %. Cracks appeared in the low-phosphorus steel above a loading of 65 % of the yield strength and the endurance times decreased with increasing load. The specimens of P-doped steel also showed stress corrosion cracking at a load of 65 % of the yield strength, while the endurance times at higher loads were markedly lower than those of the low-phosphorus steel.

Further detailed investigations that describe the negative influence of higher amounts of phosphorus in steel or of heat treatment (which promotes the formation of P precipitates on the grain boundaries) on the stress corrosion cracking behaviour of NiCrMoV steels are given in [70–74].

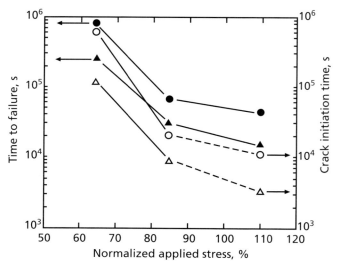

Figure 22: Time to crack development and to fracture in dependence on the load in boiling 40% NaOH [69]

If there is a leakage in sodium-cooled reactors, the sodium/water heat exchangers made of the frequently used low-alloy CrMo or NiCrMoV steels can come into contact not only with the NaOH solution, but molten NaOH can form in the evaporators. Two related reports describe in detail the investigations on the electrochemical behaviour of NaOH melts [75] and on the stress corrosion cracking tests [76] of 2.25% Cr-1% Mo steel (Table 29).

% C	% Si	% Mn	% P	% S	% Cr	% Mo	% Ni	% Al	% Cu	% Sn
0.12	0.31	0.50	0.014	0.014	2.36	0.96	0.13	0.009	0.19	0.16

Table 29: Chemical composition of the investigated steel [75]

With regards to the electrochemistry, the molten NaOH system behaves analogously to that of water; the acid-base properties are based on following dissociation reaction:

$$2\ OH^- \leftrightarrow H_2O + O^{2-}$$

with H_2O as the acid (proton donor) and O^{2-} as the base (proton acceptor). At normal atmospheric pressure, the most acidic region in the melt is in equilibrium with the vapour (1 bar) while the strongest basic region is in equilibrium with solid NaOH. The neutralisation value of a melt can be described by the pH_2O value, analogously to the well-known pH value of aqueous systems [$pH_2O = -\log c(H_2O)$, where $c(H_2O)$ is the concentration of water].

In analogy to the known Pourbaix diagrams (pH–potential diagram) for aqueous systems, a corresponding potential–neutralisation value diagram can be drawn up for iron in NaOH melts. According to this diagram, passive behaviour is expected for iron and low-alloy steels in the acidic region and active behaviour in the basic region.

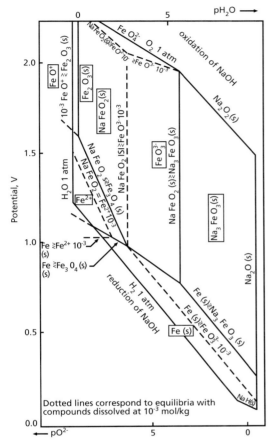

Figure 23: Potential–neutralisation value diagram for iron in NaOH melts at 623 K (350°C) [75]

As shown in Figure 24 and Figure 25 and in agreement with the above hypothesis, the typical behaviour of a passivatible material was found in the recordings of potentiokinetically determined current density–potential curves of the 2.25 % Cr-1 % Mo steel in the acidic range of the melt. The active/passive transition peak was only observed in acidic NaOH melts.

The investigations on the stress corrosion cracking behaviour were carried out on specimens with constant extension (U-bending specimen) and with constant strain rate (slow strain-rate test, CERT test). The steel of composition given in Table 29 was tested in the following heat-treated states:

1. normalised and tempered: 0.5 h at 1,198 K (925°C) + 2 h at 973 K (700°C) with air cooling

2. water-quenched: 0.5 h at 1,323 K (1,050°C) with water quenching
3. as 2. with subsequent ageing for 100 h at 673 K (400°C) and air cooling

The following test parameters were varied during the experiments:

- strain rate: 5.8×10^{-5} cm/s; 2.3×10^{-6} cm/s; 5.8×10^{-7} cm/s
- acidic NaOH melt: $pH_2O \sim 0$ free corrosion potential
  $pH_2O \sim 1$–2 $U_H = 1{,}000$ mV
  $pH_2O \sim 1$–2 $U_H = 2{,}000$ mV
- basic NaOH melt: $pH_2O \sim 10$ free corrosion potential
- in some cases, 1,000 ppm zinc was added to the NaOH melt.

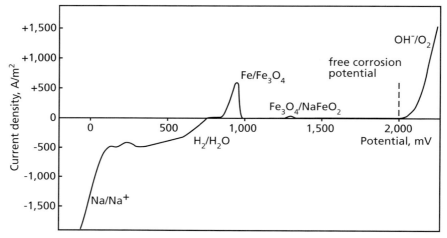

Figure 24: Current density – potential curve (5 mV/s) of 2.25 % Cr-1 % Mo steel in an acidic NaOH melt ($pH_2O \sim 1$–2) at 623 K (350°C) under argon [75]

The results can be summarized as follows.

- In the water-hardened state, the steel is susceptible to SCC in NaOH melts in a wide pH_2O range in both acidic and in basic melts.
- In acidic melts (pH_2O up to 2) the susceptibility is dependent on the potential. The greatest crack susceptibility was observed in the slightly anodic range at the active/passive transition.
- The cracks in acidic melts are generally formed in connection with prolate manganese sulphide inclusions. The distribution, morphology and composition of such inclusions appear to have a greater influence than the other test parameters, such as heat-ageing, strain rate or zinc content of the melt.
- In basic melts there is massive general surface corrosion. Intergranular crack corrosion was only found at certain strain rates and not in combination with inclusions, as found in acidic melts.
- The zinc content of the melts did not exhibit an effect in the CERT tests; however, a zinc content under 10 ppm already promoted crack formation in U-bending specimens.

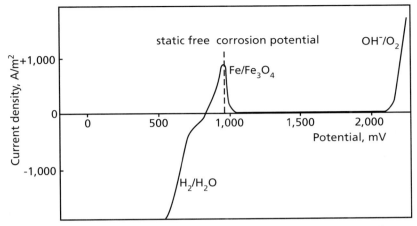

Figure 25: Current density – potential curve (5 mV/s) of 2.25 % Cr-1 % Mo steel in an acidic NaOH melt (pH$_2$O ~ 0) at 623 K (350°C) under a water vapour atmosphere of 1 bar [75]

Ferritic chromium steels with ≥ 13 % Cr

Chromium steels in boiling 48 % NaOH (413 K (140°C)) only then show a satisfactory corrosion resistance (corrosion rate of < 0.1 mm/a (< 3.94 mpy)) if the chromium content is above 18 % and if 0.1 to 0.2 % sodium chlorate is added as an inhibitor to the caustic solution [77].

According to Figure 26, chromium steels with 30 % chromium in 48 % NaOH are resistant up to 413 K (140°C) (0.01 mm/a (0.39 mpy)), even without additives which favour passivation (such as e.g. NaClO$_3$). The corrosion that occurs above 413 K (140°C) cannot be reduced by addition of sodium chlorate to the NaOH solution.

The results presented in Figure 27 on similar investigations confirm that the resistance of chromium steels with 13 to 25 % Cr in 48 % NaOH at 413 K (140°C) is drastically increased by the addition of 500 ppm and 1,000 ppm sodium chlorate, respectively. The addition of 1 % sodium chloride to the NaOH solution only has a slight effect on the corrosion behaviour [78].

Figure 28 shows the temperature dependence of the corrosion rate for steels with differing chromium contents in 48 % NaOH [79]. Thus, resistance (0.01 mm/a (0.39 mpy)) is already attained at 373 K (100°C) and a Cr content of 24 %. However, at 473 K (200°C) only the steel with a Cr content of 30 % and a corrosion rate of approx. 0.1 mm/a (approx. 3.94 mpy) can be used.

The corrosion rate of a 18 % Cr-2 % Mo steel in NaOH solution at 373 K (100°C) is dependent on the concentration of the caustic solution and amounts to

- in 25 % solution: 0.19 mm/a (7.48 mpy)
- in 35 % solution: 0.5 mm/a (19.7 mpy)
- in 50 % solution: 0.6 mm/a (23.6 mpy) [80].

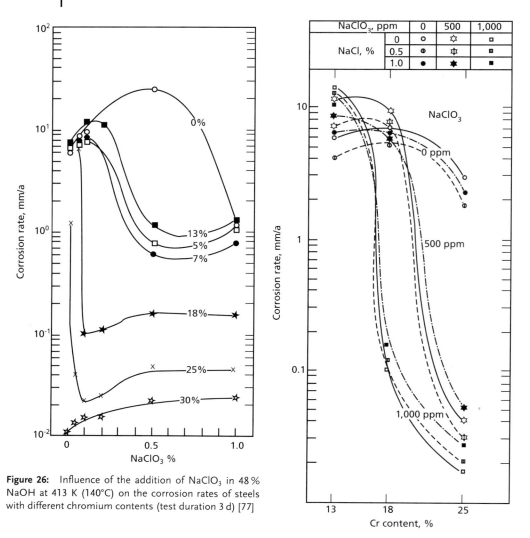

Figure 26: Influence of the addition of $NaClO_3$ in 48% NaOH at 413 K (140°C) on the corrosion rates of steels with different chromium contents (test duration 3 d) [77]

Figure 27: Influence of $NaClO_3$ and NaCl on the corrosion rate of FeCr alloys in 48% NaOH at 413 K (140°C) in dependence on the chromium content [78]

Compared to the CrNi steels (SAE 304 and 310), the chromium-molybdenum steels with similar chromium contents exhibit surface corrosion which is higher by a factor of 4 to 8; however, they are essentially resistant to stress corrosion cracking. The 26% Cr-1% Mo steel with a markedly higher chromium content exhibits the lowest corrosion rate of 0.0025 mm/a (0.098 mpy) with good stress corrosion cracking resistance in 25% NaOH at 373 K (100°C) [81].

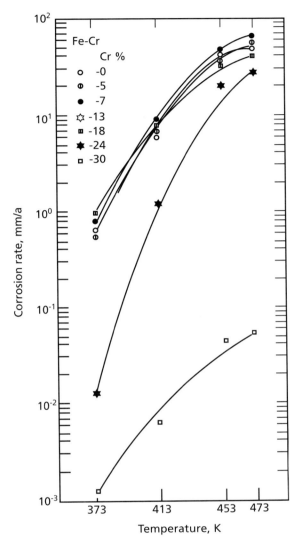

Figure 28: Temperature dependence of the corrosion rate of steels with chromium contents up to 30 % in 48 % NaOH solutions [79]

Stress corrosion cracking is also an important factor for the reliability of materials used as construction materials for steam generators and heat exchangers in power stations. Corrosion investigations were carried out on stainless steels with ferritic, martensitic, austenitic and ferritic-austenitic microstructures in various aqueous media that are liable to occur in steam generation, including NaOH solutions [82]. The major alloying elements of the investigated steels and the microstructural state are listed in Table 30.

Steel	Micro-structure	% C	% Si	% Mn	% Cr	% Ni	% Mo	% Cu	% Ti	% N
SAE 405	ferritic	0.06	0.42	0.58	12.76	0.33	–	–		
SAE 410	martensitic	0.09	0.37	0.46	12.3	0.48	0.14	0.15		
CrMoTi	ferritic	0.04	0.63	0.42	18.33	0.33	2.30	0.10	0.36	0.007
E-Brite26-1	ferritic	0.002	0.30	0.03	25.4	0.09	1.07	–		0.010
3RE60	duplex	0.025	1.64	1.58	18.2	4.9	0.24	0.11		
SAE 304	austenitic	0.046	0.47	1.67	18.92	9.14	0.24	0.11		

Table 30: Major alloy contents and microstructure of the investigated steels [82]

The stress corrosion cracking behaviour of the materials was determined by means of C-ring samples with a loading of 90 % of the yield strength in 10 % NaOH at 605 K (332°C). The materials were tested in the as-delivered state and after a sensitizing heat treatment as well as after annealing at 748 K (475°C), i.e. in the temperature range in which embrittlement of steels with higher Cr contents is liable to occur. The test duration was 200 days. The results are summarised in Table 31.

Steel	State	Result 10 % NaOH / 605 K (332°C)/ 200 d / 90 % R_e
SAE 405	as-delivered sensitized: 1,283 K (1,010°C)/1 h	no fracture / strong general corr. no fracture / strong general corr.
SAE 410	as-delivered 922 K (649°C)/1 h 838 K (565°C)/1 h 755 K (482°C)/1 h	no fracture / strong general corr. no fracture / strong general corr. no fracture / strong general corr. cracks
CrMoTi	as-delivered sensitized: 1,283 K (1,010°C)/1 h 747 K (474°C)/1 h 747 K (474°C)/20 h	no fracture no fracture / intergran. attack up to 15 μm no fracture / intergran. attack up to 5 μm no fracture / intergran. attack up to 10 μm
E-Brite26-1	as-delivered sensitized: 1,283 K (1,010°C)/1 h 747 K (474°C)/1 h 747 K (474°C)/300 h	no fracture no fracture / intergran. attack up to 10 μm no fracture fracture after 70 days
3RE60	as-delivered sensitized: 922 K (649°C)/1 h 747 K (474°C)/1 h 747 K (474°C)/300 h	no fracture / strong general corr. no fracture / intergran. attack up to 3 μm no fracture / general corr. fracture after 5 days
SAE 304	as-delivered sensitized: 922 K (649°C)/1 h	no fracture no fracture

Table 31: Results of the stress corrosion cracking tests [82]

Only the ferritic steel 304 and the austenitic steel 405 showed no signs of stress corrosion cracking. The high-Cr ferritic steels are only susceptible in the sensitised and embrittled state while the duplex steel is only susceptible in the embrittled state. The steels 304 and 405 were also tested in the as-delivered state in 50 % NaOH at 589 K (316°C). Steel 304 was susceptible to SCC and the austenitic steel showed such a strong surface corrosion that the test could not be carried out.

Free NaOH solution can be formed in vapour systems of power generators by the hydrolysis of alkaline salts or from a NaOH treatment of the feed water to the boiler. The generation of fatigue cracks in the 12 % Cr steel SAE 403, which is used for turbine rotors, was investigated at 373 K (100°C) in NaOH solutions with pH values of 10, 12 and 14 [83]. The composition and mechanical properties of the steel are given in Table 32.

C %	Si %	Mn %	P %	S %	Mo %	Cr %	R_e N/mm²	R_m N/mm²	A %	Z %	HRC
0.10	0.30	0.48	0.02	0.01	0.10	11.71	692	785	22.3	67.3	22

Table 32: Chemical composition and mechanical properties of steel 403 in the heat-treated state [83]

The corrosion fatigue tests were carried out in the form of cyclic bending tests with a sinus-shaped loading of ± 448 N/mm² and a frequency of 90 1/min. The number of stress cycles to failure obtained in the three NaOH solutions in air were:

- 5×10^4 at pH 10
- 1.2×10^5 at pH 12
- 1.7×10^5 at pH 14

and are shown in the Wöhler stress-cycle (S-N) curve in Figure 29.

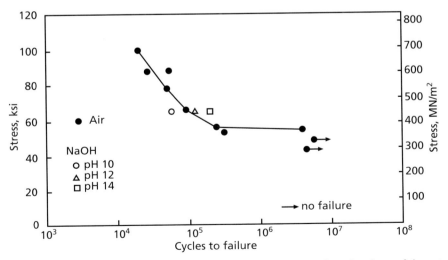

Figure 29: S-N curve of steel 304 in air at 373 K (100°C) and number of cycles to failure at 448 N/mm² in NaOH of pH 10, 12 and 14 at 373 K (100°C) [83]

For the given test conditions, the resistance to corrosion fatigue increases continuously with increasing pH value. In NaOH solutions of pH 10, the number of stress cycles endured with a loading of 448 N/mm^2 was somewhat less than that obtained in air, while at higher pH values a higher number of stress cycles was endured than in air.

High-alloy multiphase steels
Ferritic/perlitic-martensitic steels

Because of their high strength, heat-treated martensitic stainless steels with 12 or 17 % Cr are used in the construction of highly loaded components in steam turbines, particularly for turbine rotor blades. The stress corrosion cracking behaviour of these steels was investigated in various media which may occur e.g. from leaks in steam or cooling water systems.

Samples of the steel 17-4PH (X5CrNiCuNb17-4-4, mat. no. 1.4548 (1.4542)) with 0.038 % C, 0.57 % Si, 16.3 % Cr, 3.2 % Ni, 3.38 % Cu and < 0.001 % Al were tested by means of slow tensile tests in 20 and 40 % NaOH solutions at 363 K (90°C) for both free corrosion as well as under anodic or cathodic polarisation. No stress corrosion cracking was observed for this steel [84].

The 12 % Cr steel SAE 410 (composition given in Table 33) was also tested for stress corrosion cracking by means of slow strain-rate tests in NaOH solutions at 363 K (90°C): this steel was practically resistant to stress corrosion cracking [85]. The ratio of reduction of area of tensile specimens in NaOH solutions to that of specimens in oil is plotted against the NaOH concentration of the test solution in Figure 30. Even in those samples which were tested in the more concentrated solutions and which showed a small decrease of the reduction of area, no indications of secondary cracks or brittle cracking were found in the metallographic examinations.

% C	% Cr	% Ni	% Mo	% Mn	% S	% Si	% Cu
0.15	11.56	0.45	0.37	0.68	0.013	0.16	0.11

Austenitizing temperature: 1,253 K (980°C); oil quenching; tempering at 973 K (700°C)/1 h

Table 33: Chemical composition and heat treatment of the investigated steel SAE 410 [85]

Ferritic-austenitic steels/duplex steels

The differing corrosion resistance of austenite and ferrite as well as carbides in 10 M NaOH solutions is used to recognise these microstructural components by etching. This principle was used to visualise e.g. the damage to a burner nozzle made of cast steel G-X40CrNi27-4 in the blast heater of a blast furnace by etching with NaOH solution. The austenitic areas can be differentiated from the ferritic regions because they are only weakly etched and have a brown colouration [86]. In

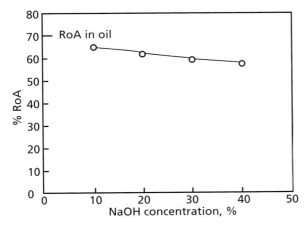

Figure 30: Influence of the NaOH concentration on the reduction of area (RoA) of tensile samples of SAE 410 at 363 K (90°C) [85]

austenitic-ferritic chromium-nickel steels, the addition of 2 to 3 % molybdenum increased the susceptibility to corrosion in NaOH solutions [87]. Precipitations of δ-ferrite also promote the corrosion of these steels.

The corrosion behaviour of the duplex steels listed in Table 34 was tested in boiling NaOH solutions of concentrations from 20 to 70 % [88]. The tests were carried out on commercially available sheets and welding materials in the solution-annealed state.

Steel	% C	% Si	% Mn	% P	% S	% Cr	% Mo	% Ni	% N	% Cu
1.4462	0.02	0.55	1.74	0.022	0.002	22.05	2.97	5.95	0.13	–
Uranus 50	0.02	0.54	1.84	0.023	0.004	20.99	2.33	6.58	–	1.4
Ferralium 255	0.03	0.24	1.52	0.023	0.002	25.93	3.15	5.6	0.18	1.71
1.4362 – 1	0.02	0.52	1.22	0.022	0.003	22.98	0.21	4.8	–	–
1.4362 – 2	0.02	0.52	1.84	0.027	0.001	23.05	–	4.2	–	–

The steel 1.4362 was present in 2 melts

Table 34: Chemical composition of the investigated duplex steels [88]

The dependence of the corrosion rates on the NaOH concentration in static solutions is shown in Figure 31.

Although the corrosion rate of all steels increases with increasing concentration of the caustic solution, the steels with values < 0.01 g/(m² h) were resistant to corrosion in boiling 20 and 30 % NaOH. At higher concentrations, the corrosion rates increased noticeably and reached a maximum at 60 % NaOH. Under these conditions, steel 1.4362 (X2CrNiN23-4) is inferior to the other steels.

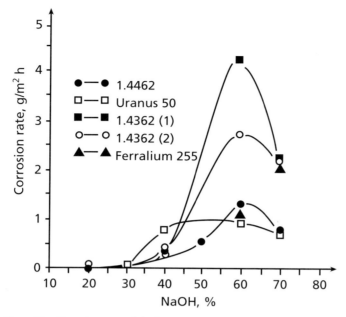

Figure 31: Corrosion rates of the ferritic-austenitic steels in boiling NaOH solutions [88]

The influence of laminar flow on the corrosion rate was tested on rotating discs of steel 1.4462 (X2CrNiMoN22-5-3). Even at very low flow rates of approx. 0.08 m/s, the rate of mass loss was increased by a factor of 6 compared to the static solutions. Further increases in the flow rate up to 1 m/s did not lead to a significant increase in the corrosion. Thus the corrosion products formed in 60 % and 70 % NaOH solutions only have a protective effect under static conditions.

At the free corrosion potential, these steels were not susceptible to stress corrosion cracking in 20 % to 70 % NaOH solutions, neither under the conditions of constant deformation (U-bending samples) nor at constant slow strain rates in the CERT test. Even in potentiostatically controlled CERT tests in 60 % NaOH at 433 K (160°C) in the potential range of $U_H = -1$ to -0.2 V, no stress corrosion cracking could be produced.

According to these results, the use of ferritic-austenitic steels in NaOH solutions up to a concentration of 35 %, even at the boiling point, could be an alternative to nickel-based materials. However, in spite of their very good resistance to stress corrosion cracking, the use of these steels at higher concentrations of caustic solutions is limited by the increasing rates of mass loss.

The studies reported in [89] also confirm that the decisive factor for the limits of utilisation of the duplex steels is general surface corrosion and not stress corrosion cracking. The corrosion rates and SCC behaviour in NaOH solutions in concentration range of 7 to 70 % with and without additional 3 % NaCl at temperatures between 120 and 473 K (−153 and 200°C) were investigated for the three steels listed in Table 35.

Steel grade	% C	% Si	% Mn	% P	% S	% Cr	% Ni	% Mo	% Cu	% W	% N	% Fe
22Cr-5Ni-3Mo	0.016	0.32	1.12	0.021	0.005	21.71	5.52	3.22	–	–	0.18	rest
Zeron®	0.017	0.16	0.48	0.003	0.001	25.34	7.02	3.56	0.66	0.53	0.23	rest
Ferralium®	0.025	0.39	1.08	0.029	0.001	25.91	5.80	3.30	1.65	–	0.23	rest

Table 35: Chemical composition of the investigated duplex steels [89]

The tests to determine the mass loss rates were carried out according to the ASTM Standard G31 [90]. The stress corrosion cracking behaviour was studied by means of CERT tests with a strain rate of 1.0×10^{-6} 1/s. The mechanical properties of tensile specimens of the three steels, determined by a CERT test in silicon oil, are given in Table 36.

Steel grade	Temperature K (°C)	$R_{e\,0.2}$ MPa	R_m MPa	Percentage elongation after fracture %
22Cr-5Ni-3Mo	393 (120)	374	656	0.50
	413 (140)	371	642	0.49
	473 (200)	359	688	0.50
Zeron®	393 (120)	494	750	0.44
	413 (140)	496	769	0.44
	473 (200)	497	799	0.43
Ferralium®	393 (120)	503	763	0.38
	413 (140)	–	–	–
	473 (200)	485	812	0.43

Table 36: Mechanical properties of the investigated tensile specimens at various testing temperatures (CERT test, 1.0×10^{-6} 1/s) [89]

The corrosion rates, calculated from the measured mass loss after a test period of 5 days, are given in Figure 32 and in Figure 33. In addition to the evaluation of the three ranges (low = < 0.01 mm/a (< 0.39 mpy); medium = 0.1–1.0 mm/a (3.94–39.4 mpy), high = > 1.0 mm/a (> 39.4 mpy)), which are labelled with different symbols, the values are also given for the values determined for each of the parallel specimens.

The 22Cr steel showed a clear dependence of the corrosion rate on the temperature and the NaOH concentration. There was no recognisable influence of the NaCl addition on the corrosion behaviour. In the superduplex steels, it is conspicuous that the highest corrosion rates were found for a NaOH concentration of 50 %, even at the lowest testing temperature of 393 K (120°C). As the individual values show, the addition of NaCl increased the corrosion rate by a factor of 3 at this concentration, while it did not have any effect at a NaOH concentration of 40 %.

Sodium Hydroxide

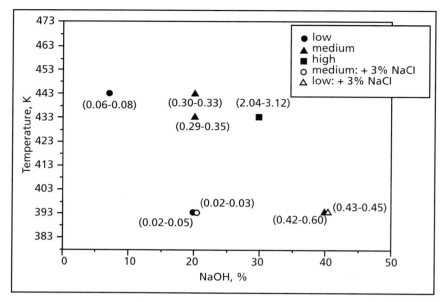

Figure 32: Corrosion rates of 22Cr duplex steel in dependence of the temperature and NaOH concentration [89]

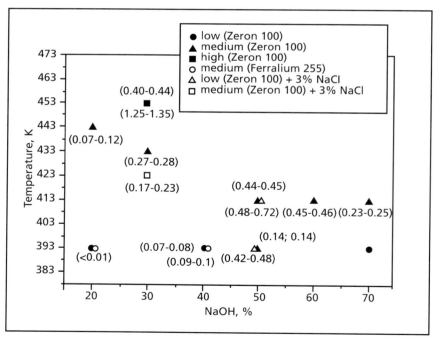

Figure 33: Corrosion rates of 25Cr superduplex steels in dependence on the temperature and NaOH concentration [89]

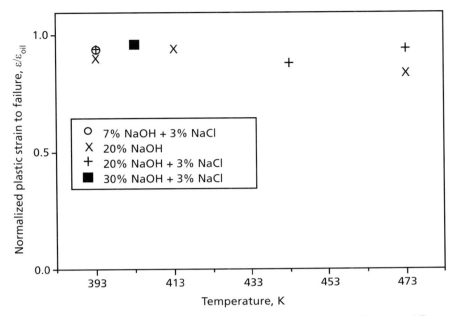

Figure 34: Results of the CERT tests of 22Cr duplex steel in various NaOH solutions at different temperatures [89]

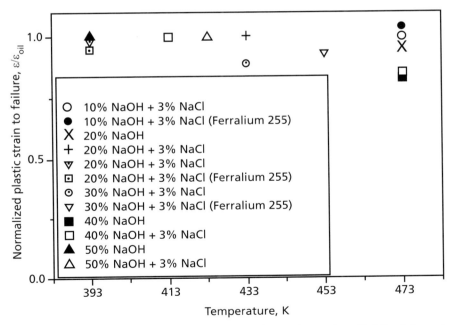

Figure 35: Results of the CERT tests of 25Cr superduplex steels in various NaOH solutions at different temperatures [89]

The results of the stress corrosion cracking tests are given in Figure 34 and in Figure 35. The percentage elongation after fracture of the tensile specimens in the CERT test in NaOH solutions is given relative to that determined in silicon oil $\varepsilon/\varepsilon_{oil}$.

For all specimens under the given test conditions, almost the same elongation values were obtained as those in oil so that these three steels are essentially not susceptible to stress corrosion cracking under these conditions.

The superiority of the stress corrosion cracking behaviour of duplex steels compared to austenitic steels was confirmed in [91] which deals with the SCC behaviour of 3 austenitic and 3 austenitic-ferritic steels in NaOH solutions that contained NaCl. Table 37 lists the contents of the major alloying elements of the investigated steels.

Steel grade	% Cr	% Ni	% Mo	% C	% N	micro-structure
18Cr-10Ni-2.5Mo	18	10	2.5	0.03		a
20Cr-25Ni-4.5Mo	20	25	4.5	0.03		a
27Cr-31Ni-3.5Mo	27	31	3.5	0.02		a
23Cr-4Ni	23	4	–	0.03	0.10	a-f
22Cr-5Ni-3Mo	22	5	3.0	0.03	0.14	a-f
25Cr-7Ni-4Mo-N	25	7	4.0	0.02	0.28	a-f

a = austenitic
a-f = austenitic-ferritic

Table 37: Nominal chemical composition and microstructure of the investigated steels [91]

The tests were performed at 473 K and 523 K (200 and 250°C) on U-bending samples in both aerated and deaerated (flushed with nitrogen) solutions that contained 200 g/l NaOH + 10 g/l NaCl. The results obtained after a test period of 500 h are summarized in Table 38.

Steel grade	473 K (200°C)		523 K (250°C)	
	aerated	deaerated	aerated	deaerated
18Cr-10Ni-2.5Mo	–	–	+	+
20Cr-25Ni-4.5Mo	–	+	+	+
27Cr-31Ni-3.5Mo	–	–	–	–
23Cr-4Ni	–	–	–	–
22Cr-5Ni-3Mo	–	–	+	+
25Cr-7Ni-4Mo-N	–	–	+	–

– = no cracks
+ = intergranular SCC cracks

Table 38: Test results of U-bending samples after 500 h in NaOH/NaCl solutions [91]

While there was only one case of stress corrosion cracking at 473 K (200°C) (steel 20Cr-25Ni-4.5Mo in oxygen-free solution), at 523 K (250°C) the steels 18Cr-10Ni-2.5Mo and 20Cr-25Ni-4.5Mo were particularly susceptible to cracking in both the aerated and in the deaerated solution. Only the high-alloy austenitic steel 27Cr-31Ni-3.5Mo and the molybdenum-free duplex steel 23Cr-4Ni remained crack-free under all test conditions.

The same solution at 473 K (200°C) was used to study round tensile specimens in slow strain-rate tests with extension rates of 5.0×10^{-6} 1/s and 9.2×10^{-7} 1/s. The values measured for the reduction of area after the slow strain-rate test in caustic solution and in glycerine are given in Table 39 along with a fractographic description.

Steel grade	Strain rate 1/s	Z_L %	Z_G %	Z_L/Z_G	Fractographic examination
18Cr-10Ni-2.5Mo	5.0×10^{-6}	62	75	0.83	+
20Cr-25Ni-4.5Mo	5.0×10^{-6}	54	69	0.78	+
27Cr-31Ni-3.5Mo	5.0×10^{-6}	82	83	0.99	(+) < 7 µm
	9.2×10^{-7}	82		0.99	(+) < 40 µm
	9.2×10^{-7}	64		0.77	+
23Cr-4Ni	5.0×10^{-6}	87	88	0.99	(+) < 15 µm
	9.2×10^{-7}	76		0.87	(+) < 15 µm
22Cr-5Ni-3Mo	5.0×10^{-6}	82	82	1.00	(+) < 15 µm
	9.2×10^{-7}	79		0.96	(+) < 50 µm
25Cr-7Ni-4Mo-N	5.0×10^{-6}	82	81	1.01	(+) < 20 µm
	9.2×10^{-7}	55		0.67	+

ZL = Reduction of area in the caustic solution; + = intergranular SCC cracks outside the fracture surface
ZG = Reduction of area in glycerine; (+) = microcracking (max. crack depth)

Table 39: Results of the slow strain-rate tests [91]

In the two austenitic steels 18Cr-10Ni-2.5Mo and 20Cr-25Ni-4.5Mo, stress corrosion cracking already occurred at a strain rate of 5.0×10^{-6} 1/s, as shown by the reduced reduction of area, the occurrence of intergranular secondary cracks and brittle fracture on the fracture surface. These results are in agreement with the SCC susceptibility found for the bending specimens of these two steels. The more highly alloyed austenitic steel 27Cr-31Ni-3.5Mo exhibited stress corrosion cracking only at slower strain rates. The following order of SCC resistance can therefore be derived for austenitic steels:

20Cr-25Ni-4.5Mo < 18Cr-10Ni-2.5Mo < 27Cr-31Ni-3.5Mo

Overall, duplex steels show a better SCC resistance to hot NaOH solutions than austenitic steels; steel 23Cr-4Ni showed the most favourable behaviour. The following order of SCC resistance can be derived for duplex steels:

25Cr-7Ni-4Mo-N < 22Cr-5Ni-3Mo < 23Cr-4Ni

For both types of steel, lower Cr contents and higher Ni and Mo contents have an unfavourable effect. The susceptibility to stress corrosion cracking in NaOH solution can be represented as a function of the nickel and the molybdenum contents [91].

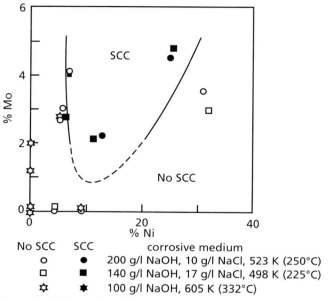

Figure 36: Influence of nickel and molybdenum on the susceptibility to stress corrosion cracking in caustic solutions [91]

This confirms the results of other authors in which the stability of duplex steels to stress corrosion cracking in NaOH solutions increases with decreasing nickel and molybdenum contents [92].

Austenitic chromium-nickel steels

The resistance of the simple austenitic 18-8-CrNi standard steels to uniform surface attack in NaOH solutions strongly depends on the concentration and the temperature of the NaOH solution. Table 40 gives the corrosion rates of some of these steels from older literature sources [93].

The sufficient stability of the simple 18-8 steels in NaOH solution only at low concentrations and temperatures up to 343 K (70°C) was also confirmed by investigations of wires made of standard 18-8 steel. These wires were noticeably attacked in 50 % solutions at temperatures above 343 K (70°C) and were very strongly attacked above 373 K (100°C) after a short period [94]. As the nickel content increased, the corrosion resistance increased. Table 41 gives the corrosion rates of a CrNi steel with 19.7 % Cr and 36.6 % Ni measured at 373 K (100°C) [95].

Steel grade SAE	Cr %	Ni %	NaOH %	Temperature K (°C)	Corrosion rate mm/a (mpy)
302	17–19	8–10	10	348 (75)	0.005 (0.197)
			20	323–333 (50–60)	0.0025 (0.098)
				378 (105)	0.05 (1.97)
			50	413 (140)	0.25 (9.84)
			70	343 (70)	0.08 (3.15)
				383 (110)	0.69 (27.2)
				438 (165)	1.9 (74.8)
			75	393 (120)	0.97 (38.2)
304	18–20	8–12	20	323–333 (50–60)	0.0025 (0.098)
			75	393 (120)	1.14 (44.9)
309	22–24	12–15	20	323–333 (50–60)	0.0025 (0.098)
			75	383 (110)	0.51 (20.08)
310	24–26	19–22	20	323–333 (50–60)	0.0025 (0.098)
			75	383 (110)	0.54 (21.3)
330	14–16	33–36	75	383 (110)	0.36 (14.2)

Table 40: Corrosion behaviour of austenitic CrNi steels in NaOH solutions at various concentrations and temperature [44, 93]

NaOH concentration %	Corrosion rate mm/a (mpy)
5	0.025 (0.098)
10	0.040 (1.57)
15	0.070 (2.76)

Table 41: Corrosion rates of a 20% Cr-36% Ni steel in NaOH solutions of various concentrations at 373 K (100°C) [95]

The chromium-nickel steels of types SAE 304 L, SAE 303, SAE 303 super; SAE 316 L and SAE 305 Cu, that are frequently used in the food industry, exhibit no noticeable corrosion in 6% NaOH solutions after 200 h [96]. The good cleaning effect of NaOH solutions and other alkaline solutions can therefore be used for plants and vessels made of chromium-nickel steel in the food industry.

Steels of type SAE 304 and 316 are susceptible to stress corrosion cracking at higher temperatures in NaOH solutions [91, 97, 98]. According to [99], the limiting temperatures for stress corrosion cracking of steel SAE 304 is 363 K (90°C) in 10% NaOH and 323 K (50°C) in 50% NaOH. In agreement with this, the steel SAE 304 L has proved resistant in a plant used for the recovery of cresol from waste-water that employs 10% NaOH at 343 K (70°C) [100]. In [101], the limits are given for the occurrence of stress corrosion cracking of the austenitic steels SAE 304 and SAE 316 (Figure 37).

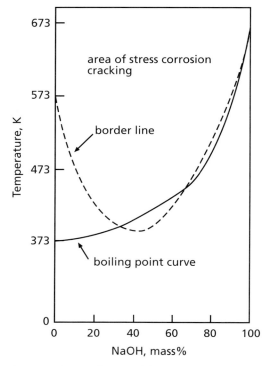

Figure 37: Range of susceptibility to stress corrosion cracking of steels SAE 304 and SAE 316 in dependence on the concentration and temperature of the NaOH solution [101]

The high-nickel, low-carbon steel Incoloy® 800 (31–34 % Ni, 20–21 % Cr, mat. no. 1.4876), that is used for steam-boiler pipes, is resistant in 35 to 70 % NaOH solutions in a wide potential range from $U_H = -0.65$ to $U_H = 0$ V and up to 458 K (185°C) against both uniform surface attack and stress corrosion cracking [102]. At 623 K (350°C) in 10 % NaOH, Incoloy® 800 developed cracks more than 100 mm deep over a period of 6,000 h [103].

Molybdenum-free chromium-nickel steels are not recommended for use in the pulp and paper industries for plants in which solutions with higher contents of sodium hydroxide and sulphur compounds are liable to occur at elevated temperatures. Instead, the nickel-chromium alloy Inconel® 600 (mat. no. 2.4816) or chromium-molybdenum steels (26 % Cr, 1–4 % Mo) should be preferred [104].

Figure 38 shows the crack susceptibility of steels Incoloy® 800 and SAE 316 as well as the Ni alloy Inconel® 600 in air-free NaOH solutions with 100 g/l NaOH at 623 K (350°C) in an autoclave in dependence on the mechanical loading [105]. Under these reaction conditions, and for a tensile stress of less than 350 MPa and a concentration of less than 150 g/l NaOH, Incoloy® 800 is clearly superior to the other two alloys with regards to the minimum time to the appearance of 100 μm-deep cracks. This is confirmed by a comparative investigation with ring and loop specimens of Incoloy® 800, SAE 304 and Inconel® 600 in 50 % NaOH at 589 K (316°C) in which, according to Table 42, Incoloy® 800 was partially superior with respect to the resistance to stress corrosion cracking than the other FeCrNi alloys [106].

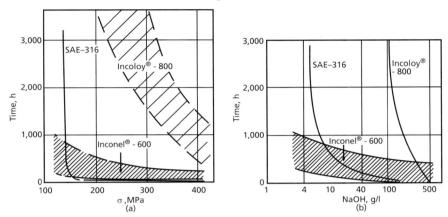

Figure 38: Minimum time until a 100 μm-deep crack had appeared in the various alloys
(a) in a deaerated NaOH solution with 100 g/l NaOH at 623 K (350°C) as a function of the mechanical loading
(b) at constant mechanical loading as a function of the NaOH concentration [105]

Test period months	SAE 304	Incoloy® 800	Inconel® 600
1	(+)	(+)	(+)
3	(+)	(+)	(+)
6	(+)	(−)	(−)

(+) = no cracks, (−) = cracks

Table 42: Stress corrosion cracking behaviour of FeCrNi alloys (C-ring specimens at a loading of 110 % of the yield strength) in 50 % equimolar KOH/NaOH solutions with additives at 603 K (330°C) after several months of storage [106].

For very low extension rates (7.7×10^9 cm/s to 1.5×10^5 cm/s), both Incoloy® 800 as well as Inconel® 600 are susceptible to cracks in 50 % NaOH above 413 K (140°C) and at a potential above $U_H = 0.25$ V. The susceptibility to stress corrosion cracking reaches a maximum at 1.7×10^{-6} cm/s [107]. Both alloys exhibited intergranular cracks at low strain rates and transgranular cracks at faster strain rates.

Cathodic polarisation at $U_H = -0.8$ V suppressed the stress corrosion cracking of C-ring samples of Incoloy® 800 in 50 % NaOH at 418 K (145°C) [108].

For Incoloy® 800 after 20 h at a temperature of 298 K (25°C) and a loading of 130 N/mm², damage caused by stress corrosion cracking was observed after there had been leakages in the steam boiler of a sodium-cooled reactor when 16 % sodium hydroxide was present in the sodium. For 1 % sodium hydroxide, there were still no cracks after 100 h [109].

According to Figure 39, the anodic dissolution of chromium-nickel steels decreases with increasing nickel and chromium contents. A nickel addition of 8% in chromium steels already noticeably decreases the corrosion rate in 48% NaOH at temperatures between 413 K and 473 K (140 and 200°C) [110].

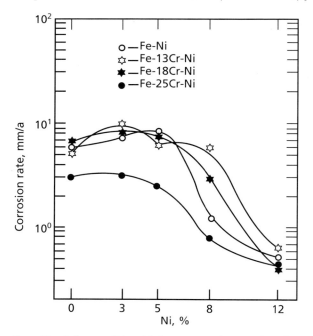

Figure 39: Influence of the nickel content on the corrosion of some CrNi steels in 48% NaOH at 413 K (140°C) (test duration 3 d) [110]

The risk of stress corrosion cracking of chromium-nickel steels in 50% NaOH at high temperatures (543 K to 603 K (270 to 330°C)) is reduced if both the chromium and the nickel contents in the steel are increased [111].

Electrochemically controlled investigations of stress corrosion cracking with tensile specimens of steel 304 L (mat. no. 1.4306) showed that the stress corrosion cracking in NaOH solutions is dependent on the potential and is caused by the preferential dissolution of the chromium in the transpassive region [112]. The behaviour of austenitic steels (composition in Table 43) was investigated by means of slow strain-rate tests in hot concentrated NaOH solutions as a function of the electrochemical potential.

C	Mn	Si	P	S	Cr	Ni	Mo	Fe
0.015	1.50	0.62	0.027	0.011	18.40	10.50	0.36	bal.

Table 43: Composition (mass %) of the investigated steel [112]

The tests were carried out in a solution with 40 g NaOH/100 g H$_2$O at 363 K (90°C). The same solution was used for the electrochemical investigations on the anodic dissolution behaviour of the steel and its individual alloying elements, Fe, Cr and Ni. The results indicate that both the occurrence of stress corrosion cracking as well as the type of crack are dependent on the potential (Table 44) and can be used directly for the dissolution and passivation of the steel.

Potential mV (Ag/AgCl)	Potential mV (NHE)	Elongation %	Type of crack
−1,000	−712	57.7	D
−825	−537	62.1	D
−650	−362	62.3	D
free corrosion	free corrosion	58.6	D
−200	+88	56.2	IG + D
−145	+143	36.7	IG + TG
0	+288	32.5	TG
+200	+488	37.1	IG + TG

D = ductile; IG = intergranular; TG = transgranular

Table 44: Results of the potential-controlled, slow strain-rate tests of steel 304L in 40 g NaOH/100 g H$_2$O at 368 K (95°C) [112]

The steel remains unsusceptible to cracks as long as the potential lies in the passive range. If the potential increases to higher values, there is crack formation. It was found that the cracks are related to the collapse of the primary passive layer and the active-passive transition to a second passive range at higher potentials. Investigations of the anodic behaviour of iron, chromium and nickel showed that the collapse of the primary passive layer is caused by a transpassive dissolution of the chromium component and that the subsequent second passivation is essentially caused by the nickel component which is the only one to remain passive.

Other investigations of the stress corrosion cracking of austenitic CrNi steels in NaOH solutions also showed that the highest susceptibility to SCC lies at potentials close to $U_H = 0.15$ V [113]. Studies of fracture-mechanics specimens of steels SAE 316 and SAE 310 in 3.35 M NaOH solutions (11.8 mass %) and in 12 M NaOH (32.4 mass %) at 365 K (92°C) gave the results presented in Figure 40.

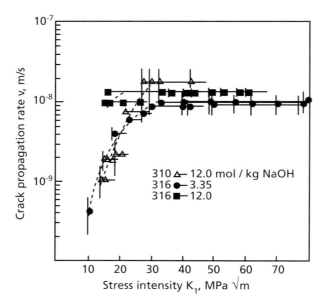

Figure 40: Stress corrosion cracking behaviour of steels SAE 316 and 310 in NaOH solutions at 365 K (92°C) and $U_H = 0.15$ V [113]

For an increase in the concentration from 3.35 to 12 M NaOH solutions, the crack propagation rate of steel 316 increased, particularly in the range of smaller stress intensities. In 12 M NaOH solutions, the steel 310 exhibited better behaviour than steel 316, although at higher stress intensities the differences are only very slight.

In summary, chromium-nickel steels with higher chromium and nickel contents are also resistant at higher temperatures in diluted and concentrated NaOH solutions; however, they can be susceptible to stress corrosion cracking and particularly to corrosion fatigue.

In NaOH melts, austenitic CrNi steels can be strongly attacked. This is given as the reason for the frequently observed damage on the furnace side of steam boilers that is caused by NaOH-containing deposits from waste gases on components made of these steels. Investigations of steels 304L and 310 show that both steels have corrosion rates of approx. 0.8 mm/a (31.5 mpy) in NaOH melts at 653 K (380°C) [114]. Figure 41 shows the linear time-dependence of the mass losses obtained for these steels.

Electrochemical investigations of carbon steel pipes clad with steel 304 showed that, under molten deposits of NaOH, steel 304 has a less noble potential than that of the unalloyed steel. In the case of cracks or discontinuities in the cladding, the unalloyed steel thus has cathodic protection [114].

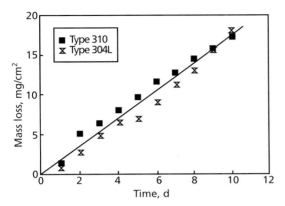

Figure 41: Mass losses of austenitic CrNi steels 304L and 310 in NaOH melts at 653 K (380°C) [114]

Austenitic chromium-nickel-molybdenum(N) steels

An addition of 2 to 5% molybdenum to austenitic CrNi steels can noticeably decrease the uniform surface corrosion in hot concentrated NaOH solutions. For comparison, Figure 42 shows the favourable effect of molybdenum on the behaviour of chromium-nickel steels and nickel-chromium alloys in boiling 50% NaOH [115]. Thus steel Incoloy® 901 (42% Ni, 13% Cr, 6% Mo) is more resistant than steel Incoloy® 825 (42% Ni, 21% Cr, 3% Mo, mat. no. 2.4858), in spite of its lower Cr content, because of its higher molybdenum content.

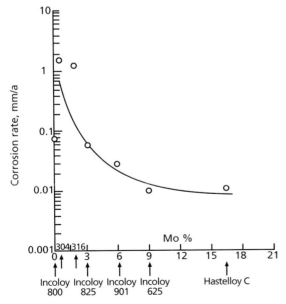

Figure 42: Influence of the molybdenum content on the corrosion of chromium-nickel steels and nickel alloys in boiling 50% NaOH [115]

If, in the case of an accident, moisture or water is able to penetrate into the sodium system in sodium-cooled reactors, then NaOH solutions are formed which can lead to stress corrosion cracking of the components made of austenitic steels at high temperatures. Hollow tensile specimens (see Figure 43) of austenitic CrNiMo steel type 316 (mat. no. 1.4401) were tested for stress corrosion cracking at various strain rates and at 823 K (550°C) in sodium with various additions of NaOH [116, 117].

The steel had the chemical composition given in Table 45. The samples were subjected to the heat treatments listed in Table 46 under an argon atmosphere.

C	Si	Mn	P	S	Cr	Ni	Mo	Cu	Ti	N	Al	Nb/Ta
0.06	0.58	1.85	0.024	0.019	17.10	13.70	2.35	0.09	0.01	0.032	0.023	< 0.01

Table 45: Chemical composition of the investigated steel [116, 117]

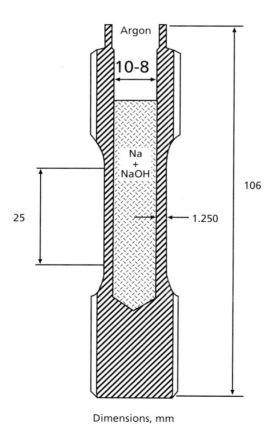

Dimensions, mm

Figure 43: Hollow tensile specimen subjected to stress corrosion cracking testing in Na/NaOH [116]

Designation	Duration	Temperature
ST	1 h	1,323 K (1,050°C) + air quenching
ST + PWHT	ST + 16 h	1,073 K (800°C)
ST + SENS	ST + 16 h	898 K (625°C)
ST + AGE	ST + 6,650 h	898 K (625°C)

Table 46: Heat treatments of the samples [116, 117]

The state ST corresponds to the as-delivered state (solution-annealed and quenched) of the steel and the subsequent treatment at 1,073 K (800°C) (PWHT) corresponds to stress-free annealing e.g. after welding. After treatment at 898 K (625°C) for 16 h (SENS), a sensitized state is expected, while tempering at 898 K (625°C) for 6,650 h simulates long-term ageing.

The hollow samples were filled with the sodium/sodium hydroxide combinations listed in Table 47 under exclusion of air and sealed under argon by autogenous welding.

Na	reactor pure	
NaOH	analysis pure	
Na + 10 % NaOH	NaOH = 10 % Na (vol.)	
Na + 5 % NaOH	NaOH = 5 % Na (vol.)	
Na + 1 % NaOH	NaOH = 1 % Na (vol.)	

Table 47: Composition of the Na/NaOH testing media [116, 117]

Table 48 gives a summary of the test results. The testing state of the material, the composition of the testing media, the applied stress and the measured endurance times to fracture of the samples are listed. For some of the samples, the tests were terminated after very long test periods even though no fracture had occurred.

In Na + 10 % NaOH with a loading of 100 MPa, all the specimens were susceptible to stress corrosion cracking. Particularly those samples with subsequent heat treatments had very short endurance times which were lower by a factor of 4 compared to the specimens in the as-delivered state (ST). In the stress-free annealed state (PWHT), an increase in the loading to 120 MPa only slightly decreased the endurance time, while for a reduction in the loading to 80 MPa, the sample did not fracture, even after 7,000 h.

The highest corrosion obviously occurred in pure NaOH, in which the PWHT specimen only showed an endurance time of 25 h at 100 MPa. In pure sodium or for lower NaOH concentrations, no fractures occurred even after very long test periods. The cracks were always intergranular.

Sodium Hydroxide

Material state	Testing medium	Loading MPa	Endurance time h
ST	Na + 10% NaOH	100	973
ST + PWHT	Na + 10% NaOH	100	179
ST + SENS	Na + 10% NaOH	100	282
ST + AGE	Na + 10% NaOH	100	213
ST + PWHT	Na + 10% NaOH	120	145
ST + PWHT	Na + 10% NaOH	80	> 7,000
ST + PWHT	NaOH	100	25
ST + PWHT	Na	100	9,935 (terminated)
ST + PWHT	Na + 1% NaOH	100	9,935 (terminated)
ST + PWHT	Na + 5% NaOH	100	15,168 (terminated)
ST	Na + 5% NaOH	100	15,168 (terminated)

Table 48: Results of the stress corrosion cracking tests on hollow tensile specimens of steel SAE 316 in Na/NaOH at 823 K (550°C) [116]

When compression-loaded components are endangered by stress corrosion cracking, the type of failure, whether it be a leakage or sudden fracture, plays a decisive role in the safety of the system. Stress corrosion cracking experiments were carried out on welded and notched capsules of austenitic CrNiMo steel SAE 316 in NaOH solutions at 573 and 623 K (300 and 350°C) [118]. Additionally, specimens were investigated which contained the following mixed materials in the welded transition joints:

316 / CrMo steel (10CrMo9-10)
316 / Inconel® 600
Inconel® 600 / CrMo steel

The composition of the 3 materials is given in Table 49. The appearance of the capsules is illustrated in Figure 44 and the type of welded transition joints are illustrated in Figure 45.

The capsules were subjected to a constant pressure or were loaded with increasing pressure. They were surrounded by an autoclave made of nickel-based material Nimonic® 75 (NiCr20Ti, mat. no. 2.4951) which contained the test solution. Figure 46 shows the typical pressure-time curves as obtained in the test under constant pressure, one in the case of a leakage with slow pressure decrease and one for a fracture with a sudden pressure decrease.

Both the crack propagation rates as well as the mode of failure are strongly dependent on the concentration of the NaOH solution, as shown in Figure 47.

Material DIN-Mat. no.	Abbreviation	% C	% Si	% S	% P	% Mn	% Cr	% Ni	% Mo	% Fe
316 1.4401	X5CrNiMo17-12-2	0.05	0.67	0.007	0.018	1.65	16.10	13.60	2.15	
10CrMo9-10 1.7380	G12CrMo9-10	0.12	0.39	0.005	0.024	0.45	9.14	0.21	0.97	
Inconel® 600 2.4816	NiCr15Fe	0.06	0.43			0.53	14.91	74.12	0.01	9.91

Table 49: Analysis values of materials in the transition joints [118]

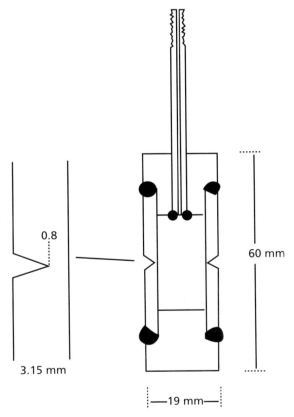

Figure 44: Notched capsule specimen [118]

While at low NaOH concentrations (0.3 M) the crack propagation rate were approx. 0.01 mm/h, they increased to 25 mm/h at high concentrations (20 M). Thus, a direct dependence of the crack propagation rate on the concentration of the caustic solution can be derived:

$$\text{Crack propagation rate (mm/h)} = \frac{c(\text{NaOH}) \ (\text{mol/l})}{16}$$

For NaOH concentrations of 5 M and greater, i.e. for high crack propagation rates, a leakage was always observed prior to fracture, while for concentrations below 5 M, the specimens fractured. The cracks were transgranular in every case. Typical pressure-time curves obtained in the tests carried out with increasing pressure are shown in Figure 48. These tests also exhibited "leakage before fracture" at high NaOH concentrations, and fracture at lower concentrations.

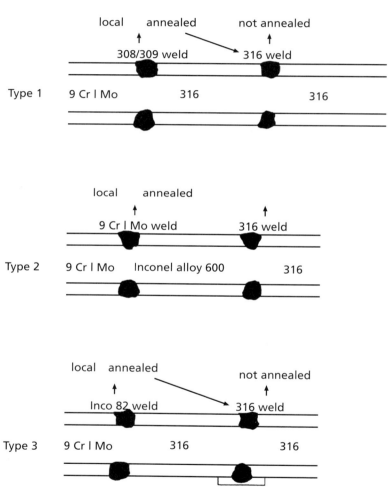

Figure 45: A schematic diagram showing the three different transition joints [118]

Austenitic chromium-nickel-molybdenum(N) steels | 355

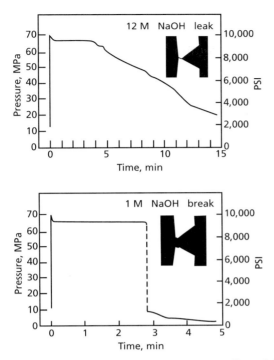

Figure 46: Typical pressure-time curves for the different failure modes of the capsules [118]

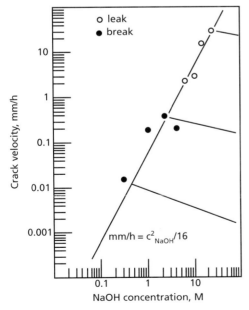

Figure 47: Influence of the NaOH concentration on the crack propagation rate and the mode of failure of notched specimens made of SAE 316 [118]

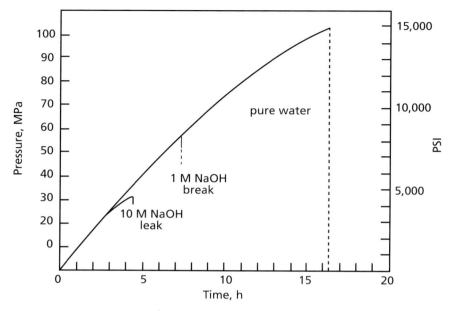

Figure 48: Pressure-time curves for capsules subjected to increasing pressure [118]

Another report describes the investigations on whether the presence of sulphides in the NaOH solution has an influence on the stress corrosion cracking behaviour [119]. For this, slow strain-rate tests were carried out on three austenitic CrNiMo steels and one austenitic-ferritic steel (compositions given in Table 50) in NaOH solutions with a concentration of 300 g/l NaOH at 473 K (200°C) in an autoclave made of a nickel-based alloy C-276 (Mat. No. 2.4819).

Steel grade	DIN-Mat. No.	% C	% Si	% Mn	% P	% S	% Cr	% Ni	% Mo	% N
UNS S31603	1.4404 (1.4435)	0.018	0.58	1.63	0.032	0.012	16.53	11.54	2.18	0.019
ASTM F138		0.025	0.59	1.80	0.022	0.002	17.29	13.99	2.73	0.074
UNS N08904	1.4539	0.009	0.59	1.51	0.028	0.002	19.61	24.54	4.12	0.077
UNS S31803	1.4462	0.012	0.48	0.87	0.021	0.001	22.15	5.62	3.09	0.19

Table 50: Chemical composition of the investigated steels [119]

Four strain rates were selected for the CERT tests:

$\dot{\varepsilon} = 5 \times 10^{-6}$ 1/s $\qquad \dot{\varepsilon} = 1 \times 10^{-6}$ 1/s
$\dot{\varepsilon} = 2 \times 10^{-7}$ 1/s $\qquad \dot{\varepsilon} = 42 \times 10^{-8}$ 1/s.

In some of the tests, 20 g/l sodium sulphide (Na$_2$S × 9H$_2$O) was added to the NaOH solution. The susceptibility to stress corrosion cracking was evaluated by means of the reduction of area Z measured on the specimens. The measured values of the four steels for each of the test conditions are given in Table 51. The given Z values are referenced against the values of reduction of area of the steels obtained in an inert media and are designated as Z_N.

Test conditions	S31603	F138	N08904	S31803
	Z_N	Z_N	Z_N	Z_N
NaOH: $\dot{\varepsilon} = 5 \times 10^{-6}$ 1/s	0.90	0.97	0.70	1.00
NaOH: $\dot{\varepsilon} = 1 \times 10^{-6}$ 1/s	0.72	0.77	0.20	0.92
NaOH: $\dot{\varepsilon} = 2 \times 10^{-7}$ 1/s	0.31	≈ 0	0.06	0.72
NaOH: $\dot{\varepsilon} = 4 \times 10^{-8}$ 1/s	0.09	≈ 0	≈ 0	0.32
NaOH +Na$_2$S: $\dot{\varepsilon} = 5 \times 10^{-6}$ 1/s	≈ 0	0.04	0.20	0.24
NaOH +Na$_2$S: $\dot{\varepsilon} = 1 \times 10^{-6}$ 1/s	≈ 0	≈ 0	0.03	0.08
NaOH +Na$_2$S: $\dot{\varepsilon} = 2 \times 10^{-7}$ 1/s	≈ 0	≈ 0	≈ 0	0.07

Table 51: Results of the CERT tests on steels given in Table 50 [119]

In both test solutions, the susceptibility to stress corrosion cracking increased at lower strain rates in all steels. The only specimens to show no signs of stress corrosion cracking were those of the austenitic-ferritic steel at the fastest strain rate of $\dot{\varepsilon} = 5 \times 10^{-6}$ 1/s. On the basis of these results, the following order can be drawn up for the increasing resistance in pure NaOH solutions:

N08904   S31603   F138   S31803

The susceptibility to stress corrosion cracking was noticeably increased by the presence of sulphides in the NaOH solution. All tested steels showed strong stress corrosion cracking in this solution, even at the highest extension rate.

The results of the investigations on the influence of the type, the concentration and the temperature of hydroxide solutions on the corrosion behaviour of Russian high-alloy stainless steels as well as a nickel grade are reported in [120]. The tested materials, the testing media and testing conditions are given in Table 52.

The influence of chloride or chlorate ions in the caustic solutions was also studied by adding the corresponding chloride up to the saturation limit at 298 K (25°C) or adding 0.05 to 5.0 g/l NaClO$_3$ to the solutions. Tests were carried out to determine the corrosion rates up to a test duration of 100 h. Stress corrosion cracking tests of welded samples were carried out for up to 255 h. The corrosion rates of the stainless steels that were determined in the various testing media are given in Figure 50 to Figure 53 (Nos. 1, 2, 3, 4 and 6).

No.	Material	Testing medium	Concentration
1	ferritic steel 15X25T	LiOH KOH NaOH	2.75–6.0 M 2.75–19.25 M 2.75–19.25 M
2	austenitic steel 12X18H10T	LiOH KOH NaOH	2.75–6.0 M 2.75–19.25 M 2.75–19.25 M
3	austenitic steel 10X17H13M2T	LiOH KOH NaOH	2.75–6.0 M 2.75–19.25 M 2.75–19.25 M
4	austenit.-ferrit. steel 08X22H6T	LiOH KOH NaOH	2.75–6.0 M 2.75–19.25 M 2.75–19.25 M
5	austenit.-ferrit. steel 08X21H6M2T	LiOH KOH NaOH	2.75–6.0 M 2.75–19.25 M 2.75–19.25 M
6	austenitic alloy 06XH28MΠT	LiOH KOH NaOH	2.75–6.0 M 2.75–19.25 M 2.75–19.25 M
7	NiCr alloy XH78T	LiOH KOH NaOH	2.75–6.0 M 2.75–19.25 M 2.75–19.25 M
8	Nickel Π2	LiOH KOH NaOH	2.75–6.0 M 2.75–19.25 M 2.75–19.25 M

Table 52: Materials, testing media and testing conditions for the investigations on the corrosion behaviour of various metals in hydroxide solutions (Temperatures 348–448 K (75–175°C) [120])

Austenitic chromium-nickel-molybdenum(N) steels

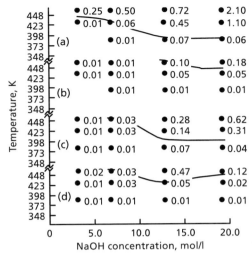

Figure 49: Corrosion rates (mm/a) and isocorrosion curves (0.1 mm/a)) for material no. 1 of Table 52 in the various testing solutions [120]
a = NaOH; b = KOH; c = NaOH + 5 g/l NaClO$_3$;
d = KOH + 5 g/l NaClO$_3$

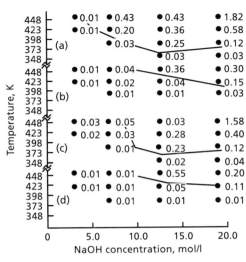

Figure 50: Corrosion rates (mm/a) and isocorrosion curves (0.1 mm/a)) for material no. 2 of Table 52 in the various testing solutions [120]
a = NaOH; b = KOH; c = NaOH + 5 g/l NaClO$_3$;
d = KOH + 5 g/l NaClO$_3$

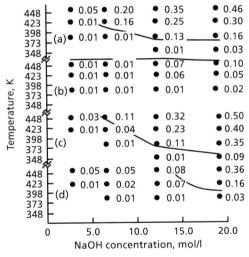

Figure 51: Corrosion rates (mm/a) and isocorrosion curves (0.1 mm/a) for material no. 3 of Table 52 in the various testing solutions [120]
a = NaOH; b = KOH; c = NaOH + 5 g/l NaClO$_3$;
d = KOH + 5 g/l NaClO$_3$

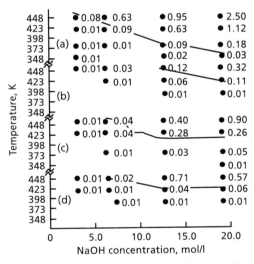

Figure 52: Corrosion rates (mm/a) and isocorrosion curves (0.1 mm/a) for material no. 4 of Table 52 in the various testing solutions [120]
a = NaOH; b = KOH; c = NaOH + 5 g/l NaClO$_3$;
d = KOH + 5 g/l NaClO$_3$

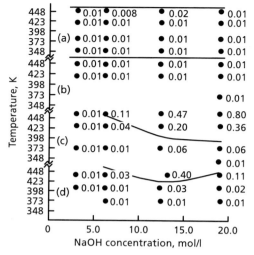

Figure 53: Corrosion rates (mm/a)) and isocorrosion curves (0.1 mm/a) for material no. 6 of Table 52 in the various test solutions [120]
a = NaOH
b = KOH
c = NaOH + 5 g/l NaClO$_3$
d = KOH + 5 g/l NaClO$_3$

Independent of the concentration or the temperature, the corrosion rates of the nickel-based materials (nos. 7 and 8) were less than 0.1 mm/a (3.94 mpy). In the lithium hydroxide solutions, the corrosion rates of all materials were less than 0.01 mm/a (0.39 mpy). Therefore, for stainless steels and nickel-based alloys, the corrosiveness of the hydroxide solutions increases in the following order:

LiOH < KOH < NaOH

The behaviour of the materials in NaOH and KOH solutions is clearly dependent on the test conditions. As the concentration and temperature increase, the corrosion rate of the stainless steels increases from 0.01 mm/a (0.39 mpy) to less than 2.5 mm/a (98.4 mpy) in NaOH and to 0.3 mm/a (11.8 mpy) in KOH. In 2.75 M KOH, the investigated steels are practically resistant, even at temperatures up to 453 K (180°C), whereas at the same concentration in NaOH the corrosion rates were approx. 0.1 mm/a (3.94 mpy). While chlorides do not have any effect, the oxidizing agent chlorate has a marked effect on the corrosion behaviour: the corrosion rate of some steels is lowered by a power of ten in the presence of 5 g/l chlorate.

As a further result of the stress corrosion cracking tests, the limits for the susceptibility to stress corrosion cracking are given for three of the investigated materials in Figure 54 in dependence on the temperature and concentration of the caustic solution.

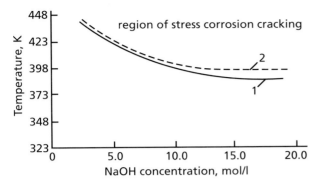

Figure 54: Regions of stress corrosion cracking for materials 2 and 3 (1) as well as 6 (2) in KOH and NaOH solutions [120]

The ferritic steel and the two austenitic-ferritic steels did not exhibit stress corrosion cracking in the testing media. However, the austenitic-ferritic steels were susceptible to selective corrosion of the austenite phase. The welded nickel-based materials exhibited a tendency to intergranular corrosion and crack formation.

Austenitic CrNiMoCu(N) steels

The high-alloy material X1CrNiMoCuN33-32-1 (Mat. No. 1.4591, alloy 33, Nicrofer® 3033) composition given in Table 53 is not only very resistant in strongly oxidizing media, various acids and chloride solutions, it is also very resistant to NaOH solutions [121, 122].

% C	% Mn	% Si	% P	% S	% Cr	% Ni	% Mo	% Cu	% N	% Fe
≤ 0.015	≤ 2.0	≤ 0.5	≤ 0.020	≤ 0.010	31.0–35.0	30.0–33.0	0.5–2.0	0.3–1.2	0.35–0.60	bal.

Table 53: Chemical composition of alloy 33 [121]

While for the austenitic steels of type SAE 316 Ti or X1CrNiMoN25-25-2 (1.4466), which are frequently used for tanks, vessels or pipelines for the storage or transport of NaOH solutions, the temperature of the medium must not exceed 363 K (90°C) in order to avoid a drastic increase in the corrosion rate. The values given in Table 54 demonstrate that alloy 33 has still quite good resistance in NaOH solutions of 25 or 50%, even at the corresponding boiling temperature [122].

Steel	25% NaOH			50% NaOH			
	348 K (75°C)	373 K (100°C)	377 K (104°C) boiling point	348 K (75°C)	373 K (100°C)	398 K (125°C)	419 K (146°C) boiling point
316 Ti	< 0.01 (< 0.39)	0.12 (4.72)	0.63 (24.8)	0.08 (3.15)	0.35 (13.8)	1.60 (63.0)	7.99 (315)
1.4466	< 0.01 (< 0.39)	0.03 (1.18)	0.02 (0.79)	< 0.01 (< 0.39)	< 0.01 (< 0.39)	0.26 (10.2)	1.35 (53.2)
alloy 33	< 0.01 (< 0.39)	< 0.01 (< 0.39)	< 0.01 (< 0.39)	< 0.01 (< 0.39)	< 0.01 (< 0.39)	< 0.01 (< 0.39)	< 0.01 (< 0.39)

Table 54: Corrosion rates (mm/a (mpy)) of austenitic steels in NaOH solutions under various conditions. Test duration 28 days [122]

Special iron-based alloys

As shown in Table 55, only higher nickel contents in iron-nickel alloys lead to a noticeable decrease in the corrosion rates in NaOH solutions [123].

Nickel content %	Corrosion rate mm/a (mpy)
0	2.3 (90.6)
5	1.2 (47.2)
15	0.75 (29.5)
20	0.15 (5.9)
30	0.01 (0.39)

Table 55: Corrosion rates of iron-nickel alloys in 50% to 65% NaOH at room temperature [123]

The corrosion resistance of manganese-aluminium steels with approx. 20 to 30% Mn and 7 to 10% Al is noticeably lower in 1 M NaOH at 298 K (25°C) than the common austenitic CrNi steels [124].

Bibliography

[1] Hubbe, M. A.
Polarization Resistance Corrosivity Test with a Correction for Resistivity
Brit. Corrosion J. 15 (1980) p. 193

[2] Neufeld, P.; Bromley, A. F.
Anodic Passivation of Mild Steel in Hot, Concentrated NaOH Solution
Brit. Corrosion J. 7 (1972) p. 285

[3] Mann, G. W.
History and Causes of On-Load Waterside Corrosion in Power Boilers
Brit. Corrosion J. 12 (1977) p. 6

[4] Burakov, M. R.; Guselnikov R. G.; Lebedev, A. N.
Untersuchung der Korrosion von Kohlenstoffstahl in heißen, konzentrierten Alkalien
Metallschutz Moskau 8 (1972) p. 452

[5] Wensley, D. A.; Charlton, R. S.
Corrosion Studies in Kraft White Liquor: Potentiostatic Polarisation of Mild Steel in Caustic Solutions Containing Sulfur Species
Corrosion NACE 36 (1980) p. 385

[6] Charlton, R. S.
A Look at Corrosion in Pulpmills
Canad. Chem. Process 62 (1978) 10/11, p. 46

[7] Berk, A. A.; Waldeck W. F.
Caustic Danger Zone
Chemical Engineering, 57 (1950) p. 235–238

[8] Herbsleb, G.
Korrosionsschutz von Stahl
(Corrosion protection of steel) (in German), p. 24
Verlag Stahleisen, Düsseldorf (1997)

[9] Ashbaugh, W. G.
Stress Corrosion Cracking of Process Equipment
Chem. Eng. Progr. 66 (1970) 10, p. 44

[10] Friedrich, H. J.; Schönewald, F.
Spannungsrißkorrosion an einer Natronlaugeleitung
Korrosion Dresden 10 (1979) p. 166

[11] Bohnenkamp, K.
Über die Spannungsrißkorrosion weicher Stähle in konzentrierter Natronlauge
(Stress corrosion cracking of mild steel in concentrated caustic soda) (in German)
Archiv Eisenhüttenwesen, 39 (1968) 5, p. 361–368

[12] Mazille, H.; Uhlig, H. H.
Effect of Temperature and Some Inhibitors on Stress Corrosion Cracking of Carbon Steels in Nitrate and Alkaline Solutions
Corrosion NACE 26 (1972) p. 427

[13] Jin, Z.; Zuo, J.; Lu, Y.
Mechanism of SCC for mild steel in 33% NaOH solution at 115°C
Conference: Corrosion Control – 7th APCCC.
Vol. 1, China (1991)
Publ.: International Academic Publishers, Xizhimenwai Dajie, Beijing Exhibition Centre, Beijing 100044, People's Republic of China (1991)

[14] Langenscheid, G.
Beitrag zur Korrosion weicher, unlegierter Stähle in Natronlauge
steel research 60 (1989) 6, p. 281–287

[15] Krautschik, H. J.; Bohnenkamp, K.; Grabke, H. J.
Einflüsse des Phosphors auf die interkristalline Spannungsrißkorrosion von Kohlenstoffstählen
Werkst. Korros. 38 (1987) 3, p. 103–110

[16] Vereshchagin, K. I.; Oskolkova, L. K.
Die Entstehung und Ausbreitung von Rissen in den Stählen St-3 und 16-GS bei der Untersuchung in alkalischen Medien
Phys.-chem. Mech. d. Werkstoffe, Kiew 11 (1975) 1, p. 106

[17] Vereshchagin, K. I.; Nechaev, V. A.; Podlipskij L. A.; Rubenchik, Yu. I.; Karpenko, G. V.
Erhöhung der Beständigkeit von geschweißten Apparaten aus St-3 und 16-GS gegen Spannungsrißkorrosion
Phys.-chem. Mech. d. Werkstoffe, Kiew 8 (1972) 2, p. 3

[18] Artemev, V. I.; Seregin, V. I.; Zholobova, E. P.; Belgaev, V. P.
Spannungsrißkorrosion von Kohlenstoff-Stahl in Alkalialuminat-Lösungen
Metallschutz Moskau 15 (1979) p. 62

[19] Breault, R.; Simard, A.
Conference: Light Metals 1992, San Diego, California, USA, 1.–5. Mar. 1992
Publ.: The Minerals, Metals & Materials Society, 420 Commonwealth Dr., Warrensale, Pennsylvania 15086, USA (1992) p. 101–107

[20] Liu, S.; Zhu, Z.; Guan, H.; Ke, W.
Stress corrosion cracking of pressure vessel steels in high-temperature caustic aluminate solutions
Metallurgical and Materials Transactions A 27A (1996) 5, p. 1327–1331

[21] Le, H. H.; Ghali, E.
Stress corrosion cracking of carbon steel in caustic aluminate solutions of the Bayer Process
Corrosion Science 35 (1993) p. 435–442

[22] Le, H. H.; Ghali, E.
Active-passive behaviour and stress corrosion cracking of A516 steel in Bayer solution
J. Appl. Electrochem. 19 (1989) 3, p. 368–376

[23] Le, H. H.; Ghali, E.
The electrochemical behaviour of pressure vessel steel in hot Bayer solution as related to the SCC phenomenon
Corrosion Science 30 (1990) p. 117–134

[24] Le, H. H.; Ghali, E.
Slow strain rate and constant load tests of A 285 and A 516 steels in Bayer solutions
J. Appl. Electrochem. 22 (1992) 4, p. 396–403

[25] Sriram, R.; Tromans, D.
Stress corrosion cracking of carbon steel in caustic aluminate solutions – Slow strain rate studies
CORROSION 41 (1985) 7, p. 381–385

[26] Sriram, R.; Tromans, D.
Stress corrosion cracking of carbon steel in caustic aluminate solutions – Crack propagation studies
Materials Transactions A 16A (1985) 5, p. 979–986

[27] Arsenault, B.; Ghali, E.
Etude de l'influence du soudage sur la resistance des recipients sous pression a la fragilisation caustique par methodes electrochimiques et essais de corrosion a taux de deformation constant
Conference: 8th European Congress of Corrosion. Nice, France, Nov. (1985) Vol. 2, p. 19–21
Centre Francais de la Corrosion, Societe de Chimique Industrielle, rue Saint-Dominique, F75007 Paris, France, 1985

[28] Arsenault, B.; Ghali, E.
Stress corrosion cracking of pressure vessel welded carbon steels
International Journal of Pressure Vessels and Piping, 45 (1991) 1, p. 23–41

[29] Bombara, G.; Bernabai, U.
The Caustic Stress Corrosion Cracking of High-Strength Steel Wire
Corrosion Sci. 21 (1981) p. 409

[30] Pöpperling, R.; Schwenk, W.; Venkateswarlu, J.
Wasserstoffinduzierte Spannungsrißkorrosion von Stählen durch dynamisch-plastische Beanspruchung in Promotor-freien Elektrolytlösungen
Werkstoffe u. Korrosion 36 (1985) p. 389

[31] Frömberg, M.
Schäden durch interkristalline Spannungsrißkorrosion an Natronlaugerohrleitungen
Korrosion, 20 (1989) 6, p. 322–330

[32] Tromans, D.; Ramakrishna, S.; Hawbolt, S.
Stress corrosion cracking of ASTM A516 steel in hot caustic sulfide solutions – Potential and weld effects
CORROSION 42 (1986) 2, p. 63–70

[33] Shah, R. S.; Desai, C. S.
Effect of Hydrogen Peroxide on Corrosion of Brass by Sodium Hydroxide
J. Electrochem. Soc. India 25 (1976) p. 181

[34] Singbeil, D.; Garner, A.
Anodic protection tot prevent the stress corrosion cracking of pressure vessel steels in alkaline sulfide solutions
Materials Performance 26 (1987) 4, p. 31–36

[35] Singbeil, D.; Tromans, D.
Caustic stress corrosion cracking of mild steel
Metallurgical Transactions A 13A (1982) 6, p. 1091–1098

[36] DIN 50915: Prüfung von unlegierten und niedriglegierten Stählen auf Beständigkeit gegen interkristalline Spannungsrißkorrosion in nitrathaltigen Angriffsmitteln – Geschweißte und ungeschweißte Werkstoffe.
Beuth-Verlag, Berlin Okt. (1993)

[37] Diekmann, H.; Drodten, P.; Kuron, D.; Herbsleb, G.; Pfeiffer, B.; Wendler-Kalsch, E.
Zum Spannungsrißkorrosionsverhalten niedriglegierter Stähle in Nitrat- und Carbonatlösungen sowie in Natronlauge.
Stahl u. Eisen 103 (1983) p. 895–901

[38] Drodten, P.; Herbsleb, G.; Kuron, G.; Savakis, D.; Wendler-Kalsch, E.
Potentialabhängigkeit der Korrosion Mo-freier und Mo-haltiger Stähle in Calciumnitrat-Lösung.
Werkstoffe u. Korrosion 42 (1991) p. 128–138

[39] Drodten, P.; Herbsleb, G.; Kuron, G.; Savakis, D.; Wendler-Kalsch, E.
Spannungsrißkorrosion Mo-freier und Mo-haltiger Stähle in Calciumnitrat-Lösung und Natronlauge unter zeitlich konstanter Kraft und unter CERT-Bedingungen.
Werkstoffe u. Korrosion 42 (1991) p. 473–489

[40] Drodten, P.; Herbsleb, G.; Kuron, G.; Savakis, D.; Wendler-Kalsch, E.
Der Einfluß mechanischer Wechselbelastungen auf die Beständigkeit Mo-freier und Mo-haltiger Stähle gegen Spannungsrißkorrosion in Calciumnitrat-Lösung und Natronlauge
Werkstoffe u. Korrosion 42 (1991) p. 576–583

[41] Drodten, P.; Herbsleb, G.; Kuron, G.; Savakis, D.; Wendler-Kalsch, E.
Spannungsrißkorrosion von niedriglegierten Stählen in Calciumnitratlösung und Natronlauge.
Stahl u. Eisen 111 (1991) p. 117–124

[42] Drodten, P.; Schwenk, W.; Sussek, G.
Einflüsse auf die Spannungsrißkorrosion niedriglegierter Stähle
Stahl und Eisen 120 (2000) 1, p. 55–64

[43] Hoar, T. P.; Jones, R. W.
The Mechanism of Caustic Cracking of Carbon Steel – I. Influence of Electrode Potential and Film Formation
Corrosion Sci. 13 (1973) p. 725

[44] ASM International, Member/Customer Service Center, Materials Park, OH (1994)
Corrosion in the Chemical processing industry: corrosion by alkalies and hypochlorite
Corrosion in the petrochemical industry
p. 204–210

[45] Isfort, D.
Moderne Alkalichlorid-Elektrolyse nach dem Hg-Verfahren. – Ausrüstung und Konstruktionsmaterialien der Produktaufbereitungsanlagen
Chem.-Anl. Verfahren (1972) 9, p. 65

[46] Maslov, V. A. et al.
Korrosionsbeständigkeit niedriglegierter Gußeisen in alkalischen Medien bei erhöhter Temperatur, Russ. Cast. Prod. (1968) 7, p. 317.
Giesserei 58 (1971) p. 260

[47] Surendranathan, A. O.; Hebbar, K. R.; Sudhaker Nayak, H. V.
Aqueous corrosion behaviour of ductile iron ans ductile iron containing 1.5Ni-0.3Mo
Journal of the Electrochemical Society of India, Vol. 45, 2 (1996) 61–69
ident. in: Transactions of the Indian Institute of Metals Vol. 50, 2–3 (1997) p. 191–199

[48] Donaldson, E. G.
Der Einsatz von Meehanite-Gußeisen in Chemieanlagen
Chem.-Techn. 4 (1975) 4, p. 127

[49] Rungta, R.; Leis, B. N.
Experimental methods for the evaluation of environmental assisted cracking of steel in caustics
in: Conference: Environment-Sensitive Fracture: Evaluation and Comparison of Test Methods, Gaithersburg, Maryland, USA, 26–28 April 1982
ASTM, 1916 Race St., Philadelphia, Pennsylvania 19103 USA, 1984, p. 341–367

[50] Pishchulin, V. N.; Kosintsev, V. I.
Korrosion der Elektroden- und Bauwerkstoffe beim Eindampfen von Natriumhydroxidlösungen in direkt elektrisch beheizten Apparaten
Izv. Tomsk. Politekhn. In.-Ta, (1976) 275, p. 90

[51] Yukalov, I. N.
Korrosionsbeständigkeit austenitischer Nickel-Kupfer-Chrom-Gußeisen
Russ. Cast. Prod. (1968) p. 7
Giesserei 58 (1971) p. 260.

[52] Nickel, O.
Eigenschaften der austenitischen Gußeisenwerkstoffe
Konstruieren & Giessen (1976) 3, p. 9.

[53] Donati, J. R.; Grand, M.; Spiteri, P.
Korrosionsvorgänge während des Ablaufs von Mikroreaktionen zwischen Natrium und Wasser innerhalb von Rissen, die durch Dampferzeugerrohre von schnellen Brutreaktoren hindurchlaufen
J. Nucl. Mat. 54 (1974) p. 217

[54] Cowen, H. C.; Thorley, A. W.
Caustic cracking of 2¼CrMo steel
Conference: Topical Conference on Ferritic Alloys for Use in Nuclear Energy Technologies, Snowbird, Utah, USA, 19.–23. June 1983
Publ.: The Metallurgical Society/AIME, 420 Commonwealth Dr., Warrendale, Pa. 15086, USA, (1984) p. 51–56

[55] Wozadlo, G. P.; Roy, P.
Caustic stress corrosion behaviour of 2.25 Cr-1Mo steels
Conference: Topical Conference on Ferritic Alloys for Use in Nuclear Energy Technologies, Snowbird, Utah, USA, 19.–23. June 1983
Publ.: The Metallurgical Society/AIME, 420 Commonwealth Dr., Warrendale, Pa. 15086, USA, (1984) p. 57–64

[56] McIlree, A. R.; Pizzo, P. P.; Indig, M. E.
Corrosion Fatigue of 2¼Cr-1Mo Steel in Caustic at 316 C
Mat. Performance 19 (1980) 3, p. 30

[57] Indig, M. E.
Stress Corrosion Studies for 2¼ Chromium-1 Molybdenum Steel
Proc. Conf. Ferritic Steels for Fast Reactor Steam Generators
London 1977, 2 (1978) p. 408

[58] Lyle jr. F. F.; Burghard jr. H. C.
Cracking of Low Pressure Turbine Rotor Discs in US Nuclear Power Plants
Mat. Performance 21 (1982) 11, p. 35

[59] Lyle jr. F. F.;
Stress corrosion cracking characterization of 3.5 NiCrMoV low pressure turbine rotor steels in NaOH and NaCl solutions
Corrosion, 39 (1983) 4, p. 120–131

[60] Shalvoy, R. S.; Duglin, S. K.; Lindinger, R. J. T
The Effect of Turbine Steam Impurities on Caustic Stress Corrosion Cracking of NiCrMoV Steels
Corrosion NACE 37 (1981) p. 491

[61] Burstein, G. T.; Woodward, J.
Effects of Segregated Phosphorus on Stress Corrosion Cracking Susceptibility of 3Cr-0.5Mo Steel
Metal Sci. 17 (1983) p. 111

[62] Burstein, G. T.; Woodward, J.
Aspects of the Stress Corrosion Cracking of Cr-Mo Steels
8. Intern. Congress on Metallic Corrosion, Mainz (1981) Vol 1, p. 460

[63] Zheng, When-long
An Investigation on Fracture behaviour of CrNiMo low alloy steels in NaOH solution
Conference: Microstructure and Mechanical Behaviour of Materials, Vol. I, X'ian, Republic of China, 21–24 Oct. 1985
Publ.: Engineering Materials Advisory Services Ltd., 339 Halesowen Rd., Cradley Heath, Warley, West Midlands B64 6Ph, UK, (1986) p. 353–360

[64] Poulson, B.
Caustic Cracking of 9Cr-1 Mo Steel at 300 °C
Corrosion Sci. 22 (1982) p. 473

[65] Woolsey, I. S.
Corrosion Resistance of Ferritic Steels in Alkaline Solutions
Proc. Conf. Ferritic Steels for Fast Reactor Steam Generators
London 1977, 2 (1978) p. 403

[66] El-Sayed, H. A.; El-Sobki, K. M.; Gouda, V. K.
Stress Corrosion Behavior of 410 Stainless Steel in Boiling 70% NaOH Solution
Surface Technol. 14 (1981) 245

[67] McCord, T. G.; Bussert, B. W.; Curran, R. M.; Gould, G. C.
Stress Corrosion Cracking of Steam Turbine Materials
Mat. Performance 15 (1976) 2, p. 25

[68] Bandyopadhyay, N.; Briant, C. L.
The effect of phosphorus on intergranular caustic cracking of NiCr steel
Corrosion NACE 39 (1982) 3, p. 125–129

[69] Pyun, S.-I.; Kim, J.-T.; Lee, S.-M.
Effects of phosphorus segregated at grain boundaries on stress corrosion crack initiation and propagation of rotor steel in boiling NaOH solution
Materials Science and Engineering A 148 (1991) p. 93–99

[70] Bandyopadhyay, N.; Briant, C. L.
Caustic stress corrosion cracking of low alloy iron base materials
CORROSION 41 (1985) 5, p. 274–580

[71] Briant, C. L.
On the role of phosphorus in the caustic stress corrosion cracking of low alloy steels
Corrosion Science 29 (1989) 1, p. 53–68

[72] Bandyopadhyay, N.; Briant, C. L; Hall, E. L.
The effect of microstructural changes on the caustic stress corrosion cracking resistance of NiCrMoV rotor steels
Metallurgical Transactions A 16A (1985) 7, p. 1333–1344

[73] Ohhashi, T.; Hasegawa, H.; Iwadate, T.
Effect of grain boundary phosphorus segregation on the SCC susceptibility of NiCrMoV steels in caustic wet environment
Transactions Iron Steel Inst. Japan 26 (1986) 12, p. 375

[74] Rechberger, J.; Tromans, D.; Mitchell, A.
Stress corrosion cracking of conventional and super-clean 3.5NiCrMoV rotor steels in simulated condensates
CORROSION 44 (1988) 2, p. 79–87

[75] Skeldon, P.
Environment-assisted cracking of 2¼Cr-1Mo-steel in fused sodium hydroxide at 623 K, 1 atm –
I. Electrochemistry in relation to stress corrosion cracking
Corrosion Science 26 (1986) 7, p. 485–506

[76] Skeldon, P.
Environment-assisted cracking of 2¼Cr-1Mo-steel in fused sodium hydroxide at 623 K, 1 atm – II. Results of slow strain rate and constant strain tests
Corrosion Science 26 (1986) 7, p. 507–523

[77] Sakaki, T.; Sakiyama, K.
Effects of $NaClO_3$ and NaCl on the Corrosion Behaviors of Iron-Chromium Alloys in Hot Concentrated Caustic Soda Solution
J. Japan Inst. Metals 43 (1979) p. 1186

[78] Sakaki, T.; Simizu, Y.; Sakiyama, K.
Effects of $NaClO_3$ and NaCl on the Corrosion Behavior of Iron-Chromium-Nickel Alloys in Concentrated Caustic Soda Solutions
J. Japan lnst. Metals 45 (1981) p. 296

[79] Sakaki, T.; Sakiyama, K.
Corrosion of Iron-Chromium Alloys in Hot Concentrated Caustic Soda Solution
J. Japan lnst. Metals 43 (1979) p. 527

[80] Steigerwald, R. F.
New Molybdenum Stainless Steels for Corrosion Resistance: A Review of Recent Developments,
Mat. Performance 13 (1974) 9, p. 9

[81] Davison, R. M.; Steigerwald, R. F.
The New Ferritic Stainless Steels
Metal Progress 115 (1979) 6, p. 40

[82] Wilson, I. L.; Pement, F. W.; Aspden, R. G.
Stress corrosion studies on some stainless steels in elevated temperature aqueous environments
Corrosion NACE 34 (1978) 9, p. 311–320

[83] Shalaby, H. M.; Begley, J. A.; Macdonald, D. D.
Fatigue crack initiation in 403 stainless steel in simulated steam cycle environments: hydroxide and silicate solutions
British Corrosion Journal 29 (1994) 1, p. 43–52

[84] Salinas-Bravo, V. M.; Gonzales-Rodriguez, J. G.
Stress corrosion cracking susceptibility of 17-4PH turbine steel in aqueous environments
British Corrosion Journal 30 (1995) 1, p. 77–79

[85] Gonzales-Rodriguez, J. G.; Salinas-Bravo, V. M.; Martinez-Villafane, A.
Hydrogen embrittlement of type 410 stainless steel in sodium chloride, sodium sulfate and sodium hydroxide environments at 90°C
Corrosion, 53 (1997) 6, p. 499–504

[86] Naumann, F. K.; Spies, F.
Verzunderte und deformierte Brennerdüsen
Prakt. Metallographie 17 (1980) p. 297

[87] Aleksandrov, A. G.; Lazebnov, P. P.
Effects of Ferrite on Corrosion Resistance of Austenitic-Ferritic Weld Deposits Avtom.
Svarka (1983) 10, p. 70

[88] Horn, E. M.; Savakis, S.; Schmitt, G.; Lewandowski, I.
Performance of Duplex Steels in Caustic Solutions
Conference: Duplex Stainless Steels '91, Vol. 2, Bourgogne, France, 28.–30. Oct. 1991
Publ.: Les Editions de Physique, Avenue du Hoggar, Zone Industrielle de Courtaboeuf, B.P. 112, F-91944 Les Ulis Cedex A, France, 1922, 1111–1119
auch: Korrosionsverhalten nichtrostender ferritisch-austenitischer Stähle in Natronlauge
Werkstoffe und Korrosion, 42 (1991) 10, p. 511–519
auch: Horn, E. M.; Schmitt, G.
Corrosion behaviour of stainless ferritic-

austenitic steels in sodium hydroxide
Werkstoffe und Korrosion, 41 (1990) 6,
p. 365–367

[89] Turnbull, A.; Griffiths, A.; Reid, T.
Corrosion and stress corrosion cracking of duplex stainless steels in caustic solutions at elevated temperatures
Stainless Steel World, Dec. (1999) p. 63–67

[90] Standard practice for laboratory immersion corrosion testing of materials; G31-72
Standard guide for laboratory immersion corrosion testing of metals; ASTM/NACE G 31a
American Society for Testing and Materials, PA, USA

[91] Rondelli, G.; Vicentini, B.; Brunella, M. F.; Cigada, A.
Stress corrosion cracking of austenitic and duplex stainless steels in caustic environments
Conference: CASS 90: Corrosion – Air, Sea, Soil, Auckland, New Zealand, 19.–23. Nov. 1990, paper 37
Publ.: Australasian Corrosion Association Inc., New Zealand Branch, P.O. Box 5961, Auckland, New Zealand, 1990
auch in: Werkst. Korros.44 (1993) p. 57–61

[92] Ward, I.; Berlung, G.; Norberg, P.
Caustic and chloride stress corrosion cracking of austenitic and Duplex steels in the alumina and paper industries
5th Asian Pacific Corrosion Control Conference, Melbourne, Australia (1987)

[93] Broschüre "Korrosionsbeständigkeit austenitischer Chrom-Nickel-Stähle"
Nickel-Informationsbüro, Düsseldorf (1961)

[94] Lew, S.
Corrosion of Stainless Steel Fibres
Anti-Corrosion 16 (1969) 7, p. 17

[95] Ahmad, Z.; Davami, P.
The Corrosion Resistance of 18Cr-37Ni Austenitic Steel in Alkaline, Acid and Neutral Solutions
Japan Soc. Corrosion Eng. 29 (1980) p. 595

[96] Condylis, A.; Bayon, F.; Chateau D.
Resultats d'essais de corrosion et de polution d'aciers inoxydables austenitiques dans l'industrie alimentaire
Rev. Metallurgie (1976) p. 787

[97] Wilson, I. L.; Pement, F. W.; Aspden, R. G.
Effect of Alloy Structure, Hydroxide Concentration and Temperature on the Caustic Stress Corrosion Cracking of Austenitic Stainless Steels
Corrosion NACE 30 (1974) p. 139

[98] Subrahmanyam, D. V.; Agrawal, A. K.; Staehle, R. W.
The Stress Corrosion Cracking Behavior of Type 304 Stainless Steel in Boiling 50% NaOH Solution
Proc. 7. Intern. Congr. Metallic Corrosion, Rio de Janeiro (1978) p. 783

[99] Anonym, Anti-Corrosion Methods and Materials. 4: Forms of Corrosion
Anti-Corrosion 19 (1972) 12, p. 9

[100] Baker, C. D.; Jesernig, W. V.; Huether, C. H.
Waste Water Treatment: Recovering para-Cresol from Process Effluent
Chem. Eng. Progress 69 (1973) 8, p. 77

[101] Nicodemi, W.
Gli acciai inossidabili; Corrosion sotto tensione
Acciaio Inossidabile 55 (1988) 4, p. 24–29

[102] Marsh, G. P.; Lawrence, P. F.; Carney, R. F. A.; Perkins, R.
The Corrosion Behaviour of Alloy 800 in Concentrated Sodium Hydroxide Solutions at Moderate Temperatures EUROCORR 6.
European Congr. Metallic Corrosion (1977) p. 505

[103] Blanchet, J.; Coriou, H.; Grall, L.; Mahieu, C.
The Stress Corrosion of Alloy 800 in Contaminated Water at High Temperature
Mém. Sci. Rev. Métallurgie 75 (1978) p. 237

[104] Komp, M. E.; Mathay, W. L.
Steels for the Pulp and Paper Industry
Mat. Performance 16 (1977) 6, p. 22

[105] Berge, Ph.; Donati, J. R.; Prieux, S., Villard, D.
Caustic Stress Corrosion of Fe-Cr-Ni Austenitic Alloys
Corrosion NACE 33 (1977) p. 425

[106] Wilson, I. L. W.; Pement, F. W., Aspden, R. G., Begley, R. T.
Caustic Stress Corrosion Behavior of Fe-Ni-Cr Nuclear Steam Generator Tubing Alloys
Nuclear Technol. 31(1976) p. 70

[107] Takano, M.; Teramoto, K.; Nakayama, T.; Yamaguchi, H.
Extremely Slow Strain Rate Stress-Corrosion Testing Machine and Some Experimental Results
J. Iron Steel Inst. Japan 65 (1979) p. 212

[108] Hickling, J.; Wieling, N.
Electrochemical Aspects of the Stress Corrosion Cracking of Fe-Cr-Ni Alloys in Caustic Solutions
Corrosion Sci. 20 (1980) p. 269

[109] Cappelaere, M.; Dixmier, J.; Sannier, J.; Coriou, H.
The Behavior of Iron-Chromium-Nickel Alloy-800 in Sodium Contaminated with Sodium Hydroxide
Mém. Sci. Rev. Metall. 74 (1977) p. 719

[110] Sakaki, T.; Simizu, Y.; Sakiyama, K.
Corrosion of Iron-Chromium-Nickel-Alloys in Hot Concentrated Caustic Soda Solution
J. Japan Inst. Metals 44 (1980) p. 582

[111] McIlree, A. R.; Michels, H. T.
Stress Corrosion Behavior of Fe-Cr-Ni and Other Alloys in High-Temperature Caustic Solutions
Corrosion NACE 33 (1977) p. 60–67

[112] Zheng, J. H.; Bogaerts, W. F.
Transpassive chromium dissolution – Interaction with stress corrosion cracking of austenitic stainless steel in caustic solution
in: Conference: Second International Conference on Corrosion-Deformation Interactions. CDI96', Nice France, 24–26 Sept. 1996
Institute of Materials, 1 Carlton House Terrace, London, SW1Y 5DB, UK, 1997, p. 104–116

[113] Crowe, D. C.; Tromans, D.
Caustic cracking behaviour and fractography of 316 and 310 stainless steels
Conference: Metallography and Corrosion, Calgary, Alberta, Canada, 25–26 July 1986
Publ.: National Association of Corrosion Engineers, 1400 South Creek Dr., Houston, Texas, 77084, USA (1986)

[114] Tran, H.; Katiforis, N. A.; Utigard, T. A.; Barham, D.
Recovery boiler air-port corrosion, Part 3, Corrosion of composite tubes in molten sodium hydroxide
Tappi Journal, 78 (1995) 9, p. 111–117

[115] Scarberry, R. C.; Graver, D. L., Stephens, C. D.
Alloying for Corrosion Control, Properties and Benefits of Alloy Materials
Mat. Protection 6 (1967) 6, p. 54

[116] Horton, C. A. P.; Crouch, A. G.
Corrosion and stress aided cracking in fast reactor steels exposed to sodium/sodium hydroxide mixtures
in Liquid Metal Engineering and Technology, Vol. 3; Oxford, UK, 9.–13. Apr. 1984
British Nuclear Energy Society, 1–7 Great George St. Kondon SW1P 3AA, UK, (1985) p. 279–286

[117] Horton, C. A. P.
Further observations on stress aided cracking in fast reactor steels exposed to sodium/sodium hydroxide mixtures
LIMET'88, Fourth International Conference on Liquid Metal Engineering and Technology, Vol. 3. Avignon, France, 17.–21. Oct. 1988; Pp 2 Paper No. 537

[118] Poulson, B.
Stress corrosion cracking of type 316 stainless steel in caustic solutions: Crack velocities and leak before break considerations
Corrosion Science, 33 (1992) p. 1541–1556

[119] Rondelli, G.; Vicentini, B.; Sivieri, E.
Stress corrosion cracking of stainless steels in high temperature caustic solutions
Corrosion Science, 39 (1997) 6, p. 1037–1049

[120] Levin, V. A.; Levina, E. E.
Corrosion resistance of metals in hot hydroxide solutions
Zashch. Met. 31 (1995) 3, p. 262–268
in: Protection of Metals 31 (1995) 3, p. 236–241

[121] Krupp VDM Firmeninformation
Nicrofer 3033
Alloy Digest April 1996, Ni-508

[122] Köhler, M; Heubner, U.; Eichenhofer, K.-W.; Renner, M.
Alloy 33, a new corrosion-resistant austenitic material for many applications
Stainless Steel World 11 (1999) 4, p. 38–49

[123] Palmer, J. D.
Alkalis Can Cause Unexpected Problems
Canad. Chem. Processing 53 (1969) 9, p. 7

[124] Cavallini M.; FeIli F.; Fratesi R. et al.
Verhalten von wirtschaftlich legierten MnAl-Stählen mit hohem Mn-Gehalt in wässrigen Lösungen
Werkstoffe & Korrosion 33(1982) p. 281

Key to materials compositions

Table 1: Chemical compositions of alloys according to German and other standards

German Standard		Materials Compositions	US-Standard
Mat.-No.	DIN-Design	Percent in Weight	SAE/ASTM/UNS
0.6015	EN-JL1020	–	A 48 (25B)
0.6025	EN-JL1040	–	A 48 (40B)
0.6655	GGL-NiCuCr 15 6 2	Fe-max. 3.0C-1.0-2.8Si-0.5-1.5Mn-13.5-17.5Ni-1.0-2.5Cr-5.5-7.5Cu	A 436 Type 1
0.6656	GGL-NiCuCr 15 6 3	Fe-max. 3.0C-1.0-2.8Si-0.5-1.5Mn-13.5-17.5Ni-2.5-3.5Cr-5.5-7.5Cu	A 436 Type 1b
0.6660	GGL-NiCr 20 2	Fe-max. 3.0C-1.0-2.8Si-0.5-1.5Mn-18.0-22.0Ni-1.0-2.5Cr	A 436 Type 2
0.6661	GGL-NiCr 20 3	Fe-max. 3.0C-1.0-2.8Si-0.5-1.5Mn-18.0-22.0Ni-2.5-3.5Cr	A 436 Type 2b
0.6667	GGL-NiSiCr 20 5 3	Fe-≤2.5C-3.5-5.5Si-0.5-1.5Mn-1.5-4.5Cr-18.0-22.0Ni	
0.6676	GGL-NiCr 30 3	Fe-max. 2.5C-1.0-2.8Si-0.5-0.8Mn-2.5-3.5Cr-28.0-32.0Ni	A 436 Type 3
0.6680	GGL-NiSiCr 30 5 5	Fe-≤2.5C-5.0-6.0Si-0.5-1.5Mn-4.5-5.5Cr-29.0-32.0Ni	A 436 Type 4
0.7040	EN-JS1030	–	A 536 (60-40-18)
0.7660	EN-GJSA-XNiCr20-2	Fe-≤3.0C-1.5-3.0Si-0.5-1.5Mn-≤0.080P-1.0-3.5Cr-≤0.50Cu-18.0-22.0Ni	A 439 Type D-2
0.7661	GGG-NiCr 20 3	Fe-≤3.0C-1.5-3.0Si-0.5-1.5Mn-≤0.080P-2.5-3.5Cr-18.0-22.0Ni	A 439 Type D-2B
0.7665	GGG-NiSiCr 20 5 2	Fe-≤3.0C-4.5-5.5Si-0.5-1.5Mn-≤0.080P-1.0-2.5Cr-18.0-22.0Ni	A 439 Type D-2C
0.7670	EN-GJSA-XNi22	Fe-≤3.0C-1.0-3.0Si-1.5-2.5Mn-≤0.080P-≤0.5Cr-≤0.5Cu-21.0-24.0Ni	A 439 Type D-2C
0.7679	GGG-NiSiCr 30 5 2	Fe-≤2.6C-4.0-6.0Si-0.5-1.5Mn-≤0.08P-29.0-32.0Ni	
0.7680	EN-GJSA-XNiSiCr30-5-5	Fe-≤2.6C-5.0-6.0Si-0.5-1.5Mn-≤0.080P-4.5-5.5Cr-≤0.5Cu-28.0-32.0Ni	A 439 Type D-4

German Standard		Materials Compositions	US-Standard
Mat.-No.	DIN-Design	Percent in Weight	SAE/ASTM/UNS
0.7688	EN-GJSA-XNiSiCr35-5-2; GGG-NiSiCr 35 5 2	Fe-≤2.0C-4.0-6.0Si-0.5-1.5Mn-≤0.08P-1.5-2.5Cr-≤0.5Cu-34.0-36.0Ni	A 439 (Type D-5S)
0.9625	EN-GJN-HV550	Fe-≤3.0-3.6C-≤0.080Si-≤0.080Mn-≤0.1P-≤0.1S-1.5-3.0Cr-≤0.5Mo-3.0-5.5Ni	A 532 (IA NiCr-HC)
0.9635	EN-JN3029	Fe-max. 1.8-2.4C-1.0Si-0.5-1.5Mn-0.08P-0.08S-14.0-18.0Cr-3.0Mo-2.0Ni	A 532
0.9640	EN-GJN-HV600(XCr14)	Fe-≤1.8-2.4C-≤1.0Si-≤0.5-1.5Mn-≤0.080P-≤0.080S-14.0-18.0Cr-≤3.0Mo-≤2.0Ni	A 532
0.9650	EN-JN3049	Fe-max. 2.4-3.2C-1.0Si-0.5-1.5Mn-0.08P-0.08S-23.0-28.0Cr-3.0Mo-2.0Ni	A 532
1.0030	St 00	Fe-≤0.30C-≤0.30Si-0.20-0.50Mn-≤0.08P-≤0.05S	
1.0032	St 34-2; (S205GT)	Fe-≤0.15C-≤0.3Si2.0-0.5Mn-0.05P-0.05S-0.007N	1010 (SAE)
1.0035	St 33; S 185		
1.0036	S235JRG1; USt 37-2; USt 37-2 G; (S235JRG1+CR)	Fe-≤0.17C-≤1.4Mn-≤0.0045P-≤0.045S-≤0.007N	K02502 (UNS)
1.0037	St 37-2; S235JR	Fe-≤0.17C-≤0.3Si-≤1.4Mn-≤0.045P-≤0.045S-≤0.009N	A 283; SAE 1015
1.0038	RSt 37-2; S235JR	Fe-≤0.17C-≤1.4Mn-≤0.045P-≤0.045S-≤0.009N	UNS K02502
1.0040	USt 42-2	Fe-≤0.25C-≤0.2-0.5Mn-≤0.05P-≤0.05S-≤0.007N	
1.0044	S275JR; St 44-2	Fe-≤0.21C-≤1.50Mn-≤0.045P-≤0.045S-≤0.009N	UNS K03000 A 853 (1020) SAE 1020
1.0050	E295; St 50-2	Fe-≤0.045P-≤0.045S-≤0.009N	
1.0070	E360; St 70-2	Fe-≤0.045P-≤0.045S-≤0.009N	
1.0114	S235J0	Fe-≤0.17C-≤1.40Mn-≤0.040P-≤0.040S-≤0.009N	
1.0116	S235J2G3; St 37-3 N	Fe-max. 0.17C-1.4Mn-0.035P-0.035S	UNS K02001
1.0120	S235JRC	Fe-≤0.17C-≤1.40Mn-≤0.045P-≤0.045S-≤0.009N	
1.0204	UQSt 36	Fe-≤0.14C-≤0.25-0.50Mn-≤0.040P-0.040S	SAE 1008
1.0208	RSt 35-2; (C10G2)	Fe-0.06-0.12C-≤0.25Si-0.40-0.60Mn-≤0.035P-≤0.035S-≤0.25Cu-≤0.012N	
1.0253	USt 37.0	Fe-≤0.2C-≤0.55Si-≤1.6Mn-≤0.04P-≤0.04S-≤0.007N	
1.0254	P235TR1, St 37.0	Fe-max. 0.16C-0.35Si-1.2Mn-0.025P-0.020S-0.30Cr-0.30Cu-0.08Mo-0.010Nb-0.30Ni-0.04Ti-0.02V	UNS K02501
1.0256	St 44.0	Fe-≤0.21C-≤0.55Si-≤1.60Mn-≤0.040P-≤0.040S-≤0.009N	A 106

Key to materials compositions

German Standard		Materials Compositions	US-Standard
Mat.-No.	DIN-Design	Percent in Weight	SAE/ASTM/UNS
1.0301	C10	Fe-≤0.07-0.13C-≤0.4Si-≤0.3-0.6Mn-≤0.045P-≤0.045S	SAE 1010
1.0305	St 35.8	Fe-≤0.17C-≤0.10-0.35Si-≤0.40-0.80Mn-≤0.040P-≤0.040S	UNS K01200
1.0308	E 235	Fe-≤0.17C-≤0.35Si-0.4Mn-≤0.05P-≤0.05S-≤0.007N	SAE 1010
1.0309	DX55D	Fe-≤0.16C-0.17-0.40Si-0.35-0.65Mn-≤0.05P-≤0.050S-≤0.30Cr-≤0.30Ni-≤0.30Cu	UNS K02501
1.0330	St 12; DC01 + ZN	Fe-≤0.12C-≤0.60Mn-≤0.045P-≤0.045S	A 366 (C)
1.0333	USt 13	Fe-≤0.08C-≤0.007N	
1.0336	USt 4	Fe-≤0.09C-≤0.25-0.50Mn-≤0.030P-≤0.030S-≤0.007N	
1.0338	DC04; St 14	Fe-≤0.08C-≤0.40Mn-≤0.03P-≤0.030S	
1.0345	P235GH; H I	Fe-≤0.16C-≤0.35Si-≤0.40-1.20Mn-≤0.030P-≤0.025S-0.02Al-≤0.30Cr-≤0.30Cu-≤0.08Mo-≤0.010Nb-≤0.30Ni-≤0.03Ti-≤0.02V	A 285; A 414
1.0346	H220G1	Fe-≤0.04C-≤0.40Mn-≤0.03P-≤0.02S-≤0.01-0.04Ti	UNS K02202 A 516 (55)(380)
1.0356	TTSt35 N	Fe-≤0.18C-≤0.13-0.45Si-≤0.55-0.98Mn-≤0.035P-≤0.04S	UNS K03000 A 524 (I,II)
1.0375	TH57; T 57	Fe-≤0.1C-Traces Si-0.25-0.45Mn-≤0.04P-≤0.04S-0.007N	
1.0401	C 15	Fe-≤0.12-0.18C-≤0.40Si-≤0.3-0.6Mn-≤0.045P-≤0.045S	SAE 1015
1.0402	C 22	Fe-≤0.17-0.24C-≤0.4Si-≤0.4-0.7Mn-≤0.045P-≤0.045S-≤0.4Cr-≤0.1Mo-≤0.4Ni	SAE 1020
1.0405	St 45.8	Fe-≤0.21C-≤0.10-0.35Si-≤0.40-1.20Mn-≤0.040P-≤0.040S	A 106
1.0408	St 45	Fe-≤0.25C-≤0.035Si-0.40Mn-≤0.050P-≤0.050S	A 108; SAE 1020
1.0414	C20D; D 20-2	Fe-≤0.18-0.23C-≤0.30Si-≤0.3-0.6Mn-≤0.035P-≤0.035S-≤0.01Al-≤0.2Cr-≤0.3Cu-≤0.05Mo-≤0.25Ni	UNS G10200; SAE 1020
1.0425	P265GH; H II	Fe-≤0.20C-≤0.4Si-≤0.5-1.4Mn-≤0.030P-≤0.025S-0.02Al-≤0.3Cr-≤0.3Cu-≤0.08Mo-≤0.01Nb-≤0.3Ni-≤0.03Ti-≤0.02V	UNS K01701
1.0426	P280GH	Fe-≤0.08-0.20C-≤0.4Si-≤0.9-1.5Mn-≤0.025P-≤0.015S-≤0.30Cr	A 662 (A)
1.0461	StE 255	Fe-≤0.18C-≤0.40Si-≤0.50-1.30Mn-≤0.035P-≤0.030S-≥0.02Al-≤0.30Cr-≤0.20Cu-≤0.08Mo-≤0.02N-≤0.03Nb-≤0.30Ni	UNS K02202 A 516 (55)(380)
1.0473	P355GH; 19 Mn 6	Fe-≤0.1-0.22C-≤0.6Si-≤1.0-1.7Mn-≤0.03P-≤0.025S-≤0.30Cr-≤0.3Cu-≤0.08Mo-≤0.3Ni	A 299

German Standard		Materials Compositions	US-Standard
Mat.-No.	DIN-Design	Percent in Weight	SAE/ASTM/UNS
1.0481	17 Mn 4; P 295 GH	Fe≤0.08-0.20C-≤0.4Si-≤0.90-1.50Mn-≤0.030P-≤0.025S-0.02Al-≤0.30Cr-≤0.30Cu-≤0.08Mo-≤0.010Nb-≤0.3Ni-≤0.03Ti-≤0.02V	A 414, 515
1.0482	19 Mn 5	Fe-≤0.17-0.22C-≤0.30-0.60Si-≤1.00-1.30Mn-≤0.045P-≤0.045S-≤0.30Cr	UNS K12437; A 537
1.0490	S275N	Fe-≤0.18C-≤0.40Si-0.50-1.40Mn-≤0.035P-≤0.030S-≥0.0200Al-≤0.30Cr-≤0.35Cu-≤0.10Mo-≤0.015N-≤0.050Nb-≤0.30Ni-≤0.03Ti-≤0.05V	UNS K03000
1.0501	C 35	Fe-≤0.32-0.39C-≤0.4Si-≤0.5-0.8Mn-≤0.045P-≤0.045S-≤0.4Cr-≤0.1Mo-≤0.4Ni	SAE 1035
1.0503	C 45	Fe-≤0.42-0.50C-≤0.4Si≤0.5-0.8Mn-≤0.045P-≤0.045S-≤0.4Cr-≤0.1Mo-≤0.4Ni	SAE 1045
1.0505	StE 315	Fe-≤0.18C-≤0.45Si-≤0.70-1.50Mn-≤0.035P-≤0.030S-≤0.30Cr-0.020Al-≤0.20Cu-≤0.020N-≤0.03Nb-≤0.08Mo-0.30Ni	A 573
1.0528	C 30	Fe-≤0.27-0.34C-≤0.4Si-≤0.5-0.8Mn-≤0.045P-0.045S-≤0.4Cr-≤0.1Mo≤0.4Ni	SAE 1030
1.0540	C50	Fe-0.47-0.55C-≤0.40Si-0.60-0.90Mn-≤0.045P-≤0.045S-≤0.40Cr-≤0.10Mo-≤0.40Ni	A 689 (1050) ASTM A 866 (1050) ASTM
1.0545	S355N	Fe-≤0.20C-≤0.50Si-≤0.90-1.65Mn-≤0.035P-0.030S-0.02Al-≤0.30Cr-≤0.35Cu-≤0.10Mo-≤0.015N-≤0.050Nb-≤0.50Ni-≤0.03Ti-≤0.12V	UNS K12709
1.0553	S355J0	Fe-≤0.20C-≤0.55Si-≤1.60Mn-≤0.04P-≤0.04S-≤0.009N	
1.0562	StE 355; P 355 N	Fe-≤0.20C-≤0.50Si-≤0.90-1.70Mn-≤0.03P-≤0.025S-0.02Al-≤0.30Cu-≤0.30Cr-≤0.08Mo-≤0.02N-≤0.05Nb-≤0.50Ni-0.03Ti	A 633 (C)
1.0564	N-80	Fe-≤0.030P-≤0.030S	—
1.0570	St 52-3 N; S 355 J2G3	Fe-≤0.20C-≤0.55Si-≤1.60Mn-≤0.035P-≤0.035S	SAE 1024
1.0580	E 355	Fe-≤0.22C-≤0.55Si-≤1.60Mn-≤0.025P-≤0.025S	A 513 (1024) (ASTM)
1.0589	GL-E 36	Fe-≤0.18C-≤0.50Si-0.90-1.60Mn-≤0.040P-≤0.040S-≥0.0200Al-≤0.20Cr-≤0.35Cu-≤0.08Mo-0.020-0.050Nb-≤0.40Ni-0.05-0.10V	UNS K11852
1.0601	C60	Fe-0.57-0.65C-≤0.40Si-0.60-0.90Mn-≤0.045P-≤0.045S-≤0.40Cr-≤0.10Mo-≤0.40Ni	A 830 (1060) ASTM A 713 (1060) ASTM
1.0605	C 75	Fe-≤0.7-0.8C-≤0.15-0.35Si-≤0.6-0.8Mn-≤0.045P-≤0.045S	SAE 1074
1.0616	C86D	Fe-≤0.83-0.88C-≤0.10-0.30Si-≤0.50-0.80Mn-≤0.035P-≤0.035S-≤0.01Al-≤0.15Cr-≤0.25Cu-≤0.05Mo-≤0.20Ni	SAE 1086

Key to materials compositions

German Standard		Materials Compositions	US-Standard
Mat.-No.	DIN-Design	Percent in Weight	SAE/ASTM/UNS
1.0619	GP240GH	Fe-≤0.18-0.23C-≤0.60Si-≤0.50-1.20Mn-≤0.03P-≤0.02S	A 216
1.0664	St 160/180	Fe-≤0.80C-≤0.20Si-≤0.70Mn-≤0.04P-≤0.04S	
1.0670	P-105	Fe-≤0.70C-≤0.03-0.30Si-1.0Mn-≤0.04P-≤0.04S-≤0.007N	—
1.0721	10S20	Fe-0.07-0.13C-≤0.40Si-0.70-1.10Mn-≤0.060P-0.150-0.250S	SAE 1109
1.0854	M125-35P	consult producer	
1.0912	46Mn7	Fe-≤0.42-0.50C-≤0.15-0.35Si-≤1.6-1.9Mn-≤0.05P-≤0.05S-≤0.007N	SAE 1345
1.1013	RFe 100	Fe-≤0.05C-≤0.10Si-≤0.20-0.35Mn-≤0.03P-≤0.035S-≤0.04-0.10Al	
1.1104	EStE 285; P275NL2	Fe-≤0.16C-≤0.4Si-≤0.5-1.5Mn-≤0.025P-≤0.015S-0.02Al-≤0.30Cr-≤0.30Cu-≤0.02N-≤0.5Ni	P275NL2
1.1106	P355NL2; EStE 355	Fe-≤0.18C-≤0.5Si-≤0.9-1.7Mn-≤0.025P-≤0.015S-≤0.3Cr-≤0.3Cu-≤0.3Mo-0.5Ni-≤0.02N	A 707
1.1121	C10E; Ck 10	Fe-≤0.07-0.13C-≤0.40Si-≤0.30-0.60Mn-≤0.035P-≤0.035S	SAE 1010
1.1127	36Mn6	Fe-≤0.34-0.42C-≤0.15-0.35Si-≤1.4-1.65Mn-≤0.035P-≤0.035S	
1.1136	G24Mn4	Fe-0.20-0.28C-0.30-0.60Si-0.90-1.20Mn-≤0.035P-≤0.035S	
1.1151	Ck 22; C22E	Fe-≤0.17-0.24C-≤0.4Si-≤0.4-0.7Mn-≤0.035P-≤0.035S-≤0.4Cr-≤0.1Mo-≤0.4Ni	SAE 1023
1.1166	34Mn5	Fe-0.30-0.37C-0.15-0.30Si-1.20-1.50Mn-≤0.035P-≤0.035S	G15360 (UNS) A 711 (1536) (ASTM) 1536 (SAE)
1.1176	G36Mn5	Fe-≤0.32-0.40C-≤0.15-0.35Si-≤1.20-1.50Mn-≤0.035P-≤0.035S	UNS H10380; A 830; SAE 1038
1.1186	C40E; Ck 40	Fe-≤0.37-0.44C-≤0.4Si-≤0.5-0.8Mn-≤0.035P-≤0.035S-≤0.4Cr-≤0.1Mo-≤0.4Ni	SAE 1040
1.1191	C45E; Ck 45	Fe-≤0.42-0.50C-≤0.4Si-≤0.50-0.80Mn-≤0.035P-≤0.035S-≤0.40Cr-≤0.10Mo-≤0.4Ni	SAE 1045
1.1520	C70U	Fe-≤0.65-0.74C-≤0.10-0.30Si-≤0.10-0.35Mn-≤0.030P-≤0.030S	
1.1525	C80U; C80W1	Fe-≤0.75-0.85C-≤0.10-0.25Si-≤0.10-0.25Mn-≤0.020P-≤0.020S	SAE W 108
1.1545	C105U; C105W1	Fe-≤1.0-1.1C-≤0.10-0.25Si-≤0.10-0.25Mn-≤0.020P-≤0.020S	SAE W 110
1.1730	C45U; C45W	Fe-≤0.42-0.50C-≤0.15-0.40Si-≤0.60-0.80Mn-≤0.03P-≤0.03S	A 830 (1045); SAE 1045

German Standard		Materials Compositions	US-Standard
MatNo.	DIN-Design	Percent in Weight	SAE/ASTM/UNS
1.2210	115CrV3	Fe-1.10-1.25C-0.15-0.30Si-0.20-0.40Mn-≤0.030P-≤0.030S-0.50-0.80Cr-0.07-0.12V	L 2 SAE A 681 (L2) ASTM
1.2311	40CrMnMo7	Fe-0.35-0.45C-0.20-0.40Si-1.30-1.60Mn-≤0.035P-≤0.035S-1.80-2.10Cr-0.15-0.25Mo	
1.2343	X37CrMoV5-1; X38CrMoV5-1	Fe-0.33-0.41C-0.80-1.20Si-0.25-0.50Mn-≤0.030P-≤0.020S-4.80-5.50Cr-1.10-1.50Mo-0.30-0.50V	H 11 SAE A 681 (H 11) ASTM T 20811 UNS
1.2344	X40CrMoV5-1	Fe-0.35-0.42C-0.80-1.20Si-0.25-0.50Mn-≤0.030P-≤0.020S-4.80-5.50Cr-1.20-1.50Mo-0.85-1.15V	H 13 SAE T 20813 UNS A 681 (H13) ASTM
1.2365	32CrMoV12-28 X32CrMoV33	Fe-≤0.28-0.35C-≤0.10-0.40Si-≤0.15-0.45Mn-≤0.030P-≤0.030S-≤2.70-3.20Cr-≤2.60-3.00Mo-≤0.40-0.70V	SAE H 10
1.2550	60WCrV8; 60WCrV7	Fe-0.55-0.65C-0.70-1.00Si-0.15-0.45Mn-≤0.030P-≤0.030S-0.90-1.20Cr-0.10-0.20V-1.70-2.20W	
1.2567	30WCrV17-2	Fe-0.25-0.35C-0.15-0.30Si-0.20-0.40Mn-≤0.035P-≤0.035S-2.20-2.50Cr-0.50-0.70V-4.00-4.50W	
1.2787	X23CrNi17	Fe-≤0.10-0.25C-≤1.00Si-≤1.00Mn-≤0.035P-≤0.035S≤15.5-18.0Cr-≤1.0-2.5Ni	
1.2823	70Si7	Fe-≤0.65-0.75C-≤1.50-1.80Si-≤0.60-0.80Mn-≤0.03P-≤0.03S	
1.2842	90MnCrV8	Fe-≤0.85-0.95C-≤0.10-0.40Si-≤1.80-2.20Mn-≤0.030P-≤0.030S-≤0.20-0.50Cr-≤0.05-0.20V	UNS T31502; SAE O2; A 681 (O2); SAE O2
1.3247	HS2-9-1-8 S2-10-1-8	Fe-1.05-1.15C-≤0.70Si-≤0.40Mn-≤0.030P-≤0.030S-7.50-8.50Co-3.50-4.50Cr-9.00-10.00Mo-0.90-1.30V-1.20-1.90W	UNS T11342
1.3355	HS 18-0-1	Fe-≤0.70-0.78C-≤0.45Si-≤0.4Mn-≤0.030P-≤0.030S-≤3.8-4.5Cr-≤1.0-1.2V-≤17.5-18.5W	A 600
1.3505	100Cr6	Fe-≤0.93-1.05C-≤0.15-0.35Si-≤0.25-0.45Mn-≤0.025P-≤0.015S-≤0.05Al-≤1.35-1.60Cr-≤0.30Cu-≤0.10Mo	SAE 52100; A 29
1.3551	80MoCrV42-16	Fe-≤0.77-0.85C-≤0.40Si-≤0.15-0.35Mn-≤0.025P-≤0.015S-≤3.90-4.30Cr-≤0.30Cu-≤4.00-4.50Mo-0.90-1.10V-≤0.25W	SAE M50; A 600 (M50)
1.3728	AlNiCo 9/5	Fe-≤11.0-13.0Al-5.0Co-2.0-4.0Cu-≤57.0Fe-21.0-28.0Ni-≤1.0Ti	
1.3813	X40MnCrN19	Fe-≤0.30-0.50C-≤0.80Si-≤17.0-19.0Mn-≤0.10P-≤0.030S-≤3.0-5.0Cr-≤0.08-0.12N	
1.3817	X40MnCr18	Fe-0.30-0.50C-≤1.00Si-17.00-19.00Mn-≤0.060P-≤0.030S-≤3.00-5.00Cr-≤0.100N-≤1.00Ni	

Key to materials compositions

German Standard		Materials Compositions	US-Standard
Mat.-No.	DIN-Design	Percent in Weight	SAE/ASTM/UNS
1.3914	X2CrNiMnMoNNb21-15-7-3	Fe-≤0.03C-≤0.75Si-6.00-8.00Mn-≤0.025P-≤0.010S-20.00-22.00Cr-3.00-3.50Mo-0.350-0.500N-0.100-0.250Nb-14.00-16.00Ni	
1.3940	GX2CrNiN18-13	Fe-≤0.03C-≤1.00Si-≤2.00Mn-≤0.035P-≤0.020S-16.50-18.50Cr-0.100-0.200N-12.00-14.00Ni	
1.3951	G-X 4 CrNiMoN 22 15	Fe-≤0.05C-0.80-1.10Si-0.50-1.00Mn-≤0.030P-≤0.030S-22.00-23.50Cr-1.10-1.30Mo-0.150-0.250N-17.00-18.00Ni	
1.3952	X2CrNiMoN18-14-3	Fe-≤0.03C-≤1.00Si-≤2.00Mn-≤0.045P-≤0.015S-16.50-18.50Cr-2.50-3.00Mo-0.150-0.250N-13.00-15.00Ni	
1.3955	GX12CrNi18-11	Fe-≤0.15C-≤1.00Si-≤2.00Mn-≤0.045P-≤0.030S-16.50-18.50Cr-≤0.75Mo-10.00-12.00Ni	
1.3964	X2CrNiMnMoNNb21-16-5-3	Fe-≤0.03C-≤1.00Si-4.00-6.00Mn-≤0.025P-≤0.010S-20.00-21.50Cr-3.00-3.50Mo-0.200-0.350N-≤0.250Nb-15.00-17.00Ni	NITRONIC 50
1.3974	X2CrNiMnMoNNb23-17-6-3	Fe-≤0.03C-≤1.00Si-4.50-6.50Mn-≤0.025P-≤0.01S-≤21.00-24.50Cr-≤2.80-3.40Mo-≤0.30-0.50N-≤0.10-0.30Nb-≤15.50-18.00Ni	
1.3981	NiCo 29 18; (X3NiCo29-18)	Fe-≤0.050C-≤0.30Si-≤0.50Mn-≤17.0-18.0Co-≤28.0-30.0Ni	UNS K94610
1.4000	X6Cr13	Fe-≤0.08C-≤1.0Si-≤1.0Mn-≤0.04P-≤0.015S-≤12.0-14.0Cr	SAE 403, 410S
1.4001	X 7 Cr 14	Fe-≤0.08C-≤1.00Si-≤1.00Mn-≤0.045P-≤0.030S-≤13.00-15.00Cr	429 (SAE) A240 (410S) (ASTM) 410S (SAE)
1.4002	X6CrAl13	Fe-≤0.08C-≤1.0Si-≤1.0Mn-≤0.04P-≤0.015S-≤0.10-0.3Al-≤12-14Cr	SAE 405
1.4003	X2CrNi12	Fe-≤0.03C-≤1.0Si-≤1.5Mn-≤0.04P-≤0.015S-≤10.5-12.5Cr-≤0.03N-≤0.3-1.0Ni	UNS S40977
1.4005	X12CrS13	Fe-≤0.08-0.15C-≤1.0Si-≤1.5Mn-≤0.04P-≤0.15-0.35S-≤12.0-14.0Cr-≤0.6Mo	SAE 416
1.4006	X12Cr13	Fe-≤0.08-0.15C-≤1.00Si-≤1.5Mn-≤0.04P-≤0.015S-≤11.0-13.5Cr-≤0.75Ni	SAE 410
1.4008	GX7CrNiMo12-1	Fe-≤0.10C-1.0Si-1.0Mn-≤0.035P-≤0.025S-12.0-13.5Cr-0.20-0.50Mo-1.0-2.0Ni	UNS J91150
1.4015	X8Cr18	Fe-≤0.10C-≤1.50Si-≤1.50Mn-≤0.030P-≤0.030S-≤16.50-18.50Cr	S43080 (UNS)
1.4016	X6Cr17	Fe-≤0.08C-≤1.0Si-≤1.0Mn-≤0.04P-≤0.015S-≤16.0-18.0Cr	SAE 430
1.4021	X20Cr13	Fe-≤0.16-0.25C-≤1.0Si-≤1.5Mn-≤0.04P-≤0.015S-≤12.0-14.0Cr	SAE 420

German Standard		Materials Compositions	US-Standard
Mat.-No.	DIN-Design	Percent in Weight	SAE/ASTM/UNS
1.4024	X15Cr13	Fe-≤0.12-0.17C-≤1.0Si-≤1.0Mn-≤0.045P-≤0.03S-≤12.0-14.0Cr	SAE 420
1.4028	X30Cr13	Fe-≤0.26-0.35C-≤1.0Si-≤1.5Mn-≤0.04P-≤0.015S-≤12.0-14.0Cr	A 743; UNS J91153
1.4031	X39Cr13	Fe-≤0.36-0.42C-≤1.00Si-≤1.00Mn-≤0.04P-≤0.015S-≤12.5-14.5Cr	UNS S42080
1.4034	X46Cr13	Fe-≤0.43-0.5C-≤1.0Si-≤1.0Mn-≤0.04P-≤0.015S-≤12.5-14.5Cr	
1.4057	X17CrNi16-2	Fe-≤0.12-0.22C-≤1.0Si-≤1.5Mn-≤0.04P-≤0.015S-≤15.0-17.0Cr-≤1.5-2.5Ni	SAE 431
1.4085	GX70Cr29	Fe-≤0.50-0.90C-2.0Si-1.0Mn-≤0.045P-≤0.030S-27.0-30.0Cr	
1.4104	X14CrMoS17	Fe-≤0.10-0.17C-≤1.0Si-≤1.5Mn-≤0.04P-≤0.15-0.35S-≤15.5-17.5Cr-≤0.2-0.6Mo	SAE 430 F
1.4105	X6CrMoS17	Fe-≤0.08C-≤1.50Si-≤1.50Mn-≤0.040P-≤0.15-0.35S-≤16.00-18.00Cr-≤0.20-0.60Mo	
1.4110	X55CrMo14	Fe-≤0.48-0.60C-≤1.0Si-≤1.0Mn-≤0.04P-≤0.015S-≤13.0-15.0Cr-≤0.5-0.8Mo-≤0.15V	
1.4112	X90CrMoV18	Fe-≤0.85-0.95C-≤1.0Si-≤1.0Mn-≤0.04P-≤0.015S-≤17.0-19.0Cr-≤0.9-1.3Mo-≤0.07-0.12V	SAE 440 B
1.4113	X6CrMo17-1	Fe-≤0.80C-≤1.0Si-≤1.00Mn-≤0.04P-≤0.03S-≤16.0-18.0Cr-≤0.90-1.40Mo	SAE 434
1.4116	X50CrMoV15	Fe-≤0.45-0.55C-1.0Si-1.0Mn-≤0.040P-≤0.015S-14.0-15.0Cr-0.5-0.8Mo-0.10-0.20V	
1.4117	X38CrMoV15	Fe-≤0.35-0.40C-1.0Si-1.0Mn-≤0.045P-≤0.030S-14.0-15.0Cr-0.4-0.6Mo-0.10-0.15V	
1.4120	X20CrMo13	Fe-≤0.17-0.22C-≤1.0Si-≤1.0Mn-≤0.04P-≤0.015S-≤12-14Cr-≤0.9-1.3Mo-≤1.0Ni	
1.4122	X39CrMo17-1	Fe-≤0.33-0.45C-≤1.0Si-≤1.5Mn-≤0.04P-≤0.015S-≤15.5-17.5Cr-≤0.8-1.3Mo-≤1.0Ni	
1.4125	X105CrMo17	Fe-≤0.95-1.20C-1.0Si-1.0Mn-≤0.040P-≤0.015S-16.0-18.0Cr-0.4-0.8Mo	SAE 617
1.4126	X 110 CrMo 13	Fe-≤1.05-1.15C-≤1.0Si-≤1.0Mn-≤0.040P-≤0.030S- ≤17.0-18.0Cr-≤0.8-1.0Mo	
1.4131	X 1 CrMo 26 1	Fe-≤0.010C-≤0.40Si-≤0.40Mn-≤0.020P-≤0.020S-≤0.015N-≤25.0-27.5Cr-≤0.75-1.50Mo-≤0.50Ni	
1.4133		Fe-≤0.01C-1.0Si-1.0Mn-≤0.030P-≤0.030S-26.0-30.0Cr-1.7-2.3Mo-0.010N	
1.4136	GX70CrMo29-2	Fe-≤0.50-0.90C-2.0Si-1.0Mn-≤0.045P-≤0.030S-27.0-30.0Cr-2.0-2.5Mo	
1.4138	GX120CrMo29-2	Fe-≤0.90-1.30C-2.0Si-1.0Mn-≤0.045P-≤0.030S-27.0-30.0Cr-2.0-2.5Mo	

German Standard		Materials Compositions	US-Standard
Mat.-No.	DIN-Design	Percent in Weight	SAE/ASTM/UNS
1.4153	X80CrVMo13-2	Fe-≤0.76-0.86C-≤1.00Si-≤1.00Mn-≤0.045P-≤0.030S-≤12.00-14.00Cr-≤0.40-0.60Mo-≤1.50-2.10V	
1.4300	X 12 CrNi 18 8	Fe-≤0.12C-≤1.0Si-≤2.0Mn-≤0.045P-≤0.030S-≤17.0-19.0Cr-≤8.0-10.0Ni	
1.4301	X5CrNi18-10	Fe-≤0.07C-≤1.0Si-≤2.0Mn-≤0.045P-≤0.015S-≤17-19.5Cr-≤0.11N-≤8.0-10.5Ni	SAE 304
1.4302	X5CrNi19-9	Fe-≤0.05C-≤1.40Si-≤1.90Mn-≤0.025P-≤0.015S-≤18.2-19.8Cr-≤8.70-10.30Ni	UNS S30888
1.4303	X4CrNi18-12	Fe-≤0.06C-≤1.0Si-≤2.0Mn-≤0.045P-≤0.015S-≤17-19Cr-≤0.11N-≤11.0-13.0Ni	SAE 305/308
1.4304	X5CrNi18-12E	Fe-≤0.12C-≤1.0Si-≤2.0Mn-≤0.045P-≤0.030S-≤17-19Cr-≤8.0-10.0Ni	
1.4305	X8CrNiS18-9; X10CrNiS189	Fe-≤0.10C-≤1.0Si-≤2.0Mn-≤0.045P-≤0.15-0.35S-≤17-19Cr-≤1.0Cu-≤0.11N-≤8.0-10.0Ni	SAE 303
1.4306	X2CrNi19-11	Fe-≤0.03C-≤1.0Si-≤2.0Mn-≤0.045P-≤0.015S-≤18-20Cr-≤0.11N-≤10.0-12.0Ni	SAE 304 L
1.4307	X2CrNi18-9	Fe-≤0.03C-≤1.0Si-≤2.0Mn-≤0.045P-≤0.015S-≤17.5-19.5Cr-≤0.11N-≤8.0-10.0Ni	
1.4308	GX5CrNi19-10	Fe-≤0.07C-≤1.5Si-≤1.5Mn-≤0.04P-≤0.030S-≤18.0-20.0Cr-≤8.0-11.0Ni	SAE 304 H
1.4310	X10CrNi18-8	Fe-≤0.05-0.15C-≤2.0Si-≤2.0Mn-≤0.045P-≤0.015S-≤16.0-19.0Cr-≤0.8Mo-≤0.11N≤6.0-9.5Ni	SAE 301
1.4311	X2CrNiN18-10	Fe-≤0.03C-≤1.0Si-≤2.0Mn-≤0.045P-≤0.015S-≤17.0-19.5Cr-≤8.5-11.5Ni-≤0.12-0.22N	SAE 304 LN
1.4312	GX10CrNi18-8	Fe-≤0.12C-≤2.0Si-≤1.5Mn-≤0.045P-≤0.03S-≤17.0-19.5Cr-≤8.0-10.0Ni	A 743
1.4313	X3CrNiMo13-4	Fe-≤0.05C-≤0.7Si-≤1.5Mn-≤0.04P-≤0.015S-≤12.0-14.0Cr-≤0.3-0.7Mo-≤0.02N-≤3.5-4.5Ni	UNS J91540
1.4315	X5CrNiN19-9	Fe-≤0.06C-≤1.0Si-≤2.0Mn-≤0.045P-≤0.03S-≤18.0-20.0Cr-≤0.12-0.22N-≤8.0-11.0Ni	
1.4317	GX4CrNi13-4	Fe-≤0.06C-≤1.00Si-≤1.00Mn-≤0.035P-≤0.025S-≤12.00-13.50Cr-≤0.70Mo-≤3.50-5.00Ni	UNS J91540 A 743 (CA-6NM)
1.4318	X2CrNiN18-7	Fe-≤0.03C-≤1.0Si-≤2.0Mn-≤0.045P-≤0.015S-≤16.5-18.5Cr-≤0.1-0.2N-≤6.0-8.0Ni	
1.4319	X3CrNiN17-8	Fe-≤0.05C-≤1.00Si-≤2.00Mn-≤0.045P-≤0.015S-≤16.00-18.00Cr-≤0.04-0.08N-≤7.00-8.00Ni	
1.4330	X 2 CrNi 25 20	Fe-≤0.03C-≤1.0Si-≤1.5Mn-≤0.045P-≤0.035S-≤18.0-22.0Cr-≤23.0-27.0Ni	

German Standard		Materials Compositions	US-Standard
Mat.-No.	DIN-Design	Percent in Weight	SAE/ASTM/UNS
1.4333	X 5 NiCr 32 21	Fe-≤0.07C-≤1.40Si-≤2.40Mn-≤0.045P-≤0.030S-≤19.00-22.00Cr-≤30.00-34.00Ni	S33200 (UNS) S33200 (SAE) B 710 (N 08330) (ASTM)
1.4335	X1CrNi25-21	Fe-≤0.02C-≤0.25Si-≤2.0Mn-≤0.025P-≤0.010S-≤24.0-26.0Cr-≤0.20Mo-≤0.110N-≤20.0-22.0Ni	
1.4340	GX40CrNi27-4	Fe-≤0.30-0.50C-≤2.00Si-≤1.50Mn-≤0.045P-≤0.030S-≤26.0-28.0Cr-≤3.5-5.5Ni	A 743
1.4347	GX6CrNiN26-7	Fe-≤0.08C-≤1.50Si-≤1.50Mn-≤0.035P-≤0.020S-25.00-27.00Cr-0.100-0.200N-5.50-7.50Ni	
1.4361	X1CrNiSi18-15-4	Fe-≤0.015C-≤3.70-4.50Si-≤2.00Mn-≤0.025P-≤0.010S-≤16.5-18.5Cr-≤0.20Mo-≤0.110N-≤14.0-16.0Ni	A 336
1.4362	X2CrNiN23-4	Fe-≤0.03C-≤1.00Si-≤2.00Mn-≤0.035P-≤0.015S-≤22.0-24.0Cr-≤0.10-0.60Cu-≤0.10-0.60Mo-≤0.050-0.200N-≤3.5-5.5Ni	
1.4371	X2CrMnNiN17-7-5	Fe-≤0.30C-≤1.0Si-≤6.0-8.0Mn-≤0.045P-≤0.015S-≤16.0-17.0Cr-≤0.15-0.20N-≤3.5-5.5Ni	SAE 202
1.4401	X5CrNiMo17-12-2	Fe-≤0.07C-≤1.0Si-≤2.0Mn-≤0.045P-≤0.015S-≤16.5-18.5Cr-≤2.0-2.5Mo-≤0.110N-≤10.0-13.0Ni	SAE 316
1.4404	X2CrNiMo17-12-2; X2 CrNiMo 17 12 2	Fe-≤0.03C-≤1.0Si-≤2.0Mn-≤0.045P-≤0.015S-≤16.5-18.5Cr-≤2.0-2.5Mo-≤10.0-13.0Ni≤0.110N	SAE 316 L
1.4405	X 5 CrNiMo 16 5; GX4CrNiMo16-5-1	Fe-≤0.07C-≤1.0Si-≤1.0Mn-≤0.035P-≤0.025S-≤15.0-16.5Cr-≤0.5-2.0Mo-≤4.5-6.0Ni	
1.4406	X2CrNiMoN17-11-2; X2 CrNiMoN 17 12 2	Fe-≤0.03C-≤1.0Si-≤2.0Mn-≤0.045P-≤0.015S-≤16.5-18.5Cr-≤2.0-2.5Mo-≤0.12-0.22N-≤10.0-12.0Ni	SAE 316 LN
1.4408	GX5CrNiMo19-11-2	Fe-≤0.07C-≤1.50Si-≤1.50Mn-≤0.04P-≤0.030S-≤18.0-20.0Cr-≤2.0-2.5Mo-≤9.0-12.0Ni	CF-8M
1.4410	X2CrNiMoN25-7-4	Fe-≤0.03C-≤1.0Si-≤2.0Mn-≤0.035P-0.015S-≤24.0-26.0Cr-≤3.0-4.5Mo-≤0.200-0.350N-≤6.0-8.0Ni	2507; A 182
1.4413	X3CrNiMo13-4	Fe-≤0.05C-≤0.30-0.60Si-≤0.50-1.00Mn-≤0.03P-≤0.02S-≤12.00-14.00Cr-≤0.30-0.70Mo-≤3.50-4.00Ni	UNS S42400 A 988 (S 41500) SAE S41500
1.4417	GX2CrNiMoN25-7-3	Fe-≤0.03C-≤1.0Si-≤1.5Mn-≤0.030P-≤0.020S-≤24.0-26.0Cr-≤1.0Cu-≤3.0-4.0Mo-≤0.150-0.250N-≤6.0-8.5Ni-≤1.0W	3RE60; A 789

German Standard		Materials Compositions	US-Standard
Mat.-No.	DIN-Design	Percent in Weight	SAE/ASTM/UNS
1.4418	X4CrNiMo16-5-1	Fe-≤0.06C-≤0.7Si-≤1.5Mn-≤0.040P-≤0.015S-≤15.0-17.0Cr-≤0.8-1.5Mo-≤0.020N-≤4.0-6.0Ni	
1.4424	X2CrNiMoSi18-5-3	Fe-≤0.03C-1.40-2.00Si-1.20-2.00Mn-≤0.035P-≤0.015S-18.00-19.00Cr-2.50-3.00Mo-0.050-1.00N-4.50-5.20Ni	
1.4427	X12CrNiMoS18-11	Fe-≤0.12C-≤1.0Si-≤2.0Mn-≤0.06P-≤0.15-0.35S-≤16.5-18.5Cr-≤2-2.5Mo-≤10.5-13.5Ni	
1.4429	X2CrNiMoN17-13-3	Fe-≤0.03C-≤1.0Si-≤2.0Mn-≤0.045P-≤0.015S-≤16.5-18.5Cr-≤2.5-3.0Mo-≤11.0-14.0Ni-≤0.12-0.22N	SAE 316 LN
1.4430	X2CrNiMo19-12	Fe-≤0.02C-≤1.40Si-≤1.90Mn-≤0.025P-≤0.015S-17.20-19.80Cr-2.50-3.00Mo-10.70-13.30Ni	S 31688 UNS S 31683 UNS
1.4434	X2CrNiMoN18-12-4	Fe-≤0.03C-≤1.00Si-≤2.00Mn-≤0.045P-≤0.015S-≤16.50-19.50Cr-≤3.00-4.00Mo-≤0.10-0.20N-≤10.50-14.00Ni	
1.4435	X2CrNiMo18-14-3; X2 CrNiMo 18 4 3	Fe-≤0.03C-≤1.0Si-≤2.0Mn-≤0.045P-≤0.015S-≤17.0-19.0Cr-≤2.5-3.0Mo-≤0.110N-≤12.5-15.0Ni	SAE 316 L
1.4436	X3CrNiMo17-13-3; X5 CrNiMo 17 13 3	Fe-≤0.05C-≤1.0Si-≤2.0Mn-≤0.045P-≤0.015S-≤16.5-18.5Cr-≤2.5-3.0Mo-≤0.110N-≤10.5-13.0Ni	SAE 316
1.4438	X2CrNiMo18-15-4	Fe-≤0.03C-≤1.0Si-≤2.0Mn-≤0.045P-≤0.015S-≤17.5-19.5Cr-≤3.0-4.0Mo-≤0.110N-≤13.0-16.0Ni	SAE 317 L
1.4439	X2CrNiMoN17-13-5	Fe-≤0.030C-≤1.00Si-≤2.00Mn-≤0.045P-≤0.015S-≤16.5-18.5Cr-≤4.00-5.00Mo-≤12.5-14.5Ni-≤0.12-0.22N	
1.4440	X2CrNiMo18-16-5	Fe-≤0.03C-≤1.00Si-2.00-3.00Mn-≤0.025P-≤0.025S-17.00-20.00Cr-4.00-5.00Mo-16.00-19.00Ni	SAE 317 L
1.4441	X2CrNiMo18-15-3	Fe-≤0.03C-≤1.00Si-≤2.00Mn-≤0.025P-≤0.01S-≤17.00-19.00Cr-≤0.50Cu-≤2.50-3.20Mo-≤0.10N-≤13.00-15.50Ni	
1.4442	X2CrNiMoN18-15-4	Fe-≤0.03C-1.0Si-2.0-3.0Mn-≤0.025P-≤0.010S-17.0-18.5Cr-3.7-4.2Mo-0.1-0.2N-14.0-16.0Ni	SAE 317 LN
1.4447	X5CrNiMo18-13	Fe-≤0.06C-1.5Si-2.0Mn-≤0.025P-≤0.020S-17.0-19.0Cr-4.0-5.0Mo-12.5-15.5Ni	
1.4449	X3CrNiMo18-12-3	Fe-≤0.035C-≤1.00Si-≤2.00Mn-≤0.045P-≤0.015S-≤17.0-18.2Cr-≤1.0Cu-≤2.25-2.75Mo-≤0.08N-≤11.5-12.5Ni	SAE 317
1.4457	X8CrNiMo17-5-3	Fe-≤0.07-0.11C-≤0.50Si-≤0.50-1.25Mn-≤16.00-17.00Cr-≤2.50-3.25Mo-≤4.00-5.00Ni	

German Standard		Materials Compositions	US-Standard
Mat.-No.	DIN-Design	Percent in Weight	SAE/ASTM/UNS
1.4460	X3CrNiMoN27-5-2	Fe-≤0.05C-≤1.0Si-≤2.0Mn-≤0.035P-≤0.015S-≤25.0-28.0Cr-≤1.3-2.0Mo-≤0.05-0.2N-≤4.5-6.5Ni	SAE 329
1.4462	X2CrNiMoN22-5-3	Fe-≤0.03C-≤1.00Si-≤2.00Mn-≤0.035P-≤0.015S-≤21.0-23.0Cr-≤2.50-3.50Mo-≤4.50-6.50Ni-≤0.1-0.22N	2205; A 182
1.4463	GX6CrNiMo24-8-2	Fe-≤0.07C-1.5Si-1.5Mn-≤0.045P-≤0.030S-23.0-25.0Cr-2.0-2.5Mo-7.0-8.5Ni	
1.4464	GX40CrNiMo27-5	Fe-0.30-0.50C-≤2.00Si-≤1.50Mn-≤0.045P-≤0.030S-26.00-28.00Cr-2.00-2.50Mo-4.00-6.00Ni	
1.4465	X1CrNiMoN25-25-2	Fe-≤0.02C-≤0.70Si-≤2.0Mn-≤0.020P-≤0.015S-≤24.0-26.0Cr-≤2.0-2.5Mo-≤22.0-25.0Ni-≤0.08-0.16N	SAE 310 MoLN
1.4466	X1CrNiMoN25-22-2	Fe-≤0.02C-≤0.7Si-≤2.0Mn-≤0.025P-≤0.010S-≤24.0-26.0Cr-≤2.0-2.5Mo-≤21.0-23.0Ni-≤0.1-0.16N	
1.4467	X2CrMnNiMoN26-5-4	Fe-≤0.03C-≤0.8Si-≤4.0-6.0Mn-≤0.03P-≤0.015S-≤24.5-26.5Cr-≤2.0-3.0Mo-≤0.3-0.45N-≤3.5-4.5Ni	
1.4469	GX2CrNiMoN26-7-4	Fe-≤0.03C-1.0Si-1.0Mn-≤0.035P-≤0.025S-25.0-27.0Cr-≤1.30Cu-3.0-5.0Mo-0.12-0.22N-6.0-8.0Ni	UNS J93404
1.4492	X 8 CrNiMoN 17 5	Fe-≤0.07-0.11C-≤0.5Si-≤0.5-1.25Mn-≤0.04P-≤0.03S-≤16.0-17.0Cr-≤2.5-3.25Mo-≤4.0-5.0Ni	
1.4500	GX7NiCrMoCuNb25-20	Fe-≤0.08C-≤1.50Si-≤2.0Mn-≤0.045P-≤0.03S-≤19.0-21.0Cr-≤1.5-2.5Cu-≤2.5-3.5Mo-≤24.0-26.0Ni	A 351
1.4501	X2CrNiMoCuWN25-7-4	Fe-≤0.03C-≤1.0Si-≤1.0Mn-≤0.035P-≤0.015S-≤24.0-26.0Cr-≤0.5-1.0Cu-≤3.0-4.0Mo-≤0.2-0.3N-≤6.0-8.0Ni-≤0.5-1.0W	UNS S32760
1.4502	X8CrTi18	Fe-≤0.09C-≤1.40Si-≤1.40Mn-≤0.030P-≤0.020S-16.70-18.30Cr-≤0.060N-0.35-0.65Ti	
1.4503	X3NiCrCuMoTi27-23	Fe-≤0.04C-0.75Si-0.75Mn-≤0.030P-≤0.015S-22.0-24.0Cr-2.5-3.5Cu-2.5-3.0Mo-26.0-28.0Ni-0.4-0.7Ti	
1.4504		Fe-≤0.09C-≤0.50Si-≤1.00Mn-≤0.025P-≤0.025S-≤0.75-1.25Al-≤16.00-17.25Cr-≤6.50-7.75Ni	UNS S17780 SAE S17700 A 705 (631)
1.4505	X4NiCrMoCuNb20-18-2	Fe-≤0.05C-≤1.0Si-≤2.0Mn-≤0.045P-≤0.015S-≤16.5-18.5Cr-≤2.0-2.5Mo-≤19.0-21.0Ni-≤1.8-2.2Cu	

German Standard		Materials Compositions	US-Standard
MatNo.	DIN-Design	Percent in Weight	SAE/ASTM/UNS
1.4506	X5NiCrMoCuTi20-18	Fe-≤0.07C-≤1.0Si-≤2.0Mn-≤0.045P-≤0.030S-≤16.5-18.5Cr-≤2.0-2.5Mo-≤19.0-21.0Ni-≤1.8-2.2Cu	
1.4507	X2CrNiMoCuN25-6-3	Fe-≤0.03C-≤0.7Si-≤2.0Mn-≤0.035P-≤0.015S-≤24.0-26.0Cr-≤1.0-2.5Cu-≤0.15-0.30N-≤5.5-7.5Ni-≤2.7-4.0Mo	UNS S43940
1.4509	X2CrTiNb18	Fe-≤0.03C-≤1.0Si-≤1.0Mn-≤0.040P-≤0.015S-≤17.5-18.5Cr-≤0.10-0.60Ti	
1.4510	X6CrTi17	Fe-≤0.05C-≤1.0Si-≤1.0Mn-≤0.040P-≤0.015S-≤16.0-18.0Cr	SAE 430 Ti
1.4511	X3CrNb17; X6CrNb17	Fe-≤0.05C-≤1.0Si-≤1.0Mn-≤0.040P-≤0.015S-≤16.0-18.0Cr	
1.4512	X2CrTi12	Fe-≤0.03C-≤1.0Si-≤1.0Mn-≤0.040P-≤0.015S-≤10.5-12.5Cr	SAE 409
1.4515	GX2CrNiMoCuN26-6-3	Fe-≤0.03C-≤1.0Si-≤2.0Mn-≤0.030P-≤0.020S-≤0.12-0.25N-≤24.5-26.5Cr-≤2.5-3.5Mo-≤5.5-7.0Ni-≤0.8-1.3Cu	
1.4517	GX2CrNiMoCuN25-6-3-3	Fe-max. 0.03C-1.0Si-1.5Mn-0.035P-0.025S-24.5-26.5Cr-2.75-3.50Cu-2.50-3.50Mo-0.12-0.22N-5.0-7.0Ni	UNS J93372
1.4519	X2CrNiMoCu20-25	Fe-≤0.02C-≤1.40Si-≤2.10-4.90Mn-≤0.025P-≤0.02S-≤19.20-21.80Cr-≤0.90-1.90Cu-≤4.10-5.90Mo-≤24.30-26.70Ni	
1.4520	X2CrTi17	Fe-≤0.025C-≤0.5Si-≤0.5Mn-≤0.04P-≤0.015S-≤16.0-18.0Cr-≤0.015N-≤0.3-0.6Ti	
1.4521	X2CrMoTi18-2	Fe-≤0.025C-≤1.0Si-≤0.040P-≤0.015S-≤1.0Mn-≤17.0-20.0Cr-≤1.8-2.5Mo-≤0.030N	SAE 444
1.4522	X2CrMoNb18-2	Fe-≤0.025C-≤1.0Si-≤1.0Mn-≤0.040P-≤0.015S-≤17.0-19.0Cr-≤1.8-2.3Mo-≤0.25Ni	SAE 443
1.4523	X2CrMoTiS18-2	Fe-≤0.03C-≤1.0Si-≤0.5Mn-≤0.040P-≤0.15-0.35S-≤17.5-19.0Cr-≤2.0-2.5Mo-≤0.30-0.80Ti	
1.4525	GX5CrNiCu16-4; GX4CrNi-CuNb16-4	Fe-≤0.07C-≤0.80Si-≤1.00Mn-≤0.035P-≤0.025S-≤15.00-17.00Cr-≤2.50-4.00Cu-≤0.80Mo-≤0.050N-≤0.350Nb-≤3.50-5.50Ni	
1.4527	GX4NiCrCuMo30-20-4	Fe-≤0.06C-≤1.50Si-≤1.50Mn-≤0.040P-≤0.030S-19.00-22.00Cr-3.00-4.00Cu-2.00-3.00Mo-27.50-30.50Ni	
1.4528	X105CrCoMo18-2	Fe-≤1.0-1.1C-≤1.0Si-≤1.0Mn-≤0.045P-≤0.030S-≤16.5-18.5Cr-≤1.0-1.5Mo-≤0.30-0.80Ti-≤1.3-1.8Co≤0.07-0.12V	
1.4529	X1NiCrMoCuN25-20-7	Fe-≤0.02C-≤0.50Si-≤1.00Mn-≤0.030P-≤0.010S-≤19.0-21.0Cr-≤6.0-7.0Mo-≤0.5-1.5Cu-≤0.15-0.25N-≤24.0-26-0Ni	A 249; ASTM N08926

German Standard		Materials Compositions	US-Standard
MatNo.	DIN-Design	Percent in Weight	SAE/ASTM/UNS
1.4530	X1CrNiMoAlTi12-9	Fe-≤0.015C-0.10Si-0.10Mn-≤0.020P-0.6-1.0Al-11.0-12.5Cr-1.5-2.5Mo-8.5-10.5Ni-0.25-0.40Ti	
1.4532	X8CrNiMoAl15-7-2	Fe-≤0.1C≤0.7Si≤1.2Mn≤0.04P≤0.015S-0.7-1.5Al-14.0-16.0Cr-2.0-3.0Mo-6.5-7.8Ni	UNS S15700
1.4533	X6CrNiTi18-10S	Fe-≤0.06C-≤1.00Si-≤2.00Mn-≤0.035P-≤0.015S-≤0.20Co-≤17.00-19.00Cr-≤9.00-12.00Ni	
1.4534	X3CrNiMoAl13-8-2	Fe-≤0.05C-≤0.10Si-≤0.10Mn-≤0.010P-≤0.008S-≤0.90-1.20Al-≤12.25-13.25Cr-≤2.0-2.50Mo-≤0.01N-≤7.50-8.50Ni	S 13800
1.4536	GX2NiCrMoCuN25-20	Fe-≤0.03C≤1.0Si≤1.0Mn≤0.035P≤0.01S-19.0-21.0Cr-1.5-2.0Cu-2.5-3.5Mo-24.0-26.0Ni	UNS J94650
1.4537	X1CrNiMoCuN25-25-5	Fe-≤0.02C-≤0.70Si-≤2.00Mn-≤0.030P-≤0.010S-24.00-26.00Cr-1.00-2.00Cu-4.70-5.70Mo-0.170-0.250N-24.00-27.00Ni	
1.4539	X1NiCrMoCu25-20-5	Fe-≤0.02C-≤0.70Si-≤2.0Mn-≤0.030P-≤0.010S-≤19.0-21.0Cr-≤4.0-5.0Mo-≤24.0-26.0Ni-≤1.2-2.0Cu-≤0.150N	SAE 904 L
1.4540		Fe-≤0.06C-1.0Si-1.0Mn-15.0-17.0Cr-2.5-4.0Cu-0.050N-0.15-0.40Nb-3.5-5.0Ni	UNS J92130
1.4541	X6CrNiTi18-10	Fe-≤0.08C-≤1.0Si-≤2.0Mn-≤0.045P-≤0.015S-≤17.0-19.0Cr-≤9.0-12.0Ni	SAE 321
1.4542	X5CrNiCuNb16-4	Fe-≤0.07C-≤0.70Si-≤1.5Mn-≤0.040P-≤0.015S-15.0-17.0Cr-≤3.0-5.0Ni-≤3.0-5.0Cu-≤0.60Mo	SAE 630; 17-4 PH
1.4544	X 10 CrNiMnTi 18 10	Fe-≤0.08C-≤1.0Si-≤2.0Mn-≤0.035P-≤0.025S-≤17.0-19.0Cr-≤9.0-11.5Ni	SAE 321; UNS J92630
1.4545		Fe-≤0.07C-≤1.00Si-≤1.00Mn-≤0.03P-≤0.015S-≤14.00-15.50Cr-≤2.50-4.50Cu-≤0.50Mo-≤3.50-5.50Ni	UNS S15500 SAE S15500 A 705 (XM-12)
1.4546	X5CrNiNb18-10	Fe-≤0.08C-≤1.0Si-≤2.0Mn-≤0.045P-≤0.030S-≤17.0-19.0Cr-≤9.0-11.5Ni	SAE 347
1.4547	X1CrNiMoCuN20-18-7	Fe-≤0.02C-≤0.70Si-≤1.0Mn-≤0.030P-≤0.010S-≤19.5-20.5Cr-≤0.5-1.0Cu-≤6.00-7.00Mo-≤0.18-0.25N-≤17.5-18.5Ni	254 SMO; A 182
1.4548	X5CrNiCuNb17-4-4	Fe-≤0.07C-≤1.0Si-≤1.0Mn-≤0.025P-≤0.025S-≤15.0-17.5Cr-≤3.0-5.0Cu-≤0.15-0.45Nb-≤3.00-5.00Ni	17-4 PH; SAE 630
1.4550	X6CrNiNb18-10	Fe-≤0.08C-≤1.0Si-≤2.0Mn-≤0.045P-≤0.015S-≤17.0-19.0Cr≤9.0-12.0Ni	SAE 347

German Standard Mat.-No.	DIN-Design	Materials Compositions Percent in Weight	US-Standard SAE/ASTM/UNS
1.4551	X5CrNiNb19-9	Fe-≤0.06C-≤1.40Si-≤1.90Mn-≤0.025P-≤0.015S-≤18.20-19.80Cr-≤8.20-9.80Ni	UNS S34780 S34781 S34788
1.4552	GX5CrNiNb19-11	Fe-≤0.07C≤1.5Si≤1.5Mn≤0.04P≤0.03S-18.0-20.0Cr-9.0-12.0Ni	UNS J92710
1.4557	GX2CrNiMoCuN20-18-6	Fe-≤0.025C-≤1.00Si-≤1.20Mn-≤0.03P-≤0.01S-≤19.50-20.50Cr-≤0.50-1.00Cu-≤6.00-7.00Mo-≤0.18-0.24N-≤17.50-19.50Ni	
1.4558	X2NiCrAlTi32-20	Fe-≤0.03C-≤0.70Si-≤1.0Mn-≤0.020P-≤0.015S-≤0.15-0.45Al≤20.0-23.0Cr-≤32.0-35.0Ni	
1.4561	X1CrNiMoTi18-13-2	Fe-≤0.02C-≤0.50Si-≤2.0Mn-≤0.035P-≤0.015S-≤17.0-18.5Cr-≤2.0-2.5Mo-≤11.5-13.5Ni-≤0.4-0.6Ti	
1.4562	X1NiCrMoCu32-28-7	Fe-≤0.015C-≤0.30Si-≤2.0Mn-≤0.020P-≤0.010S-≤26.0-28.0Cr-≤1.0-1.4Cu-≤6.0-7.0Mo-≤0.15-0.25N-≤30.0-32.0Ni	Alloy 31
1.4563	X1NiCrMoCu31-27-4	Fe-≤0.02C-≤0.70Si-≤2.0Mn-≤0.030P-≤0.010S-≤26.0-28.0Cr-≤3.0-4.0Mo-≤30.0-32.0Ni-≤0.70-1.5Cu-≤0.11N	B 668
1.4564		Fe-≤0.09C-≤1.00Si-≤1.00Mn-≤0.04P-≤0.03S-≤0.75-1.50Al-≤16.00-18.00Cr-≤0.50Cu-≤6.50-7.75Ni	UNS 17700 SAE S17700 A 705 (631)
1.4565	X2CrNiMnMoNbN25-18-5-4	Fe-≤0.03C-≤1.0Si-≤3.5-6.5Mn-≤0.030P-≤0.015S-≤23.0-26.0Cr-≤3.0-5.0Mo-≤0.3-0.5N-≤0.15Nb-≤16.0-19.0Ni	UNS S34565
1.4566	X3CrNiMnMoCuNbN 23-17-5-3	Fe-≤0.04C≤1.0Si-4.5-6.5Mn≤0.03P-≤0.015S-21.0-25.0Cr-0.3-1.0Cu-3.0-4.5Mo-15.0-18.0Ni-0.1-0.3Nb	
1.4567	X3CrNiCu18-9-4; X 3 CrNiCu 18 9	Fe-≤0.04C-≤1.00Si-≤2.00Mn-≤0.045P-≤0.015S-≤17.00-19.00Cr-≤3.00-4.00Cu-≤0.110N-≤8.50-10.50Ni	304 Cu (SAE) S 30430 (UNS) A 493 (S 30430) (ASTM)
1.4568	X7CrNiMoAl17-7	Fe-≤0.09C-≤0.70Si-≤1.0Mn-≤0.040P-0.015S-≤16.0-18.0Cr-≤6.5-7.80Ni-≤0.70-1.5Al	17-7 PH; SAE 631
1.4571	X6CrNiMoTi17-12-2	Fe-≤0.08C-≤1.0Si-≤2.0Mn-≤0.045P-0.015S-≤16.5-18.5Cr-≤2.0-2.5Mo-≤10.5-13.5Ni	SAE 316 Ti
1.4573	GX3CrNiMoCuN24-6-5	Fe-≤0.40C-≤1.0Si-≤1.0Mn-≤0.030P-≤0.020S-≤22.0-25.0Cr-≤1.5-2.5Cu-≤4.5-6.0Mo-≤0.15-0.25N-≤4.5-6.5Ni	SAE 316 Ti
1.4574		Fe-≤0.09C-≤1.00Si-≤1.00Mn-≤0.040P-≤0.030S-≤0.75-1.50Al-≤14.00-16.00Cr-≤2.00-3.00Mo-≤6.50-7.75Ni	S 15700 (SAE) A 579 (63) (ASTM) S 15700 (UNS)

German Standard		Materials Compositions	US-Standard
Mat.-No.	DIN-Design	Percent in Weight	SAE/ASTM/UNS
1.4575	X1CrNiMoNb28-4-2	Fe-≤0.015C-≤1.0Si-≤1.0Mn-≤0.025P-≤0.015S-≤26.0-30.0Cr-≤1.8-2.5Mo-≤0.035N-≤3.0-4.5Ni	25-4-4; A 176
1.4577	X3CrNiMoTi25-25	Fe-≤0.04C-≤0.50Si-≤2.0Mn-≤0.030P-≤0.015S-≤24.0-26.0Cr-≤2.0-2.5Mo-≤24.0-26.0Ni	
1.4580	X6CrNiMoNb17-12-2	Fe-≤0.08C-≤1.0Si-≤2.0Mn-≤0.045P-≤0.015S-≤16.5-18.5Cr-≤2.0-2.5Mo-≤10.5-13.5Ni	SAE 316 Cb UNS J92971
1.4581	GX5CrNiMoNb19-11-2	Fe-≤0.07C≤1.5Si≤1.5Mn≤0.04P≤0.03S-18.0-20.0Cr≤2.0-2.5Mo-9.0-12.0Ni	
1.4582	X4CrNiMoNb25-7	Fe-≤0.06C-≤1.00Si-≤2.00Mn-≤0.045P-≤0.030S-≤24.00-26.00Cr-≤1.30-2.00Mo-≤6.50-7.50Ni	
1.4583	X10CrNiMoNb18-12	Fe-≤0.10C-≤1.00Si-≤2.00Mn-≤0.045P-≤0.030S-≤16.5-18.5Cr-≤2.5-3.0Mo-≤12.0-14.5Ni	318 (Spec)
1.4585	GX7CrNiMoCuNb1818	Fe-≤0.080C-≤1.50Si-≤2.0Mn-≤0.045P-≤0.030S-≤16.5-18.5Cr-≤2.0-2.5Mo-≤19.0-21.0Ni-≤1.8-2.4Cu	
1.4586	X5NiCrMoCuNb22-18	Fe-≤0.07C-1.0Si-2.0Mn-≤0.045P-≤0.030S-16.5-18.5Cr-1.5-2.0Cu-3.0-3.5Mo-21.5-23.5Ni	
1.4589	X5CrNiMoTi15-2	Fe-≤0.080C-≤1.0Si-≤1.0Mn-≤0.045P-≤0.030S-≤13.0-15.5Cr-≤0.2-1.2Mo-≤1.0-2.5Ni-≤0.3-0.5Ti	UNS S42035
1.4591	X1CrNiMoCuN33-32-1	Fe-≤0.015C-≤0.5Si-≤2.0Mn-≤0.020P-≤0.010S-≤31.0-35.0Cr-≤0.3-1.2Cu-≤0.5-2.0Mo-≤0.35-0.6N-≤30.0-33.0Ni	
1.4592	X1CrMoTi29-4	Fe-≤0.025C-1.0Si-1.0Mn-≤0.030P-≤0.010S-28.0-30.0Cr-3.5-4.5Mo-0.045N	
1.4593	GX3CrNiMoCuN24-6-2-3	Fe-≤0.04C-1.5Si-1.5Mn-≤0.030P-≤0.020S-23.0-26.0Cr-2.75-3.5Cu-2.0-3.0Mo-0.1-0.2N-5.0-8.0Ni	
1.4603	X1CrTi17	Fe-≤0.02C-≤1.00Si-≤1.00Mn-≤0.040P-≤0.015S-≤16.00-18.00Cr	
1.4604	X2CrTi20	Fe-≤0.03C-≤1.00Si-≤1.00Mn-≤0.040P-≤0.015S-≤19.00-21.00Cr-≤0.40-0.80Ti	
1.4652	X1CrNiMoCuN24-22-8	Fe-≤0.02C-≤0.5Si-≤2.0-4.0Mn-≤0.03P-≤0.005S-≤23.0-25.0Cr-≤0.3-0.6Cu-≤7.0-8.0Mo-≤0.45-0.55N-≤21.0-23.0Ni	
1.4712	X10CrSi6	Fe-≤0.12C-≤2.00-2.50Si-≤1.00Mn-≤0.045P-≤0.030S-≤5.50-6.50Cr	

Key to materials compositions

German Standard		Materials Compositions	US-Standard
Mat.-No.	DIN-Design	Percent in Weight	SAE/ASTM/UNS
1.4713	X10CrAl7; X10CrAlSi7	Fe-≤0.12C-≤0.50-1.00Si-≤1.00Mn-≤0.040P-≤0.015S-≤0.5-1.0Al-≤6.00-8.00Cr	
1.4718	X45CrSi9-3	Fe-0.4-0.5C-2.7-3.3Si-≤0.6Mn-≤0.04P-≤0.03S-8.0-10.0Cr-≤0.5Ni	S65007 (UNS)
1.4720	X7CrTi12	Fe-≤0.08C-1.0Si-1.0Mn-≤0.040P-≤0.030S-10.5-12.5Cr	SAE 409
1.4722	X 10 CrSi 13	Fe-≤0.12C-≤1.90-2.40Si-≤1.00Mn-≤0.045P-≤0.030S-≤12.0-14.0Cr	
1.4724	X10CrAl13; X10CrAlSi13	Fe-≤0.12C-≤0.70-1.40Si-≤1.00Mn-≤0.040P-≤0.015S-≤0.70-1.20Al-≤12.0-14.0Cr	
1.4725	CrAl 14 4; (X8CrAl14-4)	Fe-≤0.1C-≤0.5Si-≤1.0Mn-≤0.045P-≤0.03S-3.5-5.0Al-13.0-15.0Cr	K91670 (UNS)
1.4742	X10CrAlSi18; X10CrAl18	Fe-≤0.12C-≤0.70-1.40Si-≤1.00Mn-≤0.040P-≤0.015S-≤0.70-1.20Al-≤17.00-19.00Cr	
1.4749	X18CrN28	Fe-0.15-0.20C-1.0Si-1.0Mn-≤0.040P-≤0.015S-26.0-29.0Cr-0.15-0.25N	
1.4762	X10 CrAl 24; X10CrAlSi25	Fe-≤0.12C-≤0.70-1.40Si-≤1.00Mn-≤0.040P-≤0.015S-≤1.20-1.70Al-≤23.0-26.0Cr	SAE 446
1.4765	CrAl 25 5; (X8CrAl25-5)	Fe-≤0.10C-≤1.00Si-≤0.60Mn-≤0.045P-≤0.030S-≤4.50-6.00Al-≤22.00-25.00Cr	
1.4773	X8Cr30	Fe-≤0.09C-≤1.90Si-≤1.40Mn-≤0.030P-≤0.025S-≤28.80-31.20Cr-≤2.00Ni	
1.4776	GX40CrSi28	Fe-0.30-0.50C-1.0-2.5Si-1.0Mn-≤0.040P-≤0.030S-27.0-30.0Cr-0.50Mo-1.0Ni	UNS J92605
1.4777	GX130CrSi29	Fe-1.20-1.40C-1.0-2.5Si-0.5-1.0Mn-≤0.035P-≤0.030S-27.0-30.0Cr-0.50Mo-1.0Ni	
1.4821	X15CrNiSi25-4; X20 CrNiSi 25 4	Fe-0.1-0.2C-0.8-1.5Si-≤2.0Mn-≤0.04P-≤0.015S-24.5-26.5Cr-≤0.11N-3.5-5.5Ni	
1.4828	X15CrNiSi20-12	Fe-≤0.20C-≤1.50-2.50Si-≤2.0Mn-≤0.045P-≤0.015S-≤19.0-21.0Cr-≤0.11N-≤11.0-13.0Ni	SAE 309
1.4829	X12CrNi22-12	Fe-≤0.14C-0.90-1.90Si-1.90Mn-≤0.025P-≤0.015S-20.8-23.2Cr-10.2-12.8Ni	UNS S30980
1.4833	X7CrNi23 14; X12CrNi23-14	Fe-≤0.15C-≤1.00Si-≤2.00Mn-≤0.045P-≤0.015S-≤22.0-24.0Cr-≤0.11N-≤12.0-14.0Ni	SAE 309 S
1.4835	X9CrNiSiNCe21-11-2	Fe-≤0.05-0.12C-≤1.4-2.5Si-≤1.0Mn-≤0.045P-≤0.015S-≤0.030-0.080Ce-≤20.0-22.0Cr-≤0.12-0.20N-≤10.0-12.0Ni	253 MA; A 182
1.4841	X15CrNiSi25-20; X15CrNiSi25-21	Fe-≤0.20C-≤1.50-2.50Si-≤2.00Mn-≤0.045P-≤0.015S-≤24.0-26.0Cr-≤0.11N-≤19.0-22.0Ni	3RE60; SAE 310; SAE 314
1.4845	X8CrNi25-21; X12CrNi25-21	Fe-≤0.15C-≤1.50Si-≤2.00Mn-≤0.045P-≤0.015S-≤24.0-26.0Cr-≤0.11N-≤19.0-22.0Ni	SAE 310 S

German Standard		Materials Compositions	US-Standard
MatNo.	DIN-Design	Percent in Weight	SAE/ASTM/UNS
1.4847	X8CrNiAlTi20-20	Fe-≤0.08C-≤1.0Si-≤1.0Mn-≤0.030P-≤0.015S-≤0.6Al-18.0-22.0Cr-18.0-22.0Ni-0.6Ti	334 (SAE)
1.4848	GX40CrNiSi25-20	Fe-≤0.30-0.50C-≤1.00-2.50Si-≤1.50Mn-≤0.035P-≤0.030S-≤24.0-26.0Cr-≤19.0-21.0Ni	A 297 (HK)
1.4856	GX40NiCrSiNbTi35-25	Fe-≤0.35-0.45C-≤1.00-1.50Si-≤0.5-1.50Mn-≤0.035P-≤0.030S-≤23.0-27.0Cr-≤0.9-1.5Nb-≤33.0-37.0Ni-≤0.10-0.25Ti	
1.4857	GX40NiCrSi35-25	Fe-≤0.30-0.50C-≤1.00-2.50Si-≤1.50Mn-≤0.035P-≤0.030S-≤24.0-26.0Cr-≤34.0-36.0Ni	A 297 (HP)
1.4862	X8NiCrSi38-18	Fe-≤0.1C-1.5-2.5Si-0.8-1.5Mn-≤0.03P-≤0.03S-17.0-19.0Cr-≤0.5Cu-35.0-39.0Ni-≤0.2Ti	
1.4864	X12NiCrSi35-16; X12 NiCrSi 36 16	Fe-≤0.15C-≤1.0-2.0Si-≤2.0Mn-≤0.045P-≤0.015S-≤15.0-17.0Cr-≤0.11N-≤33.0-37.0Ni	SAE 330
1.4871	X53CrMnNiN21-9	Fe-≤0.48-0.58C-≤0.25Si-≤8.00-10.00Mn-≤0.045P-≤0.030S-≤20.00-22.00Cr-≤0.35-0.50N-≤3.25-4.50Ni	S 63008 (UNS) EV 8 (SAE)
1.4873	X45CrNiW18-9	Fe-≤0.40-0.50C-≤2.00-3.00Si-≤0.80-1.50Mn-≤0.045P-≤0.030S-≤17.00-19.00Cr-≤8.00-10.00Ni-≤0.80-1.20W	
1.4875	X55CrMnNiN20-8	Fe-≤0.50-0.60C-≤0.25Si-≤7.00-10.00Mn-≤0.045P-≤0.030S-≤19.50-21.50Cr-≤0.20-0.40N-≤1.50-2.75Ni	S 63012 (UNS) EV 12 (SAE)
1.4876	X10NiCrAlTi32-21; X10 NiCrAlTi 32 20	Fe-≤0.12C-≤1.00Si-≤2.00Mn-≤0.030P-≤0.015S-≤0.15-0.60Al-≤19.00-23.00Cr-≤30.00-34.00Ni-≤0.15-060Ti	N 08332 (UNS) B 366 (N08332) (ASTM) N 8810 (SAE)
1.4877	X6NiCrNbCe32-27	Fe-0.04-0.08C-≤0.3Si-≤1.0Mn-≤0.02P-≤0.01S-≤0.025Al-0.05-0.1Ce-26.0-28.0Cr-0.11N-0.6-1.0Nb-31.0-33.0Ni	S33228 (UNS)
1.4878	X10CrNiTi18-10; X12 CrNiTi 18 9	Fe-≤0.10C-≤1.0Si-≤2.0Mn-≤0.045P-0.015S-≤17.0-19.0Cr-≤9.0-12.0Ni	
1.4903	X10CrMoVNb9-1	Fe-≤0.08-0.12C-≤0.20-0.50Si-≤0.30-0.60Mn-≤0.020P-≤0.010S-≤0.04Al-≤8.0-9.5Cr-≤0.85-1.05Mo-≤0.030-0.070N-≤0.06-0.1Nb-≤0.4Ni-≤0.18-0.25V	A 182
1.4913	X19CrMoNbVN11-1	Fe-0.17-0.23C-0.50Si-0.40-0.90Mn-≤0.025P-≤0.015S-≤0.02Al-≤0.0015B-10.0-11.5Cr-0.5-0.8Mo-0.05-0.1N-0.25-0.55Nb-0.20-0.60Ni-0.10-0.30V	

Key to materials compositions

German Standard		Materials Compositions	US-Standard
Mat.-No.	DIN-Design	Percent in Weight	SAE/ASTM/UNS
1.4919	X6CrNiMo17-13	Fe-≤0.04-0.08C-≤0.75Si-≤2.0Mn-≤0.035P-≤0.015S-≤0.0015-0.0050B-≤16.0-18.0Cr-≤2.0-2.5Mo-≤0.11N-≤12.0-14.0Ni	SAE 316 H
1.4922	X20CrMoV11-1	Fe-≤0.17-0.23C-≤0.50Si-≤1.00Mn-≤0.030P-≤0.030S-≤10.0-12.5Cr-≤0.80-1.20Mo-≤0.30-0.80Ni-≤0.25-0.35V	
1.4943	X4NiCrTi25-15	Fe-≤0.06C-≤1.0Si-≤2.0Mn-≤0.025P-≤0.015S-≤0.35Al-≤0.003-0.01B-≤13.5-16Cr-≤1.0-1.50Mo-≤24.00-27.00Ni-≤1.70-2.00Ti-≤0.10-050V	SAE HEV 7 UNS S66545 ASI S 66286 A 891
1.4944		Fe-≤0.08C-≤1.0Si-≤2.0Mn-≤0.025P-≤0.015S-≤0.35Al-≤0.003-0.01B-≤13.50-16.0Cr-≤1.0-1.50Mo-≤24.00-27.00Ni-≤1.90-2.30Ti-≤0.10-0.50V	UNS S66286; ASI 660; A 638
1.4947		Fe-≤0.07C-0.50-1.2Si-1.5-2.0Mn-≤0.035P-≤0.025S-22.0-23.0Cr-≤0.3Cu-≤0.75Mo-9.5-10.5Ni	UNS J93001
1.4948	X6CrNi18-10	Fe-≤0.04-0.08C-≤0.75Si-≤2.0Mn-≤0.035P-≤0.015S-≤17.0-19.0Cr-≤10.0-12.0Ni	SAE 304 H
1.4958	X5NiCrAlTi31-20	Fe-0.03-0.08C-≤0.7Si-≤1.5Mn-0.015P-≤0.01S-0.2-0.5Al-0.5Co-19.0-22.0Cr-≤0.5Cu-≤0.03N-≤0.1Nb-30.0-32.5Ni-0.2-0.5Ti	N08810 (UNS)
1.4959	X8NiCrAlTi32-21	Fe-0.05-0.1C-≤0.7Si-≤1.5Mn-≤0.015P-≤0.01S-0.2-0.65Al-≤0.5Co-19.0-22.0Cr-≤0.5Cu-≤0.03N-30.0-34.0Ni-0.25-0.65Ti	N08811 (UNS)
1.4961	X8CrNiNb16-13	Fe-≤0.04-0.1C-≤0.3-0.6Si-≤1.5Mn-≤0.035P-≤0.015S-≤15.0-17.0Cr-≤12.0-14.0Ni	
1.4970	X 10 NiCrMoTiB 15 15	Fe-0.08-0.12C-0.25-0.45Si-1.6-2.0Mn-≤0.03P-≤0.015S-0.003-0.006B-14.5-15.5Cr-1.05-1.25Mo-15.0-16.0Ni-0.35-0.55Ti	
1.4971	X12CrCoNi21-20	Fe-≤0.08-0.16C-≤1.00Si-≤2.00Mn-≤0.035P-≤0.015S-≤18.50-21.00Co-≤20.00-22.50Cr-≤2.50-3.50Mo-≤0.10-0.20N-≤0.75-1.25Nb-≤19.00-21.00Ni-≤2.00-3.00W	HEV 1 (SAE) 661 (SAE) R 30155 (UNS)
1.4977	X 40 CoCrNi 20 20	Fe-0.35-0.45C-≤1.00Si-≤1.50Mn-≤0.045P-≤0.030S-19.00-21.00Co-19.00-21.00Cr-3.50-4.50Mo-3.50-4.50Nb-19.00-21.00Ni-3.50-4.50W	R 30590 UNS
1.4980	X6NiCrTiMoVB25-15-2; X5NiCrTi26-15	Fe-0.03-0.08C-≤1.0Si-1.0-2.0Mn-≤0.025P-≤0.015S-≤0.35Al-0.003-0.01B-13.5-16.0Cr-1.0-1.5Mo-24.0-27.0Ni-1.9-2.3Ti-0.1-0.5V	663 (SAE)
1.4981	X8CrNiMoNb16-16	Fe-≤0.04-0.10C-≤0.30-0.60Si-≤1.50Mn-≤0.035P-≤0.015S-≤15.5-17.5Cr-≤1.60-2.00Mo-≤15.5-17.5Ni	

German Standard		Materials Compositions	US-Standard
Mat.-No.	DIN-Design	Percent in Weight	SAE/ASTM/UNS
1.4982	X10CrNiMoMnNbVB15-10-1	Fe-≤0.07-0.13C-≤1.00Si-≤5.50-7.00Mn-≤0.040P-≤0.030S-≤0.003-0.009B-≤14.00-16.00Cr-≤0.80-1.20Mo-≤0.110N-≤0.75-1.25Nb-≤9.00-11.00Ni-≤0.15-0.40V	
1.4986	X8CrNiMoBNb16-16	Fe-≤0.04-0.1C-≤0.3-0.6Si-≤1.5Mn-≤0.045P-≤0.030S-≤0.05-0.1B-≤15.5-17.5Cr-≤1.6-2.0Mo-≤15.5-17.5Ni	
1.4988	X8CrNiMoVNb16-13	Fe-≤0.04-0.1C-≤0.3-0.6Si-≤1.5Mn-≤0.035P-≤0.015S-≤15.5-17.5Cr-≤1.1-1.5Mo-≤12.5-14.5Ni-≤0.60-0.85V-≤0.06-0.14N	
1.5069	36Mn7	Fe-≤0.35C≤0.5Si≤1.6Mn≤0.025P≤0.025S	UNS H13400
1.5094	38MnS6	Fe-≤0.35-0.40C-≤0.20-0.65Si-≤1.30-1.60Mn-≤0.045P-≤0.045-0.065S-≤0.01-0.05Al-≤0.10-0.20Cr-≤0.015-0.020N	
1.5122	37MnSi5	Fe-≤0.33-0.41C-≤1.1-1.4Si-≤1.1-1.4Mn-≤0.035P-≤0.035S	
1.5219	41MnV5	Fe-0.38-0.44C-0.1-0.4Si-1.1-1.3Mn≤0.035P≤0.035S-0.1-0.15V	
1.5415	15 Mo 3; 16Mo3	Fe-≤0.12-0.2C-≤0.35Si-≤0.4-0.9Mn-≤0.030P-≤0.025S-≤0.30Cr-≤0.30Cu-≤0.25-0.35Mo-≤0.30Ni	A 204 (A)
1.5431	G12MnMo7-4	Fe-0.08-0.15C≤0.6Si-1.5-1.8Mn≤0.02P≤0.015S≤0.2Cr-0.3-0.4Mo≤0.05Nb≤0.1V	
1.5511	35B2	Fe-≤0.32-0.39C-≤0.4Si-≤0.5-0.8Mn-≤0.035P-≤0.035S-≤0.02Al-≤0.0008-0.005B	
1.5662	X8Ni9	Fe-≤0.10C-0.35Si-0.30-0.80Mn-≤0.020P-≤0.010S-≤0.10Mo-8.5-10.0Ni-0.05V	UNS K71340
1.5680	X12Ni5; 12 Ni 19	Fe-≤0.15C-≤0.35Si-0.3-0.8Mn-≤0.02P-≤0.01S-4.75-5.25Ni-≤0.05V-≤0.5Cr-≤0.5Mo-≤0.5Cu	A 2515 (SAE)
1.5736	36NiCr10	Fe-max. 0.32-0.40C-0.15-0.35Si-0.40-0.80Mn-0.035P-0.035S-0.55-0.95Cr-2.25-2.75Ni	SAE 3435
1.6354		Fe-≤0.03C-≤0.10Si-≤0.10Mn-≤0.01P-≤0.01S-≤0.05-0.15Al-≤8.00-9.50Co-≤4.60-5.20Mo-≤17.00-19.00Ni-≤0.60-0.90Ti	UNS J93150
1.6511	36CrNiMo4	Fe-≤0.32-0.40C-≤0.4Si-≤0.5-0.8Mn-≤0.035P-≤0.035S-≤0.9-1.2Cr-≤0.15-0.3Mo-≤0.9-1.2Ni	SAE 9840
1.6545	30NiCrMo2-2	Fe-≤0.27-0.34C-≤0.15-0.4Si-≤0.7-1.0Mn-≤0.035P-≤0.035S-≤0.4-0.6Cr-≤0.15-0.3Mo-≤0.4-0.7Ni	SAE 8630

German Standard		Materials Compositions	US-Standard
Mat.-No.	DIN-Design	Percent in Weight	SAE/ASTM/UNS
1.6562	40 NiCrMo 8 4	Fe-≤0.37-0.44C-≤0.20-0.35Si-≤0.70-0.90Mn-≤0.02P-≤0.015S-≤0.005-0.05Al-≤0.70-0.95Cr-≤0.30-0.40Mo-≤1.65-2.00Ni	UNS G43406 SAE 4340 UNS H 43406 A 829 SAE E 4340 H
1.6565	40NiCrMo6	Fe-0.35-0.45C-≤0.15-0.35Si-0.50-0.70Mn-≤0.035P-≤0.035S-0.90-1.4Cr-≤0.20-0.30Mo-≤1.4-1.7Ni	SAE 4340
1.6580	30CrNiMo8	Fe-≤0.26-0.34C-≤0.4Si-≤0.3-0.6Mn-≤0.035P-≤0.035S-≤1.8-2.2Cr-≤0.3-0.5Mo-≤1.8-2.2Ni	
1.6582	34CrNiMo6	Fe-≤0.3-0.38C-≤0.4Si-≤0.5-0.8Mn-≤0.035P-≤0.035S-≤1.3-1.7Cr-≤0.15-0.3Mo-≤1.3-1.7Ni	
1.6751	22NiMoCr3-7	Fe-≤0.17-0.25C-≤0.35Si-≤0.5-1.0Mn-≤0.02P-≤0.02S-≤0.05Al-≤0.3-0.5Cr-≤0.18Cu-≤0.5-0.8Mo-≤0.6-1.2Ni-≤0.03V	A 508
1.6900	X 12 CrNi 18 9	Fe-≤0.12C-≤1.00Si-≤2.00Mn-≤0.045P-≤0.030S-≤17.00-19.00Cr-≤0.5Mo-≤8.00-10.00Ni	UNS J92801
1.6903	X 10 CrNiTi 18 10	Fe-≤0.10C-≤1.00Si-≤2.00Mn-≤0.045P-≤0.030S-≤17.0-19.0Cr-≤0.5Mo-≤10.0-12.0Ni	
1.6906	X 5 CrNi 18 10	Fe-≤0.07C-≤1.0Si-≤2.0Mn-≤0.045P-≤0.030S-≤17.0-19.0Cr-≤0.50Mo-≤9.0-11.5Ni	
1.6932	28NiCrMoV8-5	Fe-0.24-0.32C-≤0.4Si-0.15-0.4Mn-≤0.035P-≤0.035S-1.0-1.5Cr-0.35-0.55Mo-1.8-2.1Ni-0.05-0.15V	
1.6944		Fe-≤0.35-0.40C-≤0.15-0.35Si-≤0.50-0.80Mn-≤0.015P-≤0.01S-≤0.65-0.90Cr-≤0.30-0.40Mo-≤1.65-2.00Ni-≤0.08-0.15V	
1.6948	27NiCrMoV11-6; 26NiCr-MoV11-5	Fe-≤0.22-0.32C-≤0.15Si-≤0.15-0.40Mn-≤0.010P-≤0.007S-≤1.20-1.80Cr-≤0.25-0.45Mo-≤2.40-3.10Ni-≤0.05-0.15V	
1.6952	24NiCrMoV14-6	Fe-≤0.20-0.28C-≤0.15-0.40Si-≤0.30-0.60Mn-≤0.035P-≤0.035S-≤1.20-1.80Cr-≤0.35-0.55Mo-≤3.00-3.80Ni-≤0.04-0.12V	K 42885 (UNS) A 649 (6, 7, 8) (ASTM) A 470 (5, 6, 7) (ASTM)
1.6956	33NiCrMoV14-5; 33NiCrMo14-5	Fe-0.28-0.38C-≤0.40Si-0.15-0.40Mn-≤0.035P-≤0.035S-≤1.00-1.70Cr-≤0.30-0.60Mo-≤2.90-3.80Ni-≤0.08-0.25V	
1.6957	27NiCrMoV15-6	Fe-0.22-0.32C≤0.15Si-0.15-0.4Mn≤0.01P≤0.007S-1.2-1.8Cr-0.25-0.45Mo-3.4-4.0Ni-0.05-0.15V	ASTM A 470
1.7005	45Cr2	Fe-≤0.42-0.48C-≤0.15-0.40Si-≤0.50-0.80Mn-≤0.025P-≤0.035S-≤0.40-0.60Cr	
1.7033	34Cr4	Fe-0.3-0.37C-≤0.4Si-≤0.6-0.9Mn-≤0.035P-≤0.035S-≤0.9-1.2Cr	UNS G51320

German Standard		Materials Compositions	US-Standard
MatNo.	DIN-Design	Percent in Weight	SAE/ASTM/UNS
1.7035	41Cr4	Fe-≤0.38-0.45C-≤0.4Si-≤0.6-0.9Mn-≤0.035P-≤0.035S-≤0.9-1.2Cr	SAE 5140; UNS H51400
1.7120		Fe-≤0.1C-≤0.25Si-≤0.45Mn-≤0.16Cu-≤0.07Ni-≤0.05Cr-≤0.035P-≤0.035S	
1.7131	16MnCr5	Fe-0.14-0.19C-≤0.40Si-1.00-1.30Mn-≤0.035P-≤0.035S-0.80-1.10Cr	G 51170 UNS A 711 (5115) ASTM
1.7147	20MnCr5	Fe-≤0.17-0.22C-≤0.40Si-≤1.10-1.40Mn-≤0.035P-≤0.035S-≤1.00-1.30Cr	UNS H51200 A 752 (5120) SAE 5120H
1.7214		Fe-≤0.22-0.29C-≤0.15-0.35Si-≤0.5-0.8Mn-≤0.02P-0.015S-≤0.90-1.20Cr-≤0.15-0.20Mo-0.30Ni	
1.7218	25CrMo4	Fe-≤0.22-0.29C-≤0.40Si-≤0.60-0.90Mn-≤0.035P-≤0.035S-≤0.90-1.20Cr-≤0.15-0.30Mo	SAE 4130
1.7219	26 CrMo 4; 26CrMo4-2	Fe-≤0.22-0.29C-≤0.35Si-≤0.5-0.8Mn-≤0.03P-≤0.025S-≤0.9-1.2Cr-≤0.15-0.30Mo	A 372
1.7220	34CrMo4	Fe-≤0.3-0.37C-≤0.4Si-≤0.6-0.9Mn-≤0.035P-≤0.035S-≤0.9-1.2Cr-≤0.15-0.30Mo	SAE 4130
1.7225	42CrMo4	Fe-≤0.38-0.45C-≤0.40Si-≤0.60-0.90Mn-≤0.035P-≤0.035S-≤0.90-1.20Cr-≤0.15-0.30Mo	UNS G41400 A 866 (4140) SAE 4140 RH
1.7242	16CrMo4	Fe-≤0.13-0.20C-≤0.15-0.35Si-≤0.50-0.80Mn-≤0.035P-≤0.035S-≤0.90-1.20Cr-≤0.20-0.30Mo-≤0.40Ni	
1.7259	26CrMo7	Fe-≤0.22-0.30C-≤0.15-0.35Si-≤0.50-0.70Mn-≤0.035P-≤0.035S-≤1.50-1.80Cr-≤0.20-0.25Mo	
1.7273	24CrMo10	Fe-≤0.20-0.28C-≤0.15-0.35Si-≤0.50-0.80Mn-≤0.035P-≤0.035S-≤2.30-2.60Cr-≤0.20-0.30Mo-≤0.80Ni	
1.7276	10CrMo11	Fe-≤0.08-0.12C-≤0.15-0.35Si-≤0.30-0.50Mn-≤0.035P-≤0.035S-≤2.70-3.00Cr-≤0.20-0.30Mo	
1.7279	17 MnCrMo 3 3	Fe-≤0.20C-0.50-0.90Si-0.70-1.10Mn-≤0.035P-≤0.035S-0.60-1.00Cr-0.20-0.60Mo-0.06-0.12V-0.06-0.12Zr	
1.7281	16CrMo9-3	Fe-≤0.12-0.20C-≤0.15-0.35Si-≤0.30-0.50Mn-≤0.035P-≤0.035S-≤2.00-2.50Cr-≤0.30-0.40Mo	
1.7335	13 CrMo 4 4; 13CrMo4-5	Fe-≤0.08-0.18C-≤0.35Si-≤0.4-1.0Mn-≤0.030P-≤0.025S-≤0.7-1.15Cr-≤0.3Cu-≤0.4-0.6Mo	A 182

Key to materials compositions

German Standard		Materials Compositions	US-Standard
Mat.-No.	DIN-Design	Percent in Weight	SAE/ASTM/UNS
1.7357	G17CrMo5-5	Fe-≤0.15-0.20C-≤0.60Si-≤0.50-1.0Mn-≤0.020P-≤0.020S-≤1.00-1.50Cr-≤0.45-0.65Mo	A 217; UNS J11872
1.7362	X12CrMo5	Fe-≤0.08-0.15C-≤0.50Si-≤0.30-0.60Mn-≤0.025P-≤0.020S-≤4.00-6.00Cr-≤0.45-0.65Mo	SAE 501
1.7375	12CrMo9-10	Fe-≤0.10-0.15C-≤0.30Si-≤0.30-0.80Mn-≤0.015P-≤0.010S-≤0.01-0.04Al-≤2.00-2.50Cr-≤0.20Cu-≤0.9-1.10Mo-≤0.012N-≤0.30Ni	UNS K21590
1.7380	10CrMo9-10	Fe-≤0.08-0.14C-≤0.50Si-≤0.40-0.80Mn-≤0.030P-≤0.025S-≤2.00-2.50Cr-≤0.30Cu-≤0.90-1.10Mo	A 182 (F22); UNS J21890
1.7383	11CrMo9-10	Fe-≤0.08-0.15C-≤0.50Si-≤0.40-0.80Mn-≤0.030P-≤0.025S-≤2.00-2.50Cr-≤0.30Cu-≤0.90-1.10Mo	
1.7386	X12CrMo9-1	Fe-≤0.07-0.15C-≤0.25-1.0Si-≤0.30-0.60Mn-≤0.025P-≤0.020S-≤8.0-10.0Cr-≤0.90-1.1Mo	SAE 504; UNS S50488
1.7388	X7CrMo9-1	Fe-≤0.04-0.09C-≤0.45-0.75Si-≤0.43-0.72Mn-≤0.015P-≤0.015S-≤8.60-9.90Cr-≤0.90-1.10Mo	S 50480 (UNS)
1.7707	30CrMoV9	Fe-0.26-0.34C-≤0.4Si-0.4-0.7Mn-≤0.035P-≤0.035S-2.3-2.7Cr-≤0.25Mo-≤0.6Ni-0.1-0.2V	G43406 (UNS)
1.7711	40CrMoV4-6; 40CrMoV4-7	Fe-≤0.36-0.44C-≤0.40Si-≤0.45-0.85Mn-≤0.03P-≤0.03S-≤0.015Al-≤0.90-1.20Cr-≤0.50-0.65Mo-≤0.25-0.35V	A 437 (B4D)
1.7715	14MoV6-3	Fe-≤0.1-0.18C-≤0.1-0.35Si-≤0.4-0.7Mn-≤0.035P-≤0.035S-≤0.3-0.6Cr-≤0.5-0.7Mo-≤0.22-0.32V	UNS K11591
1.7734		Fe-≤0.12-0.18C-≤0.20Si-≤0.80-1.10Mn-≤0.02P-≤0.015S-≤1.25-1.50Cr-≤0.80-1.00Mo-≤0.20-0.30V	
1.7766	17CrMoV10	Fe-≤0.15-0.20C-≤0.15-0.35Si-≤0.30-0.50Mn-≤0.035P-≤0.035S-≤2.70-3.00Cr-≤0.20-0.30Mo-≤0.10-0.20V	
1.7779	20 CrMoV 13 5; 20CrMoV13-5-5	Fe-≤0.17-0.23C-≤0.15-0.35Si-≤0.30-0.50Mn-≤0.025P-≤0.020S-≤3.00-3.30Cr-≤0.50-0.60Mo-≤0.45-0.55V	
1.7783	X41CrMoV5-1	Fe-≤0.38-0.43C-≤0.80-1.0Si-≤0.20-0.40Mn-≤0.015P-≤0.010S-≤4.75-5.25Cr-≤1.2-1.4Mo-≤0.40-0.60V	SAE 610
1.8070	21CrMoV5-11	Fe-≤0.17-0.25C-≤0.30-0.60Si-≤0.30-0.60Mn-≤0.035P-≤0.035S-≤1.20-1.50Cr-≤1.00-1.20Mo-≤0.60Ni-≤0.25-0.35V	

German Standard		Materials Compositions	US-Standard
MatNo.	DIN-Design	Percent in Weight	SAE/ASTM/UNS
1.8075	10CrSiMoV7	Fe-≤0.12C-≤0.9-1.2Si-≤0.35-0.75Mn-≤0.035P-≤0.035S-≤1.6-2Cr-≤0.25-0.35Mo-≤0.25-0.35V	
1.8159	51CrV4; 50 CrV 4	Fe-≤0.47-0.55C-≤0.40Si-≤0.70-1.10Mn-≤0.035P-≤0.035S-≤0.90-1.20Cr-≤0.10-0.25V	UNS G61500 A 866 (6150) SAE 6150H
1.8719	15MnCrMo3-2	Fe-0.10-0.20C-0.15-0.35Si-0.60-1.00Mn-≤0.025P-≤0.025S-0.0005B-0.40-0.65Cr-0.15-0.50Cu-0.40-0.60Mo-0.70-1.00Ni-0.03-0.08V	
1.8812	18MnMoV5-2	Fe-≤0.20C-0.20-0.50Si-1.00-1.50Mn-≤0.030P-≤0.025S-0.10-0.30Mo-≤0.02N-0.05-0.10V	A 202 (A) (ASTM) A 202 (B) (ASTM) A 302 (A) (ASTM)
1.8850	S460MLH	Fe-≤0.16C-≤0.60Si-≤1.70Mn-≤0.030P-≤0.025S-0.02Al-≤0.20Mo-≤0.025N-≤0.050Nb-≤0.30Ni-≤0.05Ti-≤0.12V	A 514 (F) (ASTM) A 517 (F) (ASTM) A 592 (F) (ASTM)
1.8901	S460N	Fe-≤0.2C≤0.6Si-1.0-1.7Mn-≤0.035P-≤0.03S-≤0.3Cr-≤0.7Cu-≤0.1Mo-≤0.05Nb-≤0.8Ni-≤0.03Ti≤0.2V	ASTM A 572
1.8905	P460N; StE 460	Fe-≤0.20C-≤0.60Si-≤1.00-1.70Mn-≤0.030P-≤0.025S-≤0.02Al-≤0.30Cr-≤0.70Cu-≤0.10Mo-≤0.025N-≤0.050Nb-≤0.80Ni-≤0.03Ti-≤0.20V	A 225 (C), A 633 (E)
1.8907	StE 500	Fe-≤0.21C-≤0.1-0.6Si-≤1.0-1.7Mn-≤0.035P-≤0.03S-≤0.02Al-≤0.30Cr-≤0.70Cu-≤0.10Mo-≤0.020N-≤0.05Nb-≤1.0Ni-≤0.2Ti-≤0.22V	6386 B; UNS K02001
1.8912	S420NL; TStE 420	Fe-≤0.2C-≤0.6Si-≤1.0-1.7Mn-≤0.03P-≤0.025S-≤0.02Al-≤0.3Cr-≤0.7Cu-≤0.1Mo-≤0.025N-≤0.050Nb-≤0.8Ni-≤0.03Ti-≤0.2V	A 737: UNS K02002
1.8924	S500Q; StE 500V	Fe-≤0.2C-≤0.8Si-≤1.7Mn-≤0.025P-≤0.015S-≤1.5Cr-≤0.5Cu-≤0.7Mo-≤0.06Nb-≤2.0Ni-≤0.05Ti-≤0.15Zr	
1.8931	S690Q; StE 690V	Fe-≤0.2C-≤0.8Si-≤1.7Mn-≤0.025P-≤0.015S-≤1.5Cr-≤0.5Cu-≤0.7Mo-≤0.06Nb-≤2.0Ni-≤0.05Ti-≤0.15Zr	
1.8935	WstE 460; P460NH	Fe-≤0.20C-≤0.60Si-≤1.0-1.70Mn-≤0.030P-≤0.025S-≤0.02Al-≤0.30Cr-≤0.70Cu-≤0.10Mo-≤0.025N-≤0.050Nb-≤0.8Ni-≤0.03Ti-≤0.2V	A 350; UNS K02900
1.8940	S890Q	Fe-≤0.2C-≤0.8Si-≤1.7Mn-≤0.025P-≤0.015S-≤1.5Cr-≤0.5Cu-≤0.7Mo-≤0.06Nb-≤2.0Ni-≤0.05Ti-≤0.15Zr	
1.8946	S355J2WP	Fe-≤0.12C-≤0.75Si-≤1.0Mn-0.06-0.15P-≤0.035S-0.30-1.25Cr-0.25-0.55Cu-≤0.009N-≤0.65Ni	K02601 (UNS)

German Standard		Materials Compositions	US-Standard
MatNo.	DIN-Design	Percent in Weight	SAE/ASTM/UNS
1.8952	L450QB	Fe-≤0.16C-≤0.45Si-≤1.60Mn-≤0.025P-≤0.020S-≤0.015-0.06Al-≤0.30Cr-≤0.25Cu-≤0.10Mo-≤0.012N-≤0.05Nb-≤0.30Ni-≤0.06Ti-≤0.09V	
1.8961	S235J2W; WTSt 37-3	Fe-≤0.13C-≤0.4Si-≤0.2-0.6Mn-≤0.040P-≤0.035S-≤0.02Al-≤0.4-0.8Cr-≤0.25-0.55Cu-≤0.015-0.060Nb-≤0.65Ni-≤0.02-0.10Ti-≤0.02-0.10V	
1.8962	9CrNiCuP3-2-4	Fe-≤0.12C-≤0.25-0.75Si-≤0.2-0.5Mn-≤0.07-0.15P-≤0.035S-≤0.5-1.25Cr-≤0.25-0.55Cu-≤0.65Ni	A 242; UNS K11430
1.8963	S355J2G1W; WTSt 52-3	Fe-≤0.16C-≤0.50Si-≤0.50-1.5Mn-≤0.035P-≤0.035S-≤0.02Al-≤0.40-0.80Cr-≤0.25-0.55Cu-≤0.3Mo-≤0.015-0.060Nb-≤0.65Ni-≤0.02-0.10Ti-≤0.02-0.12V-≤0.15Zr	A 588 (A)
1.8972	L415NB	Fe-≤0.21C-≤0.45Si-≤1.60Mn-≤0.025P-≤0.020S-≤0.015-0.060Al-≤0.30Cr-≤0.25Cu-≤0.10Mo-≤0.012N-≤0.050Nb-≤0.30Ni-≤0.04Ti-≤0.15V	API 5LX 60 (API)
1.8975	L450MB; StE 445.7	Fe-≤0.16C-≤0.45Si-≤1.6Mn-≤0.025P-≤0.02S-0.015-0.06Al-≤0.3Cr-≤0.25Cu-≤0.1Mo-≤0.05Nb-≤0.3Ni-≤0.06Ti	API 5LX65
1.8977	L485MB; StE 480.7	Fe-≤0.16C-≤0.45Si-≤1.70Mn-≤0.025P-≤0.020S-≤0.015-0.06Al-≤0.30Cr-≤0.25Cu-≤0.10Mo-≤0.012N-≤0.06Nb-≤0.30Ni-≤0.06Ti-≤0.10V	API 5LX70
2.4060	Ni 99,6	≤99.60Ni-≤0.08C-≤0.15Si-≤0.35Mn-≤0.005S-≤0.15Cu-≤0.25Fe-≤0.15Mg-≤0.10Ti	UNS N02200
2.4061	LC-Ni 99,6	Fe-≤0.02C-0.15Si-0.35Mn-≤0.005S-≤0.15Cu-≤0.25Fe-≤0.15Mg-99.6Ni-≤0.10Ti	UNS N02201
2.4066	Ni 99,2; S-Ni 99,2	≤99.20Ni-≤0.10C-≤0.25Si-≤0.35Mn-≤0.005S-≤0.25Cu-≤0.40Fe-≤0.15Mg-≤0.10Ti	UNS N02200
2.4068	LC-Ni 99	≤99.0Ni-≤0.02C-≤0.25Si-≤0.35Mn-≤0.005S-≤0.25Cu-≤0.40Fe-≤0.15Mg-≤0.10Ti	UNS N02201
2.4360	NiCu 30 Fe	≤63.0Ni-≤0.15C-≤0.50Si-≤2.0Mn-≤0.020S-≤0.5Al-≤28.0-34.0Cu-≤1.0-2.5Fe-≤0.3Ti	UNS N04400
2.4361	LC-NiCu 30 Fe	Fe-≤0.04C-≤0.3Si-≤2.0Mn-≤0.02S-≤0.5Al-28.0-34.0Cu-1.0-2.5Fe-63.0Ni-≤0.3Ti	N04402 (UNS)
2.4363	NiCu30Fe5	≤0.30C-≤0.50Si-≤2.0Mn-0.025-0.60S-≤0.50Al-28.00-34.00Cu-≤2.50Fe-63.00-70.00Ni-≤0.30Ti	
2.4365	G-NiCu 30 Nb	Ni-≤0.15C-0.5-1.5Si-0.5-1.5Mn-≤0.5Al-26.0-33.0Cu-1.0-2.5-1.0-1.5Nb	UNS N24130

German Standard		Materials Compositions	US-Standard
Mat.-No.	DIN-Design	Percent in Weight	SAE/ASTM/UNS
2.4366	EL-NiCu 30 Mn	≤62.0Ni-≤0.15C-≤1.0Si-≤1.0-4.0Mn-≤0.030P≤0.015S-≤0.5Al-≤27.0-34.0Cu-≤0.5-2.5Fe-≤1.0Nb≤1.0Ti	B 127-98
2.4368	G-NiCu 30Si4	Fe-≤0.25C-3.5-4.5Si-0.5-1.5Mn-27.0-31.0Cu-1.0-2.5Fe-60.0-68.0Ni	UNS N10665
2.4374	NiCu30Al	≤0.25C-≤1.00Si-≤1.50Mn-≤0.010S-2.00-4.00Al-27.00-34.00Cu-≤2.00Fe-≥63.00Ni-0.25-1.00Ti	
2.4375	NiCu 30 Al	Ni-≤0.20C-≤0.50Si-≤1.5Mn-≤0.015S-≤2.2-3.5Al-≤27.0-34.0Cu-≤0.5-2.0Fe-≤63.0Ni-≤0.3-1.0Ti	UNS N05500
2.4566	ACN 17	Ni-≤0.12C-≤10Si-≤1.2Mn-≤4Co-≤3Cu	
2.4600	NiMo29Cr	Ni-≤0.01C-≤0.1Si-≤3.0Mn-≤0.025P-≤0.015S-≤0.1-0.5Al-≤3.0Co-≤0.5-3.0Cr-≤0.5Cu-≤1.0-6.0Fe-≤26.0-32.0Mo-≤0.4Nb-≤0.2Ti-≤0.2V-≤3.0W	
2.4602	NiCr21Mo14W	Ni-≤0.01C-≤0.08Si-≤0.5Mn-≤0.025P-≤0.010S-≤2.5Co-≤20.0-22.5Cr-≤2.0-6.0Fe-≤12.5-14.5Mo-≤0.35V-≤2.5-3.5W	UNS N06022
2.4603	NiCr30FeMo	Ni-≤0.03C-≤0.08Si-≤2.0Mn-≤0.04P-≤0.02S-≤5.0Co-≤28.0-31.5Cr-≤1.0-2.4Cu-13.0-17.0Fe-≤4.0-6.0Mo-≤0.3-1.5Nb-≤1.5-4.0W	UNS N06002
2.4605	NiCr23Mo16Al	Ni-≤0.01C-≤0.10Si-≤0.5Mn-≤0.025P-≤0.015S-≤0.1-0.4Al-≤0.3Co-≤22.0-24.0Cr-≤0.5Cu-≤1.5Fe-≤15.0-16.5Mo	UNS N06059
2.4606	NiCr21Mo16W	Ni-≤0.01C-≤0.08Si-≤0.75Mn-≤0.025P-≤0.015S-≤0.5Al-≤1.0Co-≤19.0-23.0Cr-≤2.0Fe-≤15.0-17.0Mo-≤0.02-0.25Ti-≤0.2V-≤3.0-4.0W	UNS N06686
2.4607	SG-NiCr23Mo16	Ni-≤0.015C-≤0.08Si-≤0.50Mn-≤0.02P-≤0.015S-≤0.1-0.4Al-≤0.3Co-≤22.0-24.0Cr-≤1.5Fe-≤15.0-16.5Mo	UNS N06059
2.4608	NiCr26MoW	Fe-0.03-0.08C-0.7-1.5Si-≤2.0Mn-≤0.03P-≤0.015S-2.0-4.0Cu-24.0-26.0Cr-≤0.5Cu-2.5-4.0Mo-44.0-47.0Ni-2.5-4.0W	N06333 (UNS)
2.4610	NiMo16Cr16Ti	Ni-≤0.01C-≤0.08Si-≤1.0Mn-≤0.025P-≤0.015S-≤2.0Co-≤14.0-18.0Cr-≤0.5Cu-≤3.0Fe-≤14.0-18.0Mo-≤0.7Ti	UNS N06455
2.4612	EL-NiMo15Cr15Ti	≤0.02C-≤0.20Si-≤1.00Mn-≤0.015S-≤2.00Co-≤14.00-18.00Cr-≤3.00Fe-≤14.00-17.00Mo-at least56Ni	
2.4615	SG-NiMo27	Fe-≤0.02C-≤0.10Si-≤1.0Mn-≤0.015S-≤1.0Cr-≤2.0Fe-26.0-30.0Mo-64.0Ni	UNS N10665

German Standard		Materials Compositions	US-Standard
Mat.-No.	DIN-Design	Percent in Weight	SAE/ASTM/UNS
2.4617	NiMo28	Ni-≤0.01C-≤0.08Si-≤1.0Mn-≤0.025P-≤0.015S-≤1.0Co-≤1.0Cr-≤0.5Cu-≤2.0Fe-≤26.0-30.0Mo	UNS N10665
2.4618	NiCr22Mo6Cu	Ni-≤0.05C-≤1.0Si-≤1.0-2.0Mn-≤0.025P-≤0.015S-≤2.5Co-≤21.0-23.5Cr-≤1.5-2.5Cu-≤18.0-21.0Fe-≤5.5-7.5Mo-≤1.75-2.5Nb-≤1.0W	UNS N06007
2.4619	NiCr22Mo7Cu	Ni-≤0.015C-≤1.0Si-≤1.0Mn-≤0.025P-≤0.015S-≤5.0Co-≤21.0-23.5Cr-≤1.5-2.5Cu-≤18.0-21.0Fe-≤6.0-8.0Mo-≤0.5Nb-≤1.5W	UNS N06985
2.4621	EL-NiCr20Mo9Nb	Ni-≤0.1C≤0.8Si≤2.0Mn≤0.4Al-20.0-23.0Cr≤0.5Cu≤6.0Fe-8.0-10.0Mo-2.0-4.0Nb≤0.4Ti	
2.4623	EL-NiCr23Mo7Cu	Ni-≤0.02C≤1.0Si≤1.0Mn≤0.04P≤0.03≤5.0Co-21.0-23.5Cr-1.5-2.5Cu-18.0-21.0Fe-6.0-8.0Mo≤0.5Nb≤0.5Ta≤1.5W	
2.4627	SG-NiCr22Co12Mo	Fe-≤0.1C-≤0.5Si-≤1.0Mn-≤0.015S-0.8-1.5Al-10.0-14.0Co-20.0-24.0Cr-≤0.5Cu-≤1.0Fe-8.0-10.0Mo-50.0Ni-≤0.6Ti	N06617 (UNS)
2.4630	NiCr20Ti	Fe-0.08-0.15C-≤1.0Si-≤1.0Mn-18.0-21.0Cr-≤0.5Cu-≤5.0Fe-0.2-0.6Ti	N06075 (UNS)
2.4631	NiCr20TiAl	Ni-0.04-0.1C≤1.0Si≤1.0Mn≤0.03P≤0.015S-1.0-1.8Al≤2.0Co-18.0-21.0Cr≤0.2Cu≤1.5Fe-1.8-2.7Ti	UNS N07080
2.4632	NiCr20Co18Ti	Ni-≤0.13C≤1.0Si≤1.0Mn≤0.02P≤0.015S-1.0-2.0Al-15.0-21.0Co-18.0-21.0Cr≤0.2Cu≤1.5Fe-2.0-3.0Ti	UNS N07090
2.4633	NiCr25FeAlY	Fe-0.15-0.25C-≤0.5Si-≤0.5Mn-≤0.02P-≤0.01S-1.8-2.4Al-24.0-26.0Cr-≤0.1Cu-8.0-11.0Fe-0.1-0.2Ti-0.05-0.12Y-0.01-0.1Zr	
2.4634	NiCo20Cr15MoAlTi	Ni-0.12-0.17C-≤1.0Si-≤1.0Mn-≤0.045P-≤0.015S-4.5-4.9Al-18.0-22.0Co-14.0-15.7Cr-≤0.2Cu-≤1.0Fe-4.5-.5.5Mo-0.9-1.5Ti	UNS N13021
2.4636	NiCo15Cr15MoAlTi	Ni-0.12-0.2C≤1.0Si≤1.0Mn≤0.045P≤0.03S-4.5-5.5Al-13.0-17.0Co-14.0-16.0Cr≤0.2Cu≤1.0Fe-3.0-.5.0Mo-3.5-4.5Ti	NIMONIC alloy 115
2.4641	NiCr21Mo6Cu	Fe-≤0.025C-≤0.50Si-≤1.0Mn-≤0.025P-≤0.015S-≤0.2Al-≤1.0Co-20.0-23.0Cr-1.5-3.0Cu-5.5-7.0Mo-39.0-46.0Ni-0.6-1.0Ti	UNS N08042
2.4642	NiCr29Fe	≤58.0Ni-≤0.05C-≤0.5Si-≤0.5Mn-≤0.020P-≤0.015S-≤0.5Al-≤27.0-31.0Cr-≤0.5Cu-≤7.0-11.0Fe-≤0.5Ti	UNS N06690
2.4650	NiCo20Cr20MoTi	Ni-≤0.04-0.08C-≤0.4Si-≤0.6Mn-≤0.007S-≤0.3-0.6Al-≤0.005B-≤19.0-21.0Co-≤19.0-21.0Cr-≤0.2Cu-≤0.7Fe-≤5.6-6.1Mo-≤1.9-2.4Ti	UNS N07263

German Standard		Materials Compositions	US-Standard
MatNo.	DIN-Design	Percent in Weight	SAE/ASTM/UNS
2.4652	EL-NiCr26Mo	≤37.0-42.0Ni-≤0.03C-≤0.7Si-≤1.0-3.0Mn-≤0.015S-≤0.1Al-≤23.0-27.0Cr-≤1.5-3.0Cu-≤30.0Fe-≤3.5-7.5Mo-≤37.0-42.00Ni-≤1.0Ti	UNS S32654
2.4654	NiCr20Co13Mo4Ti3Al; NiCr19Co14Mo4Ti	Fe-0.02-0.2C-≤0.15Si-≤0.1Mn-≤0.015P-≤0.015S-1.2-1.6Al-0.003-0.01B-12.0-15.0Co-18.0-21.0Cr-≤0.1Cu-≤2.0Fe-3.5-5.0Mo-2.8-3.3Ti-0.02-0.08Zr	N07001 (UNS)
2.4658	NiCr7030; NiCr 70 30	Fe-≤0.1C-0.5-2.0Si-≤1.0Mn-≤0.02P-≤0.15S-≤0.3Al-≤1.0Co-29.0-32.0Cr-≤0.5Cu-≤5.0Fe-60.0Ni	N06008 (UNS)
2.4660	NiCr20CuMo	≤32.0-38.0Ni-≤0.07C-≤1.0Si-≤2.0Mn-≤0.025P-≤0.015S-19.0-21.0Cr-≤3.0-4.0Cu-≤2.0-3.0Mo	UNS N08020
2.4662	NiCr13Mo6Ti3	Fe-0.02-0.06C-≤0.40Si-≤0.50Mn-≤0.020P-≤0.020S-≤0.35Al-0.01-0.02B-≤1.0Co-11.0-14.0Cr-≤0.04Cu-5.0-6.5Mo-40.0-45.0Ni-2.8-3.1Ti	UNS N09901
2.4663	NiCr23Co12Mo	Ni-≤0.05-0.10C-≤0.2Si-≤0.2Mn-≤0.01P-≤0.01S-≤0.7-1.4Al≤0.006B-≤11.0-14.0Co-≤20.0-23.0Cr-≤0.5Cu-≤2.0Fe-≤8.5-10.0Mo-≤0.2-0.6Ti	UNS N06617
2.4665	NiCr22Fe18Mo	Fe-0.05-0.15C-≤1.0Si-≤1Mn-≤0.02P-≤0.015S-≤0.5Al-0.01-0.1B-0.5-2.5Co-20.5-23.0Cr-≤0.5Cu-17.0-20.0Fe-8.0-10.0Mo-0.2-1.0W	680 (SAE)
2.4667	SG-NiCr19NbMoTi	Fe-≤0.08C-≤0.40Si-≤0.40Mn-≤0.015S-0.2-0.8Al-≤0.006B-17.0-21.0Cr-≤0.30Cu-≤22.0Fe-2.8-3.3Mo-4.8-5.5Nb-50.0Ni-0.60-1.20Ti	
2.4668	NiCr19Fe19Nb5Mo3	Fe-0.02-0.08C-≤0.35Si-≤0.35Mn-≤0.015P-≤0.015S-0.3-0.7Al-0.006B-≤1.0Co-17.0-21.0Cr-≤0.30Cu-2.8-3.3Mo-4.7-5.5Nb-50.0-55.0Ni-0.60-1.20Ti	UNS N07718
2.4669	NiCr15Fe7TiAl; NiCr15Fe7Ti2Al	Ni-≤0.08C-≤0.5Si-≤1.0Mn-≤0.02P-≤0.015S-≤0.4-1.0Al≤1.0Co-≤14.0-17.0Cr-≤0.5Cu-≤5.0-9.0Fe-≤0.7-1.2Nb-≤2.25-2.75Ti	UNS N07750
2.4670	G-NiCr13Al6MoNb	Ni-≤0.08-0.20C-≤0.50Si-≤0.25Mn-≤0.015P-≤0.015S-≤5.50-6.50Al-≤0.005-0.15B-≤1.00Co-≤12.00-14.00Cr-≤0.50Cu-≤3.80-5.20Mo-≤1.50-2.50Nb-≤0.40-1.00Ti-≤0.05-0.15Zr	UNS N07713
2.4675	NiCr23Mo16Cu	Ni-≤0.01C-≤0.08Si-≤0.5Mn-≤0.025P-≤0.015S-≤0.5Al≤2.0Co-≤22.0-24.0Cr-≤1.3-1.9Cu-≤3.0Fe-≤15.0-17.0Mo	

Key to materials compositions | 399

German Standard Mat.-No.	DIN-Design	Materials Compositions Percent in Weight	US-Standard SAE/ASTM/UNS
2.4679	G-NiCr35	Ni-≤0.10C-≤1.00Si-≤0.30Mn-≤34.00-36.00Cr-≤1.00Fe-≤0.30N	
2.4680	G-NiCr50Nb	Ni-≤0.10C-≤1.00Si-≤0.50Mn-≤0.02P-≤0.02S-≤48.00-52.00Cr-≤1.00Fe-≤0.50Mo-≤0.16N-≤1.00-1.80Nb	
2.4681	CoCr26Ni9Mo5W	Ni-≤1.0C-≤1.0Si-≤1.5Mn-≤23.5-27.5Cr-≤1.0-3.0Fe-≤4.0-6.0Mo-≤0.12N-≤7.0-11.0Ni-≤1.0-3.0W	
2.4683	CoCr22NiW	Fe-0.05-0.15C-0.2-0.5Si-≤1.25Mn-≤0.02P-≤0.015S-20.0-24.0Cr-≤3.0Fe-0.02-0.12La-20.0-24.0Ni-13.0-16.0W	R30188 (UNS)
2.4686	G-NiMo 17 Cr	Ni-≤0.03C≤0.5Si≤1.0Mn≤2.5Co-15.5-17.5Cr≤7.0Fe-16.0-18.0Mo	
2.4694	NiCr16Fe7TiAl	Fe-≤0.08C-≤0.5Si-≤0.5Mn-≤0.015P-≤0.01S-0.8-1.6Al-14.0-17.0Cr-≤0.5Cu-5.0-9.0Fe-0.7-1.2Nb-70.0Ni-2.0-2.6Ti	N07031 (UNS)
2.4800	S-NiMo 30	–≤60.0Ni-≤0.05C-≤1.0Si-≤1.0Mn-≤0.045P-≤0.025S-≤2.5Co-≤1.0Cr-≤4.0-7.0Fe-≤26.0-30.0Mo-≤0.2-0.4V	UNS N10001
2.4810	NiMo 30	≤62.0Ni-≤0.05C-≤0.5Si-≤1.0Mn-≤0.030P-≤0.015S-≤2.5Co-≤1.0Cr-≤0.5Cu-≤4.0-7.0Fe-≤26.0-30.0Mo-≤0.6V	UNS N10001
2.4811		Fe-≤0.03C-≤0.05Si-≤0.80Mn-≤0.030P-≤0.015S-19.0-21.0Cr-≤0.50Cu-≤2.5Fe-14.0-17.0Mo-58.0Ni-0.35V	
2.4816	NiCr15Fe	≤72.0Ni-≤0.05-0.1C-≤0.5Si-≤1.0Mn-≤0.020P-≤0.015S-≤0.3Al-≤0.0060B-≤1.0Co-≤14.0-17.0Cr-≤0.5Cu-≤6.0-10.0Fe-≤0.3Ti	UNS N06600
2.4817	LC-NiCr15Fe	≤72.0Ni-≤0.025C-≤0.50Si-≤1.0Mn-≤0.020P-≤0.015S-≤0.3Al-≤0.0060B-≤1.0Co-≤14.0-17.0-≤0.5Cu-≤6.0-10.0Fe-≤0.3Ti	
2.4819	NiMo16Cr15W	Ni-≤0.01C-≤0.08Si-≤1.0Mn-≤0.025P-≤0.015S-≤2.5Co-≤14.5-16.5Cr-≤0.5Cu-≤4.0-7.0Fe-≤15.0-17.0Mo-≤0.35V-≤3.0-4.5W	UNS N10276
2.4831	SG-NiCr21Mo9Nb	Fe-≤0.10C-≤0.50Si-≤0.50Mn-≤0.015S-≤0.4Al-20.0-23.0Cr-≤0.50Cu-≤5.0Fe-8.0-10.0Mo-3.0-4.5Nb-60.0Ni-≤0.40Ti	UNS N06625
2.4851	NiCr23Fe	Fe-0.03-0.1C-≤0.5Si-≤1.0Mn-≤0.02P-≤0.015S-1.0-1.7Al-≤0.006B-21.0-25.0Cr-≤0.5Cu-≤18.0Fe-58.0-63.0Ni-≤0.5Ti	N06601 (UNS)
2.4856	NiCr22Mo9Nb	Ni-≤0.03-0.10C-≤0.5Si-≤0.5Mn-≤0.020P-≤0.015S-≤0.4Al-≤1.0Co-≤20.0-23.0Cr-≤0.5Cu-≤5.0Fe-≤8.0-10.0Mo-≤3.15-4.15Nb-≤0.4Ti	UNS N06625

German Standard		Materials Compositions	US-Standard
MatNo.	DIN-Design	Percent in Weight	SAE/ASTM/UNS
2.4858	NiCr21Mo	≤38.0-46.0Ni-≤0.025C-≤0.50Si-≤1.0Mn-≤0.025P-≤0.015S-≤0.20Al-≤1.0Co-≤19.5-23.5Cr-≤1.5-3.0Cu-≤2.5-3.5Mo-≤0.6-1.2Ti	B 163 UNS N08825
2.4869	NiCr80-20	Fe-≤0.15C-0.50-2.0Si-≤1.0Mn-≤0.020P-≤0.015S-≤0.3Al-≤1.0Co-19.0-21.0Cr-≤0.50Cu-≤1.0Fe-75.0Ni	UNS N06003
2.4882		Fe-≤0.12C-0.50-2.0Si-≤1.0Mn-≤0.020P-≤0.015S-≤0.3Al-≤1.0Co-19.0-21.0Cr-≤0.50Cu-≤1.0Fe-75.0Ni	UNS N10001
2.4883		Fe-≤0.03C-≤0.50Si-≤1.0Mn-≤2.50Co-15.50-17.50Cr-≤7.0Fe-16.0-18.0Mo	UNS N10002
2.4886	SG-NiMo16Cr16W; UP-NiMo16Cr16W	≤50.0Ni-≤0.02C-≤0.08Si-≤1.0Mn-≤0.015S-≤14.5-16.5Cr-≤4.0-7.0Fe-≤15.0-17.0Mo-≤0.4V-≤3-4.5W	UNS N10276
2.4887	EL-NiMo15Cr15W	≤50.0Ni-≤0.02C-≤0.20Si-≤1.0Mn-≤0.015S-≤14.5-16.5Cr-≤4.0-7.0Fe-≤15.0-17.0Mo-≤0.4V-≤3-4.5W	
2.4951	NiCr20Ti	Ni-≤0.08-0.15C-≤1.0Si-≤1.0Mn-≤0.020P-≤0.015S-≤0.3Al-≤0.0060B-≤5.0Co-≤18.0-21.0Cr-≤0.5Cu-≤5.0Fe-≤0.2-0.6Ti	UNS N06075
2.4952	NiCr20FeMo3TiCuAl	Fe-≤0.03C-≤0.5Si-≤1.0Mn-≤0.03P-≤0.03S-0.1-0.5Al-19.5-22.5Cr-1.5-3.0Cu-≤22.0Fe-2.5-3.5Mo-≤0.5Nb-42.0-46.0Ni-1.9-2.4Ti	
2.4964	CoCr20W15Ni	≤9.0-11.0Ni-≤0.05-0.15C-≤0.4Si-≤2.0Mn-≤0.020P-≤0.015S-≤19.0-21.0Cr-≤3.0Fe-≤9.0-11.0Ni-≤14.0-16.0W	UNS R30605
2.4973	NiCr19CoMo	Ni-≤0.12C-≤0.50Si-≤0.10Mn-≤1.40-1.80Al-≤10.00-12.00Co-≤18.00-20.00Cr-≤5.00Fe-≤9.00-10.50Mo-≤2.80-3.30Ti	N 07041 (UNS) 683 (SAE)
2.4975	NiFeCr12Mo	Fe-≤0.10C-≤0.60Si-≤2.00Mn-≤0.020P-≤0.010S-≤0.350Al-≤1.00Co-11.00-14.00Cr-5.00-7.00Mo-40.00-45.00Ni-2.35-3.10Ti	
2.4976	NiCr20Mo	Ni-≤0.10C-≤1.00Si-≤1.00Mn-≤0.020P-≤0.010S-0.50-1.80Al-≤2.00Co-18.00-21.00Cr-≤5.00Fe-4.00-5.00Mo-1.80-2.70Ti	
2.4983	NiCr18Co	Ni-≤0.15C-≤0.50Si-≤1.00Mn-≤0.02P-≤0.01S-≤2.50-3.20Al-≤17.00-20.00Co-≤17.00-20.00Cr-≤4.00Fe-≤3.00-5.00Mo-≤2.50-3.20Ti	UNS N07500 ASTM B637 (N07500)(864)
2.4999	MP35N	35.0Ni-≤0.01C-≤20.0Cr-≤9.5Mo	

Table 2: Chemical compositions of different American, CIS, Bulgarian and other steels

Steel	Materials Compositions, Percent in Weight	Note
000Ch16N13M2	Fe-≤0.07C-≤1.0Si-≤2.0Mn-16.5-18.5Cr-2-2.5Mo-10-13Ni-≤0.045P-≤0.015S-≤0.11N	CIS, formerly USSR, identical with SAE 316
000Ch16N13M3	Fe-≤0.07C-≤1.0Si-≤2.0Mn-16.5-18.5Cr-2-2.5Mo-10-13Ni-≤0.045P-≤0.015S-≤0.11N	CIS, formerly USSR, identical with SAE 316
000Ch16N16M4	Fe-≤0.07C-≤1.0Si-≤2.0Mn-16.5-18.5Cr-2-2.5Mo-10-13Ni-≤0.045P-≤0.015S-≤0.11N	CIS, formerly USSR, identical with SAE 316
000Ch18N10	Fe-≤0.03C-≤0.8Si-≤2.0Mn-17-19Cr-≤0.3Mo-9-11Ni	CIS, formerly USSR/Bulg.
000Ch18N11	Fe-≤0.03C-≤1.0Si-≤2.0Mn-18-20Cr-10-12Ni-≤0.045P-≤0.015S-≤0.11N	Bulg., comparable with 1.4306
000Ch20N20	Fe-≤0.03C-≤18.57Cr-≤19.40Ni-≤0.71Mn-≤0.26Si	CIS, formerly USSR
000Ch21N6M2	Fe-≤0.036C-≤21.1Cr-6.5Ni-2.4Mo	CIS, formerly USSR
000Ch21N10M2	Fe-≤0.02C-≤19.8Cr-≤10.5Ni-≤2.1Mo	CIS, formerly USSR
000Ch21N21M4B	Fe-≤0.03C-≤20-22Cr-20-21Ni-3.4-3.7Mo-≤0.6Mn-≤0.6Si-≤0.03P-≤0.02S-0.45-0.8Nb	CIS, formerly USSR
005Ch25B	Fe-0.005C-0.007N-25Cr	CIS, formerly USSR
00Ch18G8N2T	Fe-≤0.08C-≤0.8Si-7-9Mn-17-19Cr-≤0.3Mo-1.8-2.8Ni-≤0.2W-≤0.3Cu-≤0.2Ti-0.2-0.5Al	CIS, formerly USSR
00Ch18N10	Fe-≤0.015C-≤0.7Si-≤1.7Mn-≤17.3Cr-≤10.4Ni	CIS, formerly USSR
0Ch20N6M2T	not available (n.a.)	CrNiMoTi 20 6 2
0Ch21N6M2T	n. a.	CrNiMoTi 21 6 2
1Ch21N5T	n. a.	–
1H18N9	n. a.	cf. 1.4541; SAE 321
2Ch18N9	n. a.	Fe-0.2C-18Cr-9Ni; cf. 1.4310, UNS 30200
02Ch12N10S5	Fe-0.02C-12Cr-10Ni-5Si, Nb-stabilized	CIS, formerly USSR
02Ch12N10S5B	Fe-0.02C-12Cr-10Ni-5Si, Nb-stabilized	CIS, formerly USSR
02Ch12N10S5T	Fe-0.02C-12Cr-10Ni-5Si, Nb-stabilized	CIS, formerly USSR
02Ch17NS6	Fe-0.02C-4-6.5Si-0.43-0.52Mn-16.3-18Cr-10.5-18.2Ni-0.005-0.008S-0.012-0.014P	CIS, formerly USSR
02Ch8N22S6	Fe-≤0.02C-≤5.4-6.7Si-0.6Mn-≤0.030P-≤0.020S-7.5-10Cr-0.3Mo-21-23Ni-≤0.2Ti-≤0.20W	CIS, formerly USSR
02Ch8N22S6B	Fe-0.02C-8Cr-122Ni-6Si, Nb-stabilized	CIS, formerly USSR
02Ch8N22T	Fe-0.02C-8Cr-122Ni-6Si, Ti-stabilized	CIS, formerly USSR
03Ch16N15M3	Fe-≤0.03C-15.0-17.0Cr-14.0-16.0Ni-2.5-3.0Mo-Ti	CIS, formerly USSR
03Ch18N11	Fe-0.03C-≤0.80Si-≤0.70-2.0Mn-≤0.035P-0.020S-17.0-19.0Cr-≤0.10Mo-10.5-12.5Ni-≤0.20W-0.30Cu-≤0.50Ti	CIS, formerly USSR, comparable with DIN-Mat.No. 1.4306
03Ch18N14	Fe-0.03C-18Cr-14Ni	CIS, formerly USSR
03Ch21N21M4B	n. a.	CrNiMoB 21 21 4

Steel	Materials Compositions, Percent in Weight	Note
03Ch21N21M4GB	Fe-≤0.03C-≤0.60Si-1.8-2.5Mn-≤0.030P-≤0.020S-20-22Cr-3.4-3.7Mo-20-22Ni-≤0.2W-≤0.3Cu-≤0.2Ti, Nb 15 x C-0.80	CIS, formerly USSR
03Ch23N6	Fe-≤0.03C-≤0.40Si-1.0-2.0Mn-≤0.035P-≤0.020S-22-24Cr-≤5.3-6.3Ni	CIS, formerly USSR
03Ch25	Fe-about 0.03C-25Cr-0.6Ni	CIS, formerly USSR
03ChN28MDT	Fe-≤0.03C-≤0.80Si-≤0.80Mn-≤0.035P-≤0.020S-22.0-25.0Cr-2.5-3.0Mo-26.0-29.0Ni-0.50-0.90Ti-2.5-3.5Cu	CIS, formerly USSR
04Ch18N10	Fe-≤0.04C-≤0.8Si-≤2.0Mn-17-19Cr-≤0.030P-≤0.02S-≤0.3Mo-9-11Ni-≤0.2W-≤0.3Cu-≤0.2Ti	CIS, formerly USSR
04Ch18N10T	Fe-≤0.04C-≤0.8Si-≤2.0Mn-17-19Cr-≤0.3Mo-9-11Ni-≤0.2W-≤0.3Cu-≤0.2Ti	CIS, formerly USSR
05Ch16N15M3	Fe-0.05C-16Cr-15Ni-3Mo	CIS, formerly USSR
06Ch17G15NAB	Fe-0.05C-18.36Cr-16.5Mn-1.6Ni-0.31Nb-ß.12Si-0.01Ce-0.017P-0.014S	CIS, formerly USSR
06Ch23N28M3D3T	Fe-≤0.06C-≤0.8Si-≤2.0Mn-22-25Cr-≤2.4-3Mo-26-29Ni-0.5-0.9Ti-2.5-3.5Cu	CIS, formerly USSR/Bulg.
06Ch28MDT	Fe-≤0.06C-≤0.8Si-≤0.80Mn-22.0-25.0Cr-2.5-3.0Mo-26.0-29.0Ni-0.50-0.90Ti-2.5-3.5Cu	CIS, formerly USSR
06ChN28MDT	Fe-≤0.06C-≤0.8Si-≤0.8Mn-≤0.035P-≤0.02S-22-25Cr-2.5-3.0Mo-26-29Ni-0.5-0.9Ti-2.5-3.5Cu	CIS, formerly USSR
06ChN40B	Fe-0.055C-17.01Cr-39.04Ni-1.99Mn-0.50Nb-0.60Si-0.013S-0.022P	CIS, formerly USSR
06XH28M?T	n. a.	
07Ch13AG20	Fe-≤0.07C-≤0.60Si-≤19-22Mn-≤0.035P-≤0.025S-≤0.0030B-≤0.1Ca-≤0.1Ce -12.-14.8Cr-≤0.30Cu≤0.1Mg-≤0.30Mo≤0.08-0.18N-≤1.0Ni-≤0.20W≤0.20Ti	CIS, formerly USSR
07Ch16N4B	Fe-0.05-0.10C-≤0.60Si-≤0.2-0.5Mn-≤0.025P-≤0.020S-15-16.5Cr-≤0.30Cu-≤0.30Mo-0.2-0.4Nb-3.5-4.5Ni-≤0.20W	CIS, formerly USSR
07Ch17G15NAB	Fe-0.05C-18.4Cr-16.5Mn-1.6Ni-0.01Ce-0.005B-0.32N	CIS, formerly USSR
07Ch17G17DAMB	Fe-0.06C-17.6Cr-15.2Mn-0.43Mo-0.3Nb-0.005B-0.38N	CIS, formerly USSR
08Ch17N5M3	Fe-0.06C-0.10C-≤0.80Si-≤0.80Mn-≤0.035P-≤0.020S-16.0-17.5Cr-3.0-3.5Mo-4.5-5.5Ni-≤0.20W-≤0.30Cu-≤0.20Ti	CIS, formerly USSR
08Ch17N15M3B	Fe-≤0.08C-16.0-18.0Cr-14.0-16.0Ni-3.0-4.0Mo-Ti	CIS, formerly USSR
08Ch17N15M3T	Fe-≤0.08C-≤0.80Si-≤2.0Mn-≤0.35P-≤0.020S-16.0-18.0Cr-3.00-4.00Mo-14.0-16.0Ni≤0.20W-0.30Cu-0.30-0.60Ti	CIS, formerly USSR
08Ch17T	Fe-≤0.08C-≤0.80Si-≤0.80Mn-≤0.035P-≤0.025S-16.0-18.0Cr-≤0.6Ni-≤0.30Cu	CIS, formerly USSR

Steel	Materials Compositions, Percent in Weight	Note
08Ch18G8N2M2T	Fe-0.08C-18.2Cr-3.42Ni-8.9Mn-2.32Mo-0.22Ti	CIS, formerly USSR
08Ch18G8N2T	Fe-≤0.08C-≤0.80Si-7.0-9.0Mn-≤0.035P-17.0-19.0Cr-≤0.30Mo-1.80-2.80Ni-≤0.30Cu-≤0.035P-≤0.025S-≤0.20-0.50Ti-≤0.20W	CIS, formerly USSR
08Ch18N10	Fe-≤0.08C-≤0.8Si-≤2.0Mn-≤0.035P-≤0.020S-17-19Cr-0.3Mo-9.0-11.0Ni-≤0.2W-≤0.3Cu	CIS, formerly USSR
08Ch18N10T	Fe-≤0.08C-≤0.8Si-≤2.0Mn-≤0.035P-≤0.020S-17-19Cr-0.5Mo-9.0-11.0Ni-0.5Ti-≤0.2W-≤0.3Cu	CIS, formerly USSR
08Ch21N6M2T	Fe-≤0.08C-≤0.8Si-≤0.8Mn-≤0.035P-≤0.025S-20-22Cr-1.8-2.5Mo-5.5-6.5Ni-0.2-0.4Ti-≤0.2W-≤0.3Cu	CIS, formerly USSR
08Ch22N6M2T	Fe-≤0.08C-≤0.80Si-≤0.80Mn-20.0-22.0Cr-1.80-2.50Mo-5.50-6.50Ni-≤50.20W-≤0.30Cu-≤0.035P-≤0.025S-0.20-0.40Ti	CIS, formerly USSR
08Ch22N6T	Fe-≤0.08C-≤0.8Si-≤0.8Mn-≤0.035P–≤0.025S-21-23Cr-≤0.3Mo-5.3-6.3Ni-≤0.2W-≤0.3Cu, 5x% C max. 0.65Ti	CIS, formerly USSR
08ChP	Fe-0.25Cr-0.25Ni-0.25Mo	CIS, formerly USSR
08-KP	n. a.	cf. 1.0335; UNS G 10060
08X21H6M2T	n. a.	–
08X22H6T	n. a.	–
09Ch16N15M3B	Fe-≤0.09C-≤0.80Si-≤0.80Mn-≤0.035P-≤0.020S-15.0-17.0Cr-≤0.30Cu-2.5-3.0Mo-0.6-0.9Nb-14.0-16.0Ni-≤0.20Ti-≤0.20W	CIS, formerly USSR
09G2S	Fe-≤0.12C-0.5-0.8Si-1.3-1.7Mn-≤0.035P-≤0.035S-≤0.3Cr-≤0.3Ni-≤0.3Cu	Bulg.
0Ch17N16M3T	Fe-≤0.080C-≤0.8Si-≤2.00Mn-16.0-18.0Cr-3.00-4.00Mo-14.0-16.0Ni-≤ 0.035P-≤0.025S-0.30-0.60Ti	CIS, formerty USSR
0Ch18G8N3M2T	Fe-about 18 Cr-8Mn-3Ni-2Mo, Ti	CIS, formerly USSR
0Ch18N10T	Fe-≤0.08C-≤1.0Si-≤2.0Mn-≤0.045P-≤0.015S-17-19Cr-9-12Ni, Ti 5xC-0.70	CIS, formerly USSR/Bulg.
0Ch18N12B	Fe-≤0.08C-≤1.0Si-≤2.0Mn-≤0.045P-≤0.015S-17-19Cr-≤9.0-12Ni, Nb 10xC-1.00	CIS, formerly USSR/Bulg., comparable with DIN-Mat. No. 1.4550
0Ch20N14S2	Fe- about 20Cr-14Ni-2Si	CIS, formerly USSR
0Ch21N5T	Fe-≤0.08Cr-21Cr-5Ni, Ti	CIS, formerly USSR s. text HNO3
0Ch23N18	Fe-≤0.20C-≤1.00Si-1.50Mn-22.0-25.0Cr-≤0.30Mo-17.0-20.0Ni-≤0.035P-≤0.025S	CIS, formerly USSR
0Ch23N28M3D3T	Fe-≤0.06C-≤0.8Si-≤2Mn-22-25Cr-2.4-3Mo-26-29Ni-0.5-0.9Ti-2.5-3.5Cu	CIS, formerly USSR/Bulg.
0Ch25T	Fe-≤0.01C-25Cr, Ti-stabilized	CIS, formerly USSR
0H17N12M2T	Fe-≤0.05C-≤1.0Si-≤2.0Mn-16-18Cr-2-3Mo-11-14Ni-≤0.045P-≤0.030S, Ti 5xC-0.60	Poland

Steel	Materials Compositions, Percent in Weight	Note
10Ch13 (1Ch13)	Fe-0.08-0.15C-≤1.0Si-≤1.5Mn-≤0.040P-≤0.015S-11.5-13.5Cr-≤0.75Ni	CIS, formerly USSR/Bulg.
10Ch14AG15	Fe-≤0.10C-≤0.80Si-14.5-16.5Mn-≤0.045P-≤0.030S-13.0-15.0Cr-<0.60Ni-≤0.60Cu-≤0.20Ti-0.15-0.25N	CIS, formerly USSR
10Ch14G14N4T	Fe-≤0.10C-≤0.80Si-13.0-15.0Mn-13.0-15.0Cr-≤0.30Mo-2.80-4.50Ni-≤0.20W-≤0.30Cu-≤0.035P-≤0.020S-5x% C max. 0.60Ti	CIS, formerly USSR
10Ch17	Fe-0.10C-17Cr	CIS, formerly USSR
10Ch17N13M2T	Fe-≤0.10C-≤0.80Si-≤2.0Mn-≤0.035P-≤0.020S-16.0-18.0Cr-≤0.30Cu-2.0-3.0Mo-12.0-14.0Ni-≤0.20W, Ti>-5x% C	USA, comparable with DIN-Mat. No. 1.4571
10Ch17N13M3T	Fe-≤0.10C-≤0.80Si-≤2.0Mn-≤0.035P-≤0.020S-16.0-18.0Cr-≤0.30Cu-3.0-4.0Mo-12.0-14.0Ni-≤0.20W-≤0.7Ti	USA, comparable with DIN-Mat. No. 1.4573
10Ch18N9T(Ch18N9T)	Fe-0.08C-1.0Si-≤2Mn-≤0.045P-≤0.015S-17-19Cr-≤0.3Mo-9-12Ni-Ti5xC-0.70	CIS, formerly USSR/Bulg.
10Ch18N10M2T	Fe-≤0.10C-18Cr-10Ni-2Mo, Ti stabilized	CIS, formerly USSR
10Ch18N10T	Fe-≤0.10C-≤0.80Si-≤1.0-2.0Mn-≤0.035P-≤0.020S-≤17.0-19.0Cr-≤10.0-11.0Ni	CIS, formerly USSR/Bulg.
12Ch13G18D	Fe-0.12C-13Cr-18Mn-Cu	CIS, formerly USSR
12Ch17G9AN4	Fe-≤0.12C-≤0.80Si-8.0-10.5Mn-≤0.035P-≤0.020S-16.0-18.0Cr-≤0.30Mo-3.5-4.5Ni-≤0.20W-≤0.30Cu-≤0.20Ti-0.15-0.25N	CIS, formerly USSR
12Ch18N9T	Fe-≤0.12C-≤0.80Si-≤2.0Mn-≤0.035P-≤0.020S-17.0-19.0Cr-≤0.50Mo-8.0-9.5Ni-≤0.20W-≤0.30Cu-Ti = 5x % C	CIS, formerly USSR
12Ch18N10T	Fe-≤0.12C-≤0.8Si-≤2.0Mn-≤0.025P-≤0.020S-17.0-19.0Cr-≤0.30Cu-≤0.50Mo-9.0-11.0Ni-≤0.20W-≤0.70Ti	CIS, formerly USSR; comparable with DIN-Mat. No. 1.4878
12Ch2M1	Fe-0.12C-2Cr-1Mo	CIS, formerly USSR/Bulg., comparable with DIN-Mat. No. 1.7380, A 182, F 22, B.S. 1501-622
12Ch2N4A	Fe-0.09-0.15C-0.17-0.37Si-0.30-0.60Mn-≤0.025P-≤0.025S-1.25-1.65Cr-3.25-3.65Ni-≤0.30Cu-≤0.15Mo-≤0.03Ti-≤0.05V-≤0.12W	CIS, formerly USSR
12Ch21N5T	Fe-0.09-0.14C-≤0.80Si-≤0.80Mn-≤0.035P-≤0.025S-20.0-22.0Cr-≤0.30Mo-4.80-5.80Ni-≤0.20W-≤0.30Cu-0.25-0.50Ti-≤0.08Al	CIS, formerly USSR
12ChN2	Fe-0.09-0.16C-0.17-0.37Si-0.30-0.60Mn-≤0.035P-≤0.035S-0.60-0.90Cr-1.50-1.90Ni-≤0.30Cu-≤0.15Mo-≤0.03Ti-≤0.05V-≤0.20W	CIS, formerly USSR/Bulg.
13-4-1	Fe-0.043C-12.7Cr-3.9Ni-1.5Mo-0.68Mn-0.39Si-0.009P-0.013S-0.030N	CIS, formerly USSR

Steel	Materials Compositions, Percent in Weight	Note
14Ch17N2	Fe-≤0.11-0.17C-≤0.80Si-≤0.80Mn-16.0-18.0Cr-≤0.30Mo-1.50-2.50Ni-≤0.20W-≤0.30Cu-≤0.030P-≤0.025S-≤0.20Ti	CIS, formerly USSR
15Ch17N2	Fe-0.13C-0.49Si-0.52Mn-17.17Cr-1.75Ni-0.012P-0.09S	CIS, formerly USSR
15Ch25T	Fe-≤0.15C-≤1.00Si-≤0.80Mn-≤0.035P-≤0.025S-24.0-27.0Cr-≤0.30Cu-≤1.00Ni-0.09Ti	CIS, formerly USSR
15Ch28	Fe-≤0.15C-≤1.00Si-≤0.80Mn-27.0-30.0Cr-0.60Ni-≤0.30Cu-≤0.035P-≤0.025S-≤1.0Ni-≤0.20Ti	CIS, formerly USSR
15Ch2M2FBS	Fe-about 0.15C-2Cr-2Mo-V-Nb-Si	CIS, formerly USSR
15Ch5M	Fe-≤0.15C-≤0.50Si-≤0.50Mn-4.50-6.00Cr-0.40-0.60Mo-≤0.60Ni-≤0.03Ti-≤0.030P-≤0.025S-≤0.20Cu-≤0.05V-≤0.30W	CIS, formerly USSR
16GS	Fe-≤0.12-0.18C-≤0.40-0.70Si-0.90-1.20Mn-≤0.30Cr-≤0.30Ni-≤0.30Cu-≤0.035P-≤0.040S-≤0.05Al-≤0.08As-≤0.012N-≤0.03Ti	CIS, formerly USSR/Bulg., comparable with DIN-Mat. No. 1.0481, A 414, A 515, A 516
18/8-CrNi-steel	Fe-≤0.12C-1.0Si-≤2.0Mn-≤0.045P-≤0.030S-≤17.0-19.0Cr≤-8.0-10.0Ni	–
1815-LCSi	Fe-0.006C-18.3Cr-15.1Ni-1.5Mn-4.1Si-0.005S-0.010P-0.010N	UNS S30600, comparable with DIN-Mat. No. 1.4361
18-18-2	Fe-≤0.08C-1.5-2.5Si-≤2.0Mn-≤0.030P-≤0.030S-17.0-19.0Cr-17.5-18.5Ni	USA
18G2A	Fe-≤0.20C-≤0.50Si-≤0.9-1.7Mn-≤0.025P-≤0.020S-≤0.0200Al-≤0.30Cr-≤0.50Ni-≤0.30Cu.-≤0.08Mo-≤0.020N-≤0.050Nb-≤0.03Ti-≤0.1V	Poland
20ChGS2	Fe-≤0.25C-≤1.0Si-≤1.5Mn-≤0.04P-≤0.015S-≤14.0Cr	Russia
20Ch13 (2Ch13)	Fe-0.16-0.25C-≤0.8Si-≤0.8Mn-12-14Cr-≤0.6Ni-≤0.030P-≤0.025S-≤0.30Cu-≤0.20Ti	CIS, formerly USSR/Bulg., comparable with DIN-Mat. No. 1.4021, SAE 420, 420 S 29
20Ch23N18	Fe-≤0.2C-≤1.0Si-≤2.0Mn-22-25Cr-≤0.3Mo-17-20Ni-≤0.2W-≤0.3Cu-≤0.2Ti-≤0.035P-≤0.025S	CIS, formerly USSR
20Ch2G2SR	Fe-0.16-0.26C-0.75-1.55Si-1.4-1.8Mn-≤0.040P-≤0.040S-1.4-1.8Cr-≤0.30Ni-≤0.30Cu-0.02-0.08Ti-0.015-0.050Al-0.001-0.007B	CIS, formerly USSR
23Ch2G2T	Fe-0.19-0.26C-0.40-0.70Si-1.4-1.7Mn-≤0.045P-≤0.045S-1.35-1.70Cr-≤0.30Ni-≤0.30Cu-0.02-0.08Ti-0.015-0.05Al	CIS, formerly USSR
36NChTJu	Fe-≤0.05C-≤0.3-0.7Si-≤0.8-1.2Mn-≤0.020P-≤0.020S-11.5-13Cr-35-37Ni-0.9-1.2Al-2.7-3.2Ti	CIS, formerly USSR
40Ch13	Fe-0.36-0.45C-≤0.80Si-≤0.80Mn-≤0.030P-≤0.025S-12.0-14.0Cr-≤0.30Cu-≤0.60Ni-≤0.20Ti	CIS, formerly USSR, comparable with DIN-Mat. No. 1.4031

Steel	Materials Compositions, Percent in Weight	Note
45 G2	Fe-0.41-0.49C-0.17-0.37Si-1.4-1.8Mn-0.035P-≤0.035S-≤0.30Cr-≤0.30Cu-≤0.15Mo-≤0.30Ni-≤0.03Ti-≤0.05V-≤0.20W	CIS, formerly USSR, comparable with DIN-Mat. No. 1.0912
50Ch	Fe-0.46-0.54C-0. 17-0.37Si-0.50-0.80Mn-≤0.035P-≤0.035S-0.80-1.10Cr-≤0.30Ni-≤0.30Cu-≤0.15Mo-≤0.03Ti-≤0.05V-≤0.20W	CIS, formerly USSR
70G	Fe-0.67-0.75C-0.17-0.37Si-0.90-1.20Mn-≤0.035P-≤0.035S-≤0.25Cr-≤0.25Ni- ≤0.20Cu	CIS, formerly USSR
80S	Fe-0.74-0.82C-0.60-1.10Si-0.50-0.90Mn-≤0.040P-≤0.045S-≤0.30Cr-≤0.30Ni-≤0.30Cu-0.015-0.040Ti	CIS, formerly USSR
2320	Fe-≤0.08-1.0Si-1.0Mn-≤0.040P-≤0.030S-16.0-18.0Cr-≤1.0Ni	Sweden, comparable with DIN-Mat. No. 1.4016; SAE 430, 10Ch17T
ASTM A-159	Fe-3.1-3.4C-1.9-2.3Si-0.6-0.9Mn-0.15S-0.15P	USA
ASTM A-516 Gr. 70	Fe-0.27C-0.13-0.45Si-0.79-1.30Mn-≤0.035P-≤0.040S	USA, comparable with DIN-Mat. No. 1.0050 and No. 1.0481
ASTM XM-27	Fe-≤0.01C-≤0.40Si-≤0.40Mn-≤0.020P-≤0.020S-25.0-27.5Cr-≤0.20Cu-≤0.75-1.50Mo-≤0.015N-≤0.050-0.2Nb-0.50Ni, Ni+Cu≤0.50	USA, comparable with SAE XM-27
C 1204	Fe-0.20C-0.35Si-0.50Mn-0.050P-0.050S-0.30Cr	Yugoslavia, comparable with DIN-Mat. No. 1.0425, B.S. 1501 Gr. 161-400, 164-350, 164-400; 16 K
C 90	Fe-0.85-0.94C-≤-0.35Si-≤0.35Mn-≤0.03P-≤0.03S	Italy
Carpenter 20 Cb-3	Fe-≤0.06C-≤1.00Si-≤2.00Mn-19.0-21.0Cr-2.0-3.0Mo-32.5-35Ni-3.0-4.0Cu-≤0.035P-≤0.035S	USA
Ch12M	Fe-1.45-1.65C-0.15-0.35Si-0.15-0.4Mn-11-12.5Cr-0.4-0.6Mo-≤0.35Ni-15-0.3V-≤0.2W≤0.3Cu-≤0.03Ti	CIS, formerly USSR
Ch14N40SB	Fe-0.034C-4.0Si-0.05Mn-14.4Cr-38.9Ni-0.63Nb	CIS, formerly USSR
Ch15T	Fe-≤0.1C-≤0.8Si-≤0.8Mn-14-16Cr-≤0.3Mo-≤0.6Ni, 5x%C≤Ti≤0.8	CIS, formerly USSR/Bulg.
Ch16N15M3	n. a.	–
Ch17	Fe-≤0.08C-≤1Si-≤1Mn-≤0.040P-≤0.015S-16.0-18.0Cr	Bulg., comparable with DIN-Mat. No. 1.4016, SAE 430, 12Ch17T, X6Cr17
Ch17M2TL	n. a.	–
Ch17N2	Fe-17Cr-2Ni	CIS, formerly USSR
Ch17N5M3	n. a.	–
Ch17N12M3T	Fe-≤0.12C-≤1.5Si-≤2.0Mn-16-19Cr-3-4Mo-11-13Ni-0.3-0.6Ti	CIS, formerly USSR/Bulg.
Ch17N13M2T	n. a.	cf. 1.4571; SAE 316 Ti; CrNiMoTi 17 13 2
Ch17N13M3T	n. a.	cf. 1.4571; UNS S31635; CrNiMoTi 17 13 3

Steel	Materials Compositions, Percent in Weight	Note
Ch17N18M2T	Fe-0.09C-0.6Si-1.4Mn-16.9Cr-1.9Mo-12.3Ni, Ti stab. (p.a.)	CIS, formerly USSR/Bulg.
Ch17T	Fe-≤0.05C-≤1Si-≤1Mn-≤0.040P-≤0-0.15S-16.0-18.0Cr, Ti4x(C+N)+0.15-0.80	Bulg., comparable with DIN-Mat. No. 1.4510, 08Ch17T, X3CrTi17
Ch18AG14	Fe-18Cr-14Mg-0.5N	
Ch18N9T Ch18N10T	Fe-≤0.08C-≤1Si-≤2Mn-≤0.045P-≤0.015S-17.0-19.0Cr-9.0-12.0Ni, Ti 5xC-0.70	CIS, formerly USSR/Bulg., comparable with DIN-Mat. No. 1.4541, X6CrNiTi18-10
Ch18N10	Fe-0.08C-18.4Cr-10.2Ni-1.08Mn-0.3Si-0.005P-0.014S-0.005N	
Ch18N12M2T	Fe-≤0.15C-≤5 1.5Si-≤2Mn-17-19Cr-2-2.5Mo-11-13Ni, 4x%C≤Ti≤0.8	CIS, formerly USSR/Bulg.
Ch18N12T	Fe-0.08C-1.0Si-≤2.0Mn-≤0.045P-0.015S-17.0-19.0Cr-9.0-12.0Ni-Ti = 5x%C≥0.70	CIS, formerly USSR/Bulg.
Ch18N14	Fe-0.035C-18.8Cr-14.6Ni-0.35Mn-0.75Si-0.005P-0.03S-0.004N	
Ch18N40T	Fe-<0.08C-<1.0Si-<2.0Mn-<0.045P-<0.015S-17-19Cr-9-12Ni-<0.7Ti	Comparable with DIN-Mat. No. 1.4541, SAE 321
Ch20N20	Fe-0.004-0.015C-19.4-21.8Cr-19.3-20.8Ni-0.05-5.40Si-0.002-0.1P	CIS, formerly USSR/Bulg
Ch21N5	n. a.	–
Ch21N6M2	n. a.	–
Ch21N6M2T	n. a.	–
Ch22N5	Fe-0.07C-21.54Cr-5.73Ni	CIS, formerly USSR
Ch23N18	Fe-≤0.20C-≤1.00Si-≤1.50Mn-22.0-25.0Cr-≤0.30Mo-17.0-20.0Ni-≤0.035P-≤0.025S	CIS, formerly USSR/Bulg.
Ch23N27M2T	Fe-27Ni-23Cr-2Mo-Ti	CIS, formerly USSR/Bulg.
Ch23N28M3D3T	Fe-28Ni-23Cr-3Mo-3Cu-Ti	CIS, formerly USSR
Ch25T	Fe-≤0.15C-≤1.00Si-≤0.80Mn-24.0-27.0Cr-≤0.30Mo-≤0.60Ni-≤0.035P-≤025S-5x%C≤Ti≤0.9	CIS, formerly USSR/Bulg.
Ch28N18	Fe-0.16C-1.7Mn-1.1Si-22.6Cr-18Ni-0.4Ti (p.a.)	CIS, formerly USSR/Bulg.
ChN28MDT	Fe-0.03-0.046C-22.2-23.5Cr-26.55-27.88Ni-2.55-3.06Mo-2.68-3.38Cu-0.54-0.76Ti-0.15-0.30Mn-0.39-0.69Si-0.021-0.43P-0.008-0.017S	CIS, formerly USS
ChN40B	Fe-0.032C-18.2Cr-40.4Ni-0.08Si-0.05Mn-0.49Nb	CIS, formerly USSR
ChN40S	Fe-0.031C-20.0Cr-38.9Ni-4.2Si-0.05Mn-0.13Nb	CIS, formerly USSR
ChN40SB	Fe-0.04C-18.8Cr-39.4Ni-4.3Si-0.06Mn-0.63Nb	CIS, formerly USSR
ChN58W	Ni-0.03C-14.5-16.5Cr-15-17Mo-3.0-4.5W-1.5Fe-1.0Mn-<0.12Si-0.02S-0.025P	CIS, formerly USSR
ChN60V	Fe-0.01-0.02C-0.09N-max.0.05Zr-0.1Ti-0.015Ce-max.1.7Nb-max0.009B	CIS, formerly USSR
ChN77TJuR	Fe-≤0.07C-≤0.60Si-≤0.40Mn-≤0.015P-≤0.007S-0.6-1.0Al-≤0.01B-≤0.02Ce-19.0-22.0Cr-≤1.0Fe-≤0.001Pb-2.4-2.8Ti	CIS, formerly USSR

Steel	Materials Compositions, Percent in Weight	Note
ChN78T	Fe-≤0.25C-≤1.0Si-≤1.5Mn-≤0.04P-≤0.015S-≤14.0Cr	Russia
FC 20	Fe-3.92C-1.12Si-0.63Mn-0.072P-0.012S	Japan
JS 700	Fe-0.04C-≤1.0Si-≤2.0Mn-≤0.040P-≤0.030S-19-23Cr-4.3-5.0Mo-24-26Ni, Nb≥8xC≤0.40	USA
OZL-17u	Fe-0.04C-0.32Si-1.5Mn-23.2Cr-0.2Mo-29.4Ni-0.01P-0.01S-0.1Ti	CIS, formerly USSR
SAE 1008	Fe-0.10C-0.30-0.50Mn-≤0.030P-≤0.050S	USA, comparable with DIN-Mat. No. 1.0204
SAE 1018	Fe-0.15-0.20C-0.60-0.90Mn-≤0.030P-≤0.050S	USA
SAE C-1018	Fe-0.20C-0.25Si-0.58Mn-0.16Cr-0.04Mo-0.012-0.014P-0.02S	USA
S35C	Fe-0.32-0.38C-0.15-0.35Si-0.60-0.90Mn-≤0.030P-≤0.035S-≤0.20Cr-≤0.30Cu-≤0.20Ni	Japan, comparable with DIN-Mat. No. 1.0501
SIS 2333	Fe-≤0.05C-1.0Si-2.0Mn-≤0.045P-≤0.030S-17.0-19.0Cr-8.0-11.0Ni	Sweden, comparable with DIN-Mat. No. 1.4303; SAE 304, 03Ch18N11
SKH 2	Fe-0.73-0.83C-0.45Si-0.40Mn-≤0.030P-≤0.030S-3.8-4.5Cr-≤0.25Cu-17.0-19.0W-1.0-1.2V	Japan, comparable with DIN 1.3355
SKH-4A	Fe-0.80C-0.29Si-0.31Mn-4.16Cr-17.64W-9.3Co-1.1V	Japan
SKH-9	Fe-0.85C-0.16Si-0.31Mn-4.14Cr-4.97Mo-6.03W-1.88V	Japan
SS41	Fe-≤0.050P-≤0.050S	Japan, comparable with DIN-Mat. No. 1.0040
St35b-2	Fe-≤0.16C-0.17-0.40Si-0.35-0.65Mn-≤0.30Cr-≤0.30Ni-≤0.30Cu-0.050P-0.050S	Germany, formerly GDR, comparable with DIN-Mat. No. 1.0309
St35hb	Fe-≤0.18C-≤0.17Si-0.35-0.65Mn-0.050P-0.050S	Germany, formerly GDR
St38	Fe-≤0.20C-≤0.080-≤0.060S	Germany, formerly GDR, comparable with DIN-Mat. No. 1.0037
St38b-2	Fe-0.12-0.20C-0.17-0.37Si-0.40-0.65Mn-≤0.045P-≤0.050S-Cr+Cu+Ni≤0.70	Germany, formerly GDR, comparable with DIN-Mat. No. 1.0038, BS 4360-40C and A 570 Gr. 36
St5	Fe-≤0.045P-≤0.045S-≤0.009N	Poland, comparable with DIN-Mat. No. 1.0050
SUS 304	Fe-≤0.08C-≤1.0Si-≤2.0Mn-≤0.045P-≤0.030S-18.0-20.0Cr-8-10.5Ni	Japan, comparable with DIN-Mat. No. 1.4301
SUS 430	Fe-≤0.12C-≤0.75Si-≤1.00Mn-16.0-18.0Cr-≤0.60Ni-≤0.040P-≤0.030S	Japan, comparable with DIN-Mat. No. 1.4016, SAE 430, 430 S 15, 12Ch17
Sv-08	Fe-≤0.10C-≤0.03Si-0.35-0.60Mn-≤0.040P-≤0.040S-≤0.015Cr-≤0.30Ni-≤0.01Al	CIS, formerly USSR
TsL-17	Fe-0.1C-5Cr-1Mo	CIS, formerly USSR
TsL-9	Fe-0.07C-1.00Si-2.3Mn-24.1Cr-12.9Ni-1.1Nb-0.03P-0.02S	CIS, formerly USSR

Steel	Materials Compositions, Percent in Weight	Note
U7A	Fe-0.65-0.75C-0.10-0.30Si-0.10-0.40Mn-≤0.030P-≤0.030S	CIS, formerly USSR, comparable with DIN-Mat. No. 1.1520
U8A	Fe-0.75-0.85C-0.10-0.30Si-0.10-0.40Mn-≤0.030P-≤0.030S	CIS, formerly USSR, comparable with DIN-Mat. No. 1.1525
U10A	Fe-0.96-1.03C-0.17-0.33Si-0.17-0.28Mn-≤0.025P-≤0.018S-≤0.20Cr-≤0.20Ni-≤0.20Cu	CIS, formerly USSR, comparable with DIN-Mat.No. 1.1545, SAE W 1 10
X5CrNiMoCuTi18-18	Fe-≤0.07C-≤0.80Si-≤2.0Mn-≤0.045P-≤0.030S-16.5-18.5Cr-2.0-2.5Mo-19.0-21.0Ni-1.8-2.2Cu-Ti≥7x%C	Germany, formerly GDR, comparable with DIN-Mat. No. 1.4506
X8CrNiTi18-10	Fe-≤0.10C-≤1.0Si-≤2.0Mn-≤0.045P-≤0.015S-17.0-19.0Cr-9.0-12.0Ni-Ti≥5 x C≤0.80	Germany, formerly GDR, comparable with DIN-Mat. No. 1.4541

Index of materials

0
0.6655 184
0.6660 184, 218, 319
0.6661 218
0.6676 319
0Ch20N6M2T 59
0Ch21N5T 27, 37
0Ch21N6M2T 27, 37
0Ch23N28M3D3T 105–107, 194–195
00Cr20Ni25Mo4.4Cu 57
000Ch21N21M4B 105–107
03Ch21N21M4B 34, 59
06Ch23N28M3D3T 194, 196
06ChN28MDT 115, 279–282
06Kh23N28M3D3T 193
06ХН28МПТ 358
08Ch17N15M3T 115
08Ch21N6M2T 28, 279
08Ch22N6T 27, 279–281
08-KP 187
08Х21Н6М2Т 358
08Х22Н6Т 358

1
1.0036 99–100, 102, 188–189, 206, 215, 217
1.0037 145, 202–205
1.0308 295
1.0330 6
1.0333 4–6, 264
1.0335 187
1.0345 225, 232–233
1.0375 133
1.0401 149
1.0408 264
1.0425 313
1.0545 145, 240, 313, 315
1.0553 239
1.0562 232–233
1.0564 5
1.0580 236
1.0670 5
1.0671 5
1.1003 133
1.1104 133
1.1191 256
1.1623 133
1.4000 178, 277–278
1.4002 277–278
1.4003 277
1.4006 100, 102, 176, 241, 271, 277–278
1.4016 30, 176, 178, 241, 277–278
1.4021 30, 103, 148, 191, 277–278
1.4024 278
1.4028 278
1.4031 278
1.4034 278
1.4057 30, 277–278
1.4104 278
1.4110 278
1.4112 278
1.4113 29, 278
1.4120 277
1.4122 19, 278
1.4136 18
1.4301 28, 33, 35, 105, 181–183, 241, 255, 257–258, 271, 274–275, 277–278
1.4302 104–105
1.4303 277
1.4304 278
1.4305 274, 278
1.4306 274, 276–278, 346
1.4307 277
1.4310 192, 277
1.4311 278
1.4313 278

Index of materials

1.4315 278
1.4318 278
1.4361 278
1.4362 273, 335
1.4401 28, 30, 33, 107, 110, 182, 195, 241, 275, 277–278, 282, 350, 353
1.4404 169, 179, 184, 195, 272, 274, 277–278, 282, 356
1.4405 278
1.4410 195
1.4427 278
1.4429 277
1.4435 107, 182–183, 244–246, 272, 274, 278, 356
1.4436 29, 53, 110, 275, 278
1.4438 33, 278
1.4439 51–53, 107, 183, 278
1.4449 110, 182
1.4460 177, 278
1.4462 107, 150, 245–246, 273, 278–279, 282, 335–336, 356
1.4465 277–278
1.4466 278, 361–362
1.4467 278
1.4501 30
1.4503 193–194
1.4505 182
1.4507 183
1.4509 278
1.4510 27, 57, 278
1.4511 278
1.4512 278
1.4520 278
1.4521 278
1.4523 104
1.4528 278
1.4529 182, 195
1.4539 51–52, 69, 73, 182, 245–246, 277–278, 282, 356
1.4541 33–34, 36, 38, 52, 105, 160–161, 192–194, 257, 259–260, 278–279
1.4542 278, 334
1.4544 278
1.4546 278
1.4547 245–246
1.4548 271, 278, 334
1.4550 33, 278
1.4552 206
1.4561 278
1.4562 195
1.4565 277
1.4568 278
1.4571 33, 36, 38, 50–53, 55–56, 64, 66, 107, 115, 182–183, 194–196, 278–279

1.4573 55, 105
1.4575 245–246
1.4580 33, 107
1.4589 278
1.4591 285, 361
1.4841 110, 182, 271
1.4845 193
1.4876 245–246, 274–275, 344
1.5069 142
1.5415 232–233, 295, 313
1.5622 73
1.6580 142
1.7035 209
1.7218 225
1.7220 142
1.7279 232
1.7335 120, 145, 232–233, 313
1.7380 282, 313, 353
1.8719 225
1.8812 225
1.8850 241, 243
1.8905 232–233, 240
1.8907 145, 241, 243
1.8961 133
1.8962 133
1.8975 138
1Ch13 100, 102–103
1Ch21N5T 37, 106–107
1H18N9T 36
10Ch17N13M2T 34, 279, 281–282
10Ch17N13M3T 34, 55, 105, 196
10CrMo9-10 282, 313, 315, 352
10X17H13M2T 358
12Ch18N10T 34, 105, 193, 279, 281–282
12X18H10T 358
13CrMo4-4 146
13 CrMo 4 4 232
13CrMo4-5 120, 145, 232–233, 313, 315
15Ch25T 279–280
15Mo3 232
15X25T 358
16MnR 302–304
16MnRE 302, 304
16Mo3 232–233, 295, 313, 315
17-4PH 39–40
17-4-PH 271
17-4PH 334
17-7PH 39–40
17-10P 40
17 MnCrMo 3 3 232
18-8 nickel chromium steels 160
18Cr2Mo 28
18Cr-10Ni-2.5Mo 340
1008 6

Index of materials

1018   138, 141
1020   4, 189
1030   306
1042   256
17241   37

2
2.4060   111
2.4360   111, 241
2.4602   245–246
2.4610   241
2.4617   245–246
2.4619   245–246
2.4642   274
2.4662   284
2.4800   111
2.4816   241, 274–275, 282, 344, 353
2.4819   356
2.4858   245–246, 284, 349
2.4951   352
2Ch13   103, 191
2Ch18N9   192
20ChGS2   209, 218
20Ch23N18   193
20Cr-25Ni-4.5Mo   340
20Cr25Ni/Nb   152–154, 156, 158
20Cr-25Ni-Nb steel   163
20Cr25Ni/Ti   152–153, 155, 157–158
20Cr25Ni/TiN   153, 159
20Cr25Ni/TiNb   152, 156
20GS2   209
22Cr-5Ni-3Mo   340
23Cr-4Ni   340
25Cr-7Ni-4Mo-N   340
27Cr-31Ni-3.5Mo   340
201   32, 37
254 SMO®   245

3
3.7055   245
3RE60   332
30CrNiMo8   142–143
32CrNi3MoV   321
33   285, 361–362
33CrNi3MoV   321
34CrMo4   142–143
34CrNi3Mo   321
36Mn7   142–143
302   28, 37, 44, 47, 50, 98, 343
303   274, 343
303 super   343
304   28–30, 33–35, 38, 40, 42, 44, 49–50, 73–74, 98, 111–112, 114, 169–170, 181–183, 192, 241–242, 255, 257–260, 274–276, 285, 330, 332–333, 343, 345
304 H   169–171
304 L   30, 34–35, 38, 44, 48, 274–276, 346, 349
305   40
305 Cu   343
308   42
308 H   169–171
309   37, 52, 67, 169, 343
309 L   169–170
310   34–35, 42, 48, 110, 260, 276, 330, 343, 348–349
310 MoLN   73
316   28–30, 33–34, 38, 48, 52–53, 56–57, 59–62, 64–68, 73, 98, 110–112, 114, 169, 178, 182, 195, 241–242, 260, 274–276, 283, 285, 343–344, 348, 350, 352, 355
316 Cb   33, 61
316 ELC   52, 67
316 H   169–171
316 L   34, 169–171, 179–180, 182, 184, 272, 274, 283, 343
316 Ti   33–34, 36, 38, 50–51, 55, 67, 115, 182, 361–362
317   33, 52–53, 55–56, 60, 63, 65, 67, 110, 112, 114, 182
317 L   33, 69
321   33–34, 36, 38, 42, 44, 50, 257, 259–260
329   48, 177
330   343
347   33–34, 36, 42, 44, 47, 49–50, 53, 56, 98, 206, 258
349   57

4
40Ch   209
45   256
403   333
405   22, 26, 332–333
410   22, 26, 31, 37, 40, 52, 68, 74, 176, 178, 241–242, 271, 322, 332, 334
416   22, 40
420   19, 22, 39, 257
430   27, 52, 68, 176, 178, 241–242
431   19, 39
434   29, 177
439   27, 36
440 A   39
440 C   39
443   28
444   27–28
446   52, 68

Index of materials

4130 225–226

5
5LX65 138

6
600 241–242, 258, 274, 275, 282, 344–345, 352
625 285
690 274

7
702 245

8
800 274, 275, 285, 344–345
825 107, 114, 284–285, 349

9
901 284
904 L 69, 73, 182

a
A48CPR 302, 304
A 202 225
A 212 225
A 285 225, 300, 304, 306
A508C13 169–170
A 516 300, 303–306, 308–311
A 516 grade 70 305
A 517 (F) 241–242
A 517 grade F 227–228, 243
AC66 167
AE 316 73
Al 2205® Alloy 271, 273
Alloy 33 285, 361–362
AlMg2.5 241
Aloyco® 20 52, 67
aluminium alloys 209
aluminium bronze 33
aluminium bronze, 10 % Al 56
AM 350 73
Antacid® 176
Armco® iron 3, 6, 187, 206, 256–257, 259–260, 267
ASTM A 285 303
ASTM A 516 grade 70 303
ASTM F138 282, 356
austenitic cast iron 73, 244, 269–270, 319
austenitic chromium-nickel-molybdenum(N) steels 244, 349
austenitic chromium-nickel-molybdenum(N) steels and CrNiMoCu(N) steels 244
austenitic chromium-nickel-molybdenum steels 51, 100, 104, 107, 180, 192, 194, 219, 258, 349
austenitic chromium-nickel steels 32, 34, 68, 104, 115, 151, 180, 191, 219, 244, 258, 274, 277, 342
austenitic chromium-nickel steels with special alloying additions 68, 100, 107, 180, 196, 219
austenitic CrNi steels 169, 284
austenitic CrNiMo(N) steels and CrNiMoCu(N) steels 277
austenitic CrNiMo(N) steels 169, 244
austenitic CrNiMoCu(N) steels 169, 244, 361
austenitic-ferritic steels 340, 356
austenitic nickel chromium steels 159–160
austenitic stainless steels 35, 168
austenitic steel 32, 56
austenitic steels 32, 35, 56, 69, 151, 167, 171, 259, 340, 358
AVESTA® 254 SMO 51

c
C15 149
C-276 356
C-1018 311
carbon-manganese steels 239
carbon steels 6, 99, 102, 113–114, 175, 188–189, 200–201, 251, 266, 276, 295, 300
Carpenter® 7 Mo 48
Carpenter® 20 33, 53, 64, 260
Carpenter® 20 Cb 48
Carpenter® 20 Cb-3 67, 112, 114
cast chromium steel 104
cast ferrosilicon 244, 269, 316
cast iron 11–15, 95, 97, 99–100, 218, 244, 269, 317
cast steel 43, 95, 119, 199, 223, 263, 293, 334
CdSn-alloys 10
cement 213
Ch15T 191
Ch16N15M3 106–107
Ch17M2TL 27
Ch17N5M3 106–107
Ch17N13M2T 36, 55, 106–107, 195
Ch17N13M3T 36, 105–107, 194
Ch17T 27, 57
Ch18N9T 36, 259–260
Ch18N10T 36, 106–107, 192, 194, 206
Ch18N12M2T 191, 194

Index of materials | **415**

Ch21N5 37
Ch21N5T 37
Ch21N6M2 37
Ch21N6M2T 37
Ch23N28M3D3T 194
ChN78T 279–280, 282
chromium-molybdenum steels 29, 122, 146, 344
chromium-nickel-molybdenum-copper steel 195
chromium-nickel-molybdenum steels 59–64, 104, 109–110, 179, 195, 244
chromium-nickel steels 49, 74, 104–107, 180, 183, 192–193, 259, 276, 335, 343, 348
chromium steels 19–20, 103–104, 147, 177, 191, 257, 329
Circle® L-34 260
C-Mn steels 267
CN 20 74
concrete 202
constructional steels 206, 223, 225
container steel 227
copper 33, 53, 56–57, 257
copper alloys 35
copper-nickel 70/30 56
CrAl-steel 75
CrMoTi 332
CrNiMo-steel 196
CrNiMo-steels 20 10 195
CrNi-steel 17 14 259
CrNi-steel Ch18N9T 257
Cr-Ni steels 168, 257, 275
Cr steel 150–151
CuNi-alloy 111
Cu-steel 96
Custom 450® 74
Custom 455® 74
CuZn-alloys 9, 11

d

duplex steels 244, 271, 273, 334–335, 337–338, 341
Duracid® 176
Durichlor® 176
Durimet® 20 68–71, 74
Duriron® 175–176

e

E255 4
E-Brite® 26-1 29–30, 178, 332
El-417 193
El-943 193–194

EN AW-1100 111
EN AW-3003 111, 241–242
EN AW-5454 241–242
EN AW-6061 111
EN-GJLA-XNiCuCr15-6-2 184
EN-JL3011 73, 184, 218
EN-JS3011 75
EP-567 36
Esshete®1250 167

f

F138 357
F41000 73, 218
F41001 218
F41002 218
F41003 218
Ferralium® 337
Ferralium® 255 273, 335
Ferralium® alloy 183–184
ferritic-austenitic steels 27, 103, 164, 179, 219, 244, 271, 273–274, 282–283, 334, 336
ferritic-austenitic steels 244, 271, 334
ferritic-austenitic steels with more than 12% chromium 19, 103, 179, 219
ferritic chrome steels 244, 270–271
ferritic chrome steels with 13% Cr 146, 244, 270, 244, 271
ferritic chromium steels 29, 103, 177–178, 191, 219, 257, 319, 329
ferritic chromium steels with more than 12% chromium 19, 103, 176, 191, 219, 257
ferritic chromium steels with 13% Cr 319
ferritic chromium steels with ≥ 13% Cr 147, 329
ferritic/perlitic-martensitic steels 147, 244, 271, 334
ferritic steels 167, 209, 358
FeSi-alloys 18
fiber glass 202
fine grain constructional steels 224, 231–232, 240, 315
fine-grain structural steels 145

g

G12CrMo9-10 353
G10150 149, 202–205
G10060 187
G41350 142–143
G43400 142–143
G51400 209
GGG NiCr 20 3† 75
GGL-NiCr20-2 184, 319

Index of materials

GGL-NiCr30-3   319
GGL-NiCuCr 15 6 3   218
gray cast iron   14, 175, 269
grey cast iron   316–317
Guronit® GS2   18
G-X40CrNi27-4   334
GX70CrMo29-2   18

h
H17N13M2T   36
H13400   142–143
Hastelloy® A   52–53, 68
Hastelloy® B   52, 57, 67, 111, 175
Hastelloy® B 2   184
Hastelloy® C   33, 52–53, 56–57, 67, 107, 111, 285
Hastelloy® C 4   184, 241–242
Hastelloy® C 276   184
Hastelloy® D   111
high-alloy cast iron   13, 15–18, 73, 101, 175, 191, 218, 257, 269, 319
high-alloy multiphase steels   147, 334
high-silicon cast iron   13, 15–18, 101, 175, 191, 218, 257
H I   232

i
Illium®   52, 67
impermeable concrete   202
Incoloy® 625   285
Incoloy® 800   274, 275, 285, 344–345
Incoloy® 825   107, 114, 284–285, 349
Incoloy® 901   285, 349
Incoloy alloy 901   284
Inconel®   33, 52–53, 56, 68, 98, 258
Inconel® 600   241–242, 258, 274, 275, 282, 344–345, 352
Inconel® Alloy 690   274
interstitial steels   29
iron   188–189, 199, 218, 263–264
iron materials   241
iron-nickel alloys   362
iron-nickel cast alloys   270

j
J92610   274

k
K01600   232
K01700   232
K01702   254–255
K02001   145
K02502   99–100, 102, 145, 188–189, 206, 215, 217
K11547   120, 145, 232
K11820   232
K12000   239
K12004   232
K12709   145
K13050   73
Kovar® 29NK   74

l
Langalloy® 20 V   183
low-alloy cast iron   218, 223, 269, 316
low-alloy steels   95, 119–121, 125, 136, 144, 146, 188, 223, 238, 251–253, 255, 263, 269, 293
low-carbon steels   29, 190
low-phosphorus steel   299

m
malleable cast iron   12
manganese-aluminium steels   362
martensitic stainless steels   334
martensitic steels   244
Meehanite®   269, 318
Meehanite-HD   269, 318
Midvale® 2024   29
mild steel   96
Mn-steel   96
molybdenum-alloyed steel   104
Monel®   8, 33, 35, 52–53, 56, 68, 98
Monel® 400   111, 114, 241–242

n
N-80   5
N06022   245
N06985   245
N08031   195
N08800   245
N08825   245
N08904   51, 245, 356–357
N08926   195
N10665   245
N-A-XTRA 70   232
nickel   33, 52–53, 56, 68, 98, 114
Nickel 200   111
nickel alloys   35, 241
nickel-chromium alloys   284
nickel chromium steels   159
Nickel Π2   358
NiCr15Fe   241, 274, 353
NiCr21Mo   284
NiCr29Fe   274
Nicrofer® 3033   285, 361
Nicrosilal®   75
NiCu 30 Fe   241

Index of materials

NiMo16Cr16Ti 241
Nimonic® 75 352
niobium-stabilized steel 153, 161, 258
Ni-Resist® 75–76, 184
Ni-Resist® 1 99, 218
Ni-Resist® 1b 218
Ni-Resist® 2 75, 218
Ni-Resist® 2b 218
Ni-Resist® alloys 218
Ni-Resist® D-2 75
Ni-Resist® D-2B 75
Ni-Resist® type 1 73
NIROSTA® steels 277
nodular cast iron 317
non-alloy cast iron 257
Noricid® 176
Noricid® B 176

o

Orion® 26-1 30

p

P-105 5
P-110 5
P235GH 232–233
P265GH (HII) 313, 315
P355N 232–233
P460N 232–233, 240
PG-Sr 3 38
pipe steel 140, 229–230
Portland cement 203, 213–214
pure aluminium 241
pure iron 6, 95, 125, 136–137, 139, 187, 241–242
Pyrex® glass 175

r

Remanit® 1860 M 37
Remanit® 1860 MS 37
Remanit® 4521 28
Remanit® 4522 28
Remanit® 4575 177–178
Remanit® steel 22–23, 31, 50, 67

s

S235JR 145–146, 254–255
S500N 145
S30200 192
S30400 105
S30403 274–275
S30888 104–105
S31008 193
S31254 245
S31600 107, 110
S31603 107, 195, 245, 357
S31635 105, 107, 194, 196
S31640 107
S31726 51, 107
S31803 150, 245, 271, 273, 356–357
S32100 105, 160–161, 192–194
S32205 107
S32404 273
S32750 195
S32803 177, 245
S355J0 239
S355N 145–146, 240, 268, 298, 313, 315
S41000 100, 102
S42000 103, 148, 191
S43080 177
S44627 29, 178
SAE 201 32, 37
SAE 302 28, 37, 44, 47, 50, 98, 343
SAE 303 274, 343
SAE 303 super 343
SAE 304 28–29, 33–35, 38, 40, 42, 44, 49–50, 73–74, 98, 111–112, 114, 169–170, 181–183, 192, 241–242, 253, 255, 257–260, 271, 274–276, 285, 330, 332–333, 343, 345
SAE 304 H 169–171
SAE 304 L 30, 34–35, 38, 44, 48, 274–276, 343, 346, 349
SAE 305 40
SAE 305 Cu 343
SAE 308 42
SAE 308 H 169–171
SAE 309 37, 52, 67, 169, 343
SAE 309 L 169–170
SAE 310 34–35, 42, 48, 110, 260, 271, 276, 330, 343, 347–349
SAE 310 MoLN 73
SAE 316 28–30, 33–34, 38, 48, 52–53, 55–57, 59–62, 64–68, 98, 110–112, 114, 169, 178, 182, 195, 241–242, 260, 274–276, 282–283, 285, 343–344, 347–348, 350, 352, 355
SAE 316 Cb 33, 61
SAE 316 ELC 52, 67
SAE 316 H 169–171
SAE 316 L 34, 52, 169–171, 179–180, 182, 184, 272, 274, 283, 343
SAE 316 Ti 33–34, 36, 38, 50–51, 55, 67, 115, 182, 361–362
SAE 317 33, 52–53, 55–56, 60, 63, 67, 110, 112, 114, 182
SAE 317 L 33, 69
SAE 321 33–34, 36, 38, 42, 44, 50, 257, 259–260

Index of materials

SAE 329   48, 177
SAE 330   343
SAE 347   33–34, 36, 42, 44, 47, 49–50, 53, 56, 98, 206, 258
SAE 403   333
SAE 405   22, 26, 332–333
SAE 410   22, 26, 31, 37, 40, 52, 68, 74, 176, 241–242, 271, 332, 334
SAE 410 S   178
SAE 416   22, 40
SAE 420   19, 22, 39, 257
SAE 430   27, 37, 52, 68, 176, 178, 241–242
SAE 431   19, 39
SAE 434   29, 177
SAE 439   27, 36, 57
SAE 440   39
SAE 443   28
SAE 444   27–28
SAE 446   52, 68
SAE 904 L   69, 73, 182
SAE 1008   6
SAE 1020   4, 189, 264
SAE 1030   306
SAE 1042   256
SAE 4130   41, 225–226
SAE 4340   41
Sandvik® 2RK65   69, 73, 179
Sandvik® steel   69
Sanicro® 28   167
Sicromal®   75
Sicromal® 12   48
Siferrid®   176
silicon bronze   56–57
silicon cast iron   175, 319
slag cement   213–214
special iron-based alloys   73, 100, 107, 183, 362
spring steel   225
St 2   6
St 3   4–6, 99–100, 102, 188–189, 206, 215, 217, 264
St 35   295
St 37   202–205, 254–255
St 45   4, 264
St 52   236
St 52-3   239
stainless structural steels   219
StE 355   232, 240
StE 460   232, 240
structural steels   102
structural steels with up to 12% chromium   19, 102, 176, 219
superduplex steels   337–338

t

T 1   225–227
Tantiron® alloys   18
tempered steels   142, 227
Tenelon® steels   28
Thermisilid®   176
Ti 3   245–246
titanium   35, 112, 114, 209, 241–242, 246
titanium alloy   57
TP347H   167

u

UHB® 44 LN   179–180
unalloyed and low-alloyed cast iron   223
unalloyed and low-alloy steels/cast steel   119, 223, 293
unalloyed cast iron   12, 95, 175, 218, 257
unalloyed cast iron and low-alloy cast iron   119, 269, 316
unalloyed steels   5–6, 8, 12, 14–15, 95, 97, 121, 131–133, 136, 138, 149, 175, 199, 218, 241–242, 254, 257, 263, 276, 293, 295, 299, 306, 310, 316, 348
unalloyed steels and cast steel   3, 95, 175, 187, 199, 251
unalloyed steels and low-alloy steels/cast steel   263
Uniloy® 326   28
UNS F138   357
UNS F41000   73, 218
UNS F41001   218
UNS F41002   75, 218
UNS F41003   218
UNS G10150   149, 202–205
UNS G10060   187
UNS G41350   142–143
UNS G43400   142–143
UNS G51400   209
UNS H13400   142–143
UNS J92610   274
UNS N08926   182
UNS K01600   232
UNS K01700   232
UNS K01702   254–255
UNS K02001   145
UNS K02502   99–100, 102, 145, 188–189, 206, 215, 217
UNS K11547   120, 145, 232
UNS K11820   232
UNS K12000   239
UNS K12004   232
UNS K12709   145
UNS K13050   73
UNS N06022   245

UNS N06985 245
UNS N08031 195
UNS N08800 245
UNS N08825 245
UNS N08904 51, 245, 356–357
UNS N08926 195
UNS N10665 245
UNS S235JR 145–146, 254–255
UNS S500N 145
UNS S30200 192
UNS S30400 105
UNS S30403 274–275
UNS S30888 104–105
UNS S31000 182
UNS S31008 193
UNS S31254 245
UNS S31600 107, 110
UNS S31603 107, 195, 245, 356–357
UNS S31635 105, 107, 194, 196
UNS S31640 107
UNS S31726 51, 107
UNS S31803 150, 245, 271, 273, 356–357
UNS S32100 105, 160–161, 192–194
UNS S32205 107
UNS S32404 273
UNS S32750 195
UNS S32803 177, 245
UNS S355J0 239
UNS S355N 145–146, 240, 268, 298, 313, 315
UNS S41000 100, 102
UNS S42000 103, 148, 191
UNS S43080 177
UNS S44627 29, 178
Uranus 50 273, 335
USS Tenelon® 37

w

welding materials 169
Weldshyne EWC® 55
Worthite® 53, 68–71, 74

x

X1CrNiMoCuN20-18-7 245
X1CrNiMoCuN33-32-1 285, 361
X1CrNiMoN25-25-2 73, 361
X1CrNiMoNb28-4-2 177
X1NiCrMoCu25-20-5 180, 182
X1NiCrMoCuN25-20-7 180, 182
X2CrMoNb18-2 28
X2CrMoTi18-2 27–28
X2CrMoTiS18-2 104
X2CrNi19-11 44, 274
X2CrNiMo17-12-2 179, 274
X2CrNiMo18-14-3 107–108, 182–183, 244
X2CrNiMoN17-13-5 107–108
X2CrNiMoN17-13-5 183
X2CrNiMoN22-5-3 107–108, 150, 336
X2CrNiN23-4 335
X3CrNiMo17-13-3 29, 61, 110
X3CrNiMo18-12-3 182
X3CrNiMoN27-5-2 177
X4CrNiMo17-12-2 241
X4NiCrMoCuNb20-18-2 180–182
X5CrNi18-10 28, 44, 50, 105, 111, 182, 241, 255, 257–258
X5CrNi19-9 104–105
X5CrNiCuNb17-4-4 271, 334
X5CrNiMo17-12-2 28, 61, 67, 107, 111, 182, 195, 353
X6Cr13 178
X6Cr17 27, 176, 178, 241
X6CrAl13 22
X6CrMo17-1 29, 177
X6CrNiMoNb17-12-2 61, 107
X6CrNiMoTi17-12-2 107–108, 182–183
X6CrNiTi18-10 44, 50, 105
X8Cr18 177
X8CrMoTi17 104
X10 167
X10NiCrAlTi32-21 274–275
X12Cr13 22, 31, 176, 241
X12CrS13 22
X15Cr13 31
X15CrNiSi25-21 182
X17CrNi16-2 19
X20 167
X20Cr13 19, 22, 148–150
X20CrMo13 22–23
X39CrMo17-1 19
X-42 229–231
X-46 229–231
X-52 230–231
X-60 230–231
X 65 138–141
XH78T 358
XM-27 29–30

z

Zeron® 337
zinc 241–243
zirconium 35, 246

Subject index

a

acetaldehyde 4, 55
acetic acid 3–76, 95–96, 99–100, 102–105, 107–108, 112–113, 138, 140
acetic acid + acetaldehyde 70
acetic acid + acetic anhydride 6, 25, 70
acetic acid + acetone 6, 16
acetic acid anhydride 96, 108
acetic acid esters 8
acetic acid + ether 6, 16, 70
acetic acid + ethyl alcohol 6, 16, 21, 26
acetic acid + formic acid 6, 16, 18, 21, 26, 70, 72
acetic acid + hydrobromic acid 6, 16, 21, 26, 70, 72
acetic acid + hydrochloric acid 6, 17, 21, 26, 70, 72
acetic acid + hydrogen peroxide 22
acetic acid + mercury salts 6, 70, 72
acetic acid + phenylacetic acid 16
acetic acid + salicylic acid vapors 72
acetic acid solutions 3, 6
acetic acid + sulfuric acid 6, 17, 21, 26, 71–72
acetic acid vapor 6, 8–9, 21, 25, 64, 69–70, 72, 75, 96
acetic acid (water-free) 4
acetic acid with acetaldehyde 16
acetic anhydride 11–13, 20, 30, 35, 48, 55, 65, 69, 75
acetic anhydride + acetic acid 65, 74
acetic anhydride in acetic acid 14, 20
acetic anhydride vapor 11, 66
acetone cyanohydrin 48
acetylation equipment 65
acetyl cellulose production 8
acid vapors 9
active-passive transition 305–306, 309, 347

agricultural utilization 224
air (as contaminant) 234, 239–240
alkali electrolysis plants 269
alkali hydroxides 251
alkaline earth chlorides 202
alkaline earth hydroxides 199–219
alkaline solutions 265
alkanecarboxylic acids 95–115
aluminate solution 293, 304
amines 100
aminosulfonic acid 188, 192
ammonia and ammonium hydroxide 223–246
ammonia from pipeline 231
ammonia gas 224, 244
ammonia solution 244
ammonia sphere 236–237
ammonium stearate (as inhibitor) 214
ammonium sulfate 175
ammonium thiocyanate 99
anhydrous acetic acid 36
anodic dissolution 190
anodic passivation 293
anodic protection 305, 309–310
anodic stress corrosion cracking 142, 144
aqueous acetic acid solution 3
aromatic sulfonic acids 194
ash deposits 164, 167
austenite phase 282
azeotropic distillation 57

b

barium hydroxide 206–207
bauxite 306
Bayer digestion plant 300
Bayer process 300, 303
Bayer solution 302–303, 305
benzenesulfonic acid 187
benzimidazole 192

Subject index

benzonitrile 192
benzothiazole 192
benzotriazole 192
benzotriazole, benzothiazole and dibenzyl sulfoxide 192
benzylpyridine isooctylxanthate 188
biomass 164
boiler plate 232
boiler sheet 232
boiling acetic acid 19
boiling formic acid 180
boiling glacial acetic acid 35
Boudouard's equilibrium 121
Boudouard reaction 153
bridge constructions 205
butyric acids 3, 95, 99–100, 102, 104, 108, 110, 112–113
butyric anhydride 108
bypass line 275

c

calcium carbonate 219
calcium chloride 202, 215
calcium chloride solutions 210
calcium hydroxide 199–201, 203–204, 206, 209, 218–219
calcium hydroxide solutions 199, 205, 209, 211, 214, 219
calcium sulfate 206
calcium sulfide 206
canning industry 68
capric acid 97
caproic acid 95
caproic acid (hexanoic acid) 104
caprylic acid 97
carbamates 244
carbamic acid 244
carbonatization of concrete 200
carbon dioxide 241, 244
carbon dioxide corrosion 138
carbon dioxide gas 119, 153
carbon dioxide (oxygen-free) 133
carbon dioxide (ultrapure) 133
carbon dioxide (wet) 142
carbon disulfide 32
carbonic acid 119–171
carbon monoxide 32, 122
carbon tetrachloride 96
carboxylic acid 100, 102
cathodic polarisation 308, 345
cathodic protection 202, 276, 348
caustic solutions 357
cellulose acetate 65

cellulose acetate plants 55
cellulose digestion 308
cellulose processing 33
cementite 299
CERT experiment 299
CERT method 232
CERT test 145, 233–234, 273, 306–307, 311, 327, 336
chemical and petrochemical plants 151
chlorate 282
p-chloro-benzenesulfonic acid 187
chlorosulfonic acid 195
chromium-nickel steels 193
citric acid 12
cladding 348
cladding tubes of fuel assemblies 153
cold forming 136
collecting tank 97
combustion gases 151, 164
compact tension 239
contact corrosion 241
containers for liquid ammonia 227
contaminations 224, 227
contaminations with air 234
coolers 65
cooling water systems 271, 334
cracking tendency 142
crack initiation 142
crack nucleation 145
crevice corrosion 283
critical potential range 295
crotonaldehyde 55
crude acetic acid 18
CT specimen 239
cupric salts 183

d

deaerated water 259
decanoic acid 97
decomposer 300
deposition of magnetite 255
dew point 275
dibenzyl sulfoxide 188, 192
2,5-dihydroxy-1,4-dithiane 192
2,6-diphenylpyridylium sulfate 191
disodium phosphate (as inhibitor) 214
dissociation constant 95
distillation columns 35, 97
distillation equipment 65
dithiocarbamate 188
dry acetic acid vapors 38

Subject index | 423

e

electrochemical measurements   131
electrolysis plant   283
electro-polishing   190
epoxy resin coatings   202
erosion corrosion   148, 208, 256
erosion corrosion resistance   149
erosion corrosion tests   149
etching   271
ethanol   5, 35
ethyl alcohol   96
ethylidene diacetate   55
evaporators   269
evolution of hydrogen   175
exhaust gases   119

f

fatigue cracks   333
fatty acid   98–99
ferric salts   38
ferrite   299
fertilizer   226
fertilizing   224
fire extinguishers   142
flame sputtering   243
floret development   146
floret formation   119
flow rate   122
food   12
food industry   343
foodstuffs industry   109, 274
formaldehyde   8
formalin   207
formic acid   8–9, 27, 30, 33–36, 38, 54, 73, 96, 99, 108, 175–184
formic acid (boiling)   180
free corrosion potential   328
fruit acids   12
fruit juices   12

g

galvanic corrosion   35, 254
gas bottles   142
gas containers   144
gaseous ammonia   223
gas pipelines   138
gas production   148
gas sphere   240
glacial acetic acid   6, 13, 28, 30, 35–36, 42, 49, 52, 63, 65, 68
grain boundary sensitization   169

h

hard surfacing layer   169, 171
heat-affected zones   35
heat exchanger   246
heat exchanger pipe   244
heat exchanger tubes   119
heat-treatable steels   271
hexametylenediamine (as inhibitor)   5
hexanoic acid   95
hollow sample   351
hydration   131
hydrobromic acid   96
hydrochloric acid   9, 96, 188
hydrogen   122
hydrogen embrittlement   19, 74
hydrogen evolution   129, 131
hydrogen-induced brittle fracture   209
hydrogen-induced crack formation   144
hydrogen-induced cracking   229
hydrogen-induced stress corrosion cracking   142, 144
hydrogen iodide   32
hydrogen overvoltage   4
hydrogen permeation measurements   144
hydrogen (production of)   283
hydrogen sulfide   4
hydroxide solutions   357
hypophosphoric acid   193

i

industrial atmosphere   216
inert gas   283
inhibiting action   187
inhibiting effect   253
inhibiting efficiency   6, 99–100, 187
inhibition   187
inhibitors   6, 99, 102, 183, 187, 189–190, 192, **213**, 218, 254, 271
inhibitory effect   38
intercrystalline attack   192
intercrystalline corrosion   35, 192, 208
intergranular corrosion   33
intergranular stress corrosion cracking   259
Ion-nitriding   35, 64

k

Katapine   187
KOH melts   276

l

lactic acid   109
leather industry   218

Subject index

lime water   218
liquid ammonia   228–229, 237–239, 243
lithium chromate (as inhibitor)   254
lithium hydroxide   251–259
lithium hydroxide (as inhibitor)   254
lithium hydroxide solutions   251, 282, 360
lithium molybdate (as inhibitor)   254
low-pressure turbine   321
low temperature storage tank   240

m

magnetite   119
magnetite layer   119–120
manufacture of alkanecarboxylic acids   97
manufacture of butyric acid   110
manufacture of formic acid   175
mercaptobenzimidazole   192
mercury salts   96
metallography   271
metallurgical ammonia   231
methacrylonitrile   48
methanol   5, 32, 35
2-methylnaphthalene   34
mixed jointing welding   282
morpholine (as inhibitor)   5

n

NaOH melts   269, 318, 326, 328, **348**
naphthalenedisulfonic acid   187
natural gas production   141
nickel coatings   202
nitric acid   112
nitride hardening   223
N,N′-di-o-tolylthiourea   187
non-destructive testing   205
nuclear power plants   251, 255
nuclear reactors   119, 153
number of stress cycles to failure   333

o

octanoic acid   97
organic acids   3, 33, 38
organic-based inhibitors   215
organic inhibitors   215
organic solvents   3
orthophosphoric acid   190
oxalic acid   73
oxidic protective coating   252
oxidizing agent   282
oxidizing salts   50
oxygen atmosphere   283
oxygen-free carbon dioxide   133

p

packing industry   6
paper industry   33
passivation behaviour   293
passive range   299, 306, 347
passivity   102
peracetic acid   96
petrochemical plants   136
petroleum   189
Petrov's catalyst   192
phenolsulfonic acid   191
phenylacetic acid   65
phenylthiourea   192
phosphorus contents   299
phosphorus segregations   323
pH-potential diagram   326
pipelines   35, 224, 361
piperazine (as inhibitor)   5
pipe steels   230
pipe system   254
pitting   283
pitting corrosion   30, 34–35, 104–105, 108, 136, 193, 199, 202, 208, 244
pitting corrosion potential   203–204
pitting potential   191, 202
polyethoxylated alcohol   188
polymer impregnations   202
polyvinylalcohol   187
potassium chromate (as inhibitor)   214
potassium hydroxide   259, 263–285
potassium hydroxide solutions   251
potassium iodide   5
potential–neutralisation value diagram   326
potential range (critical)   266
Pourbaix diagram   326
power plant construction   270
power plants   151, 164, 251
preparation of acetic anhydride   18
preparation of cellulose acetate   65
preparation of vinyl acetate   55
pressure vessel   302
pressure vessel steel   302
pressurized container   244
pressurized water reactors   252
primary inhibitors   213
processing of crude oil and natural gas   125
production of acetic acid   4, 18, 69
production of acetyl cellulose   30
production of gas   148
production plants   224
propargyl alcohol   5

Subject index | **425**

propionic acid 18, 27, 95, 99–100, 103–104, 108, 111–113
propionic acid anhydride 108
protective efficiency 191
pulp and paper industries 293, 344
pumps 35, 65, 269
pure acetic acid 38
pure acetic anhydride 18, 49
pure alcohol 5
Pyrex® glass tube 175
pyrophoric iron 8

r
rail tank truck 227
reactor 326
reformer 275
reformer plant 275
refrigeration industry 225–226
reservoirs 125
Resistin N® 188
Reynolds number 149
road transport tank 227
rotating disk specimens 133
rust preventing agents 207

s
sacrificial anode 243
safety valves 227
scale break-offs 119
scaling 120
sea atmosphere 216
sea water 241
secondary inhibitors 213, 215
self-passivation 205
slow strain rate tensile tests 145
slow strain-rate tests 306
sodium acetate 3
sodium acetate + sodium salicylate 18
sodium benzoate (as inhibitor) 214
sodium bichromate 187
sodium chloride 201–203
sodium chloride solutions 3
sodium-cooled reactor 345, 350
sodium hydroxide 219, 259, 269, 293–362
sodium hydroxide solution 252
sodium nitrite (as inhibitor) 214
sodium sulfate 201
sodium sulfate solution 135
sodium sulfide 219
sodium thiosulfate 99
solid particle 150
solvent 224
sorbic acid 6

sour gas condensate 145
spherical container 224
spherical pressurized container 231, 232
spheroidal graphite 269
spray coatings 56
steam boiler pipes 275
steam boilers 251, 276
steam circuits 255
steam generators 252, 331
steam system 271
steam turbine construction 271
steam turbines 322, 334
steel spring 225
steel vessel 300
stirrers 34
storage 345, 361
storage container 227, 236, 243
storage of NaOH 295
stress corrosion cracking 5, 34, 37, 142, 144, 148, 151, 199, 203, 208–209, 215, 224–225, 227–236, 239–241, 243–244, 255, 265, **271**, 273–277, 280, 282–283, 293, 295, 298–300, 306, 308, 311, **330**, 336, 343, 346, 356
stress intensity 142
stress intensity factor 239
sulfonation of anthraquinone 194
sulfonic acid 187–196
sulfosalicylic acid 187
sulfoxylation 191
sulfuric acid 9, 27, 96, 112
sulfur infiltration 202
surface corrosion 145
surface layer 253
swelling 5
syngas 275
syngas plants 144
synthesis of urea 244
synthetic acetic acid 34

t
Tafel method 317
Tamman furnace 122
tank mounting unit 227
tanks 35, 361
tank vehicles 224
tar acids 73
tartaric acid 12
temperature difference 236
temperature resisting steel 120
thiocyanate 209
thiourea 99
TM0177-90A 148

p-toluenesulfonic acid   187–188
transcrystalline stress corrosion
    cracking   144
transport   266, 295, 308, 361
transport of acetic anhydride   49
transport of natural gas and crude oil   147
triethyl phosphate   75
turbine rotor blades   334
Twitchell's reagent   192, 195

u
U-bending specimens   273
ultrapure carbon dioxide   133
urea reactor   244
urea synthesis   244
urotropin derivates   6

v
valves   269, 316
vessels   318, 343, 361
vinegar   11, 13, 66, 73, 75
vinegar fermentation plants   12
vinyl acetate   55

w
wall thickness   265
waste gases   276
water (as inhibitor)   229, 234–235, 240
water circuit   254
water content   227–228
water pressure test   142
water vapor   122, 170
water vapor content   120
weld seams   224, 227, 243
wet carbon dioxide   142
wetting agent   188
wooden boxes   12

z
zinc coatings   202
zinc layer   243
zinc sputter layer   243